PLANT EQUIPMENT AND MAINTENANCE ENGINEERING HANDBOOK

ABOUT THE AUTHOR

Duncan C. Richardson, PE, has worked 43 years in engineering, construction, and commissioning of heavy industrial facilities. Prior to becoming an engineer he was a full time journeyman craftsman and technical writer for 12 years. He is licensed by NCEES as a professional engineer in six states and also has 10 construction licenses by state examination. Having participated in over 100 industrial engineering and construction projects, 40 of which were power plants and often as manager supervising all disciplines, his industry experience includes process, plant, maintenance, and infrastructure engineering in mining and desalination, water treatment engineering (including in the Red Zone in Iraq), and power engineering in the United States, South America, the Middle East, and Asia. Mr. Richardson has worked for such companies as Bechtel Corporation, El Paso Energy, and Portland General Electric; conducted formal classroom and on-the-job training for operations and maintenance plant personnel; and started and ran an independent engineering consulting firm.

PLANT EQUIPMENT AND MAINTENANCE ENGINEERING HANDBOOK

Duncan C. Richardson, PE

McGraw Hill Education

New York Chicago San Francisco Athens London
Madrid Mexico City Milan New Delhi
Singapore Sydney Toronto

1 2 3 4 5 6 7 8 9 0 QVS/QVS 1 2 0 9 8 7 6 5 4

ISBN 978-0-07-180989-4
MHID 0-07-180989-9

Sponsoring Editor
Michael Penn

Editing Supervisor
Stephen M. Smith

Production Supervisor
Pamela A. Pelton

Acquisitions Coordinator
Amy Stonebraker

Project Manager
Hardik Popli,
Cenveo® Publisher Services

Copy Editor
Susan Nodine

Proofreader
Chris Greene

Indexer
Robert Saigh

Art Director, Cover
Jeff Weeks

Composition
Cenveo Publisher Services

Printed and bound by Quad/Graphics.

CONTENTS

Chapter 5. Boilers 63

Chapter 6. Cleaning and Housekeeping 89

Chapter 7. Condensers 113

Chapter 8. Controls 123

PREFACE

This book is not intended to be used as an engineering manual. It is intended to give students, engineers, owners, and managers an idea of how the different equipment and processes in heavy industrial facilities—such as power plants—function. Therefore, it should give the casual observer an idea of how equipment in industrial facilities operate and what precautions should be taken in commissioning them. The book contains information from highly regarded sources. The reprinted material is quoted with permission, and the many authentic sources are credited. However, it is not a substitute for professional engineering nor for the codes and standards referenced in this work; therefore, it can only be used as a reference book and not a substitute for engineers', manufacturers', and code instructions.

I have given suggestions and sample computations for many routine operations commonly used in the field. They are guides and are not intended to be substitutes for professional engineering or for the many codes and standards readily available today. The codes available today are constantly expanding in depth and accuracy and must be utilized in their latest editions by field personnel. They are not intended for the sole use of design engineers. They are for everyone's use. Personal computers have made it possible to accumulate and process more information than could have been imagined 50 years ago. Code books that were almost leaflet form as little as 30 years ago are now huge, in some cases multivolume, standards that are paving the way for safer and more efficient engineering and construction. They also form the backbone for the safe and efficient commissioning of industrial facilities.

Some critics have told me, "In this day and age, by the time you finish writing your book, everything will have changed, and your information will be outdated." Well, great advances in computer technology have made possible great advances in other technologies. But these advances have not changed the fundamentals of classical mechanics. Newton's laws have not changed, nor will they ever. Therefore, although advances in computer science have led to great advances in other fields, including management, the same fundamental scientific laws apply today as they did before personal computers. The galvanic series has not changed, and equations of state, energy, and continuity have not changed. The first and second laws of thermodynamics have not changed. Kirchhoff's law will never change. What has changed is the speed and efficiency with which we can apply these laws.

What may be confusing to the lay observer is that technological changes in electronics have only changed the way we view things. For example, in the 1970s, solid-state protective relays and controllers started to appear. The solid-state relay replaced an electromechanical device in which one could actually observe, visually, its response to system disturbance. The solid-state relay gave no such reassurance that it was functioning at all. Therefore, its acceptance was slow. Now microcomputers are replacing all previous devices with wide acceptance. Protective relays have been replaced by multifunction protection units with wide acceptance. But don't be fooled: the old laws are still used to program the algorithms for the control units. Voltage is still a restraint and current the pickup in directional protection devices. The field instrument transformers remain the same—using the same old laws of classical mechanics. No doubt the name "protective device" may soon replace "protective relay" in our literature. But the function will remain the same.

As remarkable as they are, computers are no substitute for common sense. As computers were remolding our future, I observed how industries changed the acceptance of field test data and calculations. Once, slide-rule accuracy and instruments calibrated to known standards and manufactures' tolerances were acceptable. Today, computations are commonly made to the fourth decimal place while, incredibly, statistically derived uncertainty factors are applied in whole numbers. Tables are no longer available in some codes because, I suppose, we can make needlessly tedious computations effortlessly using computers. Regardless of how fast we can make computations, we must be able to make realistic assumptions. If we know from experience that an empirically derived coefficient, for example, is 0.999 and the overall uncertainty computation in the code is 0.5 percent, should we not have the confidence and common sense to assume the coefficient is 1?

There are other times the computer age can frustrate even the most casual observer. In 2008, an associate of mine was ordering electrical supplies from a local outlet of a national supplier. He told the supply clerk exactly what he wanted for a boiler and scrubber outage. When the supplier didn't understand what he was requesting, the customer quoted the supplier's catalog nomenclatures and numbers. The supplier then said they didn't publish those catalogs anymore and he, the customer, was lucky to have one. The supplier said he wished he had one too. Try finding industrial supplies on the Internet if you don't know their exact industrial nomenclature. Because of this, I've included an appendix of electrical parts for your assistance while Internet browsing. I still have some catalogs but unfortunately, thinking that I would never use them again, I threw many away. After all, you can now find everything you need on the Internet.

I am extremely happy, on the other hand, to see that ABB has republished *Applied Protective Relaying*, and has also offered an extensive digitalized file from its archives free to anyone. This was one giant leap forward for the 21st century—a combination of hard and digital publishing. Recall the 1980s and 1990s when such information was discontinued as budget-cutting administrators held more sway than common sense.

In this handbook, I explain how equipment and systems react to the old properties we are all familiar with. In doing so, I wrote many chapters covering the different categories of equipment. For example, I attempted to explain in simple, uncomplicated terms how boilers and turbines react to temperature change and thermal expansion and illustrate the criticality of clearance with regard to expansion. I go on to explain in simple terms how to start them up safely. There is so much more, but hopefully, this writing will be a good starting point for young engineers. For me, after graduating from Arizona State University, the power plant was a full-scale extension of the engineering laboratories at ASU. Commissioning of power plants afforded me the opportunity to test the theoretical structure that has formed the basis of all modern science. Those tests helped me satisfy for myself their unshakable profoundness. It is remarkable that so few theories are the foundation of all classical science. This theoretical structure will never change—only the speed and accuracy with which we use it.

Duncan C. Richardson, PE

CHAPTER 1

AIR COMPRESSORS AND COMPRESSED-AIR SYSTEMS

This chapter reviews the types of air compressors used in industrial plant design, their different characteristics, and applications. Keep in mind a quote from William O'Keefe in his publication "Compressed-Air System Design," *Power*, November 1978:

> Today, compressor type selection seems to depend more on individual preference than on analysis. There are several reasons for this. For one, developments in oil removal equipment have made it possible for systems to deliver low-oil air from lubricated compressors.
>
> Second, increased emphasis on compressed air—its cost, reliability, and quality—has led more attention to past performance records of compressors. Because few plants have enough compressors of various types to allow build-up of reliable information and cost data on them, the decision among competing types can fall to an engineer who has favorable memories of one type or another. If his auxiliary equipment is right, the type can probably be made to operate successfully in most systems.
>
> A third reason may be the confusing nature of data supplied by manufacturers. There is considerable room for differences of opinion and presentation in a data package containing purchase price, installation cost, efficiency at part and full load, cost of maintenance and repair, reliability, and life expectancy.[1]

Plant engineers must know what type of compressor they are commissioning, its characteristics and behavior, and the relationship between the machine and system. The manufacturer provides the compressor, not the system. It is the plant engineer's responsibility to make them work in harmony together, not the manufacturer nor the design engineer.

POSITIVE-DISPLACEMENT COMPRESSORS

Reciprocating

The reciprocating compressor had been, for long periods of time, the workhorse for the power industry because of its wide capacity range, ease of control, and especially its oil-free air delivery. They were common in large fossil-fuel-fired stations requiring large, reliable oil-free instrument air systems. They are, however, generally more expensive than other compressors in the medium- to high-capacity range and are more expensive to maintain for all capacity ranges. Because of the reciprocating action, they create vibration that requires special mounting considerations and corresponding higher installation cost, plus higher maintenance costs. The reciprocating compressor is a relatively constant-volume machine, making

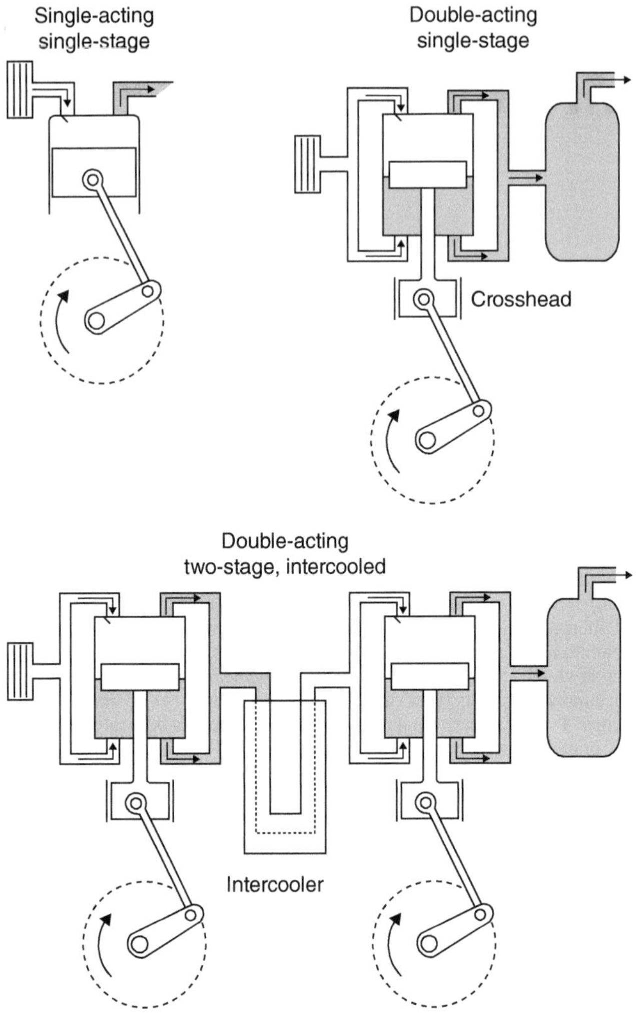

FIGURE 1-1 Typical positive displacement compressors.[2]

it ideal for instrument air systems. However, it is also a variable-pressure machine and must be controlled within a range suitable for system requirements (see Fig. 1-1).

To understand the control of recip compressors, first realize that the machines fall into either the single-acting or double-acting type and also into either the single-stage or multistage type. Figure 1-1 gives a quick review of the essential elements of this classification.

Single-acting machines are, in general, small units, inherently compact because of the trunk-piston design. Compression occurs only in one direction of stroke. In the double-acting machine, the piston has an easier time, because lateral load is less, with the crosshead's bearing surface helping to take the load. This bearing surface is outside the cylinder, so it can be given special design provisions for load-carrying and lubrication.

The number of stages in which a compressor reaches its delivery pressure is also important. For plant-air service requiring 90–125 psig air, a compressor can be single-stage or two-stage. The single-stage machine, doing all the work in a single compression, generates high heat, and is less efficient than if the work occurs in two separate stages, with cooling of the air between stages.

The general rule is that a compressor should be double-acting and water-cooled to be considered a continuous-duty unit. For plant air, such a machine will be two-stage or perhaps even three-stage, if pressures around 200 psig are needed.

Double-acting machines, with more extensive cooling and design for longer life, provide more efficient air compression, more reliability and less sensitivity to abuse and dirt, but they are relatively expensive to install and are heavy and bulky.[2]

Rotary

The rotary compressor is also a constant displacement type with variable discharge pressure.

The following sections describe the different types of rotary machines.

Sliding Vane and Screw

Screw compressor (Fig. 1-2) and sliding vane (Fig. 1-3) are positive-displacement compressors and can be provided with as high as 2000-scfm capacities. They can be oil-entrained or oil-free air delivery. The oil-entrained air compressor uses oil to lubricate the cylinder of the compressor to reduce heating. The oil-free compressor relies on material construction and jacket cooling to achieve the same results. Single-stage units will generally supply 50 psig, while two-stage units can provide up to 150 psig. The screw type is a relatively noisy machine, and therefore, noise abatement has to be taken into consideration in packaging and installation.

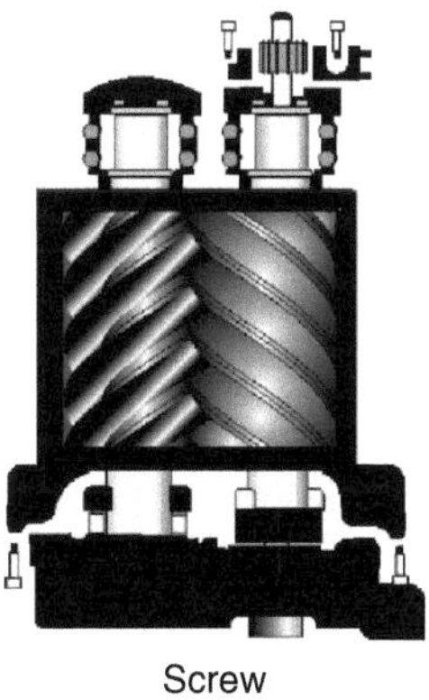

FIGURE 1-2 Screw.

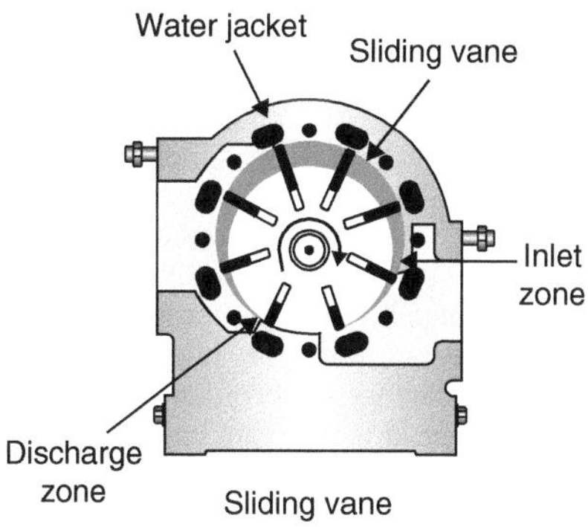

FIGURE 1-3 Sliding vane.

Rotary Tooth and Lobe

Rotary tooth or lobe compressors are used for small-capacity requirements less than 300 cfm. They are ideal for plants with relatively small air demand, such as for combustion turbine,

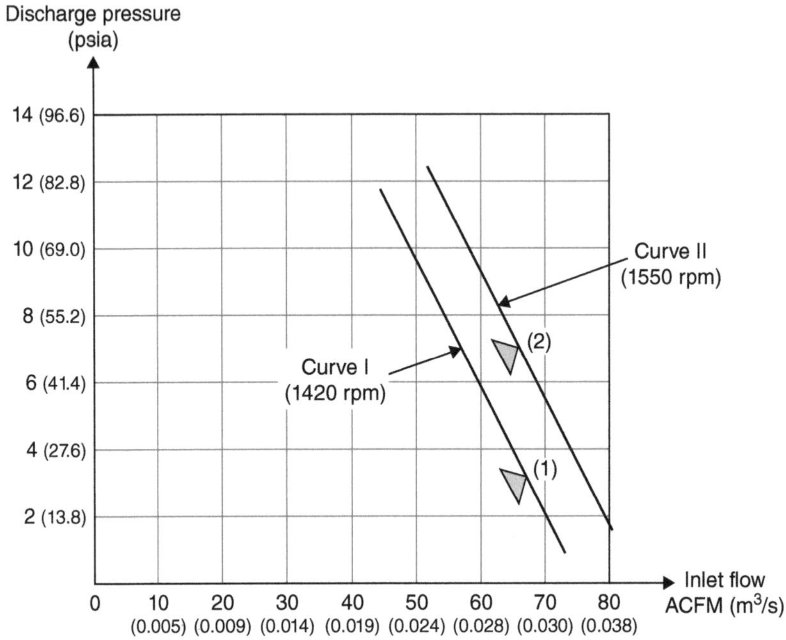

FIGURE 1-4 Speed versus capacity curves.

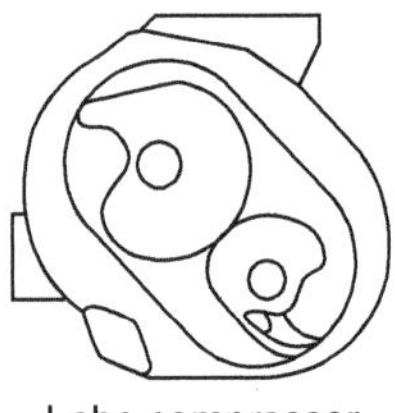

Lobe compressor

combined-cycle plants. These compressors are of relatively low cost to maintain and deliver oil-free air. Installation can be "skid mounted" from the factory, including aftercoolers, dryers, and air receivers. Changing the speed will change the capacity. See speed versus capacity curves (Fig. 1-4).

Control of a positive-displacement compressor can be obtained with the simplest of control schemes, that is, load/unload and shut down. For example, with a two-compressor scheme, a split-range pressure control can be selected for each machine. One compressor operating at a lower range will load if the lead machine cannot handle the system demand. If the unloaded standby compressor has not loaded after a predetermined period of time, the motor can be shut down for "idle standby" mode. These periods and pressure settings are preset in the factory in accordance with the engineering specifications issued by the engineer. However, it is the plant engineer's responsibility to make the necessary setting changes to meet the field situation, if necessary.

Liquid Ring

The Nash compressor used as a vacuum pump in the power industry is a liquid ring compressor that circulates water through a heat exchanger. The water, a compressant, partially fills the casing and is circulated by a radial blade impeller that is offset in an oval casing. Centrifugal force drives the water to the outer side of the casing. Figure 1-5 illustrates the process. These compressors are rugged and very quiet. In one case, a Nash pump was making

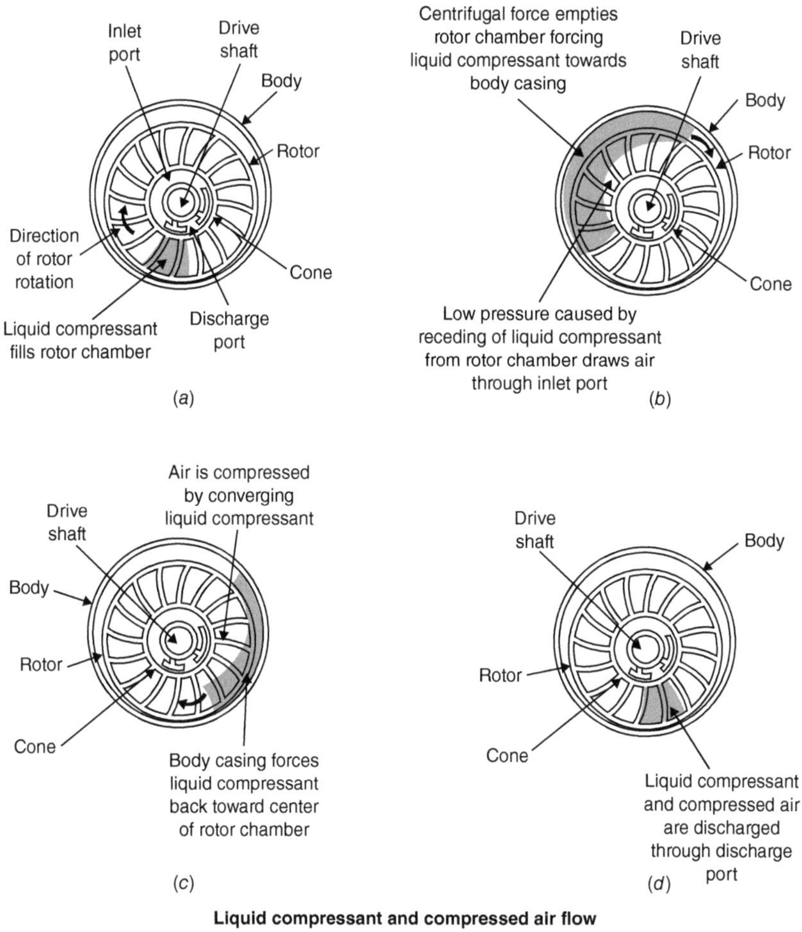

FIGURE 1-5 Nash vacuum pump.

unusual noises. During an internal inspection, the pump was found to have ingested welding rod. The impeller had flattened the rod but was undamaged.

VARIABLE-DISPLACEMENT COMPRESSORS

Centrifugal compressors become more competitive than reciprocating compressors when used for high-capacity service ranges 2000 scfm and higher. Such high-capacity ranges will be found in coal-fired power plants that use compressed air for sootblowing. Some plant locations, such as arid sites, may have limited water resources that make it prohibitive to use sootblowing steam, an unrecoverable resource. Such locations are a good candidate for sootblowing service because high pressures approaching 350 psig and capacities greater than 2000 cfm are requirements. It would not be uncommon to find a 3000-hp centrifugal per unit for sootblowing in a large, multiunit, coal-fired station.

Small machines are relatively quiet with low installation costs. However, inlet air must be free of abrasive particulate. The centrifugal compressor is a constant-pressure, variable-capacity machine. Using variable-speed control, the discharge pressure can be held constant and the capacity varied proportional to speed. See Fig. 1-6*a*.

Matching Compressor Outlet to Plant Load

Several methods exist for matching the compressor outlet to the plant load. But, in every case, the acceptable method will result in the compressor characteristic curve intersecting the load curve at the desired pressure and capacity. Speed control is the best way to obtain this match because, with variable speed, the compressor can deliver a constant capacity at variable pressure, variable capacity with constant pressure, or a combination of variable capacity and variable pressure.[1] See Fig. 1-6*b*.

Inlet Guide Vane Control

This method of control uses a set of adjustable guide vanes on the inlet to one or more of the compressor stages. By prerotation or counter-rotation of the gas stream relative to the impeller rotation, the stage is unloaded or loaded, thus lowering or raising the discharge head. The effect is similar to suction throttling, as illustrated in Fig. 1-7, but less power is wasted because pressure is not throttled directly. Also, the control is two directional, since it may be used to raise as well as to lower the band. This is more complex and expensive than throttling valves but may save 10 to 15 percent on power and is well suited for use on constant speed machines in applications involving wide flow variations.[3]

Inlet or Outlet Pressure Control

Controlling compressor capacity by varying inlet or outlet pressure is also possible to a degree. However, it is not considered a modern option for power plant applications where constant pressure is the general requirement within a reasonable range. This is especially true considering the modern advances made in the development of variable speed drives.

Surge Control

Surge control is probably the most difficult to solve of all problems relating to centrifugal compressors. It can be compared to the stall of airplane wings. Surge manifests itself in a backflow (reversed flow), usually accompanied by violent and loud noisy pressure fluctuations, and, unless reduced, may result in severe structural damage to the compressor. Surge is caused by reducing flow through the compressor to a point where the pressure restriction in the machine becomes greater than the pressure ratio developed by the compressor.

Surge limit of a compressor is the minimum flow required for stable operation at a given speed for inlet throttle valve position. This critical point for a particular compressor is determined by compressor speed and geometry, and thermodynamic properties of the gas. "Stonewall" describes the upper limit of capacity.

Although the stability range of a centrifugal compressor is commonly indicated by the design point to the surge limit, the unit can operate to the right of the design point. Stonewall occurs when the velocity of the gas approaches sonic velocity (usually at the impeller inlet). Shock waves result, restricting flow and causing either a choking effect or a rapid drop in discharge pressure with a slight increase in volume flow.[1]

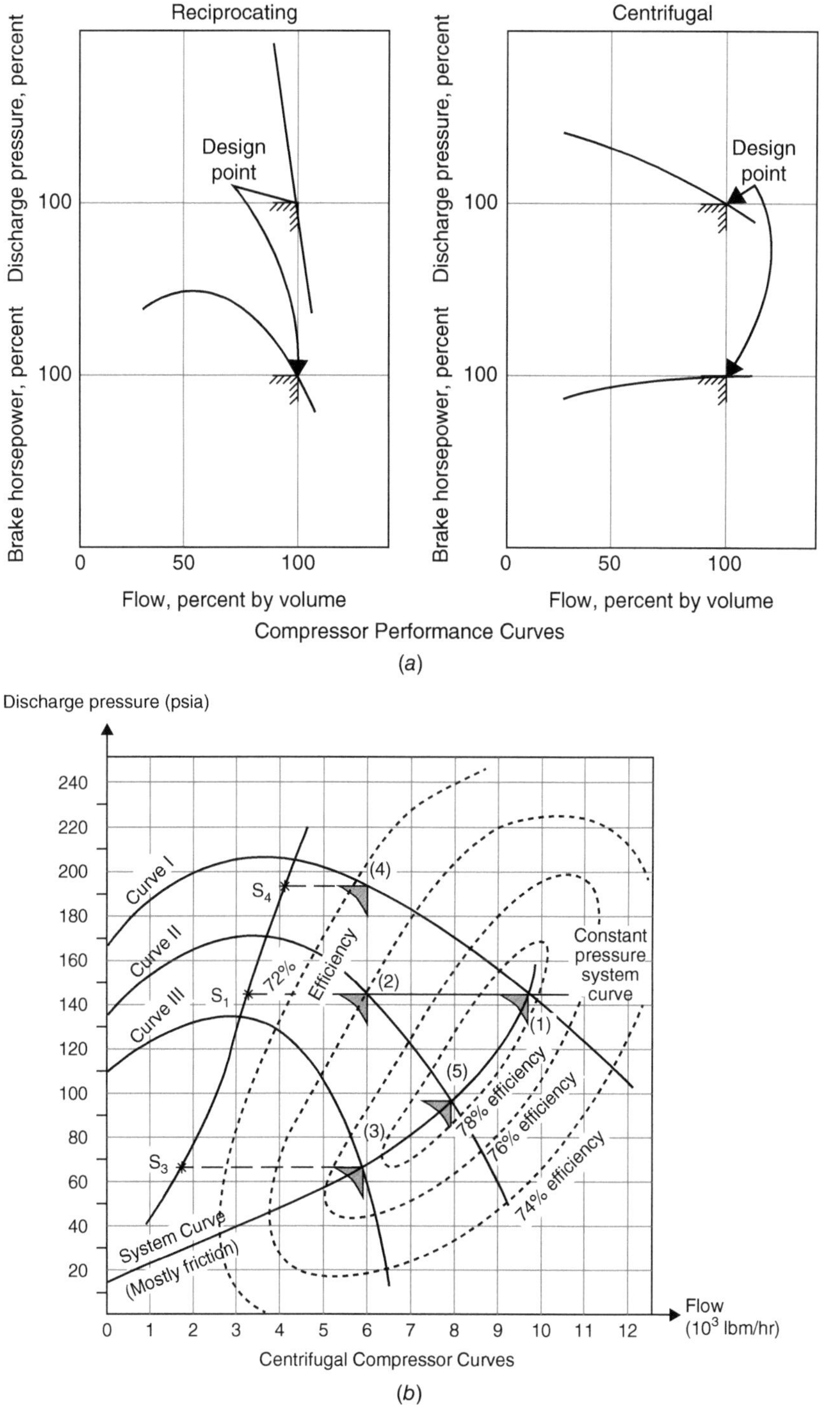

FIGURE 1-6 Centrifugal compressor curves (*a*) reciprocating versus centrifugal performance curves, (*b*) centrifugal speed-capacity curves.[3]

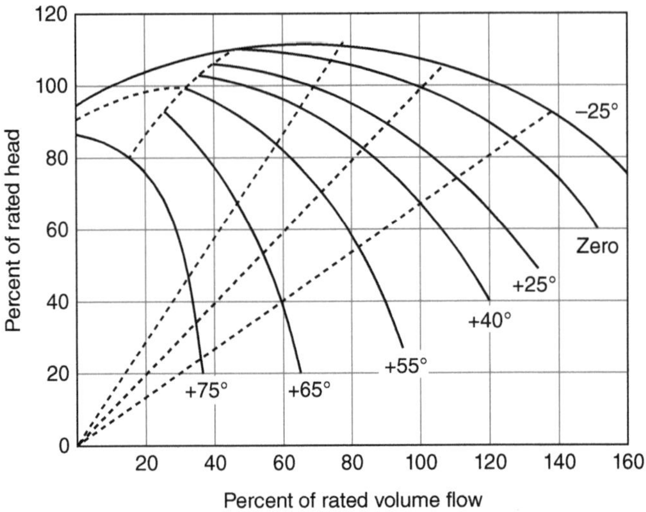

FIGURE 1-7 Effect of inlet guide vanes on capacity and head of a centrifugal compressor.[4]

It becomes obvious that a compressor operating on a constant pressure curve must be kept to the right of the surge limit curve. Early systems used inlet control and discharge dump valves to unload the compressor if it entered the surge region. If the compressor load demand suddenly drops, the inlet valve closes and the dump valve opens. From the curves in Fig. 1-7, it can be seen that the inlet guide vanes can move the compressor curves, pushing the surge limit to the left. This reduces capacity to meet the reduction in demand. From the variable-speed curves, Fig. 1-6, the same effect can be seen, but without a change in discharge pressure.

Sootblower Controls

Sootblowing operation must be controlled so that the compressors operate in their optimum load range. This is the key to the success of centrifugal compressor operation and a challenge for plant engineers. Using past experience and trial and error operation of the soot blowers, soot blower demand can be kept to the right of the surge limit of the compressor capacity curve.

> Programmed microprocessor coordinates the operation of 276 sootblowers in three boilers and one scrubber to cut air consumption and boost sootblower efficiency.... The control program is written so that each sootblower operates on a specific-time basis and can be blown more often if actual furnace conditions indicate the need. The number of sootblowers operated at any one time is automatically held within the capacity of the air compressors in service.

See Fig. 1-8.[5] The use of surge tanks will dampen system spikes generated from on/off blower operation.

> Variations in inlet-air temperature normally are not controllable, and they can have a significant effect on compressor performance. The adverse impact of warm air on efficiency emphasizes the importance of selecting a compressor with a steep curve and a high rise from working pressure to the surge point at standard conditions.[6]

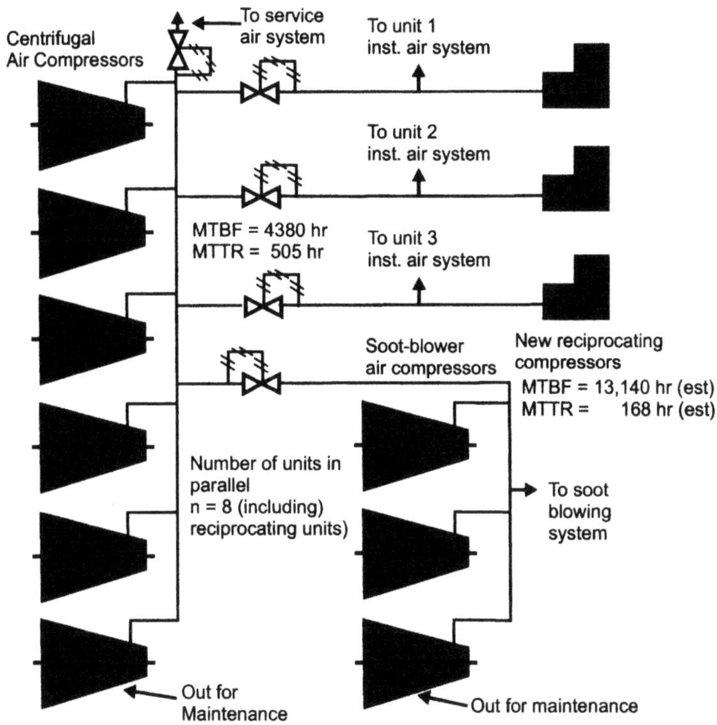

FIGURE 1-8 Plant air system with both centrifugal and reciprocating compressors.

Therefore, it is important that matching of the system capacity curve to the compressor capacity curve include ambient temperature change.

CENTRIFUGAL COMPRESSOR VIBRATION

It is imperative that centrifugal compressors operate above the minimum flow condition at all times. Flow should never drop below the surge limit/minimum flow on the performance curve. Systems that have conditions with demand requirements that can reduce flow into the surge region must be studied thoroughly by the plant engineer. This includes all system controls, compressor controls, and performance curves. Surging can be extremely violent, and catastrophic failure of the compressor can occur. Centrifugal gas compressors for combustion turbines are especially susceptible to this during startup when unit trips can be expected. This will cause the compressor to go from a loaded condition to zero flow instantly.

Constant-speed centrifugal compressors operate at full load all the time, similar to positive-displacement pumps, unless they have inlet flow control valves. To maintain constant design conditions, flow capacity that is not being used by the process can be bypassed back to the first stage inlet or dumped to atmosphere through a silencer. These machines may have high-frequency pulsations that can be in resonance with the natural frequency of the piping system and system or compressor components. It can excite high machine shaft as well as system component vibration. For variable-speed machines, the vibration can be eliminated by simply avoiding speeds at which the vibration occurs, which is usually at a low speed close to the surge limit. That is, the surge limit line should be moved to the right of the original design point. Remember, this type of vibration is caused by blade-pass-pulsation phenomena and is not turbulent flow related. Flow turbulence can cause low-frequency vibration and can be corrected by adding guide or turning vanes in the compressor inlet. This fix will not correct blade-pass-induced vibration.

Both low- and high-frequency vibrations can coexist. At the Edwardsport syngas combined cycle, an air-separation unit with a 15,000-hp compressor had a temporary suction strainer installed in the third-stage bypass to the first-stage inlet. When the strainer was removed, the fine-mesh lining was found wadded up in the bottom. Also, there was evidence of vibration wear on the coarse-mesh backing. With the strainer installed, there was no external evidence of shaft or piping vibration. Therefore, the strainer was removed because it was only temporary.

Without the strainer, severe piping vibration occurred, and the shaft vibration went from 1 to 4 mils. The inlet elbow did not have turning vanes, so a new elbow was fabricated with vanes. Also, a flow-restricting orifice was installed in the section where the strainer had been to recreate the same conditions that existed when the strainer was installed. The pipe and shaft vibration returned to the original values. However, after only 50 hours of operation, the orifice failed, and a large piece was ingested by the first stage. Similar to the strainer, it showed signs of vibrational flexing, which eventually caused embrittlement and failure. The elbow turning vanes also had severe cracking and would have eventually failed. The low-frequency problem was solved but not the blade-pass, high-frequency vibration.

Dryers

Because the water content in air does not change as the air is compressed, the given *percentage* of water increases as the air is compressed to a smaller volume. Simply stated, air at less than 100 percent humidity will compress to 100 percent humidity. Most of the free-water content produced in the process can be removed by mechanical means. However, the air will remain saturated with moisture. This moisture will condense if cooled below its dew point by ambient air. Therefore, if the system receives the saturated air at 90°F and the ambient air is at 70°F, moisture will condense because the compressed, saturated air will eventually cool to ambient conditions. Therefore, in plant instrument air systems, the dew-point temperature of the compressed air must be lowered to below the prevailing ambient air temperature. As a rule of thumb, that is at least 10°F. In freezing climates, plant air must also be dried to prevent freezing of the air lines. Drying can be accomplished either by absorption or refrigeration or by a combination of the two.

Refrigerant

> Refrigerant air dryers remove moisture from air by cooling it to within a few degrees of the freezing point of water. Resultant condensed moisture is removed in a separator and drain trap mechanism located immediately downstream of the refrigerant evaporator. The minimum pressure dew point possible with this type of dryer is 38°F because there is danger of freeze up at lower temperatures.[7]

The obvious shortcoming of this type of dryer is it has a limit to which the dew point can be lowered. This is because it can only remove aerosol moisture (mist), not vapor (gaseous) moisture that must be removed by a drying media (desiccant). In tropical climates where freezing temperatures are not present, this drying method would be ideal because of its relative inexpensive installation and operation.

Desiccant (Drying Media)

Desiccant dryers can be classified into two separate categories—regenerative and deliquescent.

The regenerative dryer uses an adsorption desiccant such as silica gel that collects moisture on the surface of the media. (Therefore, it is called an *adsorption* process.) This type of dryer can be regenerated by flushing lower dew-point air through the media to carry off the moisture to atmosphere. The air can be at ambient temperature or heated. Generally speaking, the unheated regeneration process is more expensive because the compressor must be sized for the additional volume required for regeneration. Heating substantially reduces the time and volume requirements.

The deliquescent dryer is a non-regenerative type dryer which utilizes an *absorption* process. It contains a media, such as sodium carbonate, that absorbs the moisture and dissolves in the process. They are not standard for the power industry because of daily maintenance and waste disposal considerations. Because most automatic drains will not handle the deliquescent solution, the liquid must be drained manually, at least once a day.[7]

Compressor Installation

The compression process generates considerable heat, which must be removed. Air-cooled compressors require considerably more design attention than do water-cooled compressors. The reason is that the ambient conditions are changed by the very conditions that affect the product air. A worst-case example would be a compressor located in an inadequately ventilated space. If the interstage temperature is 350°F with an ambient temperature of 90°F, and the ambient temperature is elevated by the air-cooled heat exchanger to 140°F, then the interstage temperature will rise proportionally, creating a serious cycle of events. One can see this happening if the compressor is located in a room or building and the heat exchanger exhausts into the space rather than outside. Therefore, the design engineer must consider all operating scenarios regarding air-compressor intake and cooling-air inlet and exhaust. Because two different disciplines are involved, this is too often overlooked. The process engineer must coordinate with the heating, ventilating, and air-conditioning (HVAC) engineer to marry the two into the best installation for operation. For example, the heat generated by compression can be used for space heating in the winter and exhausted during summer operation.

Air-Compressor Rooms

As in the preceding example, air compressors operating in extreme climates also need special consideration. For example, air compressors, dryers, and receivers operate more efficiently at colder temperatures but need freeze protection in the winter. Air can be exhausted from the building in the summer by ducting the compressor hot exhaust outside. Wall dampers provide air makeup. In winter months, the air can be recirculated inside by exhausting into the building. The inside air temperature is best controlled by an outside thermostat. A single diverter gate can be used as a bypass to reduce pressure drop and, therefore, back pressure on the cooling fan. Care must be taken to ensure large gates do not experience destructive vibration when operating. Harmonic oscillation can occur as the damper is moving. The inside air temperature can be controlled by ventilation air, which is distributed in the building as mix air. The mix air temperature is controlled by modulating the inlet and mix air dampers. The minimum makeup and exhaust air should be mechanically set with the damper linkages. Heat tracing is required for all drains. Radiant electric heaters can be installed for outages and extreme conditions, to warm the equipment. Also, the compressor aftercooler requires a temperature control switch so it does not subcool the air, which can cause surface condensation on downstream piping, aftercooler drains, and air-receiver external surfaces.

If the space is air-conditioned, special sizing and loading considerations are required for both the compressed air system as well as the HVAC system. In this case, the best scenario may be to exhaust the cooling air all the time or consider a separate unconditioned room within the building. Installation of the air receivers in a cold, outside environment to assist in moisture removal upstream of the dryers can be considered if the benefit is not offset by seasonal climatic change.

To improve the investment, plant standardization has always been the goal of power industry engineers. For plants firing natural gas, standardization is highly practical. However, it is sometimes impractical to enter all possible climate changes and operating conditions into an iteration that solves for optimum equipment selection for a "standard plant" design for compressed air system equipment. One exception in the late 1990s was the "standardization" of combined-cycle plants in which standards included compressed-air installations.

In short, the plant engineer must analyze the entire installation before trial operation and determine if the design properly interfaces with the environment. For example, an environmental design for Houston, Texas, may not be adequate for a plant located in Basra, Iraq. A good rule of thumb for plant engineers to follow is *there is no standard environmental design for compressor package installation.* Furthermore, if the plant engineer issues an environmentally related design change request (DCR), it is very important that he/she uses HVAC standards and equipment in the submittal.

CONCEPTUAL DESIGN STANDARD FOR COMPRESSED-AIR SYSTEMS

System Configuration

The compressed-air system conceptual design for power plants is standard for the industry because there are two basic systems integrated together. The instrument-air system and plant service-air system are standard systems for the industry. The service-air system is used for plant maintenance and the instrument-air system for plant controls. The service-air system is almost always connected to the instrument-air system through a "marshaling" valve that continuously vents full open from the service-air system to the instrument-air system. This relief-type valve closes on loss of pressure, shutting off supply to the service-air system. (In other words, it opens on increasing pressure at the minimum required for the instrument-air system.) The instrument-air system pressure is thereby maintained if overuse of the service air system occurs or a failure in either system causes a demand that is greater than compressor capacity.

Compressor Sizing

Sizing standards based on past experience are the best rule of thumb, especially for the "standard plant" such as the combined-cycle plants previously mentioned. For first-generation coal-fired plants, however, this is rarely more than an educated guess based on usage taken from device specifications, estimated maintenance demand, and past experience. This is because the vast difference in fuel quality and consistencies dictate vast differences in plant equipment selection and design. This in turn can create a vast difference in compressed-air requirements. It is not uncommon for the plant engineer to find he/she has to use the standby compressor for continuous service. This is especially true in the commissioning phase. This is not an easy problem for the design engineer because oversizing the compressed-air equipment can be as bad as undersizing from a capital investment point of view. Compressor operation at part load is inefficient and, from an energy point of view, costly.

Thus, it is extremely important that the plant engineer rigorously follow system commissioning procedures. Excessive leakage, incorrect pressure settings, and improper use of plant air can make a perfectly sized system appear inadequate. Rigorous commissioning activities include documentation that separates professional analysis from personal opinion. Without it, the design engineer should never be consulted, especially concerning the adequacy of his/her design.

Air Receiver Location and Sizing

Air receivers can act as moisture separators and sometimes are installed in cool areas upstream of the air dryers. In other cases, they are installed downstream of the dryers to eliminate the need for air-receiver water blowdown. Exposure to radiant heat should be avoided. The air receivers are generally sized for 5 to 10 percent of system demand. They are used to extend the positive-displacement compressor load/unload cycle and compensate for sporadic spikes in demand. For the plant engineer, air receivers are an excellent tool for determining system demand and compressor performance.

COMPRESSED-AIR SYSTEM PRE-STARTUP TESTING

Preparation

Read the instruction manual, and copy all sections pertaining to the specific model being commissioned. Insert the copy into a startup binder for field use. Check that all spares, consumables, and replacements are available, such as lube oil, desiccant, filter cartridges, etc. Insert the inventory information into the startup binder. Clean the compressor skids and have construction personnel, materials, machines, equipment, and debris removed from the operating area.

Inspection

Do not delegate any of the following preparations. Do it yourself so that you do not rely on word of mouth, or "coffee locker" inspection reports.

1. *Compressor controller*. Energize the compressor controller. Run through all functions of the controller prior to initial start. Be completely familiar with the compressor controller. Copy all factory settings such as alarm and trip points, load/unload ranges, measurement data logs, and scheduled advisory times and dates from the controller into your instruction manual. Also, your compressor may have functions and settings that are password-protected, such as auto restart after reestablishment of voltage after a loss of power. The factory default may prevent automatic restart and may be password-protected. In this case, you will need to obtain the password from the factory. It is essential that instrument air be reestablished automatically as soon as power is reestablished. This, of course, pertains only to the instrument-air system, not to plant or sootblowing air. There may be settings for automatic shutdown after a factory-set duration. For example, will the compressor motor shut down if unloaded for a predetermined interval? Also, will it automatically shut down at a preset time or date? Is it factory-programmed to shut down every night or weekend? It might even be set to shut down in a very short interval for the initial run. Set the internal clock to the local date and time. Call the factory to clarify any uncertainties about controller functions or settings before the initial start.
2. *Compressor package cleaning and inspection*. Have a "ShopVac" available. Remove all access panels to completely expose all compressor components for inspection. Vacuum the package thoroughly yourself or directly supervise the cleaning. You are the only one who can determine what you mean by "clean." Then, remove all desiccant bags from compressor internals. All compressors are shipped with moisture protection. Look in the interstage piping, which is a good place to put desiccant bags to protect both stages. Remove all shipping restraints. Check lubricant levels. Check the interstage and discharge automatic drains and ensure that they will operate properly. If the drains are

electronic type traps, test their operation with the electronic test option if provided. If they don't operate properly, your compressor will be damaged. Remember, compressors are water generators by nature and the water must be removed or severe damage will occur. Ensure air coolers are free of debris and the cooling fans rotate freely. For water-cooled units, ensure the closed cooling water system is available. Again, call the factory to clarify any uncertainties before the initial start. Advice should always be available, but do not call unless you have thoroughly read the manufacturer's instructions.

3. *Aftercoolers.* Check the cooler automatic functions. Do they start and stop when the compressor motor starts, or are they controlled by a temperature switch? If the room or building temperature varies with the weather, a temperature switch probably should control them. This is especially true if there is a chance of subcooling or freezing. Change this control function if field conditions dictate. Check the discharge drains and ensure they have a visual means of ensuring proper operation.
4. *Water traps.* All suspended free water generated in the compression process must be removed prior to the air reaching the air dryers. Air dryers are designed to remove saturated moisture, not free moisture.
5. *Dryers.* If not done so in the factory, desiccant dryers may have to be filled with an initial charge. Check the dryer controls to ensure that the moisture option is activated. For unheated regeneration, the compressor may operate loaded unnecessarily just for regeneration if the moisture option is not selected. This happens for extended periods if a timer presets regeneration. For heated regeneration, it can be an energy waste.
6. *Filters.* Check that the filter cartridges have been installed and differential pressure indication is available.
7. *Piping.* Check the drain piping and ensure that all drains are piped separately to an open collection point. It is essential that there is visual proof that each drain is functioning properly. If they are connected to a common header, which is piped to a single drain, remove them and re-pipe them separately. Remove permanently any isolation valves at the drain discard. Also, it is a mistaken notion that check valves installed in the separate drain legs will ensure proper operation of the header system. They are valves and, as such, are subject to failure. Failure of a drain check valve can allow water to severely damage the compressor. Check heat tracing for proper installation and operation. Remove heat tracing from piping if the local conditions don't require it. It is best to have all piping visible for routine inspection. Heat tracing requires insulation and will cover multiple pipes obscuring the origin of the individual drains.
8. *Instrumentation.* All field-installed instrumentation must be calibrated and the data sheets inserted into the startup binder for the system. For preinstalled instrumentation, it is a mistaken notion to think that "skid"-mounted indicators do not require calibration. It is a fact that factories do not calibrate indicators unless specified otherwise. They rely on vendor compliance. And, vendor quality control inspections generally require only "spot" checks for quality assurance unless specified otherwise. Nonetheless, shipping can be an extremely rough process for equipment, which can undo any rigorous calibration performed at the origin. Nevertheless, good judgment must be used. Generally, skid-mounted indicators do not require a rigorous five point calibration. The reason is they generally provide indication at a single point or narrow range. It is generally acceptable to ensure the instrument is calibrated at the design operating point. Remember also, if several indicators measuring the same parameter are reading differently, it will cause uncertainty, or worse, a sense of instrument unreliability. All instrumentation providing protection outside the compressor package, but mounted on the skid, must have a calibration check.
9. *Electrical.* The plant engineer is responsible for correct motor rotation. This must be verified and the documentation indicating correct rotation must be entered into

the startup binder. Most small compressor packages have their own motor starters; therefore, the motors must be bumped for rotation using the compressor controller. Caution should be exercised when trial operating packaged units that have their own internal "buckets." Too frequently, during construction, phasing can change at motor control center (MCC) or load center and go unnoticed because the power for the compressor originates from a disconnecting device, not a starter. Therefore, the plant engineer must investigate all MCC outages during startup operations of the equipment to ensure that phase rotation has not been changed.

PLANT- AND INSTRUMENT-AIR SYSTEM TRIAL OPERATION

Preparation

Generally speaking, the compressor installation can be trial operated with only the air system main headers available for operation. This is best for trial operation because the headers must be blown first before the individual branch lines. Initial operation will pressurize the air receivers, which will be used for performance and system leak testing. Therefore, it is recommended that test instrument devices be installed to measure the air temperature and pressure in the receivers. The receiver pressure will also be used to set the compressor load/unload range.

Line Blowing

Ear protection is required because of the high noise levels during blows. Install temporary ball valves at the end of each air system header. The valves will be used to blow the lines clean. Ball valves are used because they can be rapidly opened by hand. With the air receiver(s) at capacity blow, the headers use the expansion of air to atmosphere to provide a high-impulse velocity. Blowing for 10 s several times or until no dust or debris can be seen at the discharge should be sufficient to clean the lines. Now the device branch lines can be blown through their isolation valves to atmosphere as they become available from construction.

Testing

Positive-displacement compressor capacity can be verified by using the air receivers as a test chamber. Isolate the receiver discharge, and pressurize the receiver using a stopwatch. Mark the pressure and temperature at the end of a given pressure range. Use a range that will give at least a 5-min test. Calculate the capacity using the following equation:

$$Q_{\text{scfm}} = \frac{V_{\text{cf}}}{t}\left[\left(\frac{T_{\text{st}}}{T_{\text{act}}}\right)\times\left(\frac{P}{P_{\text{st}}}\right)\right]$$

where

Q_{scfm} = specific volumetric flow rate
V_{cf} = air receiver volume
t = timed pressure loss in seconds/60
$T_{\text{st}}/T_{\text{act}}$ = standard temperature (60°F)/actual temperature
P/P_{st} = pressure change/standard pressure (14.7 psia)

An engineering assumption can be made that the compression or expansion of the air in thc rcceiver is isentropic if an aftercooler is installed upstream. Therefore, differential temperature is ignored. The basis of the assumption is that the aftercooler heat removal will cancel the heat of compression from P_1 to P_2.

Leak test the system in the reverse manner. Time system decay for a specific range using a stopwatch. Determine the leakage with the same equation used for compressor capacity. As each area in the system is commissioned, time the unloaded cycle of the compressor. For example, when commissioning the boiler area, note the change in compressor loading. Test for leakage. Build a historic signature for system leakage and compressor loading and capacity.

Centrifugal compressor capacity and system demand can be tested by installing a flow element at the discharge of the surge tank. Or the capacity can be calculated using the equations in Fig. 1-9. The compressor must have constant flow through the surge tank at all times. To accomplish this, create a temporary flow path to atmosphere for system tuning and testing.

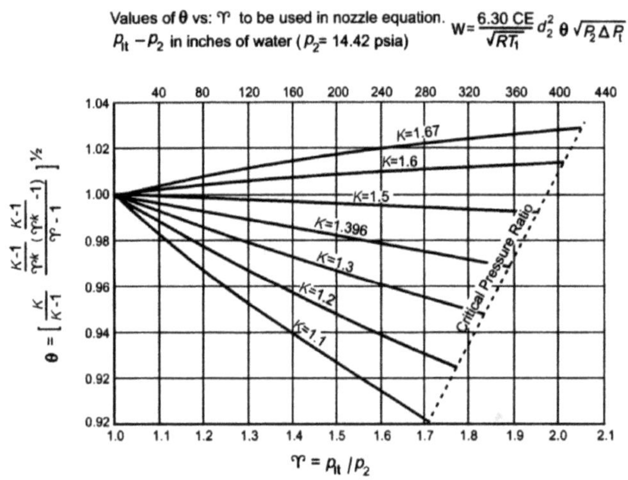

Values of θ for use in following equation:

$$W = \sqrt{\frac{2g}{RT_1}}\ d_2\ \theta \sqrt{P_2 \Delta P_t}$$

W = Weight of Gas Flow

P_1 = Initial Pressure

P_2 = Exhaust Pressure

K = Ratio of Specific Heat at Constant Pressure to that at constant volume.

P_{lt} = Total Pressure as Measured with a Pitot Tube

FIGURE 1-9 Integrated centrifugal and reciprocal compressor system.[8]

REFERENCES

1. O'Keefe, W., "Compressed-Air System Design," *Power*, November 1978.
2. Van Ormer, H., "Make Reciprocating Compressors Pay Off by Factoring Control into System Decisions," *Power*, May 1981.
3. Lipták, B. G. and Venczel, K., *Instrument Engineers' Handbook*, Chilton Book Company, 1985.
4. Avallone, E. A., Baumeister III, T., and Sadegh, A., *Marks' Standard Handbook for Mechanical Engineers*, McGraw-Hill, New York, pp. 14–56.
5. Cunningham, E. R., "Selecting a Compressed Air Dryer," *Plant Engineering*, December 12, 1974.
6. Sims, W. B. and Smith, A. A., "Automated Sootblowing Cuts Air Consumption," *Power*, April 1981.
7. Hendricks, J. F., "Understanding Centrifugal Air Compressors," *Plant Engineering*, January 22, 1976.
8. Kent, *Mechanical Engineering Handbook (Power)*, Wiley Handbook Series, pp. 1–18, "Flow of Air from a Receiver."

CHAPTER 2
AIR CONDITIONING

Combined-cycle (CC) plant engineers rarely think about heating, ventilating, and air-conditioning (HVAC) as a startup system, but coal-fired station engineers must. The reason for the lack of thought is that HVAC is thought of as a package that is subcontracted "turnkey." It generally is for most CC plants. With this in mind, this chapter's goal is to help the plant engineer understand the mechanics and fundamentals of HVAC conceptual design and testing in order to provide tools to analyze, troubleshoot, and oversee the commissioning of complex power plant HVAC systems.

AIR-CONDITIONING FUNDAMENTALS

Psychrometric Chart

The psychrometric (Fig. 2-1) chart serves as the basis for illustrating basic air conditioning. It is a good, quick reference and educational tool. HVAC engineers use equations that are more precise and very easy to use with PCs. The equations, and their application, can be found in HVAC handbooks, and the "canned" software can be purchased online. The reason engineers use the equations is that the chart is only good for a single-point

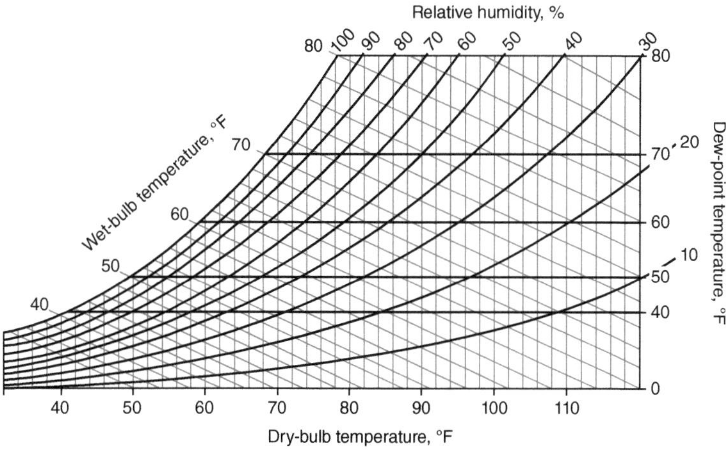

FIGURE 2-1 Psychrometric chart.

atmospheric pressure. Handbook charts are generally representative of standard pressure only. If the barometric pressure changes, a new chart is needed.

The equations used are derived from straightforward classical thermodynamics:

$$\text{Equation of state, } pv = RT$$

$$\text{Continuity equation, } W_{\text{in}} = W_{\text{out}}$$

$$\text{Energy equation, } Q_{\text{in}} = Q_{\text{out}}$$

Stated in simple terms, air conditioning is the control of both temperature and humidity. The two are inseparable. If you change one, their relation must change. For example, if we increase or decrease one variable, temperature, and hold the other, moisture content, constant, then relative humidity will change. Building the chart in steps helps clarify this relationship.

As can be seen in Fig. 2-2, the common chart is an overlay of four charts. An example of how to use the chart is as follows: We have a control room that has a packaged comfort refrigeration system. The outside air temperature is 85°F with a relative humidity of 65 percent. Find the dew-point temperature of the air. Follow the dry-bulb (DB) curve up to the 65 percent relative humidity (RH) curve, then follow the wet-bulb curve to the left and find the dew-point temperature (73°F). If temperatures are cooled to 73°F, the moisture in the air will condense on the cool surfaces.

This example is, however, a simplification. Air at the coil surface will be subcooled and promote moisture condensation. Manufacturers use empirical equations to determine the coil temperature required to cool the air to the desired room temperature. The coils will condense the moisture in the air but will nonetheless leave it just below saturation at the discharge. There it depends on the unit efficiency; the higher the efficiency, the closer to saturation. Generally speaking, cooling draws the air toward the saturation line. For control rooms and distributed control system (DCS) rooms, in cases of high humidity, the air must be dehumidified. If it is not feasible to use desiccant dryers, dehumidification can be done by a combination refrigeration-heater unit. The air is chilled to the saturation line and heated back to

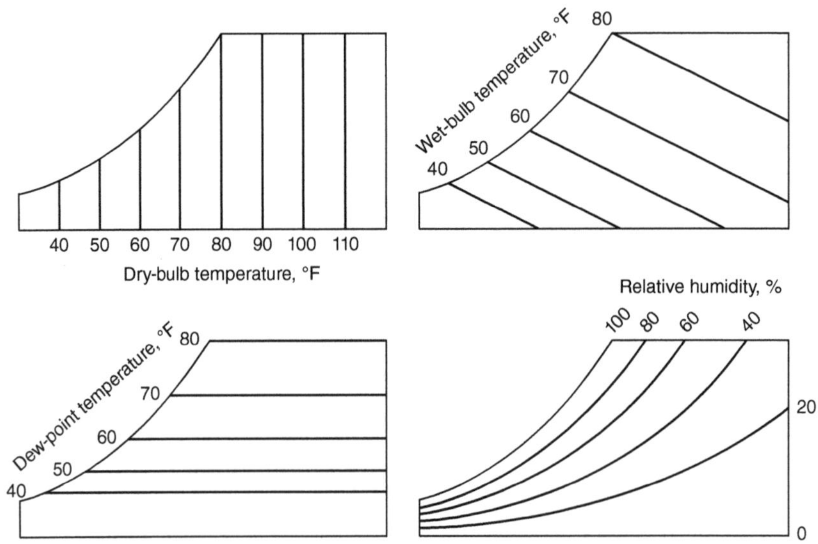

FIGURE 2-2 Four elements of a psychrometric chart.

ambient either directly or by mixing it with recirculating air. Heating causes depression of the dew point, driving down relative humidity. The air now contains less moisture and has a lower dew point (dew-point depression) and will not condense on cool surfaces. In tropical regions and regions where summer weather produces warm humid conditions, special attention to this is required.

Cabinets should never be cooled to remove moisture, for reasons illustrated in the preceding example, if the outside space is not dehumidified. If a cabinet door is opened, the outside air moisture will immediately condense on the cool surfaces inside the cabinet (the electronics and electrical components). Cabinets that require cooling must use dehumidifiers in conjunction with the air conditioners, desiccant dryers, or dehumidifiers that reheat the discharge air back to the desired air temperature for the cabinet. Nevertheless, the inside dew-point temperature must be significantly depressed. (For a relative humidity of 50 percent at 70°F, a dew-point depression of 20°F is required. Within the usual range, for each 10 percent higher relative humidity, a dew-point depression approximately 5°F less is in order, while for each 10 percent higher relative humidity, a dew-point depression approximately 6°F more should be used.)[1] For example, from the psychrometric chart, it can be seen that the intersection of the 50 percent RH curve and the 70°F DB curve yields a dew point of 50°F, or a 20°F difference.

Early Signs of Corrosion in DCS Cabinets

Look for early signs of corrosion at cabinet penetrations. Corrosion will attack the thinnest or finest metals first. Inspect connector pins for corrosion residue and circuit board surfaces for adequate corrosion protection film.

Fatal Signs of Corrosion

Example: If the DCS back panes or modules start to fail for no apparent reason, check the back-pane surfaces for adequate protective film. Check for corrosion where air can circulate through penetrations at the module connectors. Use a magnifying glass for inspection.

It is not always possible, but the best way of controlling moisture is to raise the air temperature. This is done for motors, for generators, and in outside control cabinets or cabinets in unconditioned spaces. Likewise, equipment stored under plastic tarps should have heating lamps or simple light bulbs under the tarp to raise the air space temperature to prevent condensation. The ambient moisture content remains the same, but the dew-point temperature is depressed as the air is heated.

Caution

The plant engineer must ensure that heater power supplies were not taken from the instrument power supply or that the cables were run in the same raceway. This rarely happens in design but occasionally happens during construction in outside remote cabinets or cubicles. Heater energizing can cause instrument spikes that can cause serious process control problems. That is, the heater must have its own dedicated, isolated power supply, usually from a lighting or heat-tracing panel. If systematic spiking occurs in a control system for no apparent reason, this is a good place to look for trouble.

Ventilation

Human Occupancy Design Consideration

Ventilation is used for both human occupancy, fresh-air circulation as well as for equipment heat removal. All spaces require a certain amount of ventilation in order to keep the air within the space fresh. The amount of makeup or ventilation for an air-conditioned space is given in cubic feet of air per occupant and is generally around 25 percent of the air circulation. Apart from contaminates, the carbon dioxide level is the determining factor. All workspaces must meet Occupational Safety and Health Administration (OSHA) requirements for air quality. OSHA does not dictate how. The method is the responsibility of the design engineer. The principal guideline is the threshold limit value (TLV). Values for TLV are listed in the *Industrial Ventilation Manual of Recommended Practice* published by the American Conference of Governmental Industrial Hygienists. Engineering, manufacturing, and installation are governed by the American Society of Heating, Refrigeration and Air Conditioning Engineers (ASHRAE). These requirements must be met within all spaces, including coal bunker feeder decks and ash handle enclosures, as well as office spaces. These systems must be placed in service prior to commissioning the process equipment. However, many times these are afterthoughts or are ignored completely until the main process systems are commissioned.

Equipment Heat Removal

Prior to commissioning equipment, all ventilation systems must be placed into service. Most ventilation is straightforward using a system that simply draws in outside air, which is passed over the equipment and exhausted. However, where seasonal changes in climate are severe, systems can be complex. In other cases, some spaces may be pressurized to prevent the infiltration of contaminants. Forced-draft fan air may have to be preheated in the winter for air preheater protection. Air compressors or generators may exhaust heat outside in the summer and recirculate hot exhaust air in the winter.

COMMISSIONING

Control Rooms

The control room HVAC probably gets the least field engineering attention of all areas in an industrial plant especially if it is part of the administration building. The control room final construction, however, will not be completed when commissioning activities begin. The floor penetrations will not be sealed. Craftspeople will be in and out for permits. Electricians will still be pulling and terminating cables. Other crafts will be completing the finish architectural work and others simply standing by. The occupancy loading in the space will be 10 to 20 times design with all the technical advisors, startup engineers and additional operating personnel. The infiltration of moisture and dirt can also be 10 to 20 times design.

The first thing that has to be done is restrict traffic to necessary personnel only. It is not just a matter of congestion and traffic control, air quality control is also a factor. This is especially true in electronic rooms serving the control room. The next thing to do is seal all penetrations. It may be easy to say "seal all penetrations and restrict access," but no contractor wants any penetrations sealed until he or she has completed 100 percent of the work. Nor does the contractor want any access restrictions. Nevertheless, this must be done to protect the electronic equipment.

The HVAC inlet-air automatic filter feed control should be checked for proper differential settings for feed and alarm in order to protect the fan. Chilled water and steam coils should be checked for proper operation. The mix air controller must be checked for proper operation. As the minimum/maximum (min/max) temperatures are reached ensure that the steam heating coil and chilled water inlet valves function properly. Dampers and louvers should be checked and the min/max positions locked using the linkages.

The comfort, conditioned space, air quality can be checked with CO_2 monitors. During startup, ventilation may have to be increased to maintain acceptable air quality, after which it should be returned to normal for commercial operation. Multi-meters are available that can check CO_2, humidity, and temperatures with a single monitor.

In the electronic rooms, the moisture should be checked on a regular basis. This was done in the past with mechanically aspirated psychrometers for the wet-bulb temperatures and with mercury-in-glass thermometers for dry-bulb temperatures. Today, there are electronic instruments, used in HVAC testing, that measure and record multiple parameters. If a dew-point depression of at least 20°F is not found, at a dry-bulb temperature of 70°F, corrective action must be taken. Again, the wet-bulb temperature is the dew point. Depression is the difference between the dew point and dry bulb.

Ventilated Spaces

The boiler, turbine, and the balance-of-plant areas must be ventilated and heated if necessary. In coal-fired plants, the turbine building may be under positive pressure to keep dust from entering from the boiler building. This will require air locks for personnel protection when opening the outside doors. The building ventilation system must be balanced to the correct operating pressure. Building pressure can be determined with water U-tube manometers with one leg tapped to the outside ambient air and the inside leg open to room atmosphere.

Automatic rollup filters should have a position switch that rests on the filter media; when the media runs out, the switch lever arm will be released to stop the rollers and alarm to the control room. It is also advisable to have differential pressure switches for underride and override capability. During periods of high dust conditions, a high differential pressure can override the feed timer and increase the feed rate. During periods of low dust conditions, a low differential pressure can underride the rollup controller and prevent media feed when not necessary.

A heating system must be provided in all spaces, taking into consideration winter forced outages. The heating may be by convective steam space heating, electric radiant equipment heaters, or a combination. Ventilation must take into consideration summer air ventilation as well as winter circulation when equipment heat release will reduce heating requirements.

The steam heating supply and drain system must be placed into service. Drain receivers with an outside atmospheric vent must be installed so that they don't generate icing conditions in the surrounding area. If the receiver is pressured, even slightly, during extreme winter conditions and/or if traps pass steam, a steam suppression system will be required. Figure 2-3 is an example of a steam suppression system. The heat balance can be used to determine the spray water requirements to quench steam to atmospheric pressure in the drain receiver. First find the mass flow of steam to atmosphere with a drain tank pressurized to 6 psig. (See Fig. 2-4 for an example.)

Forced-draft room air preheaters require closed-loop treated cooling water with a steam heating system. The steam side of the heater must have an automatic conductivity sampler for leak detection. It checks for leaks from the treated water side to the steam side of the cooler. Because the drain receiver will be at subatmospheric pressure, a drain ejector system will be required. The system is generally designed to automatically start at 40°F and

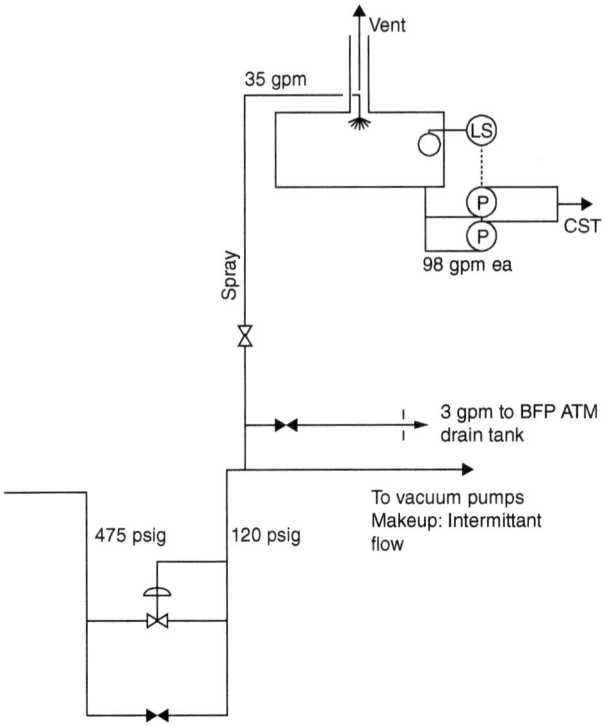

FIGURE 2-3 Spray water system for steam quenching.

shut down at 70°F. Therefore, the system drain receiver must be dedicated and not receive drains from other sources. If the system were to be shut down, drains from other sources could raise the heater tube film temperature above the acidic point of the water treatment chemicals. Glycol becomes acidic at 350°F.

Boiler House Ventilation

Some systems may incorporate complex ventilation designs for winter operation. Boiler houses located in regions of extreme seasonal weather change may have roof and lower-wall ventilators that are reversible. That is, for summer use, the roof ventilators exhaust the heat from the building, pulling fresh air in from the lower levels. In the winter, recirculation ventilators take hot, upper-level air and exhaust it at the lowest level for heating in a recirculation process. In the winter, roof ventilators can be reversed to provide makeup air. Louvers must be integrated into the system for building pressure relief and ventilation control. Some of the louvers may be balanced to open on negative pressure, while others are balanced to open on positive pressure. Louvers may have on-off-auto control so that they can be opened and close automatically when ventilators are started but can also be operated manually if required. There are other obvious startup considerations that need to be addressed. First, in the winter if balanced louvers have not completely closed because of maladjustment, cold air can infiltrate areas not designed for freeze protection when the

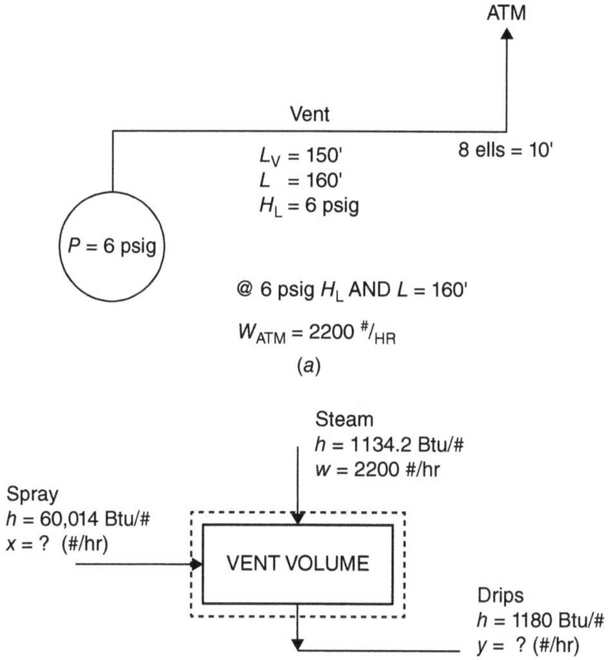

Continuity:

$$x + 2200 = y$$

Energy:

$$60\,x + 1134(2200) = y180$$

$$60\,(y - 2200) + 1134(2200) = y180$$

$$y180 + 60\,(2200 - y) = 1134(2200)$$

$$y180 + 132{,}000 - 60y = 1134(2200)$$

$$y120 + 132{,}000 = 1134(2200)$$

$$y = \frac{1134(2200) - 132{,}000}{120}$$

$$y = 19{,}750$$

$$x = y - 2200$$

$$x = 17{,}550 \text{ \#/hr}$$

$$\cong 35 \text{ gpm}$$

FIGURE 2-4 Steam heating system drain tank vent quenching calculation.

boiler is in operation. This can be serious at the drum level where vital instrumentation is located. Second, air distribution must be controlled so the subzero air is not directly impinging on equipment and instruments. The two modes of operation must provide a balanced draft, which requires a well-designed control system and considerable attention. Therefore, the plant engineer must first review the location of all dampers, ventilators, and distribution systems. Special attention should be given by checking all instrument tubing, devices, piping, and equipment in close proximity to exterior walls. After reviewing the installation, the plant engineer should also consult with the design engineer to determine the design philosophy of the systems.

Crusher Buildings, Feeder Decks, and Conveyor Tunnels

Many areas will require air-cleaning systems to capture dust. Dust is captured in dust collectors, and the air is cleaned with high-efficiency filters. These systems must be integrated into the fire protection system and freeze protection design. Deluge stations must be freeze protected and located where spillage will not cover the equipment. They will be located in a confined protective space with electric radiant heaters or in protective open areas with heat tracing. Conveyor tunnels may have piping that will require freeze protection. Inclined conveyor tunnels will require natural draft and fire control doors.

Air-Compressor Rooms

As stated in Chap. 1, air compressors operating in extreme climates also need special consideration. For example, air compressors, dryers, and receivers operate more efficiently at colder temperatures but need freeze protection in the winter. Air can be exhausted from the building in the summer by ducting the compressor hot exhaust outside. Wall dampers provide air makeup. In winter months, the air can be recirculated inside by exhausting into the building. The inside air temperature is best controlled by an outside thermostat. A single diverter gate can be used as a bypass to reduce pressure drop and, therefore, backpressure on the cooling fan. Care must be taken to ensure large gates do not experience destructive vibration when operating. Harmonic oscillation can occur as the damper is moving. The inside air temperature can be controlled by ventilation air, which is distributed in the building as mix air. The mix-air temperature is controlled by modulating the inlet and mix-air dampers. The minimum makeup and exhaust air should be mechanically set with the damper linkages. Heat tracing is required for all drains. Radiant electric heaters can be installed for outages and extreme conditions to warm the equipment. Also, the compressor aftercooler requires a temperature control switch so it does not subcool the air, which can cause surface condensation on downstream piping, aftercooler drains, and air-receiver external surfaces.

Hydroelectric Turbine Buildings

Small hydroelectric plant buildings can be heated and ventilated in a manner similar to the air-compressor rooms. In this case, the generator hot-air exhaust is used for heating in a similar manner as the air compressors.

Startup Advisory

As can be seen by these simple examples, even ventilation can be a complex process requiring a dedicated effort in order to balance the systems and ensure safe operation before the process systems are commissioned. Furthermore, the HVAC system must be fully integrated into, and commissioned with, the fire protection system and, in some cases, the plant security system. Therefore, in large installations, it is highly recommended that the same startup person be dedicated to the commissioning of both the HVAC and fire protection systems.

STARTUP TESTING

Preparation[2]

Proper preparation is necessary if testing and balancing of the air distribution system is to be accurate. Testing reveals the equipment operating conditions versus design specification. Balancing adjusts the air distribution in each zone and at each outlet to match the system design specification.

Engineering Preparation

1. Obtain the latest edition of the ASHRAE[a] Handbook.
2. Obtain the latest edition of the SMACNA[b] Testing, Balancing, and Adjusting of Environmental Systems Standard.
3. Obtain the latest edition of the SMACNA Guidelines for Seismic Restraints of Mechanical Systems.
4. Obtain the equipment operations and maintenance manuals.
5. Become familiar with the layout of the system.
6. Review the equipment and its design specifications.
7. Prepare a single-line diagram showing all duct sizes; duct traverse points; zone design cfm; supply diffuser and return grille sizes, locations, design volumes, and identification numbers; and locations of zone balancing dampers.
8. Prepare elevation and plan zone diagrams for turbine, boiler, and crusher buildings, etc.
9. Formulate a step-by-step testing method.
10. Prepare field data collection worksheets.
11. Prepare field data analysis spreadsheets for fan performance tests, duct velocity traverses, and diffuser velocity readings with corresponding specific flow results.
12. Obtain testing instrumentation for air velocity and temperature measurement. Electronic meters are available that measure all parameters for HVAC testing. These include CO_2 for air quality, the standard for air quality in occupied spaces. Some manufacturers supply a single, multifunction meter with all parameters required for HVAC testing.

System Preparation

1. Inspect and clean all equipment and ductwork prior to trial operation of the equipment.
2. Check the system for proper seismic restraint installation. The mechanical contractor may have shop drawings, but in many cases, the restraints are installed by craftspeople without using the guidelines provided.
3. Check all zone balancing dampers. Check that all fire dampers are installed in the full-open position.

[a]American Society of Heating, Refrigeration and Air Conditioning Engineers.
[b]Sheet Metal and Air Conditioning Contractors National Association, Inc.

4. The outside air dampers should be set to the minimum position, and return air dampers should be set to the maximum position, unless specified otherwise by the design engineer.
5. All probe holes must be predrilled for duct traverses and temperature and static pressure readings.
6. Close all doors and windows.

EQUIPMENT TESTING

A traverse of the main ducts to determine equipment capacity is sufficient for trial operation. The air handling equipment should be tested in accordance with the latest ASHRAE standards if trial operational tests indicate the equipment may be deficient. The traverses should be divided into not less than four equal areas, with a simultaneous temperature and velocity measurement taken at the center of each area or in accordance with the ASHRAE standard. A single U-tube water manometer is sufficient for the static pressure measurement. The following equation can be used in the analysis spreadsheet for determining specific volumetric flow rate.

$$W_{\text{scfm}} = vA\left[\left(\frac{T_{\text{st}}}{T_{\text{act}}}\right)\times\left(\frac{\Delta P}{P_{\text{st}}}\right)\right]$$

where: W_{scfm} = specific volumetric flow rate
v = air velocity
A = traverse cross-sectional area
$T_{\text{st}}/T_{\text{act}}$ = standard temperature (60°F)/actual temperature
$\Delta P/P_{\text{st}}$ = pressure change/standard pressure (14.7 psia)

SYSTEM TESTING

System testing involves balancing the distribution system. The first step is to balance the zones. This is done by measuring the airflow in the zone supply ducts. It can be done either by traverse or diffuser outlet. If a zone has less than six *diffusers*, it is generally easier to take the diffuser readings. The sum of the diffuser reading equals the zone supply. After the readings have been taken in the zones, the balancing dampers should be adjusted to supply each zone with the design flow.

Readings should then be taken at every diffuser in the zone, starting from the first branch take-off and moving out. Comparison of these readings with the original design flow will reveal how the air is being distributed and which outlets require adjustment.

In new installations, the outlets nearest the main duct will have air volumes above design values, and outlets at the end will be below design values. After the system is balanced, the air-handling equipment should be retested to ensure it is still operating as designed.

High-velocity systems (>2000 cfm) operate at over 2 in w.g. static pressure. Usually automatic control devices are used to maintain constant pressure or air volume in the system. The plant engineer must understand the automatic control for overall operation before taking the first readings. Pressures lower than design, in the mixing boxes, will cause them to malfunction, usually with excessive volume, fluctuations, and noise.

The following is a list of minimum data required when reporting HVAC deficiencies to the design engineer:

1. Date and time.
2. Name of space and location.
3. Outdoor dry-bulb and wet-bulb temperatures.
4. Volumetric air flow into space.
5. Dry-bulb temperature of air being supplied to space at air outlet.
6. Indoor dry-bulb and wet-bulb temperatures.
7. For cooling coils, heating coils and air washers:
   a. Air entering dry- and wet-bulb temperatures.
   b. Air leaving dry- and wet-bulb temperatures.
8. Where applicable, return-air, outside-air, and mixed-air temperatures taken within any given half hour.
9. For chilled water systems, water entering and leaving temperatures:
   a. Water chiller inlet and outlet.
   b. Chilled water coil inlet and outlet.
10. Was an air balance performed? If so, by whom?
11. Obtain a copy of the air and water balance report from the contractor with all the data sheets.

REFERENCES

1. Baumeister, T. and Marks, L. S., *Standard Handbook for Mechanical Engineers*, 7th ed., McGraw-Hill, New York, 1976, p. 12–130.
2. Adapted from Kisner, F., "Testing and Balancing the Plant Air Distribution System," *Plant Engineering*, August 7, 1975.

CHAPTER 3
ASH-HANDLING SYSTEMS

George A. Shultz[1] stated: A plant engineer "installing a pneumatic system for the first time may be faced with the following odds:

- 5–10 percent chance that the system will never work.
- 10–15 percent chance that the system will be abandoned.
- 15–25 percent chance that the system will never perform up to his expectations."

The reason for these high failure probabilities is the lack of existing materials data and/or the lack of testing of material-flow properties. If the fuel has never been burned before, there will be nothing more than crucible ash to analyze. This is especially true of waste fuels that have a high variation in material properties mixed with sorbents and circulating inert material such as sand. Using empirical data derived from parent coal for waste fuels could prove the preceding statistics correct.

> The variables affecting the minimum conveying velocity are material-to-air loading (type of system), absolute density, particle size and shape and physical characteristics such as tackiness, abrasiveness, moisture content, and so on. Because of the unknown effect of these variables on the conveyability of a material, the best method of determining conveying velocity is to test the material-flow characteristics in a pilot plant using the type of equipment that is being considered for selection.[2]

During the 1980s, engineers were faced with designing fuel- and ash-handling systems for unproven fuels fired in boilers designed for proven fuels. In the 1990s, engineers were faced with the same problems in new, and not yet fully proven, boiler designs. Pilot plants could not be built large enough to provide much-needed empirical data needed to guarantee 100 percent trouble-free design and, therefore, design-problem-free commissioning. The majority of design problems encountered during these periods were fuel and ash-handling related.

The plant engineer should always assume the devil's advocate role. Do not assume the manufacturer received accurate information from the design engineer or that the engineer received accurate information from the owner, or even that tests were performed if the information was outside his or her experience. Furthermore, it may have been impractical to test the ash-conveying characteristics if there were no similar facilities available to produce it in quantities required for pilot plant testing.

Pneumatic ash-handling systems can be designed from existing tabulated information available for standard common fuel sources. This information can only be used as a guess, however, when conveying ash derived from untested fuel sources such as waste solid fuels or fuels from previously unknown sources. The reason for this is simple: Every equation used in fluid mechanics has constants that are derived empirically—that is, the fluid flow was observed in a testing facility with proving vessels and test instrumentation from which equations for flow were derived.

In view of such information, short of writing a complete material conveying handbook, this chapter attempts to give the plant engineer the tools by which he or she can understand the mechanics and fundamentals of ash-handling conceptual design, from a startup perspective, and provide the tools to analyze, troubleshoot, and commission power plant ash-handling systems.

ASH-HANDLING FUNDAMENTALS

Pneumatic Systems

Pneumatic systems fall into two categories: dilute phase and dense phase.

The *dilute phase* is actually a term used in industry that refers to material flowing in suspension, sometimes referred to as "suspension-phase flow." The material travels in suspension within the conveying air. Superficial velocity is the velocity of the air within the system, irrespective of the material velocity, and is used for all flow calculations. Superficial flow velocities are in the order of 13 to 15 m/sec (43 to 50 ft/sec).[3] In the dilute phase, the flow must be large enough to maintain the material in suspension. The point at which the material can no longer stay in suspension is the velocity at which flow stagnation occurs. As material falls out of suspension, it partially restricts the flow, causing a higher velocity at the restriction. This, in turn, causes a reversal of flow, causing more material to drop out. The dropout and reentrainment of the material is called *saltation*. We can tell if the flow was at, or approaching, the saltation velocity by looking down the inside of the pipe. An indication of saltation is wave formations of material in the length of pipe. To differentiate between stagnation pocket deposits formed by directional changes in the system, we must be at least 20 ft downstream of the flow disturbances. Saltation can eventually cause complete blockage of the line. Dilute phase, therefore, is a high-velocity system of conveying material.

The curve in Fig. 3-1 illustrates how critical velocity and carrying capacity are related. If the pickup velocity is not known, within a narrow band, the carrying capacity will only be a guess. This curve only applies to one material specification and is used here as a typical plot of capacity versus velocity.

When a material is conveyed in a system that is filled with dense but fluidized material, the process is called *dense phase*. The material has to be completely fluidized in order to convey it. If it is not, the system will plug. The outlet of a transporter tank, used to fluidize a material, can be pulsed such that the material in the system conveyer piping will flow in slugs so as not to block the line. Or, internal or external air nozzles can be used to keep the mass fluidized and moving in the system. Dense phase is a low velocity system generally less than 33 ft/sec (2000 ft/min)[5] using small-bore piping systems. These are not typically used in power plant ash-handling systems.

Types of Dilute-Phase Pneumatic Systems

There are three types of dilute-phase pneumatic systems: positive-pressure, vacuum, and combination vacuum/pressure ("pull-push" system). The velocity relationship between pressure versus vacuum systems depends on density of the conveying air (Fig. 3-2). Which system design is used, however, depends on distance, operation, and construction

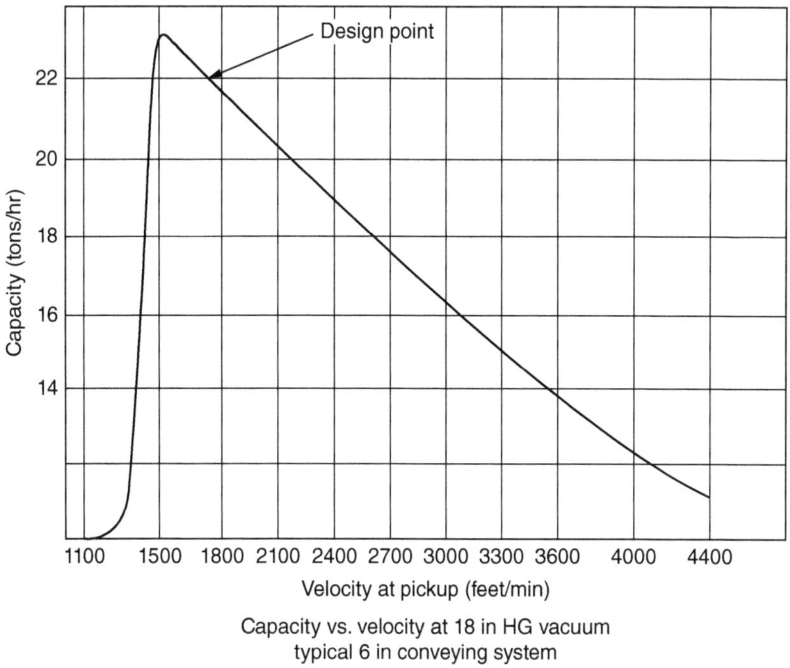

FIGURE 3-1 Capacity vs. velocity curve.[4, 5]

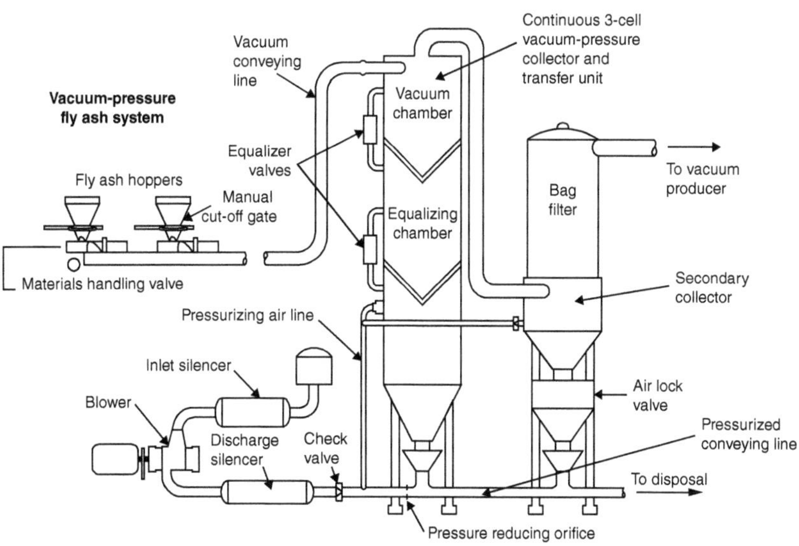

FIGURE 3-2 Combination positive pressure and vacuum system.

considerations rather than density, velocity, or erosion. For illustration purposes, a comparison of air density for the two systems is given here.

For the pressure system, the temperature rise due to compression is calculated by:

$$T_s\, p_d^{(k-1)/k} = T_d\, p_s^{(k-1)/k}$$

$$\text{Assume } T_s = 70°\text{F}$$

where T_s, p_s = supply air temperature and pressure
T_d, p_d = discharge air temperature and pressure
K = ratio of specific heats, C_p and $C_v = 1.4$

$$\begin{aligned} T_d &= T_s(p_d/p_s)^{(k-1)/k} \\ &= 70°\text{F (25 psia/14.7 psia)}^{.286} \\ &= 120°\text{F} \end{aligned}$$

Assume 10-psig pressure because 15 psig would require that the system components comply with the American Society of Mechanical Engineers (ASME) code for unfired pressure vessels.

$$\rho_d \mathrm{v} = \mathrm{RT}_\rho = \rho_d/\mathrm{RT}$$

where $R = 53.3$
ρ = density in a pressurized pneumatic conveyor
v = density in a vacuum pneumatic conveyor
$T = T_d + 460$
$\rho_p = 25\ (144)/(120 + 460)53.3$
$= 0.115$ lb m/cu ft, for the pressure system

Assume a vacuum of 12 in Hg, a nominal value, and air temperature of 70°:

$$\begin{aligned} \rho_v &= (29.9 - 12)0.491/(70 + 460)53.3 \\ &= 0.0448 \text{ lb m/cu ft, for the vacuum system} \end{aligned}$$

For reference $\rho = 0.0749$ for standard atmospheric pressure and temperature.

From the comparison, it can be seen the *pressure* system density is the larger of the two, and therefore, the suspension velocity can be lower than that of the vacuum system. As a consequence, the erosion in the vacuum system may be higher than that of the pressure system. *This should dispel the notion that vacuum systems are less susceptible to erosion.* Of course, the degree of erosion is also directly related to the material being transported. Vacuum systems are, however, preferred by plant personnel because air leakage does not blow ash into the plant.

Positive Pressure versus Vacuum System

The positive pressure system is generally used for conveying long distances and/or to multiple delivery points. "Because there is no pressure limitation, theoretically there is no conveying length limitation on pressure systems."[6] The downside is that because the system is pressurized, additional equipment, such as air locks, is required and is more expensive. On the other hand, air separation at the delivery point is much simpler than for vacuum systems. Likewise, the air-handling units are located on grade, whereas the vacuum system equipment has to be elevated at the delivery point, which requires additional structural design and construction considerations.

Overall, vacuum systems are generally preferred over pressurized systems, especially by operations, because the noise-producing equipment is away from the main operating floor, the controls are much simpler, and air leakage does not produce fugitive ash problems.

Erosion Mitigation

Important Note Regarding Erosion

Piping system erosion is a major concern when using high-velocity pneumatic conveyors. System layout should minimize piping run length and directional changes. Also, the shortest radius bends are preferred in regard to pressure loss when using long radius elbows. "Comparison of the short radius and long radius bends is in Fig. 3-3. This shows that the short radius bends are clearly the best of those tested in terms of minimum pressure drop. Only at the very highest material flow rates are the long radius bends slightly higher."[3]

The preceding reference compared elbow bend ratios of 6 versus 12. Exxon used a radius ratio of 12 for their pressure drop experiments.[6,7] Standard short radius pipe elbows have a radius equal to the diameter. In other words, they have a radius-to-diameter ratio of 1. Standard long-radius pipe elbows have a radius 1.5 times the diameter, or a 1.5 radius-to-diameter ratio. Friction loss in pipe elbows decreases as the bend radius increases up to

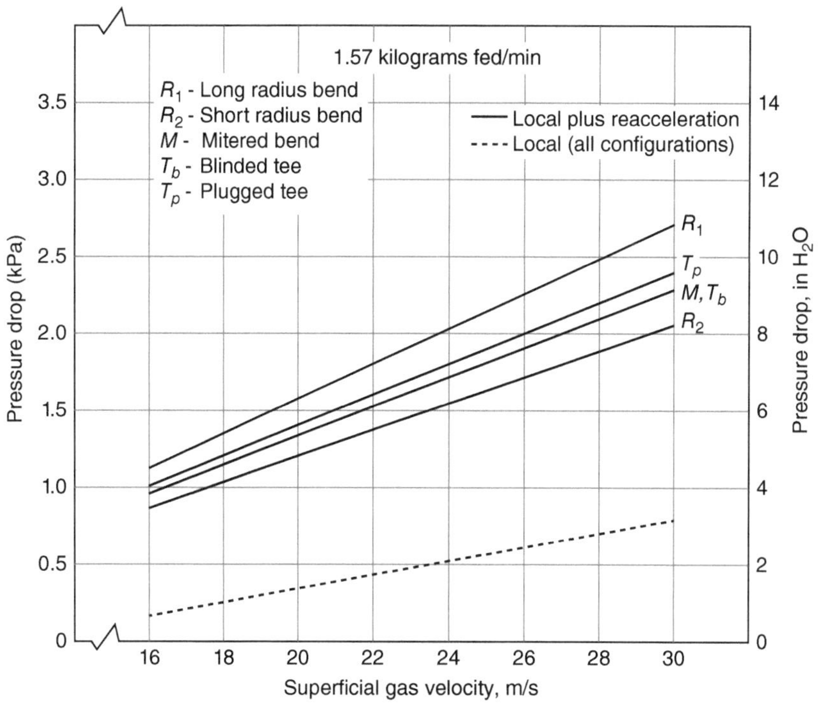

FIGURE 3-3 Pressure drop for bend configurations as a function of superficial gas velocity.

a radius/diameter ratio of approximately 2.5. For ratios greater than 2.5, the friction loss actually increases for long radius elbows.[8]

Also, tests have shown that the preferred directional change is the 90° blinded elbow, or lateral, which has been demonstrated to be the least susceptible to erosion. The author confirmed the erosion results with onsite field testing. The original tests were conducted by Exxon Nuclear for the U.S. Department of Energy (DoE). Elaborate laboratory tests were used for multiple line and fitting configurations with an assortment of materials. The tests were performed to determine flow patterns and erosion characteristics. "*These tests have shown that the use of blinded tee or lateral configurations instead of radius bends will significantly improve the erosion resistance of a dilute-phase pneumatic system without adversely affecting other performance parameters.*"[6] Also, a wear pad should be installed, fitted with an air line to the seal annulus, between the fitting and the pad. Failure of the fitting due to erosion will be indicated by air flowing through the proofing line.[6] This will provide time to plan a system outage to replace the fitting. Figure 3-4 illustrates pipe-fitting configurations with expected failure points.

Needless to say, if the plant engineer encounters failures due to erosion before commercial operation, system configuration changes will have to be made. If the configurations appear adequate, then material specifications should be checked. All materials should be highly erosion resistant. Fitting material should be cast iron or other high-wear-resistant material as well as the first pipe joint downstream of a flow disturbance. Figure 3-5 illustrates the ideal versus a worst-case configuration. Ironically, the 90° configuration is better than the double 45°s, from an erosion point of view.

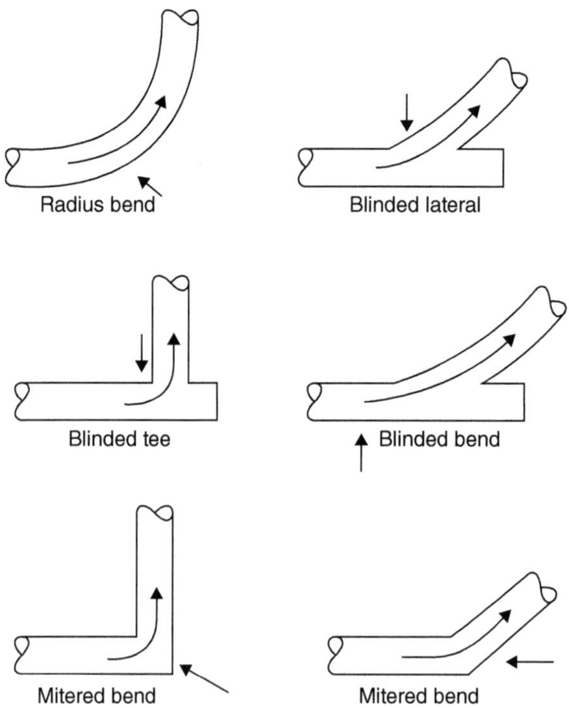

FIGURE 3-4 Typical elbow configurations.

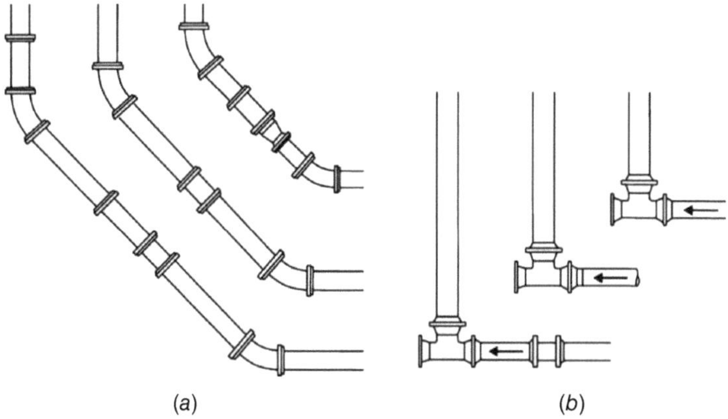

FIGURE 3-5 Poor (*a*) vs. best (*b*) elbow configuration.

If severe erosion occurs and the configuration and material specification are adequate, then the flow must be checked. Higher-than-design velocities will yield lower-than-design capacities. This can be attributed to an incorrect line or air mover sizing.

MECHANICAL AND HYDRAULIC SYSTEMS

Long-Distance Conveying

Both mechanical and hydraulic systems are used to convey material for long distances. They are often used to convey ash from a collection silo to a disposal area. Neither belt nor hydraulic conveyors are restricted by changes in direction.

The belt conveyor can utilize a roller system to transform the belt from a flat, directionally limited, conveyor to a tube, directionally unlimited, conveyor. The belt conveyor has the advantage of providing a choice between wetting at the supply end or at the disposal area. It also has an advantage in that there is no conveying water to treat and dispose of. The disadvantages are the higher cost of the equipment and its installation. Also, dust suppression can tax the maximum capabilities of the design engineer and the manufacturers. At best, it will work as designed, but it will always be a high maintenance item.

The hydraulic system has an advantage in that there is no need for dust suppression systems. Also, transport will not require transfer stations that the belt conveyor would require for extra-long distances. The cost of the equipment and the installation is less than that for the mechanical system. However, these costs could very well be increased significantly, depending on the water supply and disposal requirements. This could offset the cost advantage the hydraulic system has over the belt system. Also, because hydraulic conveyors are high-pressure, high-velocity systems, they cannot convey abrasive materials, such as sand found in fluidized bed boilers, without extra consideration for erosion. "The silica and alumina content in the limestone added to the combustion process, as well as the silica and alumina content in the rocks usually mixed in these fuels are transformed during the combustion and subsequent circulating process, thus enhancing the erosion capabilities of the bed ash. The end result is a bed ash with coarse, laminar flat particles, and fine, spherical particles."[7] Also, if the pump water supply contains sand, the sand should be removed with cyclone separators before it is pumped, or severe pump erosion problems will occur.

Hydraulic conveyors are also used for bottom ash removal. Bottom ash is collected in wet-bottom ash hoppers with clinker grinder feeders. The ground ash is conveyed to a distant dewatering silo or a disposal pond by an ash sluice system. Great care must be taken when designing controls for such systems so that clinker grinder or sluice system failure will not cause the furnace ash hoppers to fill up into the boiler beyond capacity. If this occurs, the unit must be taken off line. The extra weight of the ash, for which the boiler was not designed to hold, caused the bottom of one furnace in China to fail and fall, releasing tremendous stored energy of both fire and steam and causing an extremely high death toll.

Short-Distance Conveying

Although all systems can be used for short-haul service, the following are by far the most economical for short distances. The scraper, often called "drag" or "drag-chain," conveyor and screw conveyors are the most common considerations for short-distance ash transfer. The scraper is generally used for wet-bottom ash removal and the screw for dry-ash conveying when the deposit site is adjacent to the boiler. The wet-scraper conveyors are generally used for smaller boilers (around 150 MW thermal or less). Available also are the metallic-belt and shaker-table conveyors. Most bottom-ash conveyors, in general, have sealing requirements. The sealing design depends on whether the hoppers are wet or dry and on boiler design such as the negative furnace pressure of a pulverized coal (PC) boiler or the positive wind box pressure of a circulating fluidized bed (CFB) boiler.

Wet-bottom pits do not lend themselves well for collecting fines such as would be found in CFB boilers or stokers firing wood using drag chains. Fines require much more residence time to settle to the bottom where the conveyor is located. Also, if the conveyor speed is too fast, it will cause a turbulence that causes the fines to stay in suspension above the conveyor. An ideal match between settlement time and conveyor speed is directly proportional to conveyor width. Nevertheless, some particle settlement rates will prohibit the use of this type of conveyor. If the designer failed to make the match correctly, the system will require a major field design change, usually requiring a conversion to a sluice system. This phenomenon has been observed where the fines, concentrating in suspension, above the conveyor blades, eventually dewatered the ash and created a solid bridge between the conveyor and the conveyor ash hopper inlet (See Fig. 3.6).

The Magadli Ash Conveyor (MAC) bottom ash system does not require ash coolers (Fig. 3-7). Ambient air drawn over the ash into the boiler cools the ash; the hot air enters the boiler as a heat recovery mechanism.

Fly-Ash Hoppers

Precipitator fly ash hoppers must have heaters to prevent plugging. Fly-ash hoppers can plug when the temperature is allowed to drop below the dew point. The dew point temperature is determined by the moisture content and the sulfur dioxide and trioxide content of the flue gas. Typically, it is between 180°F and 280°F. Changes in dew-point temperature can be caused by changes in the moisture content of the ambient air feeding the furnace, changes in coal moisture and sulfur content, boiler tube leaks, sootblower operation, and flue-gas-conditioning systems.[9]

Note that the theoretical calculation of heating needs is, at best, only an approximation. Textbook values of heat-transfer coefficients may be off by as much as 20–40 percent. Test installations must be made, and measured temperature profiles taken, to determine accurate heat-transfer coefficients and actual heating needs. Companies that manufacture hopper

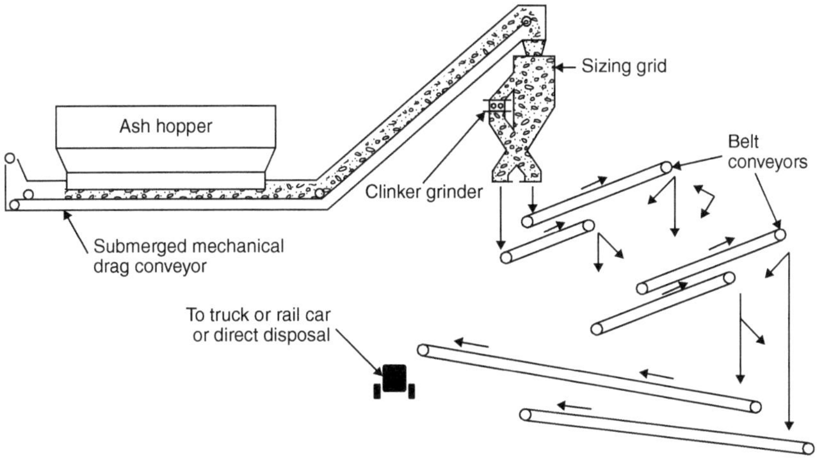

FIGURE 3-6 Wet-bottom ash drag chain conveyor.

FIGURE 3-7 Dry-bottom ash belt conveyor, Magadli ash conveyor (MAC).

heaters have such test data on file and have developed techniques to calculate realistic heating needs.[9]

DUST SUPPRESSION SYSTEMS

Dust suppression, or ash conditioning, is done by wetting ash for transport or disposal. Ash is not easily moistened because the surface tension of water rejects the process. If the surface tension could be reduced by using demineralized, or even softened, water and treated with wetting agents, the process would be greatly enhanced and water consumption reduced.

However, conventional methods utilize untreated service water that is sprayed into mixers as the primary dust control stage. Generally, the output is a mixture of water, wet ash, dry fugitive ash dust, and water vapor. The latter makes a difficult combination for fugitive dust control because the dust and vapor coat duct and fan surfaces, which make up the secondary system. Also, the wet ash falling from trucks carried by free water makes a third system necessary to process fugitive sludge.

Explosion Prevention

High-carbon ash should never be conveyed by pneumatic systems due to the explosion potential. Sometimes, flue gas is used in lieu of air because of its low oxygen content. However, flue gas should not be used because the moisture in the gas will cause system plugging. If the system is heated, the gas must be cooled before entering conveying compressors. Once in a storage silo, the explosion danger remains, regardless of how the high carbon ash was conveyed.

Mechanical systems such as screw or scraper conveyors should be used to feed high-carbon-content ash to watering ash conditioners. The water quenches cinders that will cause the carbon to ignite when it comes in contact with fresh air. It also suppresses hot carbon dust, which is highly volatile when exposed to air. Expect the same fugitive dust and sludge problems for the same reasons described earlier for dust suppression.

STARTUP TESTING

Preparation

During initial trial operation, plugging of the system is by far the most frequently encountered problem for any type system. Knowing the equipment and system operating parameters will help in preparation for solving these problems. If climates vary, pneumatic system performance will vary and, therefore, the plant engineer must know what effect increases or decreases in conveying air temperature have on system performance.

Engineering Preparation

1. Obtain all the equipment operations and maintenance manuals.
2. Obtain all ash system operational specifications, such as storage capacities, conveyor capacities, design cycles, velocities, temperature ranges, operating pressures, pressure drops, and capacity curves.
3. Prepare a test and proceed to check the system versus air mover capacity (refer to Chap. 1 or Chap. 18).
4. Prepare a test data worksheet.
5. Prepare a test calculation spreadsheet for expedient processing of the test data.

System Preparation

1. Install a flow nozzle at the compressor inlet or vacuum pump exhaust. Install pressure gauges and manometers at the compressor discharge or vacuum pump inlet and silos and bag houses.

2. Cycle all feeders and log the feed cycles.
3. Check all hopper level switches and heaters.

TESTING

Run the air movers and test for capacity and system operating pressure gradients using ambient air. Use the methods discussed in Chap. 1 and Chap. 18 to determine inlet air or sluice water velocities. The actual inlet air velocity should be checked against the design data and compared to theoretical changes in air temperature.

TROUBLESHOOTING[a]

Cold system, during startup or cold conveying air, can cause moisture to form in the piping causing the ash to plug the system. The system should be started and cycled before solid fuel burning begins and ash introduced. The dry cycling will let hot, boiler, warm-up gas heat the piping.

High inlet air velocity, due to low ambient air temperature, will reduce the carrying capacity of the air and may cause blocking of the lines if the air reaches the saltation velocity. This must be determined by the tests performed earlier. If this occurs, flow controls must be applied as described in Chap. 1. This can be a problem if the design velocity was specified 20 percent oversized as a design buffer recommended in some guides.

Overfeeding can cause plugging if cycling is not properly timed or feeder speeds are excessive. This can create worse problems in sluice systems than in pneumatic systems in that the blockage in sluice water piping will cause the feeders to be pressurized and can force conveying water into the feeder and the hoppers. If this occurs, all components will require removal and cleanout. A rotary valve speed of 1 rpm off design can cause this problem.

Purging of the system is required on a unit outage before the system is shut down. If it is not purged of ash, the ash left in the system can cause blockage when the system is restarted.

System surges caused by initial feed can cause blockage if the system is operating close to the saltation velocity. This can be determined by the testing described earlier.

Sluice water system imbalance can cause the sluice water pumps to run out reducing system pressure and branch flow below the carrying velocity of the water. The sluice water system must be balanced so that cycling of different feed points such as bottom ash, economizer ash, and silo ash do not operate at the same time, reducing the velocities to stagnation. Running the pump out beyond its design capacity can also cause severe pump erosion and cavitation damage.

Ash pit sediment can cause sump pump suction blockage. A recirculation line from the pump discharge to a suction sparger can be used to keep the ash in suspension during pump operation.

Minimum carrying velocity for slurries must be maintained. The minimum carrying velocity is dependent on the percent of solids in the water used for conveying. The minimum velocities have been determined empirically by many experts in the field of fluid mechanics. One such relationship is as follows:[10]

$$V = (k_L \, \psi \, D_P)^{0.816} \, (D \, \rho_m / \mu_L)^{0.633}$$

[a]Reference 3 is an excellent book for solving design and operating problems of pneumatic systems.

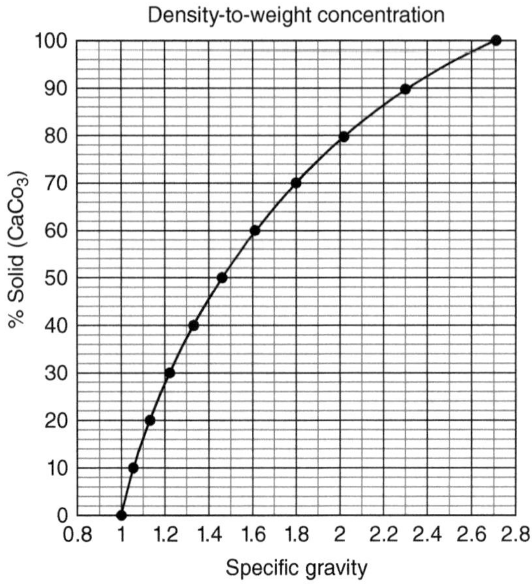

FIGURE 3-8 Material density for carrying capacity determination.

where $\psi = (\rho_P - \rho_L)/\rho_L$

k_L = 232 for minimum flow velocity

D_P = *mean* particle diameter, ft

D = pipe diameter, ft

ρ_P = particle density, in

ρ_L = liquid density, lb/ft^3

Density can be determined roughly from Fig. 3-8 for scrubber slurries.

REFERENCES

1. George A. Schultz, *Plant Engineering*, June 12, 1974.
2. Milton N. Kraus, *Pneumatic Conveying Systems for Bulk Materials*, Prentice Hall, Upper Saddle Level, New Jersey, 1991.
3. David Mills, *Pneumatic Conveying Design Guide*, Warren Springs Laboratory, 1990.
4. Thomas C. Elliot, *Standard Handbook of Powerplant Engineering*, McGraw-Hill, New York, 1989.
5. Allen-Sherman-Hoff, *A Primer on Ash Handling Systems*.
6. S.S. Bodner, "Effects of Bend Configuration on Dilute-Phase Pneumatic Transport," Exxon Nuclear Idaho for the DOE, May 1981.
7. Jose N.R. Trejo, "Bed Ash Conveying System, Case Study for American Bituminous Power," Technicon Enterprises Inco., January 1994.

8. K.H. Beij, "Pressure Losses for Fluid Flow in 90 Degree Pipe Bends," *Journal of Research of National Bureau of Standards* 21: July 1938.
9. Robert Bylin, "Hopper Heating Systems: Key to Minimizing Flyash Plugging," *Power*, June 1981.
10. Lovenstien, "Settling Phenomena in Slurries," in *Piping Handbook*, Crocker and King (Eds.), McGraw-Hill New York, 1973, Ch. 19.

CHAPTER 4

BEARINGS AND LUBRICATION

A highly recommended monthly publication for power as well as for industrial plant engineers is *Plant Engineering*. It provides extensive technical information and problem solutions. At the end of this chapter is a partial list of bearing terminology from an extensive special report from the magazine. It is free and can be acquired at www.plantengineering.com.

Bearings fall into two categories: antifriction and plain (sleeve) bearings. Antifriction bearings are either ball or variations of roller bearings. Sleeve-bearing friction is by sliding action, whereas antifriction bearings are by rolling friction. Therefore, they derive the name *antifriction*. The manufacturer of the equipment in which the bearing is installed, whether antifriction or plain, usually makes the selection of what type of bearing is used. Generally speaking, cost is an important factor in the selection process. However, this is not always the case. The late Mr. Karassik, a former great of the pump industry, gave us his views while at Dresser Pump Company:[1]

Some of the advantages in favor of the antifriction bearings are the following:

Greater interchangeability of parts. Replacement bearings are generally available throughout the world as off-the-shelf items, and bearings made by one manufacturer are interchangeable with those made by any other manufacturer, size for size and type for type. They are manufactured to a high degree of precision, so that interchangeability and performance are assured.

Shorter shaft spans. The results in a lower shaft deflection. Although this factor may not be of significant importance in single-stage pumps...it becomes quite important for multistage designs. But even if it is not too significant, the effect is in the right direction.

Accurate alignment throughout the life of the bearings. Wear will not cause the shaft center to drop, preventing eccentricity of internal fits. This is not true of sleeve bearings, which may still operate satisfactorily after considerable wear has taken place; this wear, however, will impair alignment and may lead to other mechanical damage.

Lower starting torque. This is an important advantage, because there can be no damage at start-up in the event of a momentary absence of lubrication.

Low fiction operation. The power loss of antifriction bearings is considerably lower than that of sleeve bearings. This reduces the amount of heat generated and permits satisfactory operation with a lower volume of lubricant.

Simpler lubrication. The lubrication of antifriction bearings requires less attention on the part of the operators. Relubrication of bearings is generally required as infrequently as every 3 to 6 months. Loss or deterioration of the lubricant will shorten the life for the bearing but does not generally result in the immediate failure of the bearing.

Bearing failure is not necessarily accompanied by damage to the shaft. Maintenance procedures are available for remetalizing a shaft after it has been damaged in the journals if sleeve bearings are used, but these procedures are neither simple nor inexpensive.

Antifriction bearings are capable of carrying higher momentary overloads without failure or seizure than sleeve bearings.

I would be less than factual if I failed to mention certain advantages that sleeve bearings have over antifriction bearings:

1. *They are frequently quieter in their operation.*
2. *Split bearings are more readily accessible for inspection or for removal without disturbing the coupling alignment.*
3. *They are less sensitive to small amounts of dirt or to excessive lubrication.*
4. *If necessary, they can be relined in the field by the maintenance personnel without the need of purchasing replacement bearings.*

All of these factors must be evaluated and compared when reaching a decision in favor of one or the other type. The overall evidence, however, seems to indicate that the advantages in favor of antifriction bearings outweigh those in favor of the sleeve type, because most manufacturers have standardized on antifriction bearings. It is also my impression that the majority of specifications are written today either for antifriction bearings or so as to permit either construction. Certainly I would recommend that you do so.

Earlier, I had referred to specific cases in which factors other than personal preference dictate the choice of bearings to be used. There are two areas in which this is particularly true:

1. *When shaft diameters are large and operating speeds are high, antifriction bearings are not applicable, and the standard practice is to use sleeve bearings. This is the case, for instance, with high speed boiler feed pumps.*
2. *When overhung cantilever construction is used, as in radially split casing-type process pumps, it is not practical to use sleeve bearings and only antifriction bearing design are available.*

Another point is, for most normal applications antifriction bearings do not require spares to be stocked. An investment does not need to be made for warehousing parts that the local supplier has on shelf. On the other hand, spares for both large plain and antifriction bearings should be available during startup.

Large antifriction bearings can be split if required for maintenance:

Split bearings offer the plant engineer an option that may be of considerable value when straight roller bearings are needed on a shaft that is difficult to access. These bearings, which are not to be confused with split pillow blocks, are split completely—all the way through

the inner race. This feature permits replacing the bearing without dismounting the shaft. Because the number of parts involved is much greater, the time required to install a split bearing is several times greater than that required to install a standard bearing on an easily accessible shaft. Installation procedures differ considerably, and even personnel who frequently deal with large bearings should go over the manufacturer's instructions carefully. It is vital that the shaft be slightly *larger* than the inner race, for example.[2]

Both plain and antifriction bearings can be designed as thrust bearings. Antifriction thrust bearings are radial, straight-roller bearings. Plain-thrust bearings can be fixed-land or tilted-pad (Kingsbury) bearings. Obviously, straight-roller journal bearings cannot withstand thrust loads. Therefore, when thrust loads are present, tapered roller bearings are used. Ball bearings can withstand some thrust loading, but if thrust loading is a requirement, the race cups are designed for axial contact. This is the case for double-suction centrifugal pumps where the axial load is not always completely balanced or the pump is slightly thrust-loaded off motor magnetic center to prevent the pump from oscillating axially.

For large steam turbine and boiler feed pump applications, plain-thrust bearings are used. They will be either a fixed-land (such as multi- or taper-land) or tilting-pad design. "Because taper land bearings cost less to manufacture than tilting-pad bearing, designers choose them whenever possible. But, operating conditions such as varying speeds, imperfectly aligned thrust collars, or instantaneous acceleration to full speed and load can make taper-land bearings fail too frequently, replacing them with tilting pad bearings will normally end the problem."[3]

Metal Temperature Ranges for Plain Journal and Thrust Bearings[4]

Shaft Speed	Bearing Type	Range of Bearing Sizes (in)	Range of Normal Bearing Metal Temperature (°F)[a]
3600	Elliptical	10–15	150–180
		16–22	180–200
	Tilting pad	10–15	170–200[b]
		16–22	180–210
1800	Elliptical	12–22	130–170
		23–32	150–190
	Tapered land	All	130–170
	Pivoted shoe	All	140–170

[a]Ranges of normal bearing metal temperatures presented in the table assume that the bearing and journal/collar are in good condition and aligned properly and that the lube oil flow and temperature are at design values.

[b]Temperature will be 10–20°F higher at the no. 1 and no. 2 bearings under conditions of maximum-admission steam forces.

The equivalent journal bearing to the tilting pad is the pivoted-pad plain bearing. The following is an interesting lesson on pivoted-pad journal bearings (PPJBs):

Two cases of improper design: 1300-MW steam turbine. The PPJB shown Fig. 4-1(*a*) was the result of a considerable research effort[5] to minimize power losses. It is probably the most energy-efficient bearing in the large-turbine industry today. The unfortunate feature is that most of the load is supported on the bottom pad (the largest of the three). As a result, these machines have been difficult to balance.

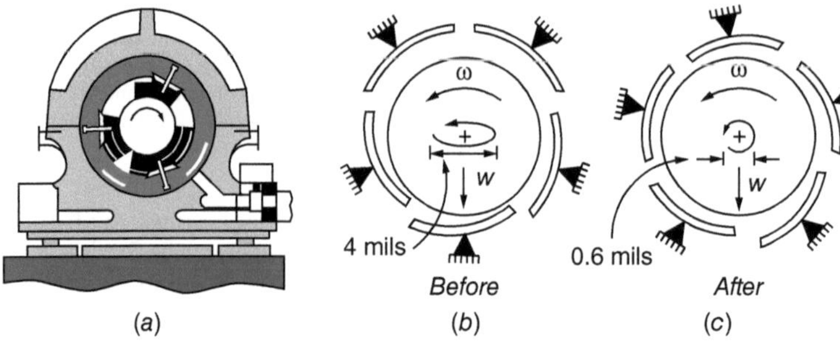

FIGURE 4-1 PPJB development.

> The field fix applied by the manufacturer to some units has been to reorient the inner bearing shell by 15 deg to distribute the load more evenly on two pads. However, this fix probably compromises the original energy-efficient feature. The manufacturer has recently changed to four equally sized pads with deadweight load between the two bottom pads. None of these redesigned units are yet operational.
>
> Boiler-feed pump. Recent field experience on a large feedwater pump demonstrates the importance of load direction. Five-pad bearings were used, with one pad directly on the bottom and no pre-load [see Fig. 4-1(*b*)]. Large horizontal vibrations were a continuous operational problem. These vibrations were eventually reduced to a small circular orbit by rotating the bearing shell one-half of a pad pitch (36°) and adding preload [see Fig. 4-1(*c*)].[6]

By the time of this writing, problems such as those that occurred in the late 1970s and early 1980s have been resolved. But these examples are good lessons learned in the development of this type of bearing.

BEARING IDENTIFICATION

The chart on the following page is used to identify antifriction bearings. Reference No. 9. A copy of Standard 20 can be acquired from the AFBMA, 1101 Connecticut Ave, NW, Ste 7000, Washington, DC; Phone 202-429-5155:

> *The most common method of identifying bearings on a motor nameplate is a standard bearing designation defined by the AFBMA in "Radial Bearings of Ball, Cylindrical Roller and Spherical Roller Type, Metric Design" (Standard 20). The standard uses alternating sets of numbers and letters to completely identify a bearing. The accompanying chart "AFBMA Standard Bearing Designation" explains the system and includes the specific information encountered on many motor nameplates.*
>
> *Older bearing designations*
> *Bearings in older motors may be more difficult to identify because motor nameplates may carry manufacturers' numbers. These designations typically begin with four digits that identify the bearing type and size. The numbers are followed by an alphanumeric identifier that designates hardware additions and clearance information.*
>
> *Bearing type*
> *The first digit of the main number (or the first two digits) identifies the type of bearing. The most common antifriction bearing used in motors is the single row, radial, deep groove, ball bearing, sometimes referred to as a CONRAD bearing. Primarily used to handle radial loads;*

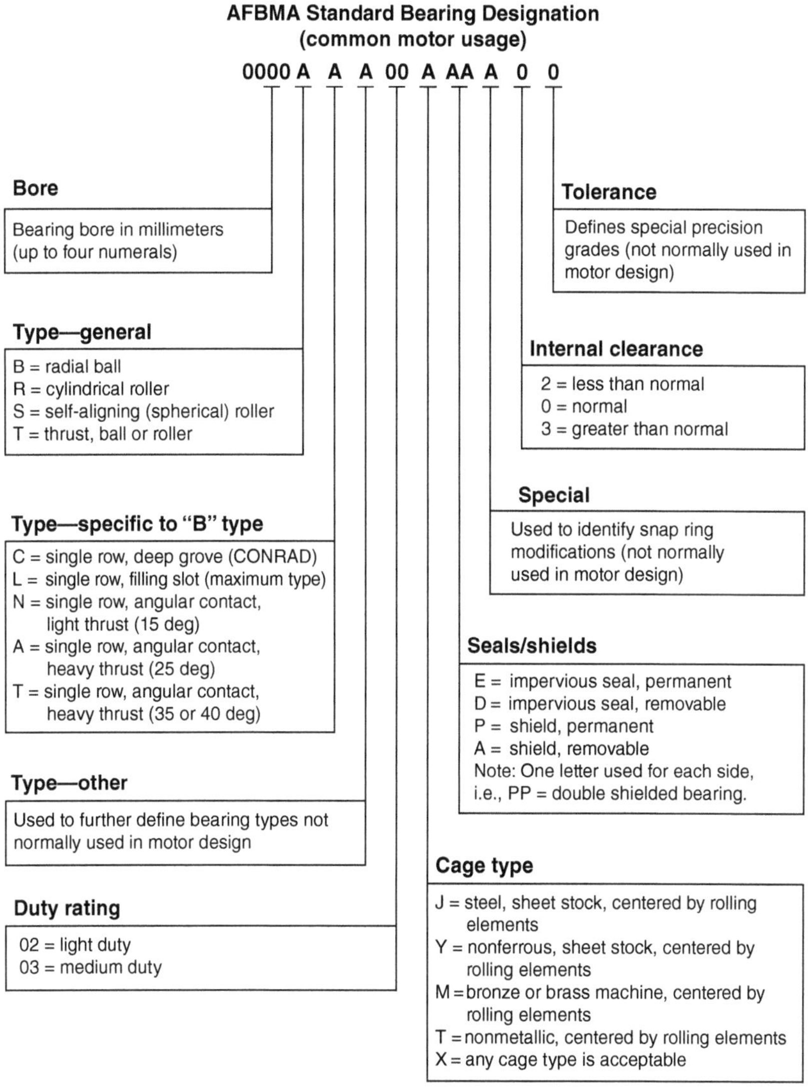

this type of bearing also can handle moderate amounts of thrust. In such applications, no additional thrust can be transferred to the motor bearings by the driven machinery. Most manufacturers use the number 6 to designate this type of bearing.

Vertical motors with hollow shafts are usually required to carry high thrust (for example, the weight of the rotating assembly in a multistage vertical pump) in addition to the weight of the motor rotor. In these motors the upper (thrust) bearing most often employed is a single row, angular contact ball bearing. It has a high contact angle (35 or 40 deg) and is identified as number 7.

Series or duty

The next digit in the main bearing number identifies the series, or duty rating, of the bearing. Common duty ratings are: 100, extra-light duty; 200, light duty; 300, medium duty; and 400, heavy duty. Series 200 and 300 are common in motors.

Because radial loads are higher at the drive end of a motor, the drive end bearing has a higher load rating. A typical motor nameplate might depict bearing duties separately as: 6309 (medium duty); opposite drive shaft bearing = 6207 (light duty).

Bore size

The last two digits of the main bearing number relate to the bore of the bearing in millimeters (mm). Up to a 20 mm bore bearing, the following applies: 00 = 10 mm; 01 = 12 mm; 02 = 15 mm; 03 = 17 mm. Multiply the last two digits of the bearing number by 5 to obtain the bore size for all larger bearings. The bearing rated 6207 in the previous example has a bore of 07 × 5, or 35 mm; number 6309 has a bore of 09 × 5, or 45 mm.

Bearing hardware

Alphanumeric characters at the end of a bearing number indicate hardware such as shields, seals, and snap rings. Designations on some nameplates include Z, one shield; 2Z, two shields; RS, one seal; 2RS, two seals; RSZ, seal and shield; and NR, snap ring.

Internal clearance

All ball bearings have an internal clearance dimension that refers to the space between a ball and the outer race. This space is partially closed when the bearing is press-fit onto a shaft. Thermal expansion of the bearing during operation takes up the remaining clearance. Clearance is often shown on the bearing, away from the bearing number. C1 and C2 bearings have less than normal internal clearance; C3, C4, and C5 have greater than normal. Normal internal clearance bearings do not have a "C" number.

Electric motors usually require a C3 internal clearance. The rotor heats up rapidly during operation, passing heat along the shaft to the inner race of the bearings, causing the inner race to expand faster than the outer race. If a C3 internal clearance bearing is not used, uneven expansion of the inner and outer races can cause bearing damage.[7]

Arcing in Turbine Generator Bearings

The exciter end bearings of generators are insulated from ground, and the turbine shaft of turbine generator sets are grounded to prevent current from intermittently discharging through the bearings. Turbine shaft currents are caused by steam flow, which causes electrostatic charge to build-up on the rotor. The oil film creates a dielectric across which a potential can build and intermittently discharge. The damage usually occurs over a long period of time. The damage is in the form of pitting, generally forming bands rather than patches. It is sometimes confused with fatigue failure, in which patches of bearing metal dislodge. Figure 4-2 is an example of fatigue failure.

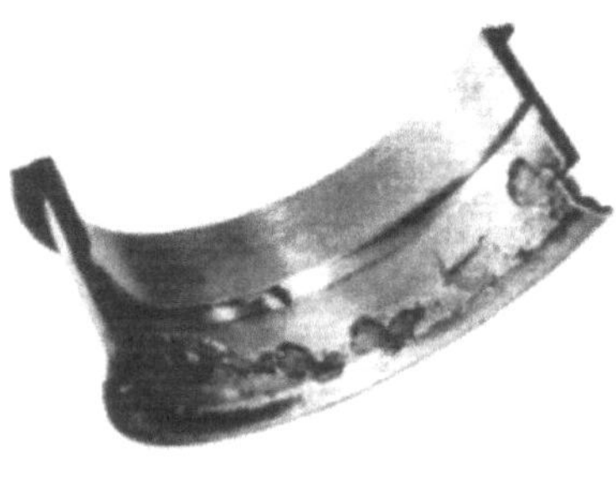

FIGURE 4-2 Fatigue failure.

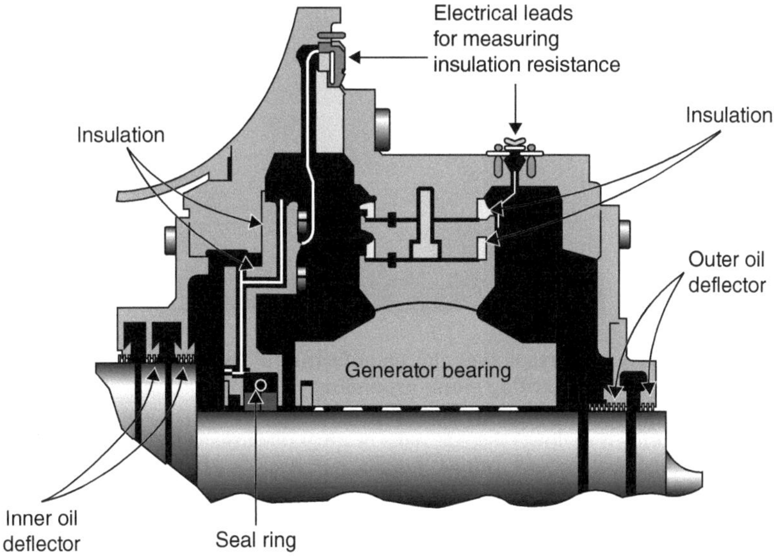

FIGURE 4-3 Generator bearing with hydrogen seal.

In Fig. 4-3, the bearing and hydrogen seal assembly is double-insulated to prevent current flow, which can cause damage where the rotating and stationary components are in contact.[4]

> A megohmmeter should be used periodically for testing insulation resistance. The minimum acceptable resistance: 100,000 ohms at normal operating temperature. If the resistance is too low, remove bolts, dowels, pipe connections, etc, one at a time, while watching the megger reading. Location of the ground is detected by the process of elimination.[4]

Generator rotors are ungrounded; therefore, even if a ground were to occur, it would not be possible to detect. The reason is that it would take two grounds for detection because the first ground is not detectable.

Arcing in Large Motors

Large motors with variable-speed drives *must be grounded* properly to prevent current damage to antifriction bearings.

> Bearing currents come in several different guises. However, while modern motor design and manufacturing practices have nearly eliminated the low frequency bearing currents induced by the asymmetry of the motor, the rapid switching in modern AC drive systems may generate high frequency current pulses through the bearings. If the energy of these pulses is sufficiently high, metal transfers from the ball and the races to the lubricant. This is know as *electrical discharge machining* or EDM.
>
> The effect of a single pulse is insignificant, but a tiny EDM pit is an incontinuity that will collect more pulses and expand into a typical EDM crater. The switching frequency of modern AC drives is very high and the vast number of pulses cause the erosion to quickly accumulate. As a result, the bearing may need replacing after only a short time of service.[8]

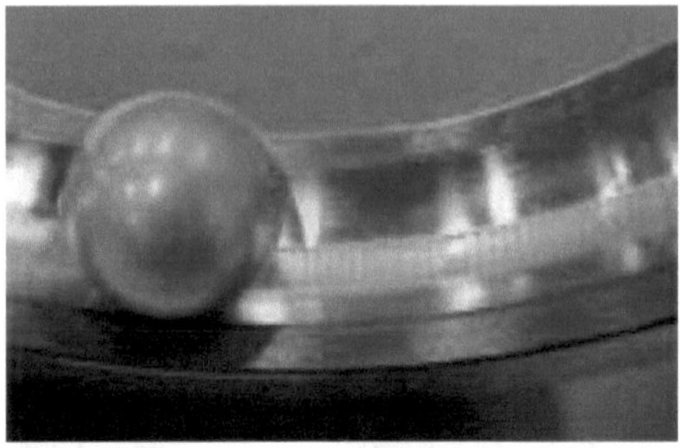

It is extremely important that the manufacturers' installation instructions are read and understood regarding motor grounding.

Fatigue failure in plain bearings is usually the result of misalignment. For example, if one turbine bearing is set lower than the other bearings, the load on that misaligned bearing will be transferred to the other bearings. If this occurs, it may *not* be accompanied by a bearing temperature rise, which would alarm a problem. Therefore, after the initial 72-hr full-load turbine run, a bearing inspection should be made. It is a very important startup activity.

Wiping is a term used to describe the loss of lubricant on a plain bearing surface (see Fig. 4-4); that is, the shaft "wipes" the bearing causing a loss of bearing metal due to overheating. This also may occur without a telltale rise in bearing temperature. It is generally caused by improper bearing fit, provided that the lube oil temperature is within design limits. Babbited journal bearings are scraped to fit the journal for an arc of 120° at the bottom of the lower bearing. The inside diameter is always 0.001 to 0.003 per inch of the journal diameter larger than the journal itself. If the bearing is not scraped and "blued" properly, you will find a polished wear pattern during bearing inspection that matches the irregular surface area. It will be on the upward side of the bearing in the direction of rotation. A proper wear pattern for bearings, without lift pumps, will be regular, approximately 60° inside and slightly off center opposite to the direction of journal rotation. Both patterns will appear if the fit is imperfect. The initial rolloff, where there is initial metal-to-metal contact, causes the normal pattern. A hydraulic wedge is immediately formed after a turbine rolls off turning gear and hydrodynamic lubrication begins. As the turbine speed increases, the shaft moves slightly upward in the direction of rotation.

Before a bearing inspection, inform the mechanics not to wipe the bearings clean. They must be inspected as found. Black oil residue is always the indicator that the bearing wiped.

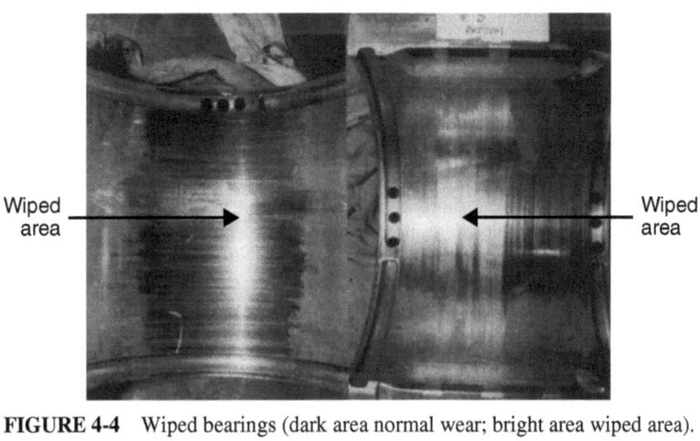

FIGURE 4-4 Wiped bearings (dark area normal wear; bright area wiped area).

This is the deciding factor if heavy scoring occurred also. If the oil is black, the bearing also wiped. If the oil is not black, it only scored and didn't wipe. Scoring is caused by foreign material in the oil, not high temperature.

Oil whip causes high vibration while a turbine is being brought up to speed. The vibration is caused by the hydraulic oil wedge in the bearing, which is loading and unloading. Whip occurs at the average of the oil velocity and the journal full-speed velocity. Because the oil boundary layer velocity is zero, the vibration will occur at half the turbine speed. The usual cause is cold lube oil.

Scoring is caused by large particles carried in the lube oil or foreign material in the bearing. This causes scratches and rough grooves in the bearing and journal.

Lapping is caused by very fine particles carried in the lube oil. It creates a very smooth, wavy, erosion pattern on the journal. This generally does not appear during startup because it requires a long time period to be noticeable. The root cause is usually the lube oil conditioning equipment not filtering the fines due to misoperation or lack of maintenance. The source of the fines can be the corrosion in the lube oil return guard pipe. The corrosion is caused by steam from the steam seal system leaking off in the bearing and return line over a period of years.

Varnishing is the formation of a hard varnishlike buildup in a bearing and the bearing journal. It is caused by the breakdown of the lubricant and requires a long period of time to occur. It is a sign of improper lube oil maintenance.

BEARING SEALS

Seals—*More progress has been made in sealing technology during the last 10 years than in any other area of roller bearing development. That is as it should be, because no area is more important. Although most bearing manufacturers provide data on the* L_{10} *life of their products—the time required for one-tenth of a representative group to fail through fatigue—almost all will admit privately that the average bearing in the field fails well before its* L_{10} *life is up. These failures do not result from fatigue, but from lack of lubrication and from the incursion of foreign matter. It is the seals' task to retain lubricant and keep out moisture and dirt. Seals may be broadly divided into contacting and noncontacting types.*

Contacting seals *include lip, felt, and face designs. Their common characteristic is that somewhere in the seal one surface rubs on another. Lip seals, employing O-ring or chevron-shaped lips, are the most commonly used type of seal. Leather was the material originally employed; it is still excellent in applications in which little heat is generated. Synthetic elastomers have replaced leather in most applications today. Plastics and laminated materials are also used. Depending on the seal material and the finish of the shaft, acceptable lip-seal speeds range from 2000 to 3000 fpm. Even when speeds are well within the acceptable range for the seal, good shaft finish and concentricity are highly desirable. The better the shaft, the better the lip seal will be able to do its job.*

Felt seals *employ wool or synthetic fibers in place of an elastomeric lip. The synthetics are more resistant to heat and corrosive materials. Provided that concentricity is good enough to prevent the seal from losing contact with the shaft, felt seals exclude dirt satisfactorily. Because the felt soaks up liquids, however, these seals cannot effectively exclude them or retain oil. Felt seals are generally limited to shaft speeds of 500 to 1000 rpm.*

Face seals *are very similar to the mechanical seals used in pumps; they are the most effective—and also, unfortunately, the most expensive—single-seal design. In these seals, a spring-loaded washer (typically carbon) with a smooth, lapped face runs against a seat (typically cast iron) that also has a very smooth, lapped face. These seals can be purchased as cartridge units. Concentricity is a must, as is installation with the face at an exact right angle to the shaft. High speeds are seldom a problem for face seals.*

Noncontacting seals *comprise labyrinth seals, grease seals, and flingers. They are frequently used together. The flinger is a rotating shield, attached to the shaft, that tends to throw off dirt by centrifugal force; it is inoperative when the shaft is not rotating, and most effective when the shaft speed is high.*

Grease seals *are cavities—two or more are often used together—packed with fibrous grease.*

__Labyrinths__ consist of interlocking circumferential fins—an inner, revolving set attached to the shaft, and an outer, stationary set fastened to the housing. The clearances between the fins are close, but the fins do not touch; instead, they act to greatly increase the length of the path that dirt must traverse to reach the bearing and that lubricant must traverse to escape. Because there is no contact between the moving and stationary parts, labyrinth seals generate very little heat, and speed is not a problem. Both plastic and all-metal designs are available.

The term taconite seals is frequently used to designate compound seals in which an inner contact seal, a grease seal, and an outer seal are combined in series. (Taconite is a finely powdered, highly abrasive material mined as a low grade iron ore; it tends to be extremely destructive to bearings.) These seals first gained popularity in the taconite mines, where bearings with L_{10} *lives of many thousands of hours frequently failed within a few weeks. A configuration consisting of an inner lip seal followed by a labyrinth seal (or a grease seal) and an outer felt seal is typical, although other combinations are also used. It is not necessary for the plant engineer to put together his own taconite seals—they are offered as complete assemblies by several manufacturers.*

Seal maintenance is one of the most neglected, yet potentially one of the most fruitful, areas of bearing technology. The split seals frequently used with heavy duty bearings render seal maintenance relatively simple, a point to keep in mind when specifying the bearing. (Preventive maintenance operations instructions for the seals can be added to the maintenance operations list for the machine utilizing the bearings.) Clean the shaft adjacent to the seal. Check to make certain they are still in contact with the shaft, and replace the felt if the old strips seem to have lost their elasticity or if grit has penetrated beyond the outer edge.

Inspect the lips of lip seals. Replace the seal if the lip shows excessive wear. Hardness indicated that the seal has been running hot; take corrective measures if the heat is coming from the bearing. High speeds may overheat the seal through friction—if this is the problem, the best answer may be to change to a high-temperature seal material.

Disassemble labyrinth seals and clean out all the old grease. This step is extremely important—pumping clean grease into the bearing until it comes out the seal does not purge it completely. When reassembling the seal, make certain the parts do not rub.

Most heavy-duty roller bearings are supplied without integral seals; however, some are so equipped. If a bearing with integral seals is to be heated for mounting make certain that the heat will not destroy the effectiveness of the seals. Viton—used in some lip seals—will tolerate 400°F (higher temperatures will draw the temper of the bearing steels in any case). All-metal seals present no problem, but may fail to provide sufficient protection unless used with an outboard auxiliary seal.[2]

LUBRICANTS

The lubricants discussed here are oils and greases. Eighty percent of all antifriction bearings are lubricated with grease, and all large plain, high-speed bearings are lubricated with oil. Oil has the ability to carry away the heat absorbed from the journal. Antifriction bearings, because of their low friction design, do not generate significant heat, but if the bearing is absorbing heat from a hot rotor, they may have to be lubricated with oil.

Lubricating Oil

Lube oils are refined from many different crude oil bases. For example, Pennsylvania crude oil is *paraffin*-based, while Texas oil is *asphalt*-based crude. Oils refined from different base crudes should never be mixed. All oil companies will have compatibility charts that must be consulted when changing oil suppliers.

Viscosity is the principle property of lube oil used to determine the temperature range of use. A machine's starting and running temperatures are crucial. If the starting and running temperature ranges are not within the viscosity range of a particular lubricant, the oil can be blended to reduce the range. For example, if we need a motor oil with a viscosity classification of SAE 10 at a starting temperature of 0°F (~7500 ssu) and an SAE 50 at 200°F (~100 ssu), the oil can be blended to give a midrange of, say, 350 ssu at 100°F. Blending oil with a low ssu with one that has a high ssu can be done to yield a viscosity index with a slope compatible to the machine operating characteristics. Viscosity index (VI) is determined by plotting two extremes, such as 100°F and 210°F, on a semi-logarithmic graph. The vertical axis will represent viscosity and the horizontal temperature. Connecting the points will yield the viscosity of the oil for any other temperature on the resulting curve. The steeper the curve (low VI), the more undesirable the oil is for meeting the startup versus running temperature range. Blending can flatten (high VI) the curve to give a desirable operating range. Refiners will also use additives to decrease the deterioration of the oil.

Note: If your minimum startup temperature is 120°F and you are in an environment in which the ambient temperature is 145°F, you will *not* have to heat the oil prior to roll. But if the ambient temperature drops to 45°F, *don't forget to preheat it.* In the Arabian deserts, for example, the summer temperature can hover at 145°F for months and drop to 45°F in the winter.

Plant engineers do not get involved with the specific details of oil refinement but must know the basic oil properties. They also must understand monitoring and sampling techniques and procedures. This applies to both the oil as delivered and during operation. The as-delivered oil sample must meet the required oil specification. During operation, samples are taken for comparison. Oil can be purchased *certified* or *uncertified.* Some engineers prefer to purchase uncertified oil and certify it themselves in order to have the same laboratory perform all testing to avoid uncertainty that can be introduced using two separate facilities. For initial purchases for pre- and postturbine oil flushes, it is best to buy bulk tanker loads. Usually, this oil will not meet a turbine manufacturer's cleanliness specifications, but it can be continuously filter polished during the flushing and postflushing operations to meet the required cleanliness.

Oxidation of the oil is one of the most important factors to test for after lube oil flushing is completed and during operation. Preoperational activities, such as flushing, do not usually contribute significantly to oxidation because the oil is not yet exposed to high temperatures. Oxidation can be determined by acid content. The neutralization number is the most commonly used indicator of acid content. Acid content can also be determined by interfacial tension. The tests are used to tell when the oil requires changing. Lube oil that has turned black does not need to be tested. The blackness is produced by carbon as a result of severe oxidation. Therefore, the oil needs to be replaced and a root-cause/failure analysis performed. If this occurs, the equipment must be immediately taken out of service because failure is imminent. During startup, the most common cause is excessively high temperatures as a result of bearing misalignment or use of the wrong lubricant. This is most common in self-lubricated plain and antifriction bearings.

Additive content can be determined in a laboratory using an infrared spectrophotometer to determine phenolic oxidation inhibitor content. Additives are used to reduce the oxidation of the oil. This should be checked against the specifications.

Moisture content must be determined and kept within specifications. If the oil and water were to emulsify, the water can be separated by adding heat under vacuum. A vacuum dehydration unit must be used. *Oil that has emulsified with water turns milky white.* This can happen with large boiler feed pumps where feedwater from the seals gets into the bearings. Considering the high cost of oil, it would pay to rent a vacuum dehydrator and restore the oil. Viscosity tests will reveal if there are liquid contaminants such as water in the oil.

Particulate contamination must be kept within specification. Portable particle counters are available to measure contaminants.

Grease

Grease is lube oil thickened with a variety of "soaps." The lube oil is generally heavy base. As with oil lubricants, chemical additives are used to resist oxidation and bearing corrosion as well as adding high-load bearing properties. Grease is color coded by the manufacturer for identification. However, there is no standardization of colors. Each manufacturer will have its own identifying color. *Multipurpose grease will always be brown.* The manufacturer will color special-purpose grease so that it can be readily identified as such. For example, a special-purpose grease may be colored light blue.

Consistency (degree of softness or hardness) is the most important physical property of a standard grease. The National Lubricating Grease Institute (NLGI) has established consistency boundaries of various grades of grease. (See Table 4-1.) The softest grade on the NLGI list, No. 000, is almost fluid; the hardest, No. 6, is solid. For handgun grease applications, an NLGI No. 2 grease is by far the most popular. In centralized systems, and for winter outdoor service in the North, No. 0 grade is the most frequently used.

Greases are classified according to the type of thickener used: alumina soap, calcium (lime) soap, lithium soap, sodium (soda) soap, and nonsoap greases which use organic compounds such as arylurea or inorganic clays as thickeners ...

Since World War II, lithium soap greases have become the first multipurpose or multifunctional greases to gain wide acceptance. Previously, calcium soap greases were used when water resistance was needed, but they are not heat resistant (normally unsatisfactory above 175°F). Soda soap greases were used for extremely hot temperatures, but they have little water resistance.

A lithium soap grease resists both water and high temperatures. In addition, lithium soap greases have good mechanical stability with little loss in consistency with use. In high-speed bearings, greases are subjected to high rates of shear with a resultant thinning effect. A grease with good mechanical stability can withstand the shear with little effect in consistency; a grease with poor mechanical stability will thin and run out of the bearing.[9]

Note: When applying grease to a bearing, ensure that the applied grease is displacing the old grease from the bearing until new grease appears. If it does not displace the old grease, the bearing must be cleaned with a solvent and repacked by hand. Most high-speed rotating machinery bearings cannot be greased using a handgun. Always check the manufacturers' instructions.

TABLE 4-1 Grease Classification by Thickener Type[9]

Class	Type	Kind of Thickener	Dropping Point[a]	Characteristics
Lime soap	Cup	Calcium-fat soap	190°F	Smooth, water resistant, separates on loss of water content
	Complex	Calcium-fat soap and calcium acetate	>500°F	Smooth or slight fiber, water and heat resistant, usually some extreme pressure (E.P.) properties
Soda soap	Fiber or sponge	Soda-tallow soap	375°F	Fibrous structure, not water resistant
	Medium or short fiber	Soda-fatty acid soap	375°F	Semi-smooth to smooth, not water resistant
Lithium soap	Multi-purpose	Lithium-12 hydroxystrate soap	375°F	Smooth, water and heat resistant, good multipurpose properties
Non-soap	High-temperature multipurpose	Inorganic clay or orgânic arylureo	Usually 500°F or higher	Smooth, water and heat resistant, high temperature stability, good multipurpose properties
Residuum	Pinion or gear compound	Usually none, asphaltic base	Rapid thinning when heated	Smooth, black, adhesive, pressure and water resistant

[a]The maximum usable temperature is considerably below the dropping point due to oxidation and evaporation.

Gear couplings, gearboxes, and limited end-float couplings require special attention. Often, couplings are installed in the field by mechanics who don't understand that they have to be packed with grease to lubricate the coupling against the fretting action of the coupling and shaft teeth. The proper grease is an essential element of the assembly. Also, if a gearbox failure occurs, the first thing to look at in the root-cause/failure analysis is the lubricant.

Wear at Low Speeds
Gear coupling manufacturers and many oil companies recommend that when a grease is used for lubrication, a soft grease be used at low speeds and a stiffer grease at high speeds. Sometimes, though, a maintenance mechanics will use a stiff grease in a coupling running at low speed. Rapid wear is a consequence. In order to determine, more precisely, when to change from a stiff to a soft grease, a study was conducted in which couplings of various sizes were lubricated with stiff greases and run at gradually decreasing speeds. Below a given speed, the wear-rate changed drastically, and rapid failure of the coupling occurred. The appearance of a coupling tooth which failed under these conditions is shown below. An empirical formula was developed, which gives the minimum speed at which a coupling lubricated with a stiff grease (penetration factor smaller than 310) can operate and not suffer this type of rapid wear:

$$N = 850/(2R)^{1/2}$$

It was found, experimentally, that rapid failures at low speeds are independent of torque, misalignment, and tooth geometry, but the wearing process can be retarded by hardening of the teeth. It was also found that this type of failure is very rapid when the grease used has more than 8 percent soap.

See Fig. 4-5 wear failure scar known as "worm tracking."

FIGURE 4-5 Gear failure illustration.

Conclusion: It has been shown that the speed has a significant influence on coupling wear: High speeds reduce the wear rate; while low speeds can, with some lubricants, cause the coupling's failure. It can only be concluded that increased rotating speed helps in improving the lubrication conditions. It was also shown that worm-tracks are caused by very localized welding and, as such, the sliding velocity alone cannot be blamed for this type failure. It is likely that worm-tracks are caused by a combination of unfavorable torques, misalignments, and sliding velocities. Because little can be done about the torque, and because the alignment often cannot be improved, the best way to avoid reoccurrence of worm-tracks is to improve the lubrication ...[10]

Bearing materials and definitions were adapted from "File-Bearings,"[11] as follows:

A **bearing *PV*** number furnishes a convenient method of comparing the performance of plain bearing materials:

$$PV = P \times V$$

where P = pressure, psi
V = velocity, fpm

Babbitt is a term used to label a wide variety of soft, low-melting, nonferrous alloys. Isaac Babbitt's original formula called for 89 percent tin, 9 percent antimony, and 2 percent copper. Alloys of this type are called *tin-based babbitts.* They offer good embedability, and resistance to scoring and corrosion. But, because tin is now very expensive, they are seldom used.

The tin-based babbitts have been replaced by *lead-based babbitts,* in which the lead is hardened with antimony. These alloys are typically 70 to 90 percent lead. In some cases, tin is added to improve corrosion resistance.

Thick babbitt bearings are susceptible to fatigue wear, characterized by spalling of the bearing surface. This problem can be overcome by bonding a thin layer of babbitt to a steel, bronze, or cast iron backing—a process that also permits the economical use for tougher, corrosion resistant tin based babbitts. Such bearings have *PV* values of about 100,000 psi-fpm.

Babbitt bearings are often found on large shafts in steam engines and similar heavy machinery (such as large, low-speed Diesels). Babbitt bushings are also used in fractional horsepower electric motors.

Bronze was the first metal used in bearings, and bronze and other alloys going under that name are still among the most common plain bearing materials. *Bronze* indicates an alloy primarily of copper and tin. For example, one manufacturer offers bushings with a composition of 88 percent copper, 8 percent tin, and 4 percent zinc, which makes them true bronze. (Traditionally, brass is an alloy of copper and zinc softer than bronze, an alloy of copper and tin.) Solid bronze bearings are suited to heavily loaded, low-speed applications with *PV* values of about 50,000 psi-fpm. Bronze of this type, also called *tin bronze*, is used at times as a backing material for babbitt. If the thin babbitt layer fatigues, the bronze offers a bearing surface less resistant to seizure but considerably tougher. *Copper-leads, leaded bronzes,* and *aluminum bronzes* are often called "bronze" and find use as bearing materials. Copper-leads are similar to babbitt in performance but are somewhat stronger. Leaded bronzes are alloys of copper, lead, and tin. They perform much like bronze, but are slightly softer. Aluminum bronzes have great strength and thus are suitable for use under high loads, but they are hard and require a hard shaft and good lubrication.

Carbon is used for plain bearings in high-temperature (to 750°F) applications, and in certain applications requiring water lubrication or corrosion resistance. Unlubricated carbon bearings are limited to *PV* values of about 15,000 psi-fpm; lubrication can increase these by a factor of 10.

Cast iron is a valuable bearing material in high-temperature (to 1000°F) applications. Lubrication is provided by the graphite in the cast iron. Loads must be light and speeds low. Furnace and flue applications are typical.

TFE (tetrafluoroethylene) is perhaps the most advanced plastic bearing material. With and without addition of other polymers, graphite, glass, ceramics, silicates, and metal powders, it is sold under such trade names as Teflon, Ebolon, Rulon, and Duralon. It can be used at temperatures ranging from –400 to 550°F. In addition to steel-backed shells and the common bushing and thrust washer forms, it is available as bondable tape. In continuous service without lubrication, some TFE bearings are capable of *PV* values as high as 25,000 psi-fpm. TFE can be formed into fibers, and the molecular orientation of these fibers greatly increases its strength. Bearings of TFE fibers can carry up to 50,000 psi. Such bearings may be backed with steel shells, interwoven with polyester, and wrapped with fiberglass.

TERMINOLOGY AND NOMENCLATURE

ABEC is an abbreviation for Annular Bearing Engineers Committee. This committee has set up a system of numbers that specifies tolerances for ball bearings. The higher the number, the tighter the tolerance for ball bearings. ABEC 1 corresponds to the ISO's Normal

Class, ABEC 5 to ISO Class 5, and ABEC 7 to ISO Class 4. There is also an ABEC 9. The committee is part of the Anti-Friction Bearing Manufacturers Association.

Adapter bearings have a wide inner ring to accommodate some means of attachment to the shaft, such as an eccentric locking collar.

AFBMA is the abbreviation for Anti-Friction Bearing Manufacturers Association.

Angular contact bearings are ball bearings having high shoulders on one side of the outer race and on the opposite side of the inner race, an arrangement that permits them to carry high thrust loads in one direction, as well as radial loads.

Basis dynamic capacity or basic load rating indicates the load a bearing should be able to carry for 1 million revolutions without failure.

Bath oiling is a system in which the lowest rolling element in a bearing passes through a pool of oil supplied by a reservoir. It is limited to moderate speeds.

Brinelling is the indentation of the races by the rolling elements. It occurs when the static load rating is exceeded.

Cam followers are needle bearings having a threaded shank instead of an inner ring bore.

Cartridge blocks are premounted bearings intended to be pressed into a cavity.

Circulating oil is a system in which oil from a reservoir is fed through the bearing, or to the bearing holes in the outer race; filtered; and returned to the reservoir.

Complement is a term used to refer to the number of rolling elements in a bearing.

Cone is a special name for the inner ring of a tapered roller bearing.

Conrad bearings are radial ball bearings in which balls running in deep grooves in both races are separated by a cage or spacer. The deep grooves permit the bearing to carry some thrust load in either direction, and spacing the balls permits the bearing to hold a great deal of lubricant, although its load capacity is not as high as that of a full-compliment bearing. Conrad bearings are probably the most important type of rolling-element bearing in industrial applications.

Cuff seal is another name for lip seal.

***dn* numbers** are often used to specify bearing speed. Applications with *dn* numbers of 1,000,000 or more are considered high speed.

$$dn = d \times n$$

where d = bore diameter, mm
n = speed, rpm

Drop-feed oiling is a system in which oil drips from a reservoir into the bearing. The rate of feed is controlled by a needle valve.

End play is looseness that permits the inner ring to move parallel to the shaft. All radial rolling element bearings have some end play.

False brinelling results from vibration experienced while the bearing is stationary.

Foil bearings, developed by AiResearch, (initially for aircraft air-handling units), are now being tested for aircraft power plants. The bearings are oil free, so aircraft engines will be able to operate without lube oil systems.

Foil air bearings are self-acting hydrodynamic bearings that use ambient air as the lubricant instead of oil. Both air and oil are fluids that can perform the job of separating moving metal parts. The very high rotational speeds of turbine engine shafts (>10,000 rpm) allow the replacement of oil with air. At these speeds, foil air bearings maintain the air film between moving parts by actually pumping air between the rotating and stationary surfaces. Air is drawn in and sticks to the bearing surface, no matter how fast it is moving.[12]

Flange blocks are premounted bearings provided with flanges so that they can be mounted on thin, usually vertical, surfaces.

Fretting corrosion results when wear particles remain in or around the bearing and rust. A mixture of black and red oxides is often apparent. Poor shaft and housing fits are common causes of fretting corrosion.

Full-complement bearings are ball bearings in which the balls are so closely spaced that there is no room to add another one. These are also called *max bearings, maximum capacity bearings,* and *filling slot bearings.* They can carry more radial load than Conrad bearings, but less thrust load, and they are more limited in speed capacity.

Heat checks are surface cracks caused by overheating metal.

Interference fit is the term used to describe the dimensional relationship between an inner ring as it is pressed on the shaft when the inner ring must stretch slightly, or to describe the relationship when the outer ring must compress slightly as it is pressed into the bore. Interference fits are often achieved by heating or chilling the parts to reduce the force required. They are intended to prevent relative motion between the mating parts, even under heavy load.

Internally self-aligning bearings are ball or roller bearings in which the outer race is geometrically a section of a sphere, permitting the inner ring to turn out of the plane of the outer ring. This type of bearing is extremely valuable in compensating for shaft misalignment.

L_{10} life, also called *rating life*, is the time required for 10 percent of a group of bearings of given size and design to fail. If a bearing's basic load rating is known, its L_{10} life for a given load and speed will be:

$$L_{10} = (C/F)^k \times 106/60N$$

where L_{10} = rating life, hr
N = speed, rpm
C = basic dynamic load
F = equivalent dynamic load
k = life expectancy
Exponent, $k = 3$ for ball bearings, 3.33 for roller bearings.

Lobe bearings are plain bearings with two or more lobes. A two lobe bearing is an **elliptical bearing.**[13]

> *Distributed across the entire shaft diameter, there are as many individual hydrodynamic carrying forces directed at the center of the shaft as there are lobes. The strength of the individual hydrodynamic force is, among other things, dependent on the width of the wedge gap. The vector total of all the individual carrying forces represents the effective load capacity of the "multilobe [sic]" bearing towards the outside. This results in a strong centering effect being*

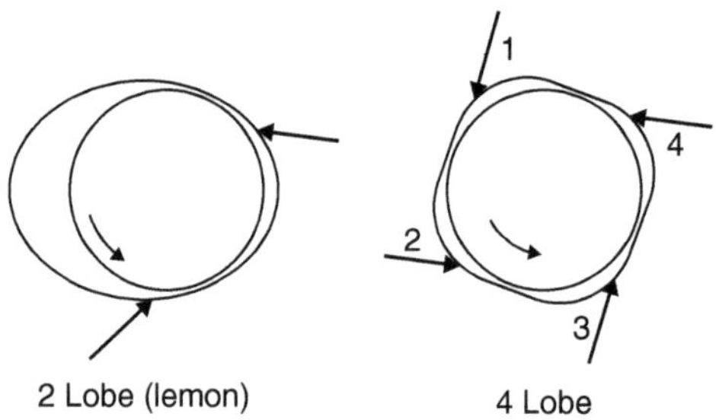

FIGURE 4-6 Two vs Four Lobe Bearing Hydralics (arrows indicate oil wedges).

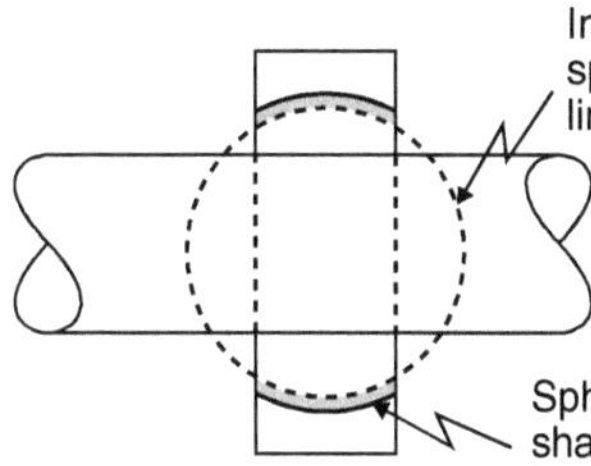

FIGURE 4-7 Spherical Bearing.

applied to the shaft which produces good concentricity and generates a defined shaft position. By matching the lubricant viscosity to the shaft's peripheral speed and the wedge gap shape, the degree of the hydrodynamic carrying force and the bearing friction can be varied to meet individual requirements.

An elliptical bearing is sometimes referred to as a **lemon bearing** because of its shape. A **spherical bearing** (see Figs. 4-3 and 4-7) is a bearing whose sphere is the outer bearing case which allows the bearing to self align with the shaft.

Mist oiling is a system in which compressed air picks up atomized oil from a filter-regulator-lubricator unit and delivers it to a bearing. The air assists in cooling the bearing. The system is particularly well suited to high-speed applications.

Naked bearings are bearings that are not premounted bearings.

Needle bearings are bearings having long slender rollers.

Pads are the feet of a pillow block.

Pillow blocks are premounted bearings provided with feet so that they can be mounted on flat surfaces. It should be noted that the shaft through a pillow block is parallel to the plane of the mounting surface. The shaft through a cartridge block or flange block passes through the plane of the mounting surface. Cartridge blocks and flange blocks are often erroneously called pillow blocks.

Preloading is used to take up end play and hold the shaft in the correct position, to quiet a bearing and prevent skidding and to make certain several bearings share the load equally.

Prelubricated bearings have been packed with grease at assembly and are closed by integral seals. Because no provision is made for relubrication, the installed cost of these bearings is low. They are satisfactory in many applications in which only moderate life is required.

Radial play is looseness that permits the inner ring to move toward the outer ring. It may be required to permit thermal expansion or to allow for slight misalignment between the shaft and the housing bore or for other reasons.

RBEC is an abbreviation for Roller Bearing Engineers Committee. This committee has set up two tolerance numbers for roller bearings. The higher the number, the tighter the tolerance for ball bearings. RBEC 1 corresponds to the ISO Normal Class and RBEC 5 to ISO Class 5. The committee is part of the Anti-Friction Bearing Manufacturers Association.

Retainer is another name for cage.

RMS is the abbreviation for root-mean-square. It indicates a system for expressing surface finish roughness; the higher the number, the rougher the surface. RMS values are often used to designate shaft and housing finishes. An ordinary milled steel surface is normally about 125 RMS, a buffed surface 16 RMS or less.

Scoring is galling resulting in furrows in the bearing metal.

Scuffing is the scratching resulting when one bearing part slides on another under load. It often occurs in large bearings because of poor handling before or during installation. If the scuff marks can be rubbed out with 240-grit emery cloth, they are not serious. Galling is also called scuffing at times.

Skidding is caused by rapid starts and insufficient preloading. The rolling elements slide on the races—often leaving characteristic smear marks—instead of rolling.

Slingers are regular metal pieces attached to the shaft, either to spray a bearing with oil or to throw off foreign matter before it can enter the bearing.

Smearing is galling without scoring.

Spalling is the breaking away of metal in flakes.

Spherangular roller bearings are similar to tapered roller bearings but employ spherical rollers. They can carry substantial thrust loads in one direction, as well as allow for misalignment.

Spherical roller bearings are internally self-aligning roller bearings.

Standard designs are those available from a number of manufacturers. The use of standard designs permits plant engineers to substitute one brand of rolling-element bearings for another. *IBI Guide*, published by International Bearing Interchange Inc., of St. Louis Park, Minnesota, and *Bearing Manual*, published by Bearing Manual Inc., of Broadview, Illinois, list thousands of possible substitutions. Some bearing makers include substitution data in their catalogs.

Tapered roller bearings employ conical rollers set at an angle to the shaft, an arrangement that permits them to carry substantial radial and thrust loads. They are one of the most valuable industrial types.

REFERENCES

1. Karassik, I. J. *Centrifugal Pump Clinic*, 2nd ed., Marcel Dekker, Inc, 1981.
2. Wolfe, G. "Selecting and Maintaining Heavy-Duty Roller Bearings," *Plant Engineering*, April 1, 1976.
3. Tyson, L. "Tilting Pad Thrust Bearings," *Plant Engineering*, June 12, 1976.
4. Missana, A., Barton, S. C., and Reck, F. H. "Achieve Higher Availability through Better Lubrication and Shaft Sealing," *Power*, April 1976.
5. Varga, Z. "900-mm pivoted-pad journal bearings for steam turbosets—characteristics and testing," *Brown Boveri Review*, 6 (1977), pp. 325–336.
6. Adams, M. L. and Makay, E. "How to Apply Pivoted-Pad Journal Bearings," *Power*, October 1981.
7. Bos, M. H. and Brown, M. V. "Identifying Motor Bearings," *Maintenance Technology*, January 1993.
8. "Bearing Currents in Modern AC Drive Systems," ABB Technical Guide No. 5, 1999.
9. Lindenbaum, H. J. "Grease Is...What?" *Plant Engineering*, September 19, 1974.
10. Calistrat, M. M. "Wear and Lubrication," *Mechanical Engineering*, October 1975.
11. File-Bearings, *Plant Engineering*, July 9, 1981.
12. "Creating a Turbomachinery Revolution—Research at Glenn Enables an Oil Free Turbine Engine," FS-2001-07-014, GRC.
13. John Crane Corporation "Multilobe Bearings," October 2008, http://www.kupujucpavky.cz/e-catalogue/DataSheets/Bearings/Multilobe_bearings.pdf

CHAPTER 5
BOILERS

This chapter covers boilers, including boiler auxiliaries that are also covered in separate chapters (e.g., Chaps. 3, 13, and 15). The design philosophies presented here do not reflect any particular design or manufacturer. The principal goal is to clarify manufacturing and engineering literature that quite often assumes the plant engineer understands how the researchers got from point A to the writing. In other words, often the ABCs, or what is "obvious" to the engineer is left out.

For design details and fundamentals of steam generators and auxiliaries, we can find excellent reference in manufacturers' manuals such as *Steam* by Babcock & Wilcox, and *Combustion* by Combustion Engineering. A must for any power industry professional is *The Control of Boilers* by Sam G. Dukelow, Instrument Society of America.

DESIGN ABCs

Boilers, simply stated, are water and steam *cooled* pressure vessels that include a saturated steam generation section and steam superheating sections. The steam generation section is the furnace enclosure containing the waterwalls. Its source of energy is primarily radiant heat. The steam superheating sections are the *radiant and convection*–type heating sections downstream of a saturated steam collection point. Their source of energy is primarily convective, with *finishing or final heating* by radiant heat. The steam generation section of a boiler generates steam following the saturation curve of the Mollier diagram. The steam heating section curves are above the saturation line of the Mollier diagram.

If there is no flow from the steam generator, pressure cannot be controlled, unless the heating source is removed. Flow, as well as firing rate, is directly related to pressure control. In actuality, if there is no flow there will be no *cooling* of the furnace enclosure, and the boiler will eventually fail due to overheating and pressurization. Also, in the superheating sections, if flow is not established, there will be no *cooling* of the heating elements, and they can also fail.

The steam generation section, and superheating sections in line of sight of the fires, absorb heat by radiant heat transfer. The reheater is generally placed between the primary and secondary superheaters. The superheating sections in the dark area, so to speak, of the boiler absorb heat by convective heat transfer and are referred to as the convection pass. This presents special problems for plant engineers during commissioning. Because there is more radiant heat available than convective heat at low loads, overheating of the radiant heaters is a prime concern. That is, there can be more heat available than there is steam flow to cool the elements. Furthermore, at low firing rates, there is less convective heat available because of the low energy (temperature and mass flow) of the combustion gas. Increasing airflow does not increase heat energy, and increasing firing rate overheats the radiant heaters. The net result is pressure rises faster than superheat temperature, which can start a deteriorating cycle difficult to control during startup. This situation comes to the forefront when firing to achieve proper superheat temperatures required for turbine roll.

To remember the concept of heat absorption in the different sections of the boiler, imagine cooling effects rather than heating. For example, during startup and after pressure is established in the boiler, using the upper burners will provide more heat to the convection pass. The reason is that there is less cooling of the gas in the furnace by the waterwalls than if the lower burners were used. Boiler engineers would say less heat is absorbed in the radiant section, and therefore, there is more heat available when the gas reaches the convection pass. This practice of using the upper burners helps control pressure rise at the same time the engineer is trying to raise superheat temperature in the convection pass after steam flow is established. For initial firing, the lower burners are used to raise boiler pressure and reduce radiant absorption in the radiant heaters until steam flow is established.

One method in which boiler physical design is used to control superheat is to place the finishing, or secondary, superheater at the radiant section and the primary superheater in the convection section, with the reheater in between. The result is illustrated in the diagram below.

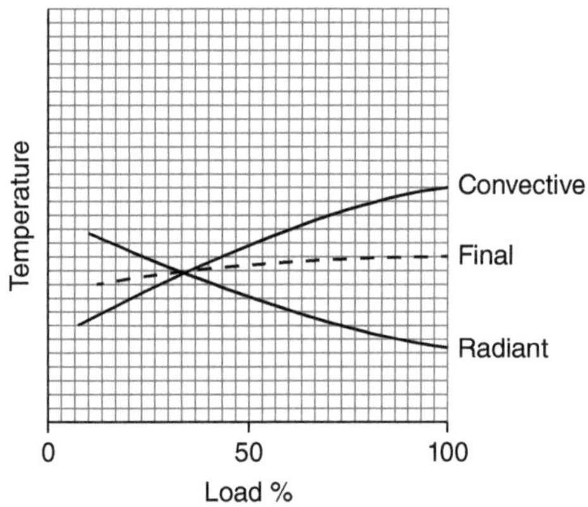

Ideal physical superheater arrangement.

STARTUP STEAM FLOW

If the unit does not have a turbine steam bypass system, another means of establishing steam flow must be used. Both the boiler drum and superheaters must be vented to establish flow. The drum is vented to vent off air until approximately 10 psig is established. The superheater outlet vent must be sized sufficiently to allow enough flow to cool the heaters until flow is established through the turbine. Drains are used to remove condensate while warming the main steam lines to the turbine. When these drains are used for the purpose of establishing cooling flow for the superheaters, instead of the superheater outlet vents, serious damage to the superheaters can occur.

The photos shown in Fig. 5-1 are of a pendant superheater, which is not drainable. The distortion of the tube occurred during a hot restart after a boiler trip. After the trip, the purge cycle cooled the pendant, which condensed steam to form a loop seal. On a relight off, the water boiled, providing steam flow on the turbine side through the startup vent, but was blocked on the drum side by drum pressure. In this case, the vents were not opened and the tubes overheated, causing the distortion. Even with the vent open, the drum side will have no flow until the loop seal is evaporated. Therefore, adequate venting is required to quickly evaporate the loop.

Close up of HTSH front tube distortion

Close up of inner LTSH tube distortion

FIGURE 5-1 Pendants.

The tubes shown in Fig. 5-2 are in a horizontal superheater section and supported by sling tubes. Operators used the main steam drains to establish flow in the superheaters. Once the tubes fell off their supports, due to distortion caused by overheating, they created a loop similar to the ones in the pendant section previously mentioned. The remedy they used, ignoring the root cause, was to install clips so the tubes would not fall off the supports.

The support clips are cooled by the tubes they are attached to as well as the tubes they are supporting. The restraining tabs shown in Fig. 5-2 will receive minimal cooling because the tubes are not touching the supports. Also, this remedy did not address the root cause for

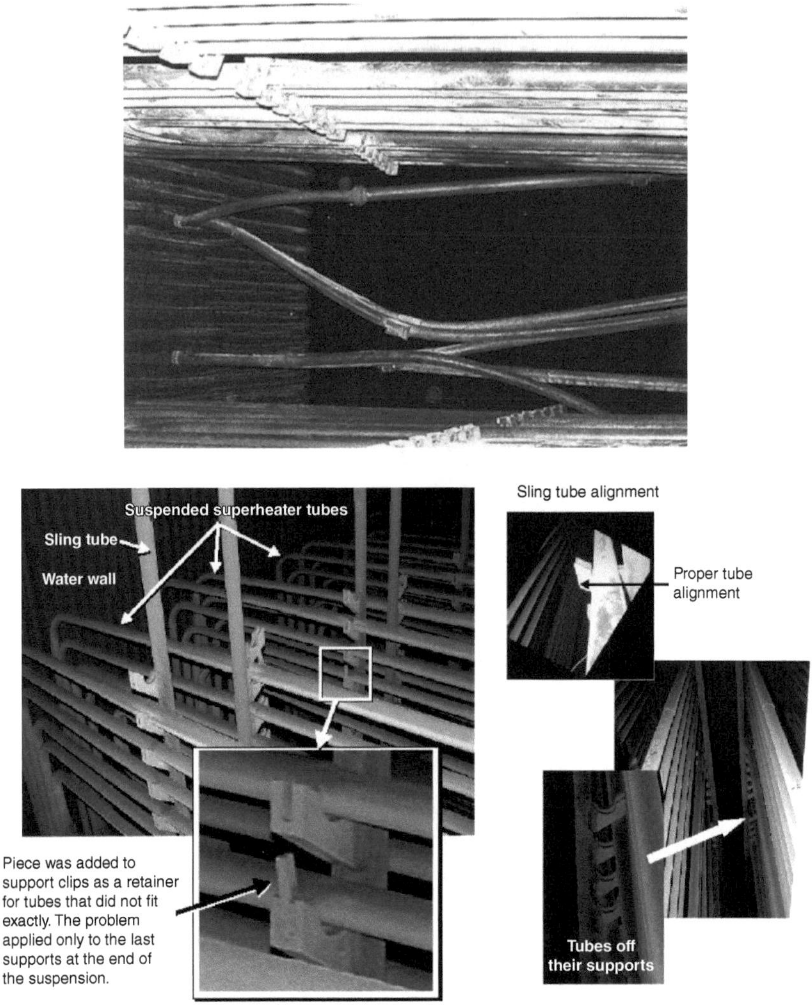

FIGURE 5-2 Sling tubes.

the failures. This unit was designed with a startup bypass system that was not in service. Therefore, because it had a bypass system, the designer did not size the superheater vents for startup; there was no need for a startup vent. The vents did not have the capacity to vent enough steam to cool the superheaters. Starting up with the bypass out of service was the root cause of the problem.

FURNANCE

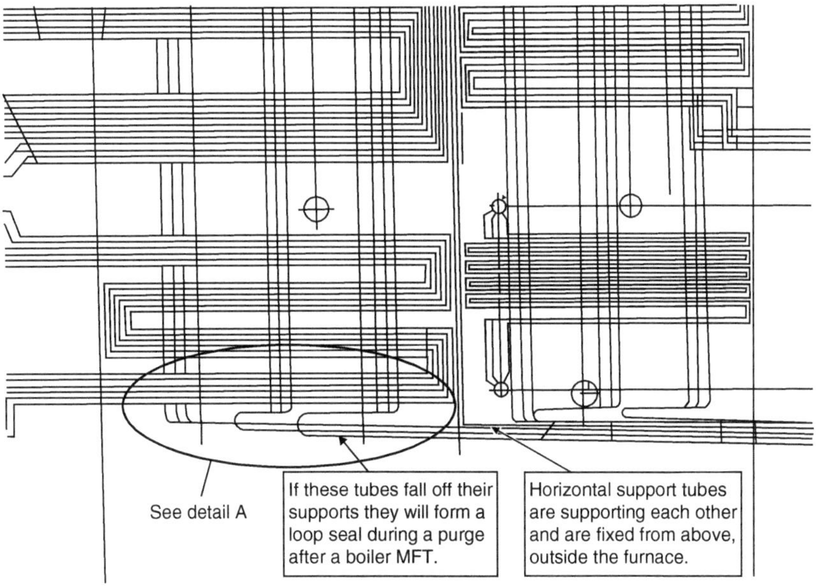

This superheater is drainable and the support tubes are inclined for drainage. However, if steam flow is not established, these tubes could warp and fall off their supports and cause a water loop to form as shown in Fig. 5-2.

Furnace detail A

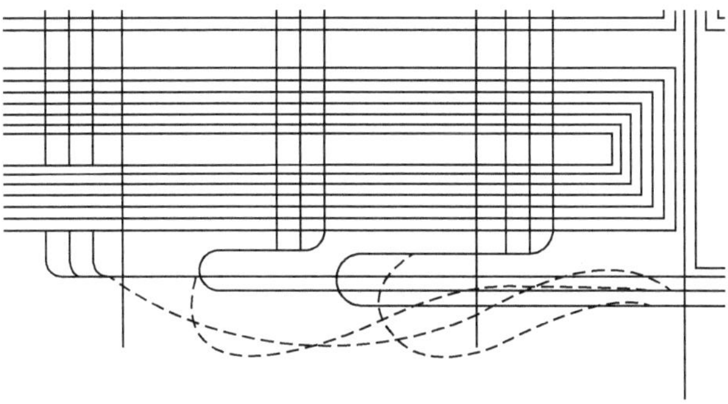

STEAM TEMPERATURE CONTROL

Designers employ different methods to control steam temperature, but all strive to achieve limited control at some optimum point. All trade-offs are driven by engineering economics; however, most designers strive for boiler designs that minimize control at full load. Therefore, there are always trade-offs while getting to full load. This generally means more complicated controls are required for low-load operation than required at high load. The ideal temperature control at low load is achieved by altering the furnace fire pattern and gas flow because of the risks of turbine steam hammer and water injection at low steam flows. Low-load operation generally requires changing furnace heat absorption and gas flow, while high-load operation requires steam attemporator water injection.

Because radiant heat transfer depends on *gas temperature and flame pattern* and convective heat transfer depends on *gas temperature and mass flow,* a designer can change one to offset the other. For example, recirculation gas from the economizer outlet to the bottom of the furnace will reheat the gas and increase the mass flow rate. Increasing the gas mass flow rate and temperature increases the steam temperature in the convection pass elements. The gas temperature in the furnace will decrease and heat absorption in the radiant section will decrease. This will have the effect of controlling drum pressure while giving more energy for steam superheating. On the other hand, the gas can be used for tempering in the convection pass by taking the cooled gas at the economizer outlet and injecting it above the fires in the top of the furnace. The furnace absorption will be relatively unchanged, and the steam temperature will be reduced by the tempering effect of the cooler gas.

Gas recirculation is a requirement on dual-fuel units that burn coal and liquid or gaseous fuels because the radiant heat in the furnace burning coal is not as intense as when burning oil or gas. Recirculation is used to lower the radiant absorption of oil or gas firing while increasing convective heat absorption. This is because the boiler would normally be designed for coal as the primary fuel which requires a much larger furnace. The difference in volume is due to the fuel residence time needed in the furnace. See Fig. 5-3.

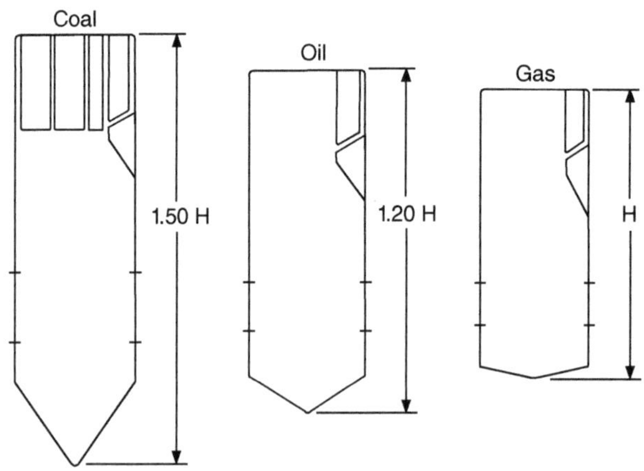

FIGURE 5-3 Furnace volume for different fuels.

Methods of temperature control vary from manufacturer to manufacturer but the overall principles are the same and must conform to the laws of classical mechanics. Therefore, the options are fixed. The only difference between one boiler and another is how the designer chooses to use them. Because of this, it is imperative that plant engineers study and understand fully the designer's intent.

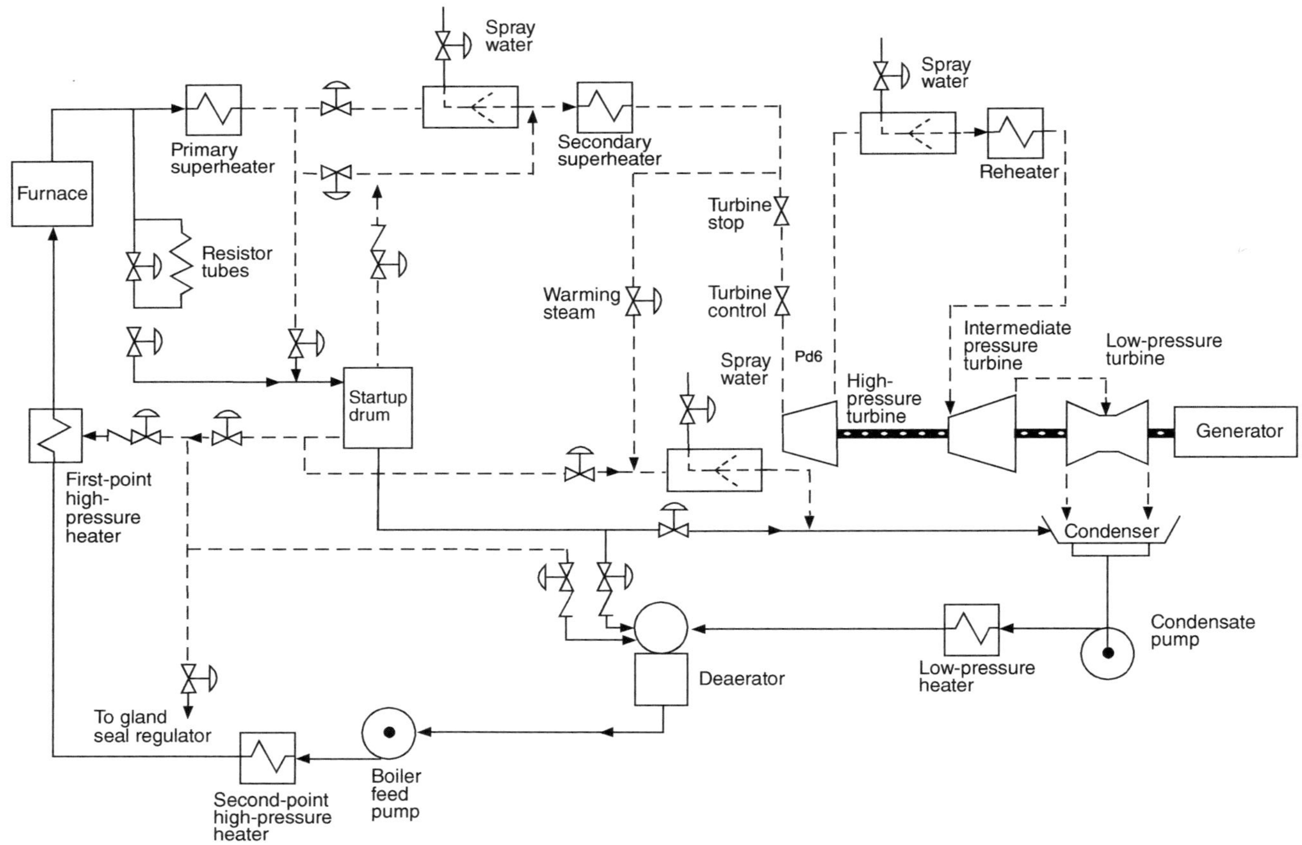

FIGURE 5-4 An early supercritical unit.

Note: Turbine steam temperature is controlled to within ±5°F. Variations larger than 10°F can cause the blades to expand and contract, while the shaft, because of its mass, does not—that is, because of the differential mass, there will be differential expansion. This causes fretting, which can lead to blade fatigue failure at the dove tails within a very short time.

SUPERCRITICAL STEAM GENERATORS

Supercritical steam generators operate above the critical pressure of steam (3208.2 psia). The saturation temperature at the critical pressure point is 705.5°F (374.2°C). At that point, steam and water have the same density therefore natural circulation is not possible. Nevertheless, these boilers must use either forced boiler water circulation or natural circulation to startup. Therefore, they require a drum, called either a startup separator or flash tank. For initial warm up, steam emission is from the separator vessels just like a subcritical boiler. As the boiler pressure approaches the critical point, furnace waterwall two phase flow is directed to both the startup separators, and through the superheaters, to the turbine. When water separation stops and the vessels run dry, all steam emission is from the furnace evaporator to the turbine with no boiler internal circulation. Therefore, supercritical boilers are called once-through. Figure 5-4 is an example of an early supercritical unit with startup transfer valves. These valves have been eliminated on newer designs.

Because these boilers operate at such high pressures and temperatures, they must have full flow demineralizer condensate polishing systems to assure the highest possible boiler water quality. Subcritical boilers only have condensate makeup demineralizers. Another major difference is the control of feedwater flow. Subcritical boilers use steam flow as the feed forward signal. For supercritical units the feed forward signal is load demand. The primary variable for drum boilers is drum level and the secondary variable is feedwater flow. For supercriticals the primary variable is steam flow with feedwater flow also the secondary variable. See Fig. 5-5 once through operating curve and Fig. 5-6 Control logic diagrams.

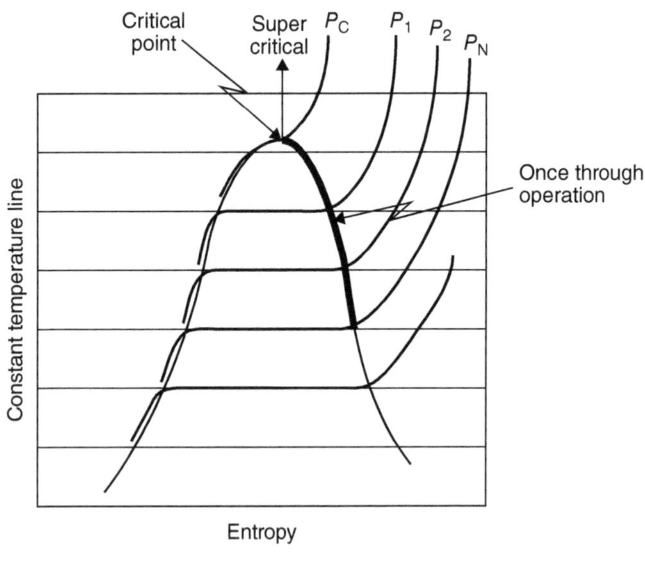

FIGURE 5-5 Once through operation curve.

Drum boilers are sensitive to drum level shrink and swell for which the steam flow feed forward signal acts as an anticipatory command. Rapid increase in steam flow anticipates a rapid drop in steam header pressure causing the drum level to swell. Drum level shrink in turn is caused by an increase in steam header pressure caused by a reduction in steam flow.

Supercritical units must anticipate load change before there is a change in steam flow in order for feedwater flow to match firing rate. Rapid increases in firing rate must be anticipated by the feedwater control system to prevent furnace over heating and tube blow out.

Supercritical units operate in the once through region from approximately 30 to 100% load. See Fig. 5-5. As load is increased from 0 to 30% the unit must operate similar to drum boilers. When approaching 30% the drum or startup steam separator

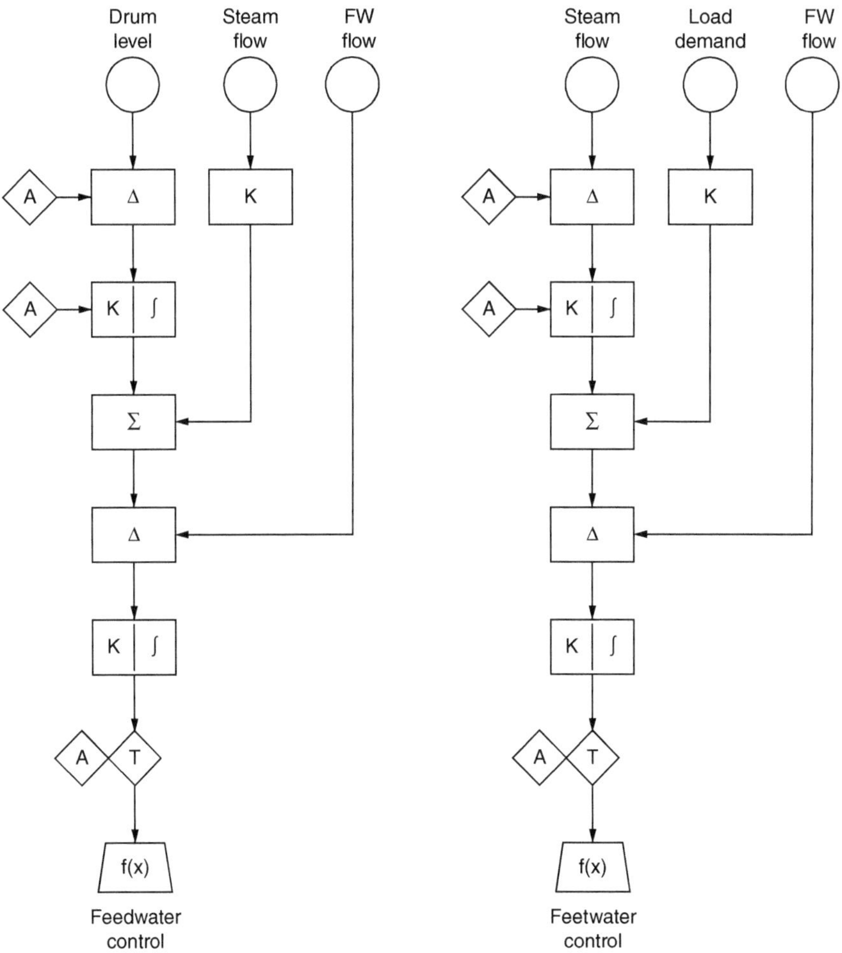

(*a*) Drum boiler FW logic (*b*) Supercritical FW logic

FIGURE 5-6 The control logic diagrams.

will go dry and then the transfer from the steam separator to once-through operation is complete. The raising of pressure, and therefore load, is done while maintaining the rated turbine steam inlet temperature constant.

GAS TEMPERATURE MEASUREMENT

Before first fires, a means of measuring the radiant section temperature must be available. Although the air preheater (APH) exit gas temperatures are irrelevant at low flows, the convection pass inlet temperature is critical because there is low steam flow in the superheaters. The most common device used to measure the gas temperature from the radiant section of the boiler is a retractable temperature probe located at the top of the furnace. After adequate steam flow is established, the probe has to be retracted to avoid damage.

Before steam pressure is established, the radiant superheaters will be subjected to high burner radiant heat without the benefit of cooling steam flow. As steam flow increases, the initial undesirable effects of radiant heat will diminish. No attempt should be initiated to use attemporation sprays to lower the temperature at low load. Low steam flow is the heart of the problem. At low steam flow, water injection for attemporation can cause water induction into the super- or reheaters, flashing, and subsequent piping system or turbine damage. The damage caused by super- or reheater water induction can be latent, surfacing much later in the startup or commercial operation.

After steam flow is established in the superheaters via a venting or bypass system, attention is now on the reheater, which will not have flow until the turbine is in service (only, of course, if a bypass system is not available). If no bypass is available, we have to rely on the superheater to cool the gas before it enters the reheater. At this time, the temperature probe becomes even more critical. As can be imagined, this becomes a juggling act until flow to the turbine is established.

Note: Overheating for even short periods, even if no metallurgical damage occurs, can deform and warp the tubes, causing them to fall off their supports. Because the situation will go unnoticed until a boiler inspection is made, this can have long-term effects that may not surface until later in startup or even after commercial operation.

WATER SPRAY FOR TEMPERATURE CONTROL

As the unit load is increased, temperature control shifts to the convection path. As the gas flow increases, convection heat transfer starts to become dominant. Because of this, excessively high superheater steam temperatures can occur. Because the steam flow is now high enough to support water injection, a superheater attemporation system can be used to control steam temperature. Some boiler manufacturers may use gas balance dampers to adjust the gas flow in the reheat–superheat sections to control the reheater–superheater steam outlet temperatures. In this case, the attemporators should only be available for emergency excursions at high steam flows.

An important inspection detail prior to startup is to verify where all the steam and gas temperature elements are located. This should be a part of the initial study and understanding of the boiler design. It will help in analyzing future problems or operating anomalies. Special consideration should be given to the location of spray nozzles in regards to temperature element location and physical layout of the piping systems. Overinjection of water due to an erroneous reading from a temperature element can have devastating consequences. Causes for erroneous readings, aside from the obvious calibration error or defective devices, are nozzle and/or element location. An element located too close to the spray nozzle is one example that could cause the element to sense an erroneously high temperature and subsequently call for more water than required, resulting in water induction. That is, the element is reading the steam temperature before the spray heat absorption is achieved. The efficiency of the nozzle, if affected by debris plugging, location, or improper design application, can also cause problems.

It is always good practice to have a blank calculation prepared to compare actual temperature readings to theoretical ones to make absolutely certain of the accuracy of the measured variables. The following is an sample calculation that can be used to determine the theoretical value of any number of questionable readings. In this case, the superheater inlet temperature downstream of an attemporation station is in question. The superheater inlet distributed control system (DCS) reading was 406°C with little variation while attemporating. However, the superheater outlet temperature was normal.

In this example, the sprays are not attemporating the steam before reaching the temperature element. In other words, the water was not completely vaporized by the time it reached the temperature sensing element causing a high reading. Therefore, more water than necessary might have been injected.

DIFFERENTIAL EXPANSION

Because the temperatures of the water and steam vary as they circulate through the various sections of the boiler, a little thought and a lot of inspection should be carried out by the plant engineer before the boiler is fired for the first time and, preferably, while the boiler is being assembled. The most important consideration of the boiler itself during precommissioning activities is how the different sections are supported and how they will expand. Expansion is a vital part of boiler design. Furthermore, the sections must expand in unison.

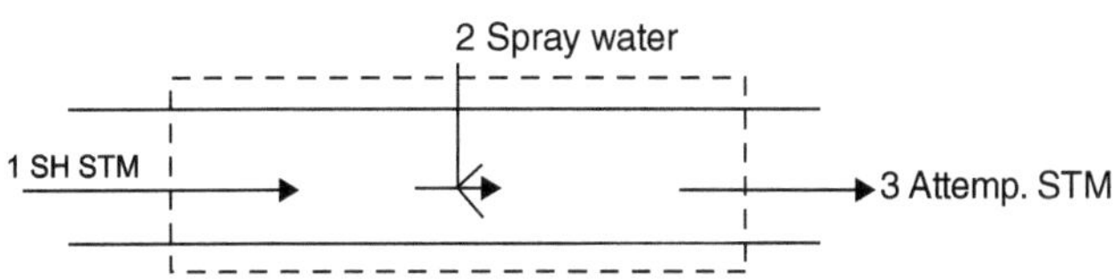

Given: $T_1 = 374°$ $\quad \dot{m}_2 = 6.1$ t/hr

$T_2 = 96.8°$ $\quad \dot{m}_3 = 44.3$ K/s

Find: T_3

NRG $\quad \dot{m}_3 \text{ Cp } T_3 = \dot{m}_2 \text{ Cp } T_1 + \dot{m}_2 \text{ Cp } T_2$

Continuity $\quad \dot{m}_3 = \dot{m}_2 - \dot{m}_1$

Solution $\quad \dot{m}_1 = 44.3 - 6.1\ (1000/36000)$

$= 44.3 - 1.7$

$= 42.6$

$$T_3 = (\dot{m}_2\, T_2 + \dot{m}_1\, T_1)\, \dot{m}_3$$
$$= [1.7\ (96.8) + 42.6\ (0.374)]\ /44.3$$

$T_3 = 363.4°$ theoretical

$T\ = 406.0°$ actual reading

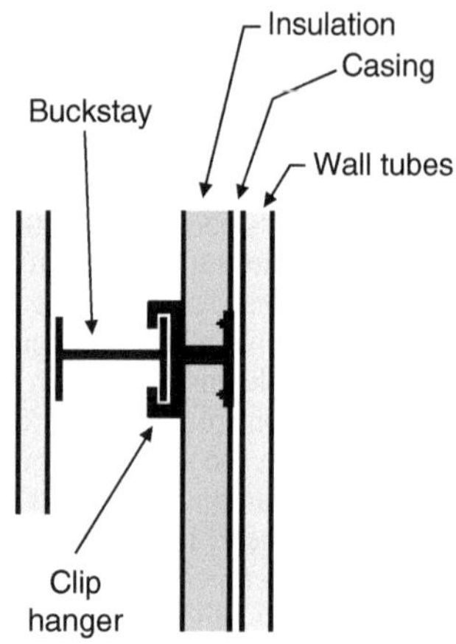

Buckstays

Large boilers are suspended from the top by the drum and/or headers because the boiler waterwalls cannot support the weight of the drum. Otherwise, the ideal direction of expansion would be upward and only the foundation would be sized to support the boiler, eliminating the need for heavy structural steel to also support the weight of the boiler. In short, the boiler must be held rigid at one point so that there is free expansion of all components in the opposite direction. Small boilers, such as marine boilers, can be supported from the bottom because the waterwalls are short and will not buckle from the weight of the drum. Heat Recovery Steam Generator (HRSG) drums are relatively light; therefore, the heat exchangers can be supported from the bottom. To guide the waterwall during expansion and protect it from explosion or implosion forces, the boiler is enclosed in a cage of structural steel. Girders, called *buckstays*, are hung on the boiler walls and follow the expansion and contraction of the walls. The lower buckstays are guided via slide plates with one side attached rigidly to the main structure. The guides prevent the boiler from swinging. While guiding and protecting the boiler, they must also allow for outward expansion of the walls. Therefore, the buckstays are hung by clips on tie bars welded to the walls. This allows the walls to move freely. The tie bars are at the same temperature as the walls, while the buckstays remain at ambient temperature.

Boiler representatives install tram points by which the movement can be measured at various points on the boiler. Regardless, the expansion can and should be checked at the buckstay guides. It is extremely important for plant engineers to check that everything attached to the waterwall is free to expand with it. That includes air and gas ducts as well as piping and conduit. If an attachment pushes an expanding furnace wall into a guide, the guide can jam the buckstay, causing the wall above it to buckle. This can happen if the obstruction causes the freely suspended boiler to swing to one side. This holds true also for free upward expansion of HRSGs. Even a small bore pipe or conduit can cause the swing.

If not provided by the manufacturer, the expansion can be calculated for any section of the boiler or piping system. It is always a good idea to know how far the boiler is going to move. The following is a sample calculation from which we can see an obvious rule of thumb (~6 in per 100 ft per 600° temperature rise).

Caution

As an example, if one side of a furnace expands 6 in and the other 9 in you have a serious problem and must trip the boiler.

Partition walls, horizontal banks, and pendants must also expand together freely. Steam partition walls and waterwalls are sometimes used to support horizontal superheater banks. The support on the wall must allow free expansion. They can consist of cleats and hooks. The tubes within a bank support each other using guides rather than rigid connections. The plant engineer must check that these guides and supports were not welded during assembly for ease of construction. Another consideration is that the walls will be cooled by water or saturated steam that will be at a lower temperature than the superheated section. Therefore, the two sections must be free to move at different expansion rates. Stress cycling due to differential expansion can cause failures after startup, during commercial operation, if they are not free to expand and contract separately.

$$\Delta L = \alpha L \Delta T$$
$$L = 100 \text{ ft}$$
$$\Delta T = 600°\text{F}$$
$$\alpha = 8 \times 10^{-6}/\text{in } °\text{F}$$
$$\Delta L = (8 \times 10^{-6})(12)(100)(600)$$
$$= 5.76 \text{ in}$$

Precise expansion for different materials can be taken from the table in App. N.

Some horizontal banks are supported by vertical tubes from the same steam source, called *sling* or *stringer tubes*. These tubes will have saddles or clips from which the tube bank is supported. The saddles or clips are, in turn, cooled by the tube they are attached to. The first of a group may support the outward suspension tubes at the bottom of the tube bank, which form a horizontal run. These supports must also allow free expansion. Pendants are normally supported by hangers attached to the pendent headers outside the furnace enclosure. They can be designed with guide tubes that are cooled from the same source as the pendant.

Care should be exercised during inspection to ensure that all tubings within any section are not rigidly attached to each other so they move together freely.

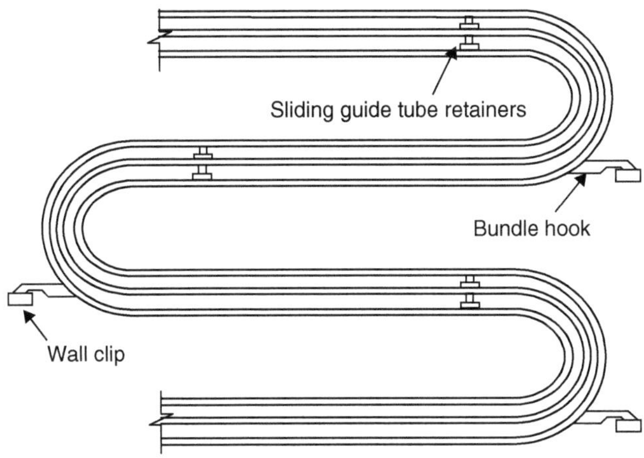

GAS RECIRCULATION DUCT EXPANSION

Gas recirculation for dual fuel oil or gas- and coal-fired units can present special problems. Because the gas recirculation system is not in service for coal firing, expansion joints and ductwork can fill with fly ash. This is caused by the lack of normal gas flow through the ducts that carries the ash when the system is in operation. Ash will accumulate any place there is no movement of gas, in stagnation areas, and in areas with pressure fluctuations (such as across closed dampers). The accumulation of ash can clog expansion joints and prevent free expansion and contraction of the jointed components. It can also cause blockage in the APH when attempting to place the system in service after a long coal-burning period. A positive pressure in the gas recirculation duct, using blowers, will provide leakage flow through dampers, preventing ash accumulation. Whenever transferring from coal operation to oil or gas, the boiler should be shut down and inspected for ash accumulation that could cause the aforementioned problems. Remember, when the boiler contracts during cooldown, boiler movement restriction may not be obvious. As the boiler contracts, it is under tension, if restrained, but it can generally move the restricting element with it. However, when the boiler expands again during the next startup, the waterwalls will be in compression and will buckle and collapse if restrained. Inward-bowed waterwalls are testimony to this.

A rule of thumb for gas recirculation systems is to disconnect them and fill the furnace openings with refractory if the unit is to use coal as the primary fuel. These systems are designed to transfer from gas or oil to coal, not the reverse.

OIL IGNITION AND WARMUP SYSTEMS

Prior to first fires, the plant engineer should review the ignition and warmup system piping and controls for both the fuel side and atomizing side. He or she must also check the system compliance with National Fire Protection Association (NFPA) standards. He or she should check the flame scanner operations and maintenance manual for scanner and scanner module startup, adjustment, and calibration. All of these checks should be done regardless of what the boiler rep is doing or has done.

The most frequent cause of air preheater fires is a malfunctioning ignition oil system. The most frequent cause of the malfunctions is a malfunctioning atomizing steam system. This occurs when water is blown into the ignition gun during light-off attempts. The water prevents adequate atomization and light-off. The failure to light-off is too often blamed on the flame scanners. Continuous attempts to achieve ignition while adjusting scanners allows large amounts of fuel to be carried over into the gas path and APH.

Before first fires, the plant engineer must check that the atomizing steam system has an adequate warmup loop and condensate drain system. The atomizing system must provide superheated steam prior to any attempts for first fires. Steam jets do not operate efficiently, if at all, using saturated steam. On the other hand, steam that is too hot can cause *petro coking*, which can plug the nozzle tips and carry raw fuel and coke into the backpasses and APH. Heavy oil-fired units switch from steam atomization to mechanical atomization after the oil becomes hot enough to flow and atomize in the nozzles without assistance. Refer to Chap. 15 for additional information.

The second-most frequent cause of APH fires is malfunctioning flame scanners during light-off. If all the obvious adjustments that can be made at the burner front are made, the plant engineer should check the sensitivity adjustments at the scanner module in the DCS room.

Caution

After startup activities are concluded each day that involve first firing, the APH should be opened and inspected for combustibles. If combustibles are found, the APH should be cooled and steam cleaned. Never introduce water into a hot APH.

Also, an APH should never be washed after firing coal no matter what material or coating the baskets have. The most frequent cause of serious APH damage is water washing. Even trace amounts of sulfur will cause the water pH to drop enough to dissolve the thin basket material. It doesn't matter if they are enameled or have some other exotic coating. If water washing must be done, the water pH must be raised until the exit water pH is at least neutral. Also remember the fuel will be deep in the basket and not readily visible. Refer to Chap. 14 for additional information.

Gas recirculation duct-pulling furnace east

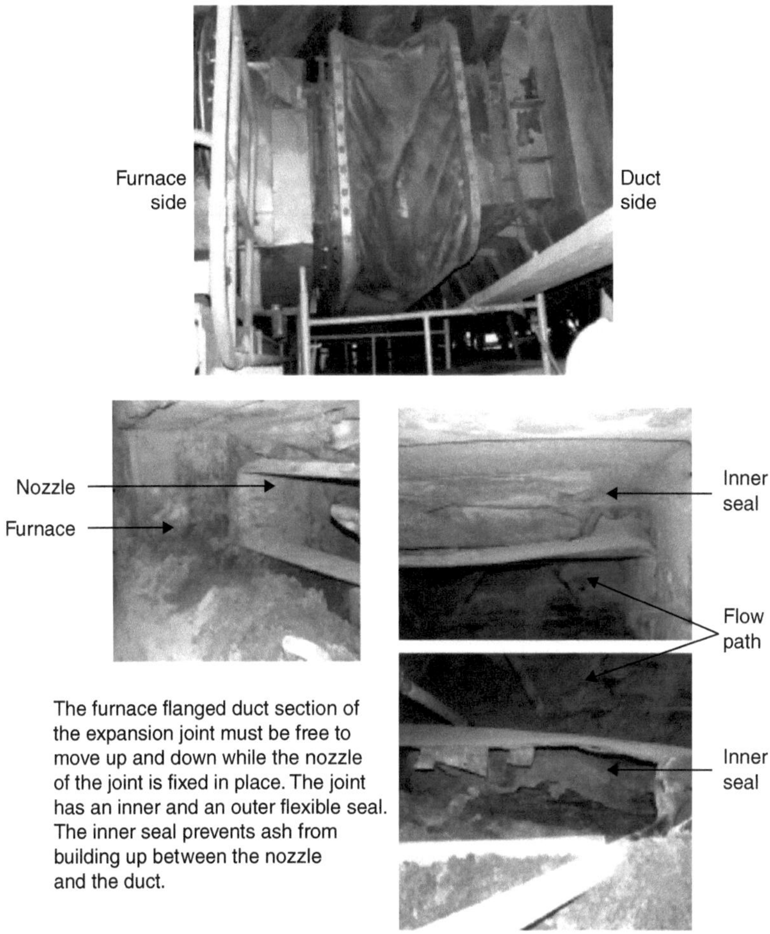

The original expansion joint was not repaired internally during rehabilitation.

It should be noted here that scanners for coal firing and liquid or gas firing fall under different guidelines. That is, the flame envelope of a coal fire is a turbulent mass with changing patterns. This generally makes it impractical to trip a boiler on loss of anything other than loss of all flame or low energy. And it is generally impractical to trip a mill on loss of flame in one burner associated with that mill.

> Upon detection of loss of flame on a predetermined number or arrangement of burners served by a pulverizer, that feeder or pulverizer and feeder shall be automatically tripped. In addition, an alarm shall sound upon detection of loss of flame on each burner involved, and the operator shall inspect flame conditions visually to determine whether burner or pulverizer operation can be continued safely. The number and arrangement of burners for each pulverizer showing loss

of flame that are necessary to automatically initiate a feeder or pulverizer and feeder trip shall be determined by the boiler manufacturer after considering the special arrangement of burners and unit load. Under circumstances of operation where adequate support between the operating burners cannot be shown, either from proven ignition or from proven adjacent burner, loss of flame on an individual burner shall automatically initiate a trip of the feeder or pulverizer and feeder.[1]

EXIT GAS TEMPERATURE

Exit gas temperatures should be checked while raising load to ensure that the design temperatures are met and the O_2 readings follow the load curve. High gas temperature is usually indicative of higher-than-design excess air. High gas temperature caused by high excess air will cause *precipitator overloading* because the temperatures are designed around volumetric flow. The induced draft (ID) fan performance may appear normal because load (amperage) is determined by mass flow.

The precipitator will generally show considerable degradation in performance before the ID fan overloads. Even 1 percent more O_2 than design can cause unanticipated downstream problems. Higher-than-design excess air is often caused by the deliberate adjustment of the air-to-fuel ratio to give conservative exit-gas O_2 readings. This often causes more problems than necessary. The O_2 monitors must be bench checked and O_2 traverse readings taken to ensure the average gas sample is read properly. Traverse readings with test equipment during startup should be the calibrated reference. If the online monitors are reading stratified gas samples rather than a true gas mixture, location modifications will have to be made. If the engineer is confident the startup checkout was complete, conservative air-to-fuel ratio control settings are not necessary.

FURNACE DRAFT CONTROL

In the mid 1970s, boiler implosion prevention standards were developed and added to the boiler explosion prevention codes because of numerous furnace implosion incidences in the early 1970s. Root cause of the failures cannot be attributed to only one source, but one factor behind the sudden emergence of these failures was the mandatory addition of air quality equipment. The addition of the air quality equipment resulted in substantial pressure drop in the system. Therefore, fans that otherwise were sized within the boiler design maximum now exceeded boiler negative pressure design. Secondly, the furnaces for balanced draft units had increased in size substantially (600- to 800-MW units replacing 150- to 300-MW units), creating an enclosure more susceptible to furnace-draft-caused damage. Office theory, which was not based on empirical evidence, offered that after a *master fuel trip* (MFT), the replacement of hot gas with cold purge air could create a negative pressure capable of collapsing the furnace, even at low loads. However, this was proven false with field empirical testing. The trends, shown in Fig. 5-7, were recorded on a 560-MW coal-fired boiler in which the newly designed furnace-draft-control systems, with implosion protection, were developed.

In the trends, the furnace pressure instantaneously dropped to -10 in H_2O, airflow was cut to purge rate, and the ID fan dampers were run back. Cold air never reached the furnace within that period. It is only logical that an instantaneous collapse of the furnace fire, from flame temperature to hot gas temperature, will instantaneously cause a collapse in gas volume and furnace pressure.

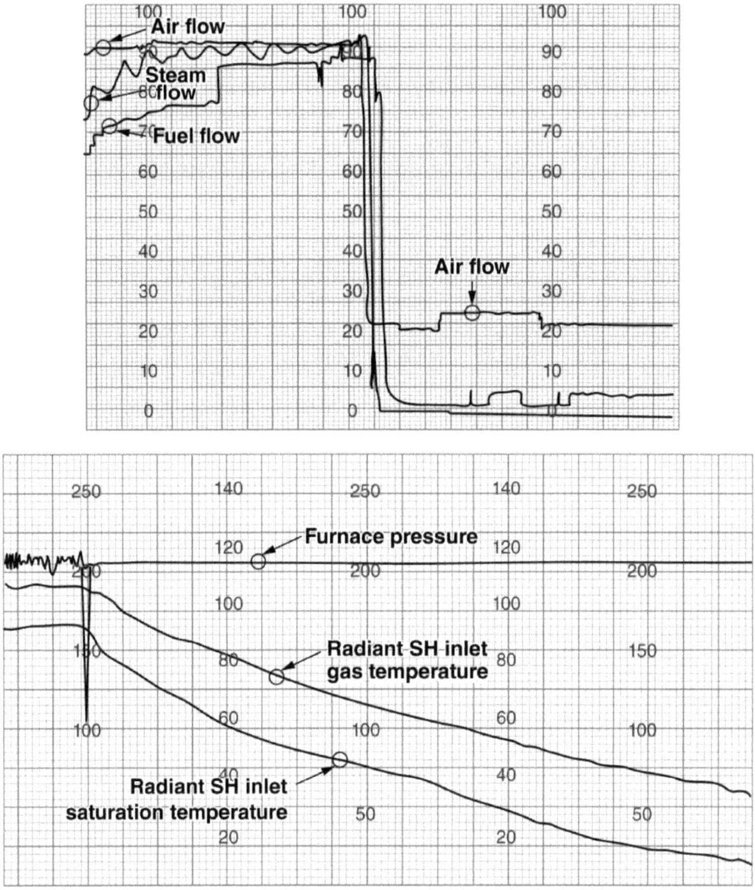

FIGURE 5-7 Furnace recorded reaction to master fuel trip (MFT).

This problem only involved oil- and coal-fired units, probably because gas-fired boilers are much smaller than oil- or coal-fired units and are not normally balanced draft.

The principal control modification incorporated to protect the boiler, ducts, etc., was the ID fan runback control and the ID fan inlet control damper feedforward positioning. That is, during normal operation, the ID fan damper position is set by a flow feedforward signal indicative of total airflow or unit load. On an MFT, a digital, timed runback is initiated to drive the ID fan dampers close to minimize the negative pressure excursion. The feedforward signal is an anticipatory signal that positions the damper in anticipation of load changes. One engineering firm included an ID fan suction pressure control loop that limited the ID fan damper position according to a position versus suction pressure curve. This effectively protected upstream equipment from collapsing due to high negative pressures caused by accidental upstream flow restriction. Furnace pressure is used to "fine tune" the furnace pressure controller by adding final adjustment for damper position. Refer to Chap. 8 for additional information.

DRUM-LEVEL CONTROL[a]

The drum is a vessel used to chemically treat boiler water and to separate (condition) the steam and water entering from the evaporation section of the boiler. It is sized only for that purpose and retains seconds of storage, not minutes. The evaporation section, consisting of waterwalls, contain two-phase flow, while the drum and downcomers are single-phase flow.

Drum level presents interesting observations. First, drum level can be an abstraction, looking into a vessel through instrumentation that can be manipulated by the very process it is monitoring. Take, for example, an instrument column that is flashing. As it is flashing, there is no true level in the column. The true level is the level in which equilibrium—or steady state, exists, but flashing causes separation of liquid and vapor. The apparent level is the level at any other state, or change of state. If, for example, the drum pressure is decreasing, the apparent level will be increasing, all other conditions remaining the same. If the operator or the control system were to react by decreasing feedwater flow, the drum level would be lower than the control point once the system returns to steady state.

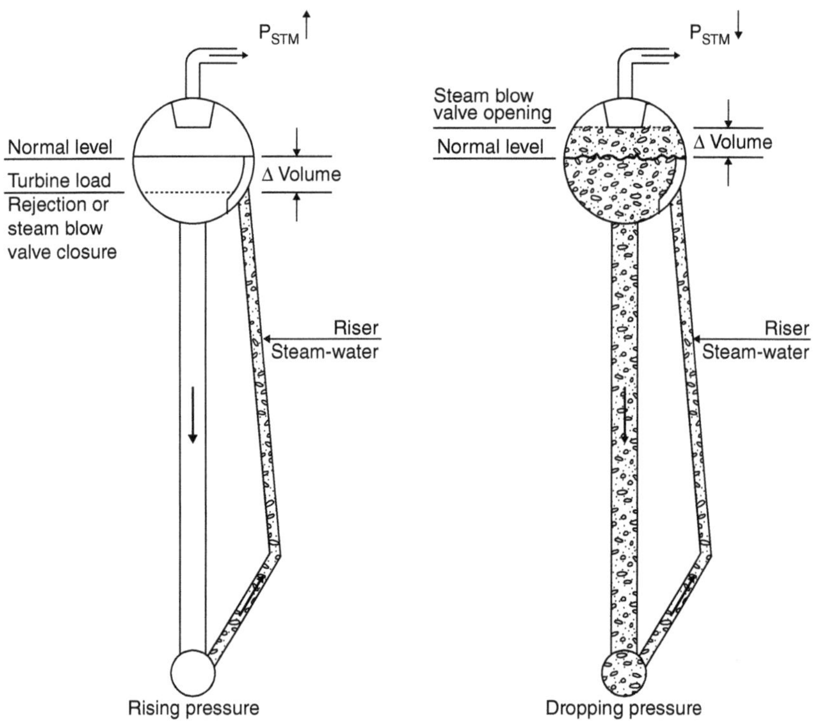

[a]Refer to Chap. 8 for additional information.

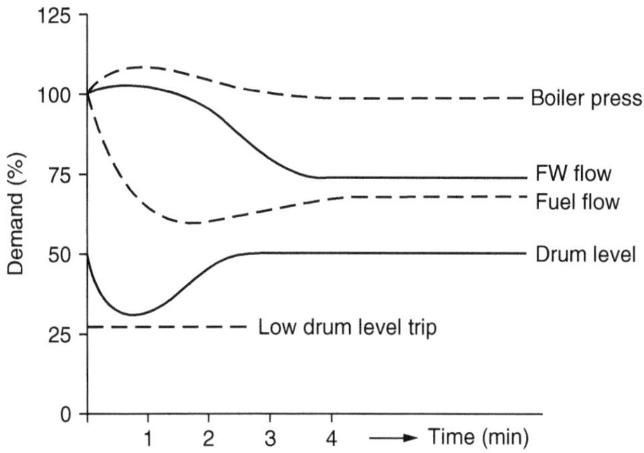

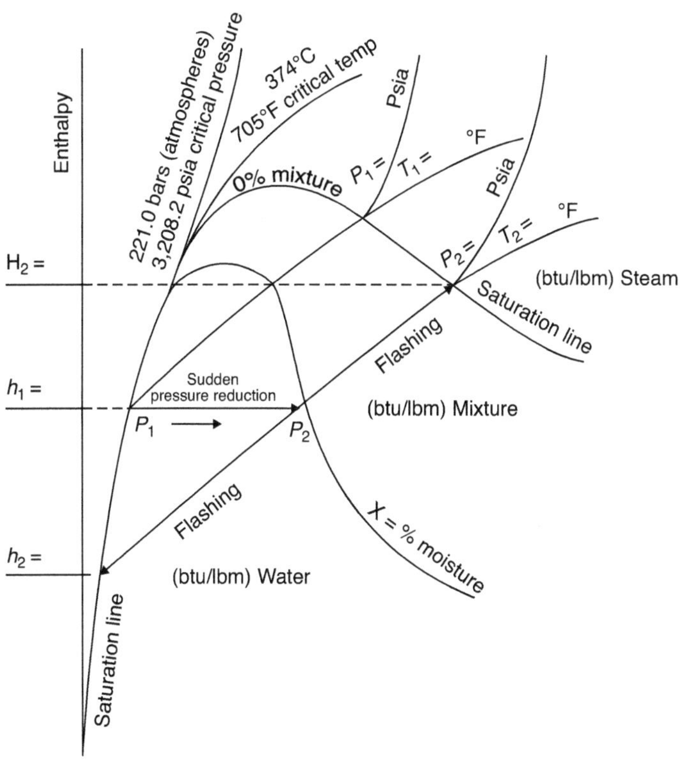

The figure above illustrates the flashing process.
Moisture in the vapor zone is determined as follows:
$h_2x + H_2(1 - x) = h_1$; $x = (H_2 - h_1)/(H_2 - h_2)$

The reason for this is decreasing pressure below the saturation line, which is the equilibrium state, into the vapor dome of the Mollier diagram generates vapor within the liquid, thus causing a temporary increase in volume of the liquid phase. In contrast, increasing pressure suppresses flashing and holds the volume constant. In other words, boiler water can fluctuate between a semi-compressible and an incompressible state. If steam flow were to suddenly increase, decreasing drum pressure, the water columns (downcomers and drum, as well as instrument tubing) would *swell*, providing a temporary high level in the drum that is not representative of the incompressible liquid state. From the preceding examples, we can see the importance of steam flow in the control of drum-level. In all drum level control systems, steam flow is used as the anticipatory signal to predict feedwater demand.

Shrink and swell are volumetric changes not mass changes. Therefore, drum level must not be allowed to chase these volumetric changes. Under normal conditions, water mass flow should follow steam mass flow. Therefore, a level rate-of-change inhibitor function block can be added to the drum level output for system stability. However, this inhibitor block must be restrained by main steam header pressure. If there is no sudden rise in header pressure the level inhibitor is restrained and the feedwater controller is allowed to respond to drum level uninhibited.

First-stage pressure is directly proportional to steam flow. In paired-unit operation, where a single boiler generates steam for only one turbine, steam flow is measured by turbine first-stage pressure. Because turbine nozzles behave as critical, or sonic flow, elements, the turbine first stage is directly proportional to pressure and can be used for steam flow measurement. A subcritical flow nozzle, for example, is proportional to the square root of the element differential pressure. When units use a steam header system by which multiple boilers can serve multiple turbines, then steam flow elements are required for each boiler main steam outlet. This is the case with combined-cycle units that use multiple HRSGs and one turbine.

The true drum level is the level at steady state and can be calculated by using equations of state. Therefore, instrument manufacturers correct for the difference in densities of water in the drum and water in the instrument columns to reflect true drum level. We can see the importance of *blowing down* an unheated, local-gauge glass to get a more realistic sight-glass reading for drum level. In blowing down the glass, the column is being heated to the saturation temperature, and the water column will more closely represent the density of the drum water, therefore, the level in the gauge-glass column. Note there is a rapid reduction of gauge-glass level shortly after blowdown because the uninsulated device cools rapidly. *The higher reading is the closest representation of true drum level. The level will quickly drop again to the false low level.*

It is good practice to incorporate a variable-drum setpoint based on unit boiler pressure to facilitate startup and restart. After an MFT, the boiler cools rapidly while purging, with the unheated cold air depressing the water level. To complicate the matter, without extraction steam, the now unheated feedwater entering the drum subcools the water and causes further suppression of drum level. Then a re-light-off causes rapid swelling in the boiler that can cause a boiler MFT on high drum level. *Drum level shrink–swell control* is a vital part of the control system. If the drum level setpoint were lowered for cold water (low pressure) and raised for hot (high pressure), the difference would serve to help compensate for swell. This will substantially reduce drum-level problems associated with reestablishing fires in the boiler.

Drum level can be calculated using the gauge glass. Calculating the difference at the design operating pressure is a good reference when checking drum level locally. That is, when you blow down the gauge glass, you will know more precisely what to expect if you have calculated the difference beforehand.

Note: Drum-level transmitter manufacturers do not use the simple algebraic expression using the water density ratio of the water in the drum and level column. They also calculate the differential head of the steam column above each water level. However, for practical purposes, the differential steam column correction is not within instrumentation accuracy.

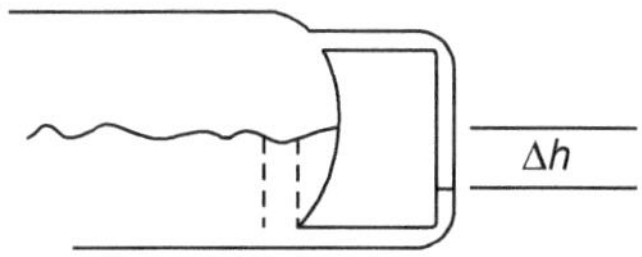

Assumption: Differential steam column (Δh) weight is negligible.
Given:
Drum: $T_D = 636°F$, $\nu = 0.02566$, $h_D = ?$
Gauge: $T_G = 300°F$, $\nu = 0.01745$, $h_G = 12$ in
Solution:

$$h_D/\nu_D = h_G/\nu_G$$

$$h_D = h_G(\nu_D/\nu_G) = 12(0.02566/0.01745)$$

$$\mathbf{h_D = 17.7\ in}$$

Sodium Monitors

An additional point that should be mentioned is that sodium monitors can act as a crucial element in drum-level monitoring during startup. Sodium is a constituent of phosphate water treatment. The presence of sodium in the main steam is an indication of high drum level, or foaming carryover. Sodium in the steam can mean uneven waterwall steam generation when all other indications are normal. That is, if the tubes entering the drum at its center have more flow than those at the end of the drum, the level will represent only the ends. Therefore, carryover can occur because of high level in the center of the drum. If not corrected, the turbine can be seriously fouled with phosphate, or worse, the emergency stop valve stems plated could prevent them from closing. Both situations have occurred in the industry.

Chemical Feed System

The chemical feed system will consist of differential pressure feed pots for periodic chemical feed and positive displacement pumps for continuous chemical feed. The feed pots are normally used for closed cooling-water systems for periodic addition of corrosion-inhibiting chemicals. This system must be in service prior to first firing. Ensure that the discharge of the chemical pumps is located at a positive pressure point in the condensate or feedwater systems. Connections to the condenser or suction of the condensate pumps can result in pulling all *chemicals* from the mixing tanks into the system. All chemical-feed diaphragm pumps require initial priming. If they are not primed, they will simply compress and expand the air within the pump casing. Follow the pump manufacturer's startup procedures. Above all, do not overfeed chemicals as a safety precaution.

BOILER BLOWDOWN

A blowdown system is designed to keep the cycles of chemical concentrations stable. Therefore, *cycles are defined in concentrations.* For example, if boiler feedwater is injected with chemicals that result in a 50-ppm feedwater concentration of the chemical and the boiler water has a concentration of 1000 ppm of the chemical, then the cyclic concentration is 1000/50 = 20, and the blowdown ratio must be 1/20 or 5.0 percent to maintain this concentration. It can be seen that the higher the chemical feed, the higher the blowdown must be to maintain a specific concentration. Therefore, the lower end of the chemical feed chart, coupled with the high limit for concentration, will yield the most economical blowdown ratio. The blowdown system can be automated by using conductivity as the control parameter. This will be important for surface (continuous) blowdown during normal operation. Remember: Blowdown wastes energy and must be set up prudently.

Both the surface and the bottom blowdown will be used during startup for silica reduction. The length of time available to blow down to achieve steam purity has a direct correlation with the quality of the preboiler and boiler cleaning process. The bottom blow is used during normal operation to periodically blow down suspended solids that collect in the bottom drum or headers. The bottom blowdown should never be used continuously at high load, and never at full load, because it reduces the cooling flow in the boiler waterwalls. That is, it is prudent to reduce load for bottom blowdowns.

STEAM BYPASS SYSTEM

Bypass steam systems were developed in the early 1970s for combined-cycle unit startups. Because the startup heat input to the boilers could not be controlled as well as for conventional-fired boilers, drum-level swell was a major problem. That is, initial pressure rise in a fired boiler is the result of radiant heat absorption and can be controlled by initially placing one burner at a time into service. With HRSGs, the evaporator heat absorption is by convection only. The evaporators receive untempered combustion turbine exhaust gas that causes immediate drum-level swell as a result of steam generation in the evaporator. The control of drum-level swell during startup must be reduced to a large extent by blowdown. This was marginally successful and wasted large amounts of water. Quickly, bypass systems developed into today's designs. This system is not to be confused with the reheat bypass systems that were used by some turbine manufacturers to control the IP turbine inlet during startup and shutdown. In those systems, reheat steam bypasses the IP turbine via a pressure control valve and is dumped to the condenser.

The heart of the bypass system design philosophy is to bypass the turbine to allow better boiler–turbine temperature coordination on startup and to prevent a boiler trip after a turbine runback or trip. Units that do not have bypass systems must use drains to warm the main steam line and stop valves to at least 100°F superheat before turbine roll. If the turbine should trip, the boiler must also be immediately tripped.

A typical large reheat unit will have a high- and low-pressure (HP, LP, respectively) bypass. Instead of flowing to the turbine, the steam flows directly to the condenser through the superheaters and reheaters via the HP and LP bypass lines. It is obvious that large amounts of energy must be absorbed in the process—first to protect the heaters, then the condenser. An extreme example would be a full-load turbine trip. Because it is impractical to absorb all the energy output from the boiler with attemporation, the energy must also be reduced by load rejection. Because the only practical means of dissipating the energy that would have been absorbed in the turbine-generator is by using attemporating water sprays, the condensate and boiler feed pumps would have to be sized not only to keep up with steam production, but also for energy absorption or reduction. Therefore, a "full-bypass" system, in most plants, does not really bypass maximum continuous-rated (MCR) steam flow, but rather allows for boiler runback from full load (full steam flow) to partial boiler load to maintain fires. Refer to Chap. 8 for additional information.

TESTING AND OPERATION

Pre-trial Operation Testing

A decision has to be made whether or not first fires will be immediately followed by boilout, steam blows, or turbine roll. However, if the unit is not ready for any of these activities but the boiler is ready for first fires, then it is still possible to proceed—as long as all considerations for cooling the boiler sections have been met, such as steam flow through the superheaters. First fires will be limited to proving flame on individual ignition and main burners until a high steam flow path is available.

1. *Furnace leak tests.* After boiler hydro, the first thing that must be done is to air leak test the gas side of the boiler. This is done by placing tarps across dampers as far downstream of the furnace and as close to the ID fans as possible. Lacking dampers, the flue duct may have to be blanked off. The purpose of an air-pressure test is to avoid air infiltration during commissioning. Air infiltration can overload the boiler fans and cause erroneous O_2 readings at the economizer outlet. Erroneously high O_2 readings can mean low combustion excess air, a dangerous situation.

   The easiest way to perform the test is to use air horns to blow air into the furnace through manways. Furnace pressure must be monitored while conducting the test to ensure positive pressure is established. As air is blown in, smoke is pulled into the furnace by the one of the air horns for furnace- and duct-leak detection.
2. *Air- and gas-side inspection.* A plant engineer, prior to operating the boiler fans, should inspect the air and gas sides of the boiler internals. Do not accept secondhand information for inspections that are required for startup.

   During the inspection, all dampers should be stroked for proper operation. Pitot tube elements and flow venturies should be checked per the manufacturers' installation specifications, and photographed. Special attention should be given to all structural supports. Some of the supports may actually have been temporary shipping or construction spreader braces. They are usually easy to identify because they will have been tack welded rather than structurally welded. Close attention to this is required near the boiler fan inlets and discharges. Photographs are extremely important because, once the boiler has been fired, access will become very uncomfortable. No one wants to climb into a dirty boiler for a verification inspection if it could have been done when the boiler was clean.

   All superheater and reheater sections should be photographed. Hangers and guides should be photographed. In the process, they should be checked for clip hanger and guide freedom of movement. In some cases, guides were welded, for ease of installation, preventing free expansion, which resulted in costly long-term outages.
3. *Boiler fan control interlock testing.* Before the fans are started, the controls must be verified without energizing the fan motors. This is done by testing the switchgear in the "racked in" and ready position. All dampers must be energized and operational. All interlocks must be checked. For example, when an induced draft (ID) fan trips,

a forced draft (FD) fan must trip. If an FD fan trips, the ID fan dampers must run back, and so on. All furnace pressure interlocks must be verified with the fans in the simulated-run condition. This is done, preferably, with U-tube water manometers, hand pumps and vacuum pumps. All tests are performed with the fans simulating the run condition.

4. *Cold-air tests*. Because the ID fans, designed for hot flue gas, will not be able to handle the "heavy" cold air for other than partial forced-draft fan load, only low-load furnace draft control testing can be done at this time. This applies, of course, to gas recirculation fans as well, which can only operate at partial load. Analog controls are tested as the FD fans are started and flow is established. The furnace draft controls should control furnace draft as the FD fans are loaded to the full-load current of the ID fans and on an FD fan trip.

   After the fan tests, a flow distribution verification test should be performed by a plant engineer inside the ducts with the fans running. Good flood lighting is required as well as scaffolding for efficient and safe testing. Two areas of importance are the section of duct containing the O_2 monitors and the inlet to precipitators. This is done by traversing the section using a velocity meter. It is necessary to ensure that these sections have equal velocity distribution for obvious reasons. If the sections where the monitors are located do not have good flow distribution, the monitors may have to be relocated. If the inlet to a precipitator does not have good flow distribution, baffles may have to be changed or added. These tests must be done with care because entering the ducts for testing after the boiler has been fired on liquid or solid fuel is not recommended. Entering the boiler ducts after the unit has been fired will require personal protection gear, including continuous air-breathing apparatuses.

   U-tube manometers should be mounted at convenient locations for measuring all gas-pass differential pressures. The manometers can be mounted on hand rails with plastic Tygon tubing running to the various boiler-gas-path taps. The measurements are used to determine the clean boiler resistance versus design and during loading to determine if fouling is occurring. Refer to flow equations in App. S.

5. *Drum inspection*. The drum internals must be inspected to ensure they have been installed and no tools, rags, or debris were left inside.

   The drum-level piping should be checked to ensure it was installed in accordance with the manufacturer's specification. The steam leg of all drum-level instruments is usually insulated, or partially insulated as required by specification, to ensure that the steam entering the condensing pot is saturated. The water legs should not be insulated to promote rapid condensation for refilling if flashed during pressure drops. Also, this ensures the variable and reference legs are at the same temperature.

6. *Economizer bypass*. The economizer bypass valve must be in service to prevent the economizer from boiling out during initial firing. If the economizer were to boil out, the lag time to fill the drum while refilling the economizer will adversely affect drum level control.

7. *Startup vents and drains*. Startup vents and drains during pretrial operation must be opened. The main steam stop valve after seat drain must be open and monitored to ensure that the stop valve is not passing steam as pressure is established. At this time, the turbine governor valves cannot be considered as secondary emergency valves. If the stop valves pass steam, there is a possibility of prematurely rolling the turbine.

8. *Turbine preparation*. The turbine-generator should be on turning gear with the lube-oil system in operation. The turbine must be in a condition that will allow accidental rolloff without damage.

9. *Air preheater*. The APH must be started to prevent warping while firing the boiler.

Trial Operation and Testing

1. *Boiler purge.*[b] Purge requirements are for 5 min minimum or the greater of five volume changes, no matter how long it takes to achieve a five volume purge of the greatest volume chamber in the system. The constant flow shall not be less than 25 percent nor more than 40 percent of the total rated volumetric airflow.

   If a trip occurs due to loss of fans, all fan dampers shall be opened gradually to the wide-open position. The natural draft purge shall be held for no less than 15 min, after which the fans can be started and the normal 5-min purge initiated.
2. *Fuel charging.*[b] The burner headers must be vented to fill them full of gas prior to first fires. A fuel leakage test must be performed after the fuel is forwarded to the individual burner fuel shutoff valves. The leak test must determine that no fuel is leaking into the furnace.
3. *Initial light-off.*[b] Registers are adjusted to the light-off position of the first burner. Energize igniters. If ignition does not occur within 10 s, the burner must be tripped and the burner register shall be opened to the purge position for at least 1 min before repeating the light-off cycle. Continue to light-off burners to prove proper operation of all ignition burners.
4. *First main burner light-off.*[b] Registers are adjusted to the light-off position. If a failure to achieve main flame ignition within 10 s (revised, 1999, from 15 s) occurs, then a master fuel trip must be effected and a mandatory furnace purge initiated.
5. *Subsequent main burner light-off.*[b] Registers are adjusted to the light-off position. If a failure to achieve main flame ignition within 10 s occurs, then all fuel to that main and ignition burner shall be shut off.
6. *Refractory dry-out.* All utility boilers will have some refractory dry-out requirements that must be adhered to. Follow the manufacturer's instructions before continuous firing is initiated and after proving a reliable burner management system is available.
7. Record all gas-pass differential pressures, O_2, APH inlet and outlet temperatures, and boiler expansion tram readings.

REFERENCE

1. Bogot, A. and Sherrill, R.C., "Principle Aspects of Converting Steam Generators Back to Coal Firing," Combustion, March 1976.
2. Monitoring with Automatic Tripping of Individual Feeder or Pulverizer and Feeder, "NFPA 8502, 1999, Standard for the Prevention of Explosions/Implosions in Multiple Burner Boilers," Section 8-4.4.3 (b).

[b]NFPA 8502, "Standard for the Prevention of Furnace Explosions/Implosions in Multiple Burner Boilers," 2011.

CHAPTER 6
CLEANING AND HOUSEKEEPING

PLANT COMMISSIONING

The responsibility for site housekeeping rests with construction. The condition of the site will give the commissioning manager a feel for what impact construction's policies will have on the startup schedule. If the site is a "mess" in which poor working conditions exist, it will affect the overall schedule and require more startup labor-hours. If it can be seen that the site is poorly kept, expect internal inspections to reveal the same. The condition of the site is directly related to construction mentality.

What is, or is not, done to keep equipment clean during construction has a major effect on the plant engineer's activities. For this reason, during the period that he or she is on the site, the commissioning engineer should assist construction in policing this effort. This includes not only the internal conditions of equipment, but also the external cleanliness and protection.

Prior to operating any equipment, the area around the equipment should be cleaned of all debris and scaffolding. All obstructions should be removed so that the operating personnel have free access to all parts of that equipment. Before firing the boiler, the burner decks should be cleared of all obstructions to give operating personnel clear access to the burner front. All planks or other combustible material should be cleared away from steam lines or other equipment that will be heated. In general, the area around equipment to be trial operated must resemble an operating plant.

CONSTRUCTION CLEANLINESS PRACTICES

The following are some good construction practices used by many companies to maintain the internal cleanliness of equipment:

Boiler

1. Tube ends should be securely capped and wrapped with aluminum-foil duct tape when received from the supplier and while in storage.
2. Tube ends received without such wrapping should be bottle brushed, air-blown, inspected for cleanliness, and capped and wrapped.
3. Caps should be pressed metal only, made to fit snugly over the tube end and wrapped with aluminum-foil duct tape.

4. Caps should be kept in place until final connection is made to the pipe systems.
5. Headers, partial headers, and piping received without caps should be cleaned and capped.
6. Superheater and reheater elements should have had sponges blown through them and have been capped in the assembly plant. However, in today's international market, that cannot be taken for granted. Therefore, if the elements are received uncapped, they should be air-blown and capped.
7. Membrane wall tubes should have been flow—and hydrostatically—tested, dried, and capped by the manufacturer. This should be included in the specification.
8. Caps should remain tight on the tube ends and stubs during erection of the boiler. Tubes and elements should be positioned for welding before the wrapping tape and caps are removed.
9. Header openings should remain tightly capped during boiler erection. Caps should not be removed until work is to be performed at the individual opening.

Equipment, Valves, and Piping Storage

1. A hard-surfaced lay-down area, or an area free from loose sand, mud, and debris with proper drainage, should be provided.
2. Keep down dust, sand, and other loose material in areas adjacent to lay-down areas.
3. Require suppliers to deliver pipe, valves, pumps, and vessels protected with dust and watertight covers over openings.
4. All valves, pumps, and cabinets should be warehoused in suitable structures until required.
5. All valves, pumps, and cabinets should be inspected as received and protected as required for storage.
6. Storage areas should be part of the daily site walkdown.
7. Prior to installation, the equipment should be internally inspected and air-blown or vacuum-cleaned.
8. All valves with weld end preparations should be made to remain closed during welding operation.
9. All valves that require opening for hydrostatic testing should remain open until after high-velocity flushing.
10. Nozzles and flanges for external connections should be flanged, plugged, or capped. Flat discs of sheet metal or 1-in-nominal dressed lumber can be used to substitute for metal flanges as covers.
11. Pipes of either random or standard-length sections should have ends flanged, plugged, or capped. All gauge and sensing element bosses should be plugged, or capped until ready for final installation.
12. Any pipe that is cut should be wire-brushed and air-blown before placing in position. Pipe that is cut in position should be wire-brushed and vacuum-cleaned at least four pipe diameters in each direction from the opening. Any plug resulting from cutting an outlet into a pipe system should be retrieved and discarded prior to welding a device or nipple into the pipe.
13. Open ends should be covered at the end of each workday. Inspect the openings for tools or personal belongings.
14. Inspection by quality control personnel should be made before making final connections to pumps.

15. Oil from threading or cutting machines should be removed with a solvent.
16. All exposed bearing surfaces should be covered with a rust-preventive compound.

Mechanical Cleaning

Mechanical cleaning is the easiest form of cleaning and should be done whenever accessibility permits.

1. Chemical tanks should be hand-cleaned and -dried after hydrostatic testing. All chemical feedlines should be air-blown clean and dry.
2. All pressure vessels should be hand-cleaned, including the deaerator and condenser hotwell. This will save time during high-velocity flushing by reducing the time required to clean suction strainers.
3. While installing startup suction strainers, hand-clean the pipe and pump suctions.
4. Mechanically clean all reservoirs. Air-blow all seal-water and seal-cooling waterlines. Air-blow all small-bore cooling waterlines for small lube oil coolers.
5. Mechanically clean all piping that can be disassembled. Bottle-brush and air-blow.

HIGH-VELOCITY FLUSHING

All systems require high-velocity flushing. High-velocity flushes are performed by flushing the systems with the normal process pumps to waste. An open, unrestricted discharge will give sufficient velocity to carry loose dirt and debris from the systems. The acceptance criterion in fossil-fuel plants is a clean discharge. The condensate and feedwater systems are flushed using the normal process condensate pumps.

The boiler can be high-velocity flushed by using the normal bypass system if so provided in the design or via a jumper from the main steam to the reheat line and to the condenser. This will be a closed-circuit flush that will require two screen-cleaning teams—one at the boiler feed pumps and one at the condensate pumps. If the boiler feed pumps cannot be used for closed circuit flushing, such as the boiler, temporary pump and filter platforms can be used. For either method, the circulating water system must be in service to cool the water as it enters the condenser. In a closed circuit, the pumps will overheat the water in a short period of time.

PREBOILER CHEMICAL CLEANING

The condensate and feedwater systems must be cleaned with a detergent to remove caked dirt, mud, and petroleum-based preservatives and coatings in the systems. Ebasco Services recommended that a combination of trisodium and disodium phosphates be used as follows:

5 lb trisodium phosphate per 1000 lb of water

2.5 lb disodium phosphate per 1000 lb of water

0.01 percent wetting agent by volume

This mixture is close to the mixture requirements for boiler manufacturers' boilout procedures, without an inhibitor. If the condensate and feedwater systems are to be cleaned with the boiler, then the boiler manufacturer's requirements must be followed, which will include an inhibitor. The wetting agent reduces the water surface tension and aides in the dispersing of oil and grease.

When these formulas were being tested in the field years ago, the commissioning engineer would add the detergent recommended and test the mixture by shaking it in a bottle. If it didn't foam, there was too little detergent. If it foamed excessively, there was too much. The enterprising individual would wipe a patch of grease in the suction spool piece of the startup strainer and then check it every time the strainer was cleaned. The engineer would also take a mixture sample and shake it to determine when the detergent was depleted. The engineer could then report to the office that the *empirical* formula for adequate preboiler detergent flushing was found as follows:

5000 ppm trisodium phosphate (Na_3PO_4–$12H_2O$)

2500 ppm disodium phosphate (Na_2HPO_4–$5H_2O$)

0.025 percent by volume (maximum) low-foam-emulsifying wetting agent

Acceptance criteria has always been by inspection of the system internals after a given time. If a more quantitative analysis is required, while the flush is in progress, the following is suggested (as always, these suggestions are not a substitute for the engineer-of-record procedures): Samples can be taken every hour and tested for phosphate, silica, pH, and turbidity. When the silica stabilization point is reached, with at least 10 h of continuous flushing at 180–200°F, cleaning of that section of the system is usually deemed complete.

Do not confuse detergent cleaning with alkaline cleaning. The latter is used commonly in a boiler after acid cleaning to neutralize any residual acid. However, at least one cleaning company lists it as an alternative to detergent cleaning and called it alkaline cleaning versus detergent cleaning.

If soda ash (otherwise known as sodium bicarbonate or baking soda) could be substituted for detergents, commissioning engineers would have been using it a long time ago. If required to use it as a substitute for a detergent, remember to wipe a greasy hand in the suction strainer spool piece. If the greasy handprint is still there after two strainer cleanings, add detergent.

The system should be rinsed by circulating properly treated demineralized water at 150°F for 1 hr.

Note: There will occasionally be questions about the need for preboiler cycle cleaning. *Experience has shown that when a preboiler detergent cleaning is performed, silica problems during boiler trial operation are greatly reduced.*

WASTE DISPOSAL

Sewer treatment facilities process detergent wastes on a routine basis. Detergent wastewater is a major part of sewer wastewater. Sewer or wastewater treatment facilities near large airports in cold-climate areas will also process glycol base, deicing solutions. When consulting with the facility that will accept the wastewater, always specify "wastewater containing detergent." Specify phosphate ppm and water pH. Do not specify oil, grease, paint, and preservatives simply because the flush was intended to remove these substances along with dirt. They know what detergents are used for. If you must be quantitative, use a lab analysis. If these substances are in the solution in measurable quantities, they will be trace, or "nil." There should be one quantity of wastewater for each rinse. Dilute the temporary storage tanks holding the detergent wastewater with the rinse water. That is, partially fill the tanks with the flush water so there is room for the rinse water.

BOILER BOILOUT

Note: Startup experience has shown that boilers assembled in a factory require far more time to clean than those field erected. This information was acquired by comparing flush pump filter plugging during high-velocity water flushing of the different assemblies; field and factory assembled.

Chemical requirements for boiler boilout usually resemble those of precycle cleaning. Because the boiler is fired, an inhibitor must be included to avoid caustic embrittlement. Sodium nitrate is commonly used for that purpose. There also are many combinations of chemicals: trisodium plus disodium, trisodium only, disodium only, and combinations of caustic soda and soda ash. The boiler manufacturer will have all these combinations so that if one chemical is not available, substitutions can be made. Chemical suppliers' handbooks also list suggestions for cleaning chemicals. The proportions used by manufacturers and recommended by engineering firms are very similar. One such recommendation, as a basic phosphate boilout, is as follows:

5 ¼ lb trisodium phosphate per 1000 lb of water *or*

2 ½ lb disodium phosphate per 1000 lb of water

1 lb caustic soda (sodium hydroxide) *or* soda ash (sodium bicarbonate) per 1000 lb of water

0.15 lb sodium nitrate per 1000 lb of water

0.10 lb wetting agent per 1000 lb of water

Note: To convert parts-per-million to pounds, multiply pounds by 1000. For example, 5250 ppm of trisodium phosphate = 5¼ pounds trisodium phosphate per 1000 lb of water. To convert to gallons of water, use 8.33 pounds per gallon.

All boiler manufacturers will have detailed procedure to follow. The preceding is to merely point out the similarities of boilout to preboiler detergent cleaning. Boilout is a detergent flush using natural circulation instead of forced circulation. Intermittent firing of the boiler is required to induce circulation. Therefore, the chemical solution will be hotter than required for cleaning. If forced circulation is used, such as cleaning the preboiler and boiler in a single cycle, the temperature requirement will be that of the preboiler flush.

Caution

Remember to isolate the main drum gauge glasses. Install a temporary-gauge glass for detergent cleaning. Because there is no flow in the glass, the detergent will leave a level watermark on the glass.

The recommended practice for boilout acceptance is a drum inspection. Remember, the main purpose of a boilout is to remove grease used to roll tubes in the drum and steam headers. Many low-pressure boilers have only welded, not rolled and welded, tubes. Therefore, oil or grease may not be found in a drum prior to boilout and, if so, it will not be in a readily accessible location. Therefore, a known reference should be used to determine acceptance. Wipe a greasy hand on the drum surface close to a manway. Check it after the boilout for acceptance.

Note: If the boiler and preboiler systems are flushed simultaneously with a high-capacity pumping platform, for high-velocity flushing, a heating boiler will not be required. Throttling the pump discharge will cause the pump to generate enough heat for the process. However, if acid cleaning follows chemical cleaning, the pumps must be changed out. If a chelate is used instead of acid, the flush pumps can be used as before. Also note, because of the limitations of natural circulation, boilout generally takes 72 hr compared to 8 to 10 hr required for forced circulation cleaning of the same boiler.

MILL SCALE REMOVAL

Mill scale removal should always be performed using the manufacturer's procedure. The following are discussions that are intended to clarify the intent of those procedures and are not intended to supplement or replace them.

After detergent cleaning of the boiler, mill scale (magnetite) and red iron oxide (hematite) must be removed. Magnetite is a black, magnetic oxide, and hematite is a nonmagnetic red iron oxide. Heating and cooling during manufacturing and fabrication form magnetite. Hematite is formed during storage and assembly, transportation, and erection. It may be years before a boiler is assembled and ready for startup trial operation. Therefore, hematite removal can be much more significant than magnetite removal.

It is very important to estimate the amount of iron oxide to be removed. Theoretically, acid can remove iron until it is spent or no more metal is available including base metal.

Iron oxide content determination

- Take a sample from the drum and lower headers. The sample should be close to a manway so inspection without entering can be made easily after cleaning.
- Measure a 1-ft-square section to be sampled.
- Scrape the section, using a square or spatula, to a clean metal surface.
- Collect the samples from all sample locations, and place them into a platinum crucible. Dry the combined sample in a drying oven.
- Record the weight of the sample.

If there is not a very significant amount of oxide, cut a boiler tube at least 1-ft in length for further sampling.

- Using a band saw, cut the tube in half then cut a strip at least 3 in across the tube.
- Remove the sample as before.
- Dry and pass the sample through a fine mesh screen to remove saw-cutting metal from the sample.
- Weigh the sample as before.

Calculations

- Deposit weight (g/ft^2) = (grams deposit/in^2 sample area)(144)

With the total boiler internal surface area provided by the manufacturer, the estimated total iron removal can be determined. The remaining tube sample can be used to determine the inhibitor effectiveness. Place one sample in a heated *uninhibited* solution and another in a heated *inhibited* solution and compare the iron concentration after 72 hr. Examine the samples for base metal loss.

Caution

If the iron content of the solution exceeds the estimate, before the acid is spent, one should be wary. One claim, by a small chemical cleaning company, that they cleaned a small boiler so corroded that they had to remove 6000 lb of iron should make one a little uneasy. Even the casual observer should suspect that most of the iron removed by the claimant was base metal.

There are many acids available today for oxide removal. Most people will use acids that the industry has the most experience with. Those are hydrochloric, citric and EDTA (ethylenediamine tetraacetic acid). Hydrochloric acid is probably the most common industrial acid cleaning solvent in use today. It is readily available and is the least expensive of all other options. It is commonly used to neutralize swimming pool water and is marketed as muratic acid for that use. Citric acid came into extensive use with the development of thin-tube, heat-recovery steam generators. It is less aggressive than hydrochloric acid, which makes it more attractive to manufacturers of smaller, lower-pressure boilers. Today, citric acid and EDTA are deemed safe enough to use in boilers in which the feedwater, boiler, and steam systems are cleaned in one circulation loop.

The following are some points to remember regarding the use of acids:

- All solvents of this nature require inhibitors.
- Hydrochloric and citric acids must be circulated using corrosion-resistant pumps.
- Where circulation is *not* obtainable, the "fill and soak" method is used. Inject nitrogen in the lower headers periodically to create mixing.
- Wetting agents reduce the amount of corrosion during the cleaning process.
- The inhibitor used determines the temperature for the solution. The temperature is usually 150 to 170°F.
- Acid concentrations range between 3 and 6 percent; citric acid is at the lower of the scale and hydrochloric at the higher.
- The solution will change from a clear to a pale yellow color as iron is dissolved. When it becomes dark brown or olive drab, it is spent. A sample check, which indicates an asymptotic (leveling-off) acid content, will verify the depletion.
- Sodium nitrite, as well as ammoniated citrate, is used for copper removal in commercially operational plants. Copper removal is never a commissioning consideration for new plants.
- The solution can be neutralized by adding ammonia, caustic soda, or soda ash, raising the pH to 9.5 to 10.
- A post-alkaline cleaning should be performed to further neutralize any acids left in the boiler.

Note: Aqueous ammonia or anhydrous ammonia must be added to citric acid to prevent the precipitation of a white salt identified as ferrous acid citrate.[1]

Passivation is usually specified without consideration of how long the boiler will be open after acid cleaning. The usual folklore is that the *raw metal* is vulnerable to oxidation. If the boiler is to be in service within a short period of time (3 days), then passivation should not be required. The normal procedure for passivation calls for the addition of sodium nitrite to create a protective layer of magnetite. However, this magnetite layer will always be replaced in time with hematite, the red iron oxide. Furthermore, the sloughing-off process of the magnetite layer can cause delays in boiler startup until the feedwater iron content and conductivity are reduced to satisfactory levels. Sometimes the boiler is overpassivated and "black water" is created, which plugs pump screens with "black mud" and causes further delays. Also, after cleaning a boiler, the magnetite layer created can cover up what the inspectors would like to see in a postcleaning inspection.

Plant engineers are accustomed to seeing black magnetite and water-level marks in drums and pressure vessels, while maintenance engineers are accustomed to seeing a thin red, satinlike surface layer of hematite with no waterline mark at all. Because plant personnel rarely inspect pressure vessels that have been in service for a long period of time, they are unaware of the normal internal surface condition.

EDTA or chelate cleaning cannot dissolve base metal; therefore, it became an attractive alternative to acid cleaning.

> It cannot directly solubilize a base metal or scale which has no dissolved ions associated with it. This is in contrast to an acid which theoretically can attack any metal above hydrogen in the electromotive series. A chelate *ties up* essentially all of the free metal ions adjacent to the deposit to be dissolved. In layman's terms, a chelate produces a "supersoft" water at this point. To give you an idea as to the order of magnitude of this "softening" action, it is quite possible for EDTA in a boiler to change the free iron concentration from 1×10^{-7} moles per liter to 1×10^{-14} moles per liter. The removal of these free ions immediately causes more of the deposit material to be ionized which, in turn, is bound by the chelate—and so on—until the deposit has been completely removed.[2]

EDTA solvents are normally heated to ~300°F (maximum 320°F), with a pH of ~9.5. The solution is approximately 6 to 10 percent EDTA. If firing is required for circulation, the temperature will be circulation driven, rather than the cleaning action. The solution is circulated ~8 to 10 hours until the iron concentration levels off.

ACID SOLUTION DISPOSAL

The disposal of neutralized acid solutions with high iron content, as well as EDTA, can be accomplished by incineration in the boiler that was cleaned. The SO_2 and NO_x emissions will decrease during the burn period.[1] See Table 6-1. The solution also can be treated for disposal in a municipal sewer system by precipitation and filtration to remove the iron.

TABLE 6-1 Representative Emissions during Cleaning Solution Incineration[1]

Data Taken during Burn Operations	Ammoniated Citric Burn	Water Flush Burn
Steam flow, lb/hr	54,000	540,000
NO_x during normal operation, ppm	353	341
NO_x during burning period, ppm	86	196
Decrease in NO_x, ppm	287	145
Decrease in NO_x, due to burn, %	81	43
SO_2 during normal operation, ppm	94	124
SO_2 during burn period, ppm	3	90
Decrease in SO_2, ppm	91	34
Decrease in SO_2, due to burn, %	97	27

Normally, the iron comes dissolved in the water in the ferrous (fe+2) state and for it to precipitate you have to oxidize it to the ferric (+3) state (pH of water ~9). Oxidize it with air and then chlorinate it, converting it to ferric (+3) chloride, and it precipitates (with flocculant) as a sludge (for high water-flow rates). Use either a greensand filter (low water flow) and regenerate it with potassium permanganate (like a softener with salt), or use a settling pond and let it precipitate using flocculant.

Also, farmers have used spent EDTA cleaning solutions successfully as fertilizers because of the high iron content. An interesting note is that EDTA is used as a stabilizing agent in the food industry. It promotes color retention in dried bananas, beans, chickpeas, canned clams, pecan pie filling, frozen potatoes, and canned shrimp. It inhibits rancidity in salad dressings, mayonnaise, sauces, and sandwich spreads. EDTA salts are used in foods at levels ranging from 33 to 800 ppm.[3]

STEAM BLOWS AND ALTERNATE SUPERHEATER AND MAIN STEAM CLEANING

Although it is a routine part of startup operations, steam blowing remains a potentially dangerous procedure that should be analyzed for each plant. Therefore, experienced commissioning engineers should be on hand to help plan and direct the operation. Steam blowing is a procedure used to clean boiler and plant steam piping. Steam is purged to atmosphere through an open-ended temporary pipe. With high-pressure purging, the open end of the pipe acts in the same manner as an under-expanded supersonic nozzle. That is, steam is not expanded to atmospheric pressure at the exit plane, but through a series of shocks well downstream. Figure 6-1 is illustrative of a typical high-pressure steam blow.

Consider using a correctly sized silencer such as the superheater safety valve silencer to suppress noise. Many times, water sprays are used to quench the steam in a transition

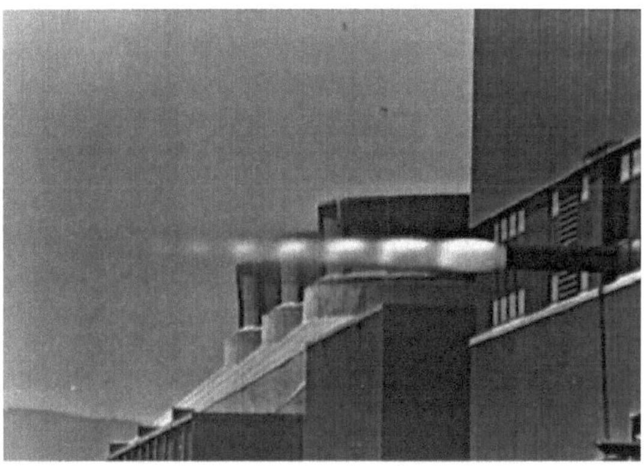

(a)

(b)

FIGURE 6-1 (a) A 1650 ft/sec blow, (b) a 1850 ft/sec blow.

piece between the blowpipe and an over sized expansion discharge section. The discharge to atmosphere will be subsonic. Continuous steam blows are also used but are limited by water consumption. In order to continuously blow, demineralized makeup water must be continuously fed to the condenser to replenish the water blown to atmosphere. It is more practical to bypass steam to the condenser, which must be under vacuum. In this manner, there is no appreciable noise, and the water is recovered. The shortfall of all alternative flushing procedures is achieving the equivalent mass and momentum conditions that exist during maximum continuous operation. Bypass systems are flow-limited because of condensate pump sizing. Steam normally extracted from the turbine and fed to the deaerator is all bypassed to the condenser. Unless the unit is equipped with additional condensate pump capacity beyond 100 percent the condenser level will rise and limit the operation. Also, the turbine and condenser both must be ready to pull vacuum, which can be a schedule restraint.

To be an effective cleaning process, the friction and inertial forces within the system must equal or exceed those which will occur during full load operation. The friction and inertia generated by mass flow act as a scrubbing device to break loose dirt, slag, and debris and carry it to the atmosphere. This is possible because, at the velocities encountered in steam blowing, the primary resistance to flow is wave drag.[4] Wave drag is predominant when the surface boundary layer is decreased, exposing the rough pipe surface to the turbulent flow outside the sublayer. This is the primary mechanism by which the system is cleaned.

Note: Experience has shown that factory-assembled boilers, such as heat recovery steam generators, contain much larger quantities of debris than field-assembled units. Flush pump and steam bypass system screens used for both have yielded as high as 10 times the amount of debris in factory assemblies. Therefore, mechanical and chemical cleaning such units should precede steam blowing. If this is done, bypass blows to the condenser can be substituted for high-pressure blowing to remove large particles not carried out by the low-velocity chemical-cleaning process. Consider both options, high mass flow or medium flow bypass blows (with chemical cleaning), for field-assembled unit. Also, it should be remembered that the main steam line is always cleaner, by a vast margin, than the boiler after they are both assembled. Because the main steam lines can be cleaned by alternate methods other than steam blows, steam blowing is primarily for the boilers.

The real test for the success of the steam blows is inspection of the machine receiving the steam after a reasonable period of full load operation. In most installations, the inspection is made at the first annual scheduled outage. However, a judgment of system cleanliness must be made prior to terminating the steam blows. To do this, a target coupon is installed in the temporary blowoff pipe. System cleanliness is then interpreted by target cleanliness. Because forces during the purging process are generally greater than with full-load unit operation, a virtually impingement-free target is rarely sought. After steam blows, fine-mesh screens are installed in turbine stop valves for the first 72 hr of full-load operation as a further precaution against damaging the turbine blades. If debris is found in the screens, further steam blows will be required. If a large quantity of small uniform pieces of metal (that resemble square shot) is detected, the steam piping must be checked for exfoliation.

Steam blowing is one of the last major procedures carried out prior to turbine roll and unit trial operation. For this purpose, practically all plant equipment is required. In general, the unit will be operated following the boiler manufacturer's operating procedures. The initial steam blow should be at reduced pressure to permit operators to test unit response and equipment and instrumentation reliability. Succeeding controlled pressure blows are then made at progressively higher pressures. Blows continue until an acceptable target coupon is obtained.

Plant Considerations

In preparing for steam blows, special attention should be directed toward pipe supports, the turbine lubricating oil system, insulation, and fire protection.

All steam line pipe supports are sometimes pinned, or otherwise restrained, prior to hydrostatic tests to prevent support overloads and possible damage. However, many manufacturers provide supports that do not require such restraining. But construction will nevertheless pin or restrain the piping.

Prior to steam blow, these restraining devices must be removed to allow proper expansion of the lines. Also prior to steam blow, the supports must be checked and adjusted, if necessary, to the cold support settings. These settings should be recorded and kept as a part of the plant record. During steam blows, the hot settings should be recorded and kept as a plant record. Pipe movement and interference should also be monitored during the initial heat-up phase of startup.

If the main-steam-to-turbine pipe connections are made, the turbine lube oil system should be operational with the turbine on turning gear. This precaution should be taken to mitigate the unlikely, but possible, rolling of the turbine. If enough leakage were to occur past the untested turbine stop and governor valves to roll the turbine, the consequence would be disastrous without an operational lube oil system.

The boiler drum must be insulated prior to operation to prevent excess inner-to-outer metal temperature differential. Control valves must be insulated to minimize distortion, which could prevent their operation. Steam lines must be insulated to prevent excess condensation and destructive water hammer during steam blows.

Severe fire hazards exist during this phase of the plant transition from construction to operation. Construction activities are still in progress with scaffolding and other combustible material in close proximity to lines and equipment that will be operated and heated, possibly for the first time. The plant staff should ensure that no combustibles are close enough to hot ignition sources, such as hot pipes, to cause fire. Close attention should be paid to the vicinity under the turbine, where insulation is generally not complete and scaffolding is dense. As a first-line precaution, fire-retardant scaffolding should be considered. Also, all fire-fighting equipment should be checked and fire-fighting procedures reviewed prior to light-off.

Design Considerations

In further preparation, design features peculiar to steam blows should not be overlooked. The most important are predetermination of boiler operating pressure, temporary blowoff valve installation, blowoff pipe supports and restraints, and temporary instrumentation requirements.

Design considerations for steam blows should include estimates for required drum operating pressure. This helps the plant operating staff determine proper equipment lineup. For example, if the required drum pressure is 500 psig and the discharge pressure of the condensate pumps is 250 psig, an alternate pump, such as a startup feed pump or boiler feed booster pump, must be considered. If this is not addressed in advance, the required drum pressure may be too great to allow water to be fed during steam blows using the desired equipment lineup. This can happen if the main feed pumps are turbine-driven and its steam lines require blowing prior to operation. Drum pressure can generally be estimated from design information already available—for example, steam line and boiler superheater losses from the plant heat balance for full-load operation—and the manufacturer's specification. For optimum cleaning, the friction and inertial forces within the pipe and superheater must equal or exceed the forces during full-load operation.

An equation relating steam parameters to the friction required for cleaning can be found from dimensional analysis or by simply using the Darcy[5] formula as follows:

$$\Delta P = (cfLW^2V)/d^5$$

where: ΔP = pressure loss, psi
c = constant
f = friction factor
d = pipe diameter,
L = pipe length, ft
W = mass flow, lb/hr
V = specific volume, ft^3/lb

The term cfL/d^5 is a constant for high velocities, that is, friction factor becomes asymptotic at high Reynolds numbers. See the Moody diagram in App. S. Because the relation can be analyzed with only two variables—mass flow and specific volume—the ratio of these variables, which is proportionate to the ratio of friction and inertia forces present, is:

$$R = (W/W_R)^2\,(V/V_R)$$

where:

R = force ratio
W_R = design steam flow, lb/hr
W = steam flow during steam blows, lb/hr
V_R = specific volume at superheater outlet design conditions, ft^3/lb
V = specific volume at superheater outlet steam-blow conditions, ft^3/lb

Steam flow is directly proportional to the steam pressure at the plane of criticality. Therefore, using the sonic velocity equation:

$$v_s = (144\,k\,g\,P_{BP}V_{BP})^{1/2}$$

where:

v_s = sonic velocity
k = 1.3, ratio of specific heats
g = 32.2 ft/s^2, acceleration due to gravity
P_{BP} = blowpipe pressure, psia
V_{BP} = blowpipe specific volume, ft^3/lb

Then:

$$W = 3600Av_s/V_{BP}$$

where:

A = blowpipe cross-sectional area, ft^2
ΔP = pressure loss, psi

Actual steam flow will be approximately 5 percent greater than calculated because steam and vapors deviate from the perfect gas laws. The Darcy formula may be used for more accurate results but is much more cumbersome for field evaluation.

If the flow does not reach critical velocity because of pipe size or use of a silencer etc. a portable ultrasonic flow meter can be used for volumetric flow determination.

The estimated pressure required at the steam drum can now be found by iteration, using assumed superheater outlet and blowpipe conditions and rated steam system pressure drops. The superheater and main steam line can be assumed to be a fixed orifice for calculating steam-blow flows and pressure drops. See App. S. First, find the flow coefficient for design conditions. Because the flow coefficient does not change, the pressure will be proportional to the change in steam conditions. Find the flow for saturated steam at a unity force ratio. Use that flow to find the drop for saturated steam across the flow element, or geometry, then add it to the assumed blowpipe pressure to find drum pressure. Assume saturated conditions and 100-psig blowpipe conditions to start with. Most blows should be around this figure. Also remember, the higher the drum pressure, the harder and longer the required firing is going to be. For a typical 2800-psig natural circulating boiler, 500 psig to 600 psig is what can be expected as the maximum before the blows become hot. High superheat will greatly reduce the force ratio. In the end, the final determination of the correct drum pressure is done by making low-pressure test blows and slowly increasing the pressure until the desired force ratio is achieved.

It is also necessary to consider whether the boiler will be fired during blows. If the boiler is tripped during blows, the boiler will cool down during purges and will greatly extend the time required between blows. The most productive steam blows are achieved by having continuous firing during blows. However, this takes experience because the steam drum water level will swell out of sight as soon as the blow valve is opened. During this period, water must be fed to the boiler or the boiler will trip on low level once the blow valve is closed again. The high drum level trip for turbine protection must be defeated for continuous firing during steam blows.

BLOWOFF VALVE

The blowoff valve should be located in the blowpipe in a place that allows complete pressurization of the entire piping system as well as the boiler. This will provide the maximum possible volume for stored energy. This should be especially considered for reheat systems that generally have larger volumes and, as a consequence, are not cleaned as well as the main steam system if the piping volume is not pressurized. Blowoff valves should be motor-operated gate valves for optimum blows and safety. Manual valves should not be considered because of the problems associated with communicating with the individuals operating them. Although all safety precautions are taken to prevent accidents, it is not considered good practice to have personnel in the vicinity of a steam blow.

Gate valves are recommended because nearly full flow can be established after only 20 percent opening of the valve (see Fig. 6-2). This allows for better utilization of the piping system volume by establishing full flow at the highest possible pressure. Slow valves,

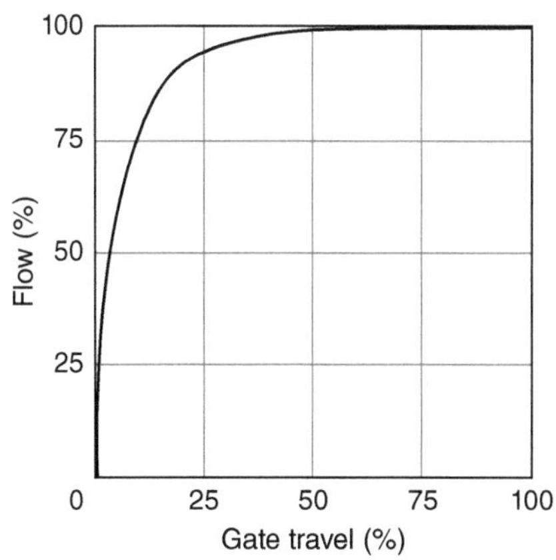

FIGURE 6-2 Gate valve flow versus position.

for example, do not allow full flow until a significant amount of pressure is bled off in the opening process. Typically, gate valves are supplied with 1800-rpm motors, which provide a 12-in/min valve stem speed. For steam blowing, the motor speed should be increased to 3800 rpm, to increase the valve speed to 24 in/min, if permitted by the valve manufacturer. This greatly improves the blow performance because the rapid reduction in pressure within the boiler takes place without a corresponding reduction in temperature, and the boiler water will provide a greater volume of steam by flash evaporation.

During steam blowing, it is especially important to remember that a gate valve should not be installed on its side. If it is not in a vertical position, excessive wear can occur or debris can collect in the bottom guides. This can cause the valve to stick open. If the valve sticks open, it can cause excessive boiler drum metal differential temperature excursions due to rapid blowing of the boiler to atmospheric pressure.

BLOWPIPE RESTRAINTS

The blowoff pipe must be considered as a free jet in calculations to determine maximum stresses imparted to the pipe. This will allow for the proper selection of pipe restraints. It must be remembered that a steam blow can easily exert 250,000 lb of thrust in large-volume, high-pressure boiler applications. See Chap. 17 for thrust calculations. For a typical horizontal application, supports are considered at three critical points: for thrust, buckling, and pipe whip (refer to Fig. 6-3).

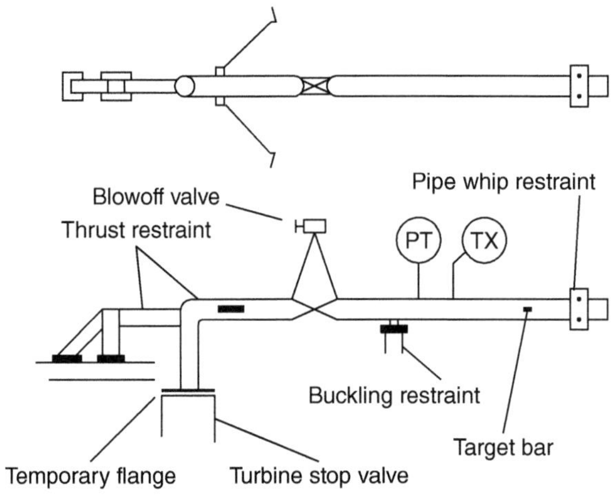

FIGURE 6-3 Typical horizontal restraints.

Considering an exit velocity of 1800 fps in a 10-in pipe, the thrust would be approximately 270,000 lbf. Torque exerted at the pipe-to-flange joint, without a thrust restraint, for a 4-ft moment arm would be >1,000,000 ft-lb.

It should be noted that the blowpipe in Fig. 6-3 is not mitered at the end. Mitering is a common practice for vent stacks in order to minimize bird resting. It is not recommended because it creates lateral thrusts besides the axial thrust. Construction often miters vent pipes and stacks without the engineer-of-record's knowledge. Unless these additional forces are considered in the design, this can be a dangerous practice. Furthermore, mitering does not increase flow or reduce noise as generally thought in power plant folklore. If a blowpipe extending a short distance from a building is mitered, the spreading flow pattern can create a low-pressure zone on the outside building wall, which can cause the pressure inside to blow out the siding. This is especially true of coal-fired plants where the turbine building is pressurized to prevent the entry of dirt, coal dust, or fly ash.

A typical blowpipe arrangement (Fig. 6-3) shows which restraints are necessary to counteract thrust, buckling, and pipe whip. The target bar should be polished mild steel, rather than aluminum or other soft metal, to avoid conflicting interpretations of target cleanliness.

Instrumentation

A recorder with variable speed capability or a DCS trend is recommended to record data used to determine the steam-blow force ratio. The following readings or trends are required: (1) blowpipe pressure and temperature, (2) superheater outlet pressure and temperature, and (3) drum pressure.

Blowpipe pressure and temperature will be used to determine the steam sonic velocity at the blowpipe exit. From the sonic velocity and cross-sectional area, steam flow can be found. Superheater outlet pressure and temperature are used to determine specific volume. From the actual steam flow and specific volume at the superheater outlet, a force ratio can be found using design steam flow and specific volume information. Drum pressure will be used as a steam-blow record and reference.

Thermocouple wires twisted and peened into a small hole in the blowpipe give satisfactory steam temperature and avoid the need for a thermal well in the flow stream. Pressure-sensing lines should be welded to the pipe to prevent loosening by vibration. Usually, spare or redundant transmitters with appropriate range can be found elsewhere in the plant or local turbine instrumentation can be used because the turbine is not commissioned yet.

Evaluating Data

Pressure in the blowpipe is the single most important reading required to accurately determine the effectiveness of a blow. However, experienced commissioning engineers know that the pressure can vary drastically depending on whether or not a target is installed in the pipe. The pressure reading taken without a target coupon installed will be much *lower* than the one with the target installed. If the higher reading is used, the indicated flow and resulting force ratio will be much greater than is the actual case.

To better understand why there is a higher pressure with a target installed, we must first review the mechanics of exhaust jet behavior. The photograph in Fig. 6-4 is of an actual steam blow, which shows the expansion and compression waves typical of underexpanded

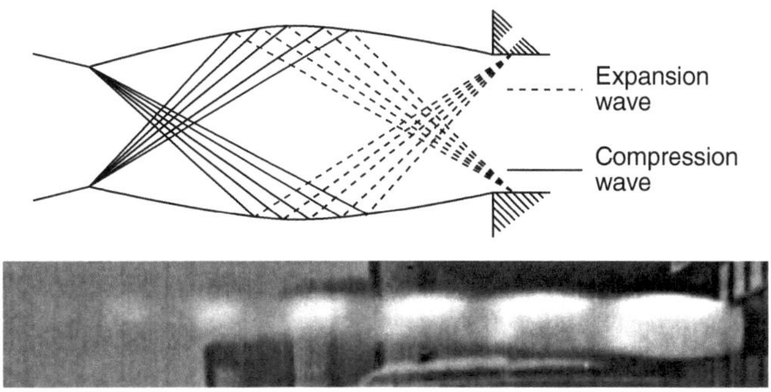

FIGURE 6-4 Sonic shock waves.[6]

supersonic nozzles. The steam is visible due to the varying luminance of the alternating compressing and expanding fluid waves. These waves are actually sudden pressure changes or shock waves.

In Fig. 6-4, theoretical expansion and compression waves are shown in the diagram above the photo of a steam blow, which shows how the waves actually appear.

Figure 6-5 shows normal pressure readings in the blowoff pipe versus pressure readings with a target installed. The pressure curve with the target installed shows the spike caused by a shock. It is of further interest to know that the shock pattern completely fills the entire pipe. Therefore, it doesn't matter where in the pipe the pressure reading is taken. Pressure will then decay slightly to the exit plane due to friction loss. This can be seen in Fig. 6-6, which shows how shock pattern fills the entire tube. The upward vertical slopes are a family of tests representing shocks, and the downward horizontal slopes represent the friction loss over the length of the tube. The curves are pressure versus time that follows the opening and closing of the blowoff valve. This phenomenon, observed with supersonic flow in constant area tubes, has also been observed in steam-blow blowoff pipes. This was verified by additional pressure readings at equididistant points along the entire pipe length. Therefore, readings at points before and after the target yield only the difference in pressure due to friction loss.

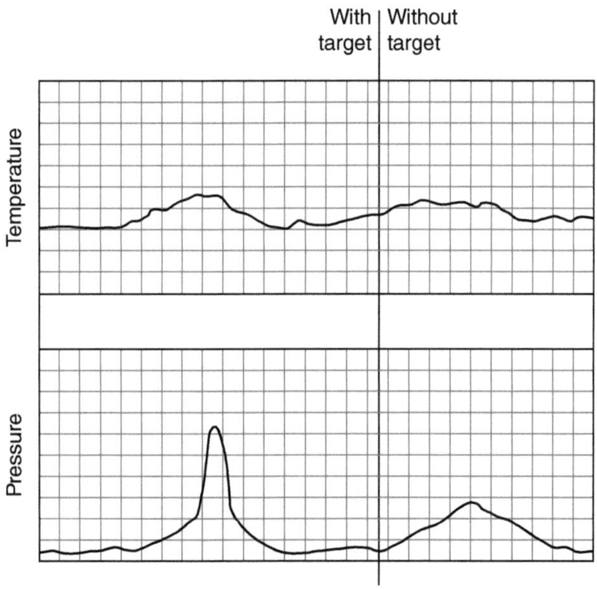

FIGURE 6-5 Pressure and temperature recorded with and without target bar.

At the station where the steam-blow discharge was photographed, flow nozzles in the main steam lines were monitored in an attempt to verify the flow obtained by sonic velocity calculation. However, the flow obtained from the flow nozzle differential pressure was so unreasonably high that it was theorized that shocks were forming there also. The flow nozzle, it is theorized, was causing the same effects seen in the pressure versus time curves.

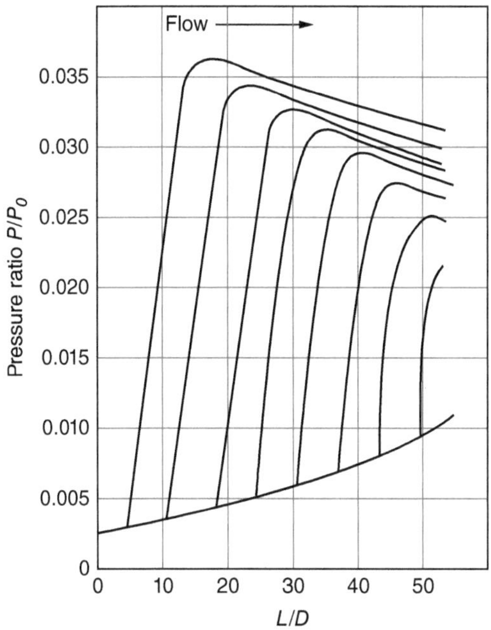

FIGURE 6-6 Plot of pressure distribution along a constant area tube (ratio of wall pressure to stagnation pressure versus ratio of length to diameter).[7]

Therefore, the sonic velocity calculation remains the most reliable method for determining flow for steam blows.

During the steam-blow phase of startup, more untested, unproved equipment and systems will be operated than during any other phase of operation. People and systems will be integrated into an operating unit for the first time on a large-scale basis. Because of this, every plant must be analyzed separately. And, commissioning engineers, with steam-blow experience, should plan and organize the activities.

AIR-BLOWING

The keys to successful air-blowing are high-capacity storage and rapid discharge. The more storage capacity, the longer the blows, and the faster you can open the discharge, the more air will discharged at maximum velocity.

Storage containers can be large-diameter pipe or pressure vessels. For example, if you are blowing a 10-in pipe and a 30-in pipe is nearby, you can connect the two and use the 30-in pipe for your air accumulator. Pressure vessels can also be used as accumulators. Remember that the velocity in the accumulator will be very low or, for all intents and purposes, zero. If you are using the pipe you are blowing as an accumulator, the velocity of the air in the back end of the pipe will be zero. Therefore, you must blow in both directions in order to clean both ends.

Fast acting valves are 90° valves such as ball valves. Rupture disks can also be used. If you use a manufactured rupture disk, use a flat disk without a disk holder to save money. However, gasket material can be used instead, which will be as much as 20 times cheaper. Order gaskets from your supplier without the center punched out. Garlock Blue-Guard Style 3000 give the best rupture pattern. Stiffer material will shatter, and softer material will pull out of the bolts instead of rupturing.

NATURAL GAS BLOWING

Natural gas lines can be blown using the normal gas supply. Natural gas is lighter than air and, therefore, will disperse into the atmosphere. Gas blows should be performed by experienced personnel only when safety regulations permit.

Some rules of thumb are as follows:

- All ignition sources must be eliminated. This means all sources, including welding, burning, heaters, and smoking, cannot be tolerated.
- High-thrust loads will occur similar to steam blows. Therefore, select a discharge point as close as possible to an anchor point. The best anchor is the ground where a pipe exits vertically. On horizontal runs, select the last anchor point with minimum pipe length extension.
- Dispersion of the gas must be vertical to the atmosphere above any structure where gas accumulation can occur. This may require temporary piping above the highest structure at the final endpoints if close to, or within, the plant.
- If there is pipe extending more than a few feet, the pipe must be supported to prevent whip, buckling, and "launching" due to thrust loading.
- The pipe open end must be grounded to prevent static electrical discharge.
- This means flames from welding, burning, propane heaters, and the like cannot be tolerated the entire day.
- Install a temporary bypass around the metering, conditioning, and compressor skids (where applicable), and blow the piping to the unit. Never disassemble a skid for cleaning or blowing. They are assembled in clean, controlled spaces in the factory and should not be disassembled by construction personnel.
- If the pipeline has a low-point loop, any water or debris will be blown clear. It is not necessary to drain the loop. Mitering the discharge will direct the blow if desired.
- The blow will obtain sonic velocity and form shock waves at the discharge. If the line is mitered, the shocks will not be visible even though critical velocity has been obtained.
- The blow should take not more than a few minutes.
- The industry standard acceptance criteria is a clean discharge. When there is no visible discharge (with the exception of shock waves), the blow is complete.
- Remember, methane is odorless and colorless and can only be detected with an analyzer. Methane that has an odor has had an odorant added to it. After the New London, Texas, school explosion on March 18, 1937, that killed more than 300 students and teachers, a federal regulation was passed to mandate odorization of fuel gas in the United States. The present federal regulation is 49CFR192.625, "Odorization of Gas,"[8] which mandates compliance unless the distribution line is in a Class 1 or 2 location. In general, Class 1 and 2 locations are relatively low population areas. Furthermore, NFPA 54, National Fuel Gas Code,[9] exempts electric utilities and gas-processing plants. If you are not sure your gas has an odorant, call your supplier. Nevertheless, do not rely on smell; use an analyzer to detect gas.

- The auto ignition temperature of methane is 1076°F and corresponding adiabatic compression pressure from the perfect gas law $Pv = RT$ is 2121 psi. Therefore, diesel ignition of natural gas is impractical. Ignition in a gasoline engine can only be achieved by spark. A diesel engine requires a seed feed of diesel fuel or a spark ignition. It cannot diesel ignite natural gas.

FUEL OIL PIPE LINES

Fuel oil pipelines should be cleaned by using petroleum industry methods, not power piping methods. The standard used in the pipeline industry is to clean the lines using pigs. Because the cleaning procedure will require preliminary planning and purchasing, the process should be started well in advance of trial operation. There are many different types of pigs supplied by many manufactures. Do not accept construction-fabricated innovations as substitutes for properly designed and manufactured pigs.

The simplest design is a polyurethane pig for light cleaning (Fig. 6-7). This pig can be wrapped with wire-brush or silicon carbide straps for removal of hard deposits. The latter is recommended for removal of iron oxide and debris. Also available are a number of mechanical pigs that can clean heavy scales and biological growth within a pipeline (see Fig. 6-8). These are used in scrubber piping or other post-startup uses. Also available are transmitters that can be installed in the pig to locate it in case it becomes stuck in the pipe.

FIGURE 6-7 Pig wrapped with silicon carbide straps.

The pipeline should be designed with a pig launcher and receiver (see Fig. 6-9). This design should be a permanent installation for future use. All thermal wells or instrument elements protruding in the pipe should be removed. Pigs can pass through valves, elbows, and other obstacles but may be damaged by sharp protrusions. The flexibility of the pig varies according to what it has to pass through. Pigs are made slightly larger than the pipe inside diameter, so they must be soaked in water, twisted, or compressed before they are inserted into the pipe. Once in the line, they will swell to normal size. Pigs are propelled by water or air. In short lines, they can be pulled. The receiver can resemble the launcher but needs a basketlike device to retain the pig after retrieval.

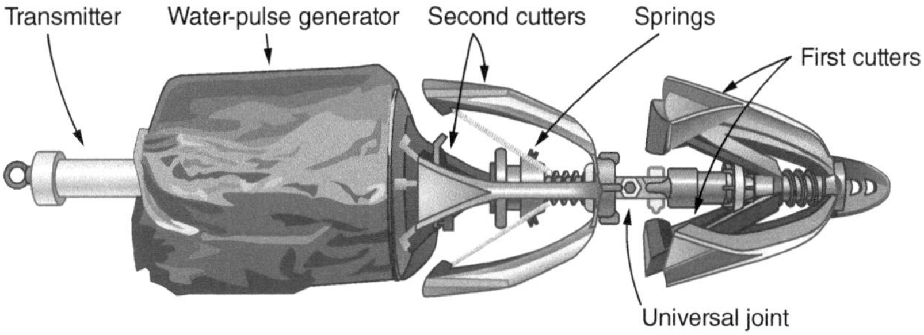

FIGURE 6-8 Mechanical pig.[10]

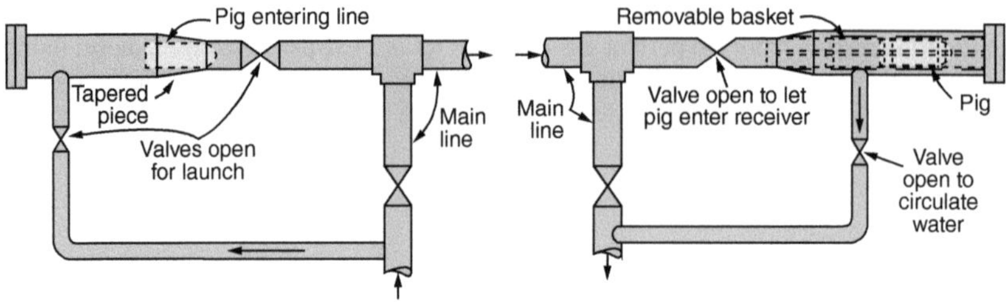

FIGURE 6-9 Pig launcher and receiver.[10]

HYDROLASER CLEANING

Hydrolaser or hydroblast cleaning was developed in the nuclear industry in the early 1970s. It was used primarily during construction as the pipe spool pieces were welded in place. Therefore, the lines had already been cleaned prior to startup.

Hydrolaser cleaning is very effective when the water can be flushed freely from the line with the debris during construction where an open-ended pipe is available. When access valves are used, they must be an unrestricted type, such as gate, ball, or stopcocks. The access usually acts as the drain; therefore, they must be large enough to flush the debris from the line.

The positive displacement pumps used can produce 8000 to 10,000 psi with about 100 gpm flow. The spray nozzle is inserted into the pipe where it cleans using a high-pressure, conical water jet. The shape of the jet keeps the nozzle centered in the pipe. The force of the spray propels the nozzle and supply hose through the pipeline. Therefore, as the hose is pulled out, the pipe is sprayed again. For a 6-in pipe, the force of the spray can propel the hose approximately 400 ft for a horizontal section and 100 ft upward in vertical section. The length of propelling capacity depends on the diameter of the pipe and obstructions in the system.

The water is collected and filtered for acceptance. Acceptance criteria will usually be the same for a target impingement acceptance. That is, there should be no particles visible to the naked eye or single particles that would cause an unacceptable impingement. All parties, prior to commencement of flushing, must agree upon the acceptance criteria.

If the lines are cleaned and pacified during construction and adequate quality control records are kept, the lines should not require further cleaning during commissioning. However, when startup performs the laser cleaning, the lines must be blown afterward because one can never be sure that all the debris was removed from the pipe if the pipe is not open ended. Even spray nozzles can become stuck in the pipe, and they are more than likely going to be left in the pipe if no one knows any different. The contractor will not want to endure the expense of cutting the line to retrieve a nozzle.

After hydrolasing, carbon steel pipe should be laid up with a nitrogen blanket to prevent corrosion.

LUBE OIL FLUSHING

The design and construction of a lube oil system will have a significant effect on the duration of a lube oil flush. If the system piping is stainless steel, the flushing can be shortened drastically if fabricated with a stringent quality control program in place. The extra cost of the material will be offset by the time and money saved during cleaning. Another argument for stainless steel is the guard-return pipe. If carbon steel, the material can also create problems during the life of the plant. Carbon steel is susceptible to corrosion because of the moisture passing the turbine steam seals. The iron oxide produced and washed to the reservoir can circulate, causing journal lapping if full-flow microfilters are not installed.

Construction of the piping system must be done with the same care as a boiler assembly. All piping must be cleaned before it is fit in place. Carbon steel should be pickled and internally coated with a preservative; capped; and stored in a covered, clean space. Stainless pipe should be bottle-brush cleaned and also capped and stored in a clean space. First-pass welds should be done by inert gas welding using *consumable inserts*.

Flushing large diesel station piping cannot be done because the piping diameters are too large. Because of the large diameters, adequate flushing velocities cannot be obtained. On the other hand, the piping spool pieces are short and all can be hand cleaned.

ASTM D6439-5, Standard Guide for Cleaning, Flushing, and Purification of Steam, Gas, and Hydroelectric Turbine Lubrication Systems,[12] provides the standards by which the lube oil flush should be performed. The latest copy of the guide should have been used by construction, and it should be studied and a detailed procedure written well in advance of flushing operations. The detailed procedure must reflex both the standard and the turbine manufacturer's flushing procedure.

TYPES OF FLUSHES

Gravity return flushing is a flush that uses the normal lube oil return lines for circulation. Flushing is done with the process lube oil pumps. This type of flush is usually limited by the flow of the oil back to the reservoir. When the oil supply is discharged into the pedestal cavity, the flow will be determined by the cavity oil level. The flow must be controlled so that the oil does not spill out. It can be injected into the return line, but the return flow will always be the supply determinant. It is advisable to use a variable-speed temporary pump when using this method unless the pump is used to heat the oil because the flow is limited by the drain system design, not the lube oil pumps. Flushing velocity of three times the design velocity can be achieved by flushing through a limited number of bearings at a time.

Check the flush velocity by timing how long it takes to fill a container of known volume at each bearing. Compare with the design flow by calculating the velocity for each. This will tell you how many bearings to flush at the same time.

Forced return circulation is used by installation of a temporary, pressurized return line. This method returns the oil to the suction of a temporary pump, providing a high-velocity closed loop. If this method is used, the lube oil coolers must be used to cool the oil.

Regardless of the method used for the main bearings, the forced circulation method must be used for the generator seal oil piping. To facilitate the temporary piping installation, flexible, fiber-reinforced, clear plastic hose can be used for the oil return branches. All seals can be flushed individually or simultaneously. (See App. T.)

Preparation Tips

- The lube oil reservoir, bearing pedestals, and front standards must have temporary covers made to limit access if the gravity flush is used. They should have lockable access doors and Plexiglas observation windows. The doors should be installed with rubber gaskets so dirt will not be drawn in.
- The lube oil reservoir, bearing pedestals and front standards must be cleaned using *lint-free* rags.

- The lube oil reservoir internal component should be removed, inspected, and cleaned. Special attention should be given to screen channels and pump piping dead-legs.
- Ensure that the lube oil coolers are clean and not a source of contamination.
- Pneumatic vibrators should be purchased well beforehand. Pipe vibration will lift debris into the flow stream.
- Provide a temporary instrument air connection, at the pump discharge, to periodically admit short busts of air to lift and carry debris.
- Install temporary emergency stop buttons at the turbine deck and under the turbine for tripping the pumps in the event of a leak or spill.
- Fabricate 60- and 100-mesh screen baskets (two of each) for each bearing supply. The surface area should be approximately 10 times the pipe diameter. See illustrations in App. T. Have the screens fabricated by an offsite shop that has experience with screens and strainers. The silver solder process used for fine-mesh screens is very delicate, and it is difficult to obtain good results in the field.
- Install throttling valves at each bearing feed, block the lines in, and air-test the piping for leaks.
- Purchase a Milipour sampler (sampler pulls solvent through filter paper under vacuum), filters (0.8-m, 47-mm diameter), and Petri slides (plastic sample containers). The fine-mesh screens should be washed in a clean, stainless-steel bowl using solvent. The solution is then drawn through the filter paper and allowed to dry. They are then placed in the Petri dishes and onto a sample board, by bearing, for comparison and progress observation.

Flushing Tips

During flushing do not leave anything open and unlocked. Suspicion of sabotage must be eliminated as a cause for contamination. Assignments should be made for around-the-clock surveillance of the flush.

- Both the AC and DC pumps must be run to flush out anything that could have been in the pump or pump discharge.
- Start a pump and check the flow at the bearings and open bearing supplies to balance return flow with full-load motor current.
- Heat the oil to obtain maximum velocity, minimize boundary layer thickness, and maximize the flow through the fine mesh screens.
- As motor current drops below full load, and pump discharge pressure increases, the filters are plugging. **Note:** Start the flush using the 60-mesh screens. Do not deadhead the lube oil pump against plugged screens. The pressure will blow the fine mesh through the perforated backing plate. Stop and replace the screens with the spares. After the flush can be maintained, change the screens every 3 hr and compare samples. Increase the changeout period until a 24-hr run can be made. Install the 100-mesh screens and repeat the process. After all bearings are cleaned, move to the generator seal oil system.
- Maintain the Petri dish board to obverse the progress and help the analysis of problem areas.

Acceptance criteria and sampling methods can be found in ASTM D6439[11] and the manufacturer's lube oil flushing procedures. Figure 6-10 is a typical use of lube oil coolers for heating during lube oil flushing. Refer to App. T for more information on setting up flushing operations.

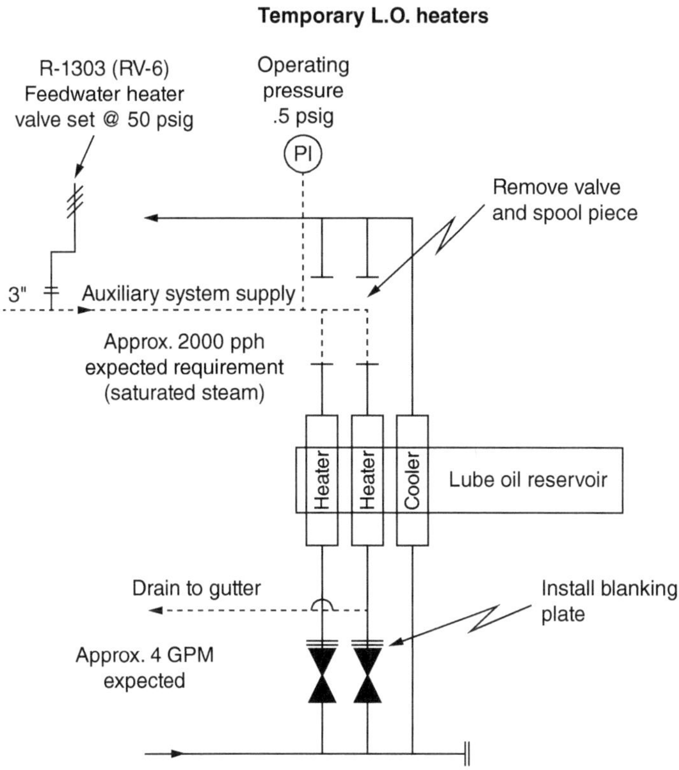

FIGURE 6-10 Temporary lube oil heaters.

REFERENCES

1. Wackenhuth, E. C., Lamb, L. W., and Engle, J. P., "Use and Disposal of Boiler Cleaning Solvent," *Power Engineering*, November 1973.
2. Osborn, O., "Theory of Chelation Chemistry and Its Application to the Utility Industry," presented to the Power Station Chemistry Subcommittee of the Edison Electric Institute at Fort Worth, TX, September 15, 1970.
3. Chemical of the Week, Chelates and Chelating Agents, http://scifun.chem.wisc.edu/chemweek.
4. Shames, I. H., *Mechanics of Fluids*, Chapter 10, McGraw-Hill, New York, 1962.
5. Crane Technical Paper No. 410, *Flow of Fluids through Valves, Fittings, and Pipe.*
6. Shapiro, Ascher H., The Dynamics and Thermodynamics of Compressible Fluid Flow, John Wiley & Sons, 1953.
7. Neuman-Lustwerk, N., "Supersonic Diffusers for Wind Tunnels," *Journal of Applied Mechanics*, June 1949.
8. "Odorization of Gas," 49CFR192.625.
9. "National Fuel Gas Code," NFPA54.
10. O'Keefe, W., "Lift Efficiency, Availability by Cleaning Piping In-Line," *Power*, May 1983.
11. "Standard Guide for Cleaning, Flushing, and Purification of Steam, Gas, and Hydroelectric Turbine Lubrication Systems," ASTM D6439-5.

CHAPTER 7
CONDENSERS

The condenser is the focal point of both the condensate and circulating water systems. Therefore, to speak of condensers the circulating water system must be considered likewise.

The condenser not only creates a vacuum by collapsing the steam from the turbine exhaust, it is also the primary collection point for all the noncondensable gases in the steam cycle. Other sources of noncondensable gases are the condensate makeup water and air leakage into the condenser. Therefore, the condenser also serves as a deaerator and is susceptible to damage from the gases. As the steam is condensed and cooled by the circulating water, the condensate is also reheated and deaerated as it falls through the steam from the turbine exhaust and the steam vents and drains. The gases collect in the air cooling section, which is the coldest part of the condenser. This is the collection point where air ejectors or vacuum pumps extract the gases. As well as air, the gas can consist of ammonia from the boiler water treatment which can cause *dezincification* of copper alloy tubes. Therefore, the plant engineer must evaluate the water treatment method to be used in relationship to the condenser tube material, especially that of the air cooler section of the condenser.

MATERIALS CONSIDERATIONS[a]

Stored demineralized water will absorb carbon dioxide from the atmosphere, which will, in turn, form carbonic acid that is corrosive to carbon steels. The stored condensate pH normally stabilizes at approximately 5.6. The condensate storage tank must be lined or coated and the makeup line to the condenser must be of stainless steel to resist corrosion. Residual CO_2 in the water will be liberated in the condenser by deaeration.

Stainless steel, such as Type 316 that contains 2.4 percent molybdenum, is often used for hardware exposed to a saltwater environment. Stainless such as Type 304 should be never used in seawater applications. Stainless with a 5 percent or greater molybdenum content is virtually immune to pitting.[1] Results of a test for a 13-month duration, gave evidence that Type 316 stainless steel corrodes in stagnant seawater where marine growth adhered to the specimens, but little corrosion occurred when the water was circulated at 4 to 5 ft/s.[2] Field surveys indicate that Type 304 and 316 stainless steels have inadequate pitting resistance in seawater applications unless extreme care is taken to keep tubes clean.[1] This is something that the plant engineer must keep in mind.

On the circulating waterside of the condenser, the tubes can be attacked by microbiological corrosion during a shutdown. Sulfate-reducing bacteria that produce hydrogen

[a]See also, in Chap. 10.

sulfide (H_2S), which in turn produce sulfuric acid (H_2SO_4), was the cause of severe pitting in stainless-steel piping in the Caribbean Islands. The source of the bacteria was from silt carried to the open sea by rivers close to the stations. These contaminants are usually suspended in the water during the Caribbean storm season and settle out during calm weather. Systems that circulate such water must be nonmetallic. This was observed in the Dominican Republic in stainless-steel, traveling-screen wash piping and in Puerto Rico in a 316 stainless-steel seawater makeup system. In both instances, extensive pitting penetrated the bottom of the piping where the silt had settled during an outage. It would be prudent for the plant engineer to take a sample of water with silts at the intake structure of circulating water pumps and immerse a condenser tube or stainless pipe sample in it for a month to determine if corrosion occurs in the stagnant sample. This applies to plants that are located on rivers, open sea locations on islands, or near river mouths.

An example of stress corrosion cracking failure occurred after an outage of a unit located on a river tributary to the Atlantic Ocean:

> The failures occurred at the inlet of the condenser tube, adjacent to the transitional zone tube joint… The unit had been in service many years before the failures… Metallurgical examination confirmed the fact that the failures were due to stress corrosion, but there was no obvious corrodent in the normal river circulating water. Furthermore, stress conditions in at least some tubes were worse during pump operation than during shutdowns. The strong correlation between failures and outages, together with the organic material remaining in the tube ends and waterboxes during outages, suggested a strong probability of local ammonia generation at the tube inlets during outages. Proteins in the decomposing fish, plant tissues, and other organic debris trapped at the tube ends break down into the constituent amino acids, and subsequent reactions could produce local concentrations of ammonia sufficient to induce stress-corrosion cracking in the admiralty metal tubes in the portion of the tube exposed to that local environment.[3]

The need for startup to be conscious of waterbox cleanliness during shutdowns is exemplified by this example.

Condenser tubesheets must be made of a material that is of the same type as the galvanic series (Table 7-1). However, the waterbox will be of cast iron or carbon steel, and the condenser casing and support sheets will also be carbon steel. Therefore, the plant engineer must ensure that cathodic protection is installed for the waterboxes if the tubes are in the noble range (see Fig. 7-1).[4]

Open recirculating water systems using cooling towers must have a water treatment system in place before turbine roll. Open recirculating water systems may have high hardness levels that can cause calcium carbonate scale to form. The high hardness is due to concentration caused by evaporative cooling. The scale will be a hard, "sandy-textured" pale green coating that will form in the distribution box nozzles of the cooling tower and the hot end tubes of the condenser.

EDTA can be used to remove calcium carbonate scale in condensers without the use of corrosive acids. But the cleanliness may never be restored to an as-delivered condition. (See Chap. 6.)

PRECOMMISSIONING INSPECTION

Condenser steam side:

- Check the turbine exhaust hood spray nozzles for proper installation.
- Check that exhaust pressure taps have been installed.

TABLE 7-1 Galvanic Series in Seawater[2]

Magnesium	18-8 stainless steel, type 304 (active)
Magnesium alloys	18-8, 3% mo stainless steel, type 316 (active)
	Lead
Zinc	Tin
Galvanized steel or galvanized wrought iron	Muntz metal
Aluminum 52SH	Manganese bronze
Aluminum 4S	Naval brass
Aluminum 3S	Nickel (active)
Aluminum 2S	78% Ni -13,5% Cr-6% Fe (Inconel) (active)
Aluminum 53S-T	Yellow brass
Alclad	Admiralty brass
Cadmium	Aluminum bronze
Aluminum A17A-T	Red brass
Aluminum 17S-T	Copper
Aluminum 24S-T	Silicon bronze
Mild steel	5% Zn - 20% NI, Bal. Cu (Ambrac)
Wrought iron	70% Cu - 2% Zn - 10% Sn (Composition G - bronze)
Cast iron	88% Cu - 3% Zn - 6.5% Sn - 1.5% Pb (Comp.
Ni-Resist	M - bronze)
13% chromium stainless steel, type 410 (active)	Nickel (passive)
50-50 lead tin solder	78% Ni -13.5% Cr - 6% Fe (Inconel) (passive)
	70% Ni - 30% Cu (Monel)
	18 - 8 Stainless steel, type 301 (passive)
	18 - 8, 3% Mo stainless steel, type 316 (passive)

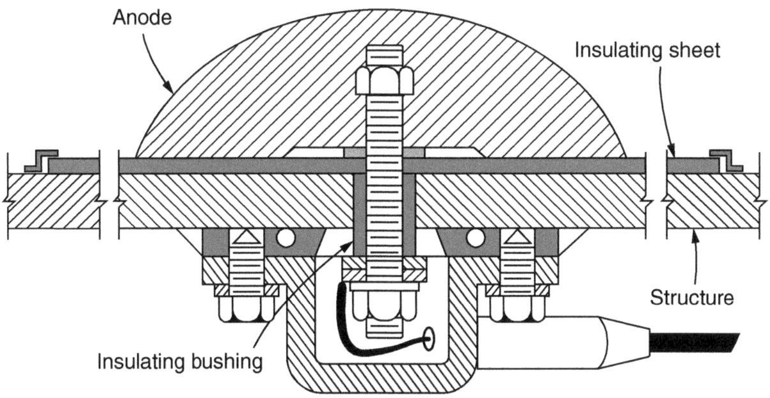

FIGURE 7-1 Impressed-current anode.

- Inspect the condenser above the tubes for construction debris such as welding rod, wood, or other foreign materials.
- Inspect the tubes for weld spatter and damage due to stepping or from fallen objects. Plug damaged tubes.

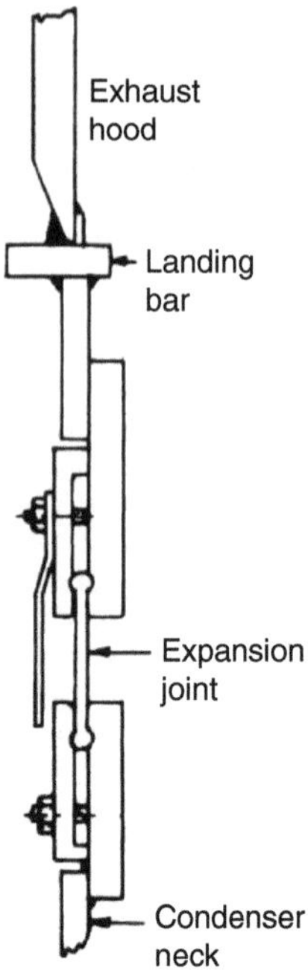

FIGURE 7-2 Typical expansion joint.[3]

- Check the expansion joint for proper installation (see Fig. 7-2).
- Check all drain and steam inlets to ensure baffles are properly installed to prevent impingement of the tubes. Figure 7-3 illustrates a properly installed baffle and one reversed by construction.
- Inspect the condenser hotwell for proper cleanliness. Supervise cleaning if necessary. Generally, if not always, construction has a different idea of condenser cleanliness than startup. The main reason for this is many supervisors are reluctant to crawl into the condenser and, therefore, rely on the word of the laborers.
- Check the condenser outlet to the hotwell pumps and ensure there is a standpipe to prevent sludge from washing into the condensate system. Ensure that the standpipe has a coarse-mesh screen that will prevent failed tubes from entering the hotwell pumps.
- Check the condensate trough, if installed, for properly located conductivity cells (see Fig. 7-4).

Condenser water side:

- Check the main circulating water piping for cleanliness. Supervise cleaning if necessary.
- Check the piping and waterbox linings for damage.
- Check cathodic protection.
- Check waterbox automatic air vents or priming system for proper installation.

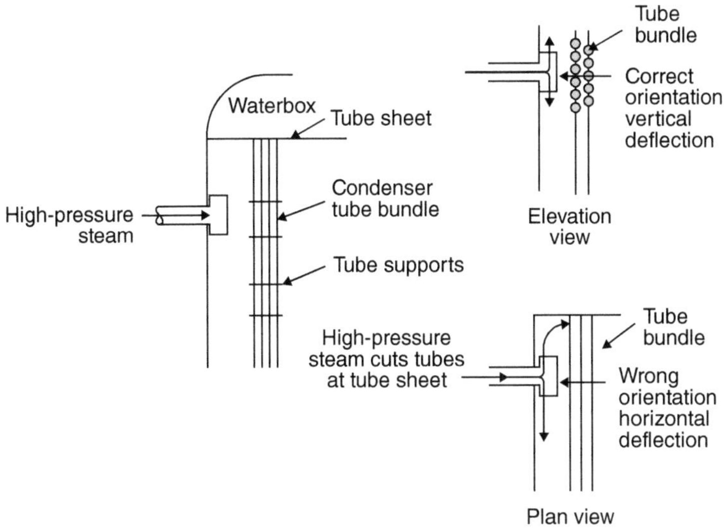

FIGURE 7-3 Steam drains to condenser.

- Check that there are properly installed thermowells for condenser performance testing. Three thermowells mounted vertically at 45° are required for each inlet and outlet pipe. The tests are required for determining benchmark and future condenser cleanliness. The Heat Exchange Institute's Standard for Steam Surface Condensers[6] governs condenser testing. The HEI's Construction Standards Addendum[7] covers design.
- Check to verify that system expansion/dialectric joint shipping restraints have been removed.

PRECOMMISSIONING TESTING

Condenser Hydro

The first condenser test is the steam-side hydro. ASME PTC 19.11, Part 11-1974, Water and Steam in the Power Cycle (Purity and Quality, Leak Detection and Measurement)[8] contains detailed procedures on how to leak test condensers as well as boilers and other power plant equipment. It was replaced by PTC 19.11-1997, Steam and Water Purity in the Power Cycle.[9]

- Obtain a tubesheet map (drawing of the condenser tube locations).
- Fill the condenser to 24 ins above the turbine expansion joint with warm water. Cold water can make the tubesheets sweat, making leak detection difficult.
- Check the tubesheets and expansion joints for leaks.

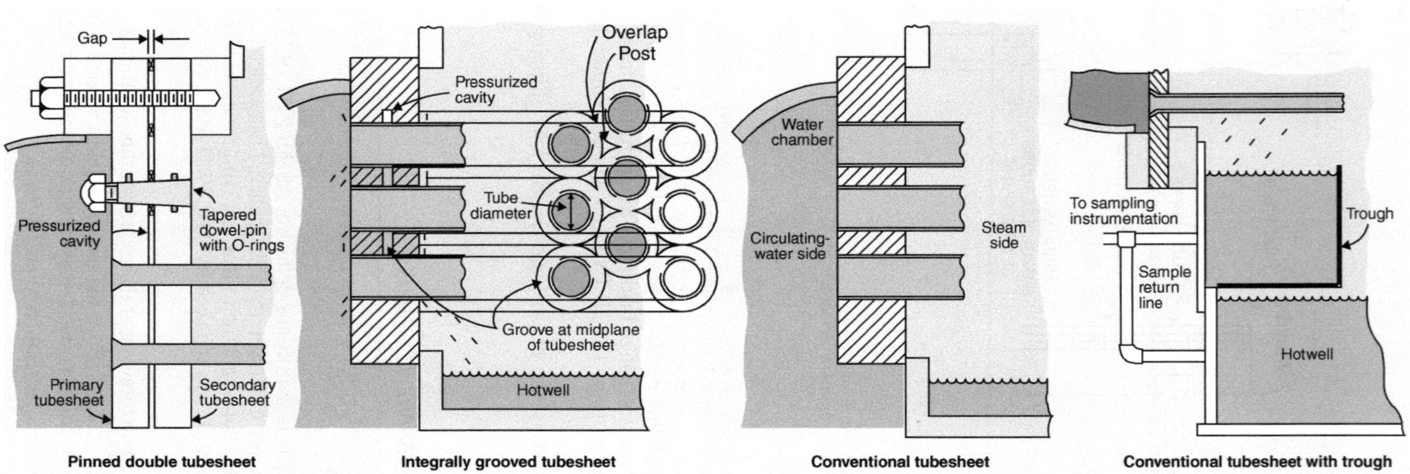

FIGURE 7-4 Typical tubesheet designs.[5]

Circulating Water Pump Pretrial Operation

The circulating water pump operation must be checked with the pump breakers in the test position.

- Initiate a pump start, and ensure the pump motor breaker closes and the pump discharge valve partially opens (generally 20 percent) and holds that position long enough for the circulating water system to prime. The discharge valve should then go to full open.
- Check the auto start of the backup pump on loss of the running pump.
- Check traveling screen and pump pits for discarded construction debris that could enter the pump or damage the screens. If the pit is full of water, install damming planks and pump out the pit. Use divers if necessary, but do not ignore this step. Do not accept secondhand information about inspections required by startup. Pits full of water are tempting sites for laborers performing final housekeeping to discard construction debris.
- Check traveling screen operation, including level differential operation. Check the level device for proper installation in turbulent flow or icing conditions.
- Check the operation of the screens on low level, and verify that the circulating water pump trips on low-low level.
- Check the screen-wash pumps for proper operation. Ensure that the controls are designed so two pumps will never operate together on the flat side of the pump curve. Test this function by reducing the flow. Generally, horizontal screen-wash pumps will have fairly flat curves. Therefore, pressure devices cannot be used in place of flow devices to detect low flow. Severe damage to a pump, such as a broken shaft, can happen to centrifugal pumps operating in parallel on the flat side of the head-capacity curve, even if it is much greater than minimum flow. The reason for this is that one pump will always lead and back the other pump down to deadhead. This can happen if the pumps are supplying washwater to more than one screen and one screen stops and its washwater valve closes.
- Check the proper operation and travel of all circulating waterbox valves.

Circulating Water System Trial Operation

- Install pressure differential gauges across the inlet and outlet waterboxes. It is very important to record the clean condenser circulating water differential pressure and monitor the pressure throughout the startup. Rising differential pressure is a sign of tubesheet fouling or, worse, tube scaling.
- Obtain the pump curves and condenser design waterbox differential pressure.
- Start the screen-wash system.
- Start a circulating water pump, and check that the opening of the discharge valve to the priming position is sufficient to prevent high pump vibration. Vent the condenser waterboxes while filling.
- Check the discharge pressure against the pump curve.
- Verify the waterbox differential pressures are within design values. Record the pressures on a clipboard at the condenser and in a startup log.

Condenser Trial Operation

Condenser air leakage testing is to be completed with the condenser under the highest achievable vacuum and only if the vacuum leakage is unacceptable. The leakage is read on the leakage rotometer located at the air discharge from the vacuum pump or ejectors.

The main discharge valve of the vacuum pump or ejector is temporarily closed and the discharge airflow is read at the bypass containing a flow meter. The reading should be in the bottom portion of the meter. If it is "pegged out" for example, a search for the leakage has to be conducted.

Before proceeding check the following:

Check the steam ejector steam temperature. A steam air ejector, steam supply must be superheated or the ejectors will not function. If the temperature control setpoint is too close to saturation, the temperature may dip in and out of saturation. Check the condensate flow through the inner- and after-condensers. Also the condensate return loop seal must be operating properly.

Note: Vacuum pumps are rarely, if ever the cause of poor vacuum.

Rent or purchase a helium detector for leak detection. The detector will be used to sample helium released around the condenser. The helium will be drawn in through the leak and be discharged by the vacuum pumps or ejectors. Practice with the detector, and get used to the type or strength of the device indication for helium detection in the office.

- Place the detector near the air discharge, and run a metal tube coil from the air discharge to an open can near the detector. The purpose of the coil is to cool the air sample. The purpose of the can is to decrease the sample velocity. Put the sample tube from the detector in the can.
- Open a vent or drain, under vacuum, at a location near the detector, and observe the reading as you did in the office. Repeat the test at a remote location. The detection of helium should be almost instantaneous.
- After you are confident about detection, spray helium around the condenser neck and low-pressure (LP) turbine exhaust. The trick is to spray general areas and then narrow your search. Start at the highest point on the turbine deck.
- Check all equipment under vacuum, such as LP heaters, by spraying helium around them. If you detect helium, narrow your search to vents, drains, and so on. Make certain that you are not causing the detector to pick up the helium outside the condenser. For example, if your detector is on the mezzanine, and you spray underneath, you will get a false reading. On the other hand, you may spray under your detector and think it is picking up the helium outside, but in reality a leak in the LP turbine casing or exhaust hood is pulling it in above you. Therefore, start your search from the highest elevation and work your way down.
- Locate condenser inlet drain headers, and walk each drain line back to the source. Spray helium around each walking towards the header. *If the header is leaking because of erosion caused by the inlet steam impingement, the vacuum will pull air from under the insulation and lagging via a break in the lagging, sometimes at a great distance upstream.*
- Check all systems, such as auxiliary steam, that will be out of service after load has increased.
- Check condensate return lines to the condenser.

REFERENCES

1. Baur, P. S., "What Condenser Materials Can Minimize Your Plant's Downtime?"*Power*, January 1982.
2. Uhlig, H. H., *Corrosion Handbook*, 24th ed., John Wiley & Sons, New York, 1948.
3. Impagliazzo, A.M. and Bell R. J., "Steam Surface Condenser Operating and Design Considerations," ASME Publication 78-JPGC-Pwr-19.

4. "Impressed Current Anode from How to Design Condenser Cathodic-Protection Systems," *Power*, June 1984.
5. Fowler, J. E., "Factors Affecting the Design of Turbine-Condenser Connections," *Combustion*, March 1977.
6. "Standard for Steam Surface Condensers," Heat Exchange Institute.
7. "Construction Standard Addendum," Heat Exchange Institute.
8. "Water and Steam in the Power Cycle (Purity and Quality, Lead Detection and Measurement)," ASME PTC 19.11, Part 11-1974.
9. "Steam and Water Purity in the Power Cycle," ASME PTC 19-11-1997.

CHAPTER 8
CONTROLS

Information on the control and startup of power plants can be found in IEEE Str 502-1985, Guide for Protection, Interlocking, and Control of Fossil-Fueled Unit-Connected Steam Stations.[1]

Logic diagrams are the key to understanding control systems. For many years instrument and control logic diagrams were referred to as SAMA drawings and still are today by many engineers. (SAMA stands for Scientific Apparatus Makers Association.) SAMA RC22-11 was the standard used for developing logic diagrams. RC (Recorder-Controller Inc.), as in RC22-11, was replaced by PMC (Process Measurement and Control Section), both of which are now identified as the Measurement, Control & Automation Association (MCAA) (www.measure.org). Nevertheless, the basic symbolism and diagram formatting are used throughout the industry and are a lasting tribute to those members of the association that developed them.

It should be pointed out that engineers customarily do not use these diagrams unless required by contract or contract negotiation. Architect-engineer firms, as well as commissioning instrument and controls (I&C) engineers, do not require logic diagrams because they are intimately familiar with the logic and work directly with engineering schematics or with distributed control system (DCS) function blocks. Therefore, it is considered very important that the owner-operator include the logic diagrams in the purchase specification. The diagrams are considered invaluable for the operations personnel to understand how the engineer designed the system controls. These diagrams are also used in conceptual design prior to detail engineering. They are also a useful tool to determine, or negotiate, the scope of work for the final engineered product. As such, these documents form standards that provide uniformity in the industry. They are also the tools by which plant engineers can convey design change requests to the design engineers. The basic control schematics are discussed next.

BASIC CONTROL SCHEMATICS

Single-Element Control

The single-element controller (Fig. 8-1) is used in processes that do not usually have upsets and are stable. For example, they are used in closed feedwater heater level control where there are no sudden changes in level. Closed feedwater heater level is maintained by condensing steam and is very stable. The feedwater heater example is for level control accomplished by controlling the vessel outlet flow. It can be used when the change in input flow does not significantly affect the response of the outlet controller. Flow of steam into a feedwater heater does not cause significant upsets in level control.

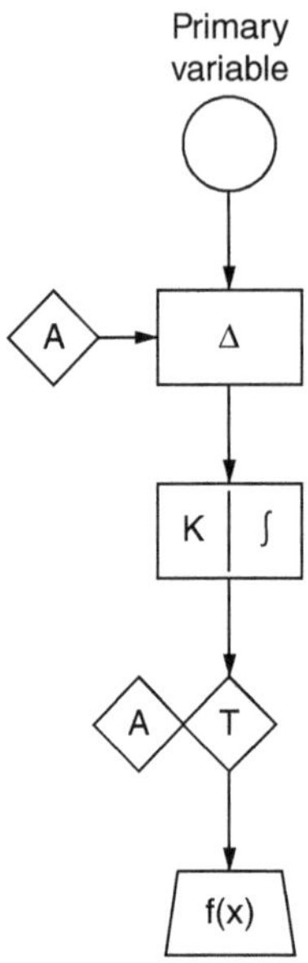

FIGURE 8-1 Single element control.

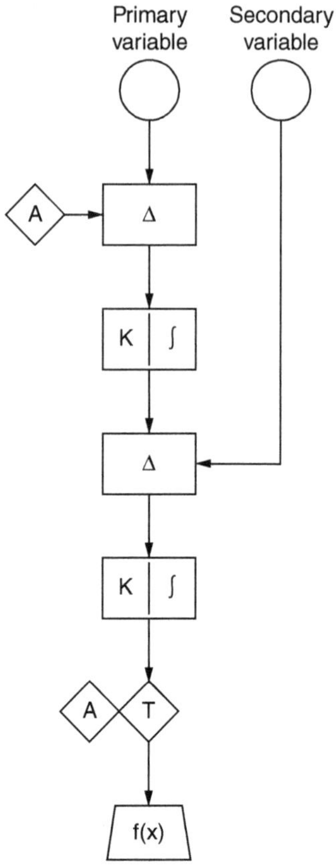

FIGURE 8-2 Cascade control.

Cascade Control

The cascade controller (Fig. 8-2) uses a second process variable to change the setpoint of the primary-variable controller. An example is attemporation steam temperature controllers that use the superheater inlet temperature as the primary variable and the outlet as the secondary variable. In doing so, outlet temperature essentially readjusts the primary (inlet) setpoint. In this case, the attemporator sprays are located upstream of the primary sensing element. Cascade can be identified immediately by observing the type controller the secondary variable feeds. It is always a differentiator.

Feedforward Control

The feedforward controller (Fig. 8-3) uses an anticipatory signal to index the primary controller. There always has to be a curve to index the setpoint. For example, a load curve is used for damper or valve position versus load. The load could be unit load, steam, or airflow. If there was an abrupt change in the load, the feedforward signal acts to anticipate a disturbance

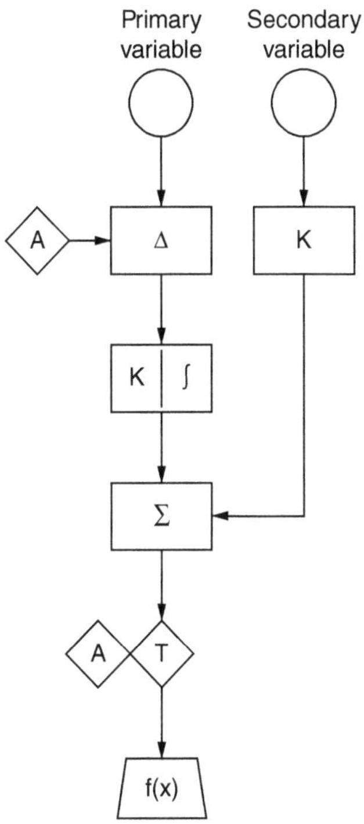

FIGURE 8-3 Feedforward control.

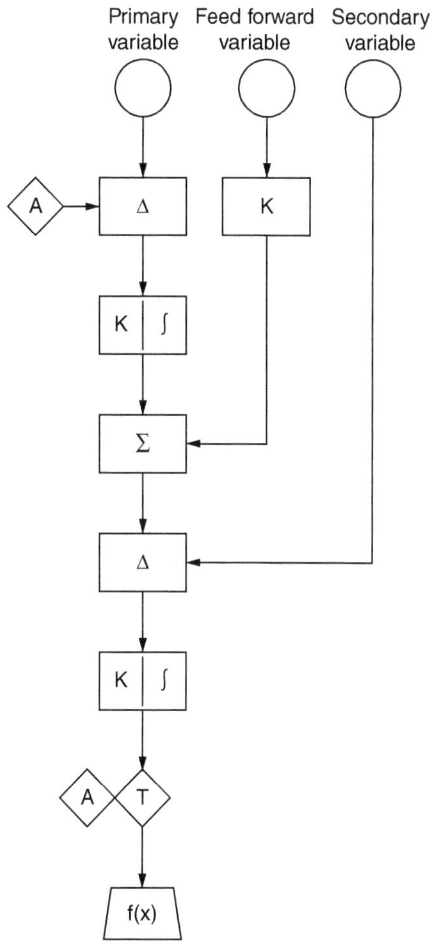

FIGURE 8-4 Cascade feedforward control.

and indexes the controller to the new point on the curve. This provides system stability. A feedforward controller always uses a summer.

Cascaded Feedforward Control

The cascaded feedforward control (Fig. 8-4) combines the cascade controller and the feedforward controller. A good example is a typical drum-level control system. The feedforward signal is turbine first stage pressure, or steam flow, which represents load and is the feedforward signal. The primary variable is drum level. The output of the primary controller (drum level) is indexed by summing its output with load. The output of the summer is an input to a differentiator. The output of the differentiator is the difference between the feedwater (–) and steam (+) flows. This control scheme provides stability for systems that are very sensitive to disturbances.

Algorithm

An algorithm is a finite sequence of instructions used to solve problems, make calculations, and so on. They are used in computer programming but have been around much longer than computers.

Derivative and Integral

The terms *derivative* and *integral* are mathematical terms used in calculus. Derivatives are expressions, in terms such as $xd(t)$, to represent a curve with respect to the variable "x" and time "t". Integrating the derivative (e.g., with $x = 1$ and time $d(t)$ with a range from 0 to 2), we can find the area under the curve (see Fig. 8-5). The average area can be illustrated by shading the area bounding the curve or the average area below the line that we can call the setpoint. A double integration will yield a three-dimensional surface area, such as for a cube. A third integration gives its moment of inertia. These mathematical expressions have been in use, and part of control system engineering, since the early part of the 20th century.

Computers cannot perform calculus functions. Differential equations, used in calculus, must be reduced to algebraic expressions. (See Table 8-1 for a list of some signal processing symbols.) Some complex nonlinear partial differential equations must be reduced by making engineering, or scientific, assumptions. This can lead to inaccuracy if the assumption error is multiplicative.

A good example is the science of climatology, where second-order nonlinear partial differential equations must be used for weather predictions.[2] But, computers cannot solve calculus equations so weather, normally, cannot be predicted for more than a week.

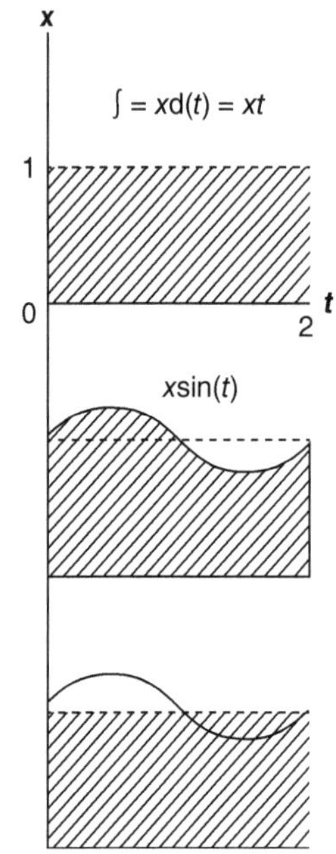

FIGURE 8-5 Simple integration.

LOGIC DIAGRAMS

The diamond with an arrow signifies a setpoint, which is compared with the process variable. The diamond without the arrow signifies a manual-auto (MA) transfer function at an MA station.

Figure 8-6 combines cascade, feedforward, and ratio control. Steam flow is the feedforward index for the boiler master. The air controller and fuel controller are cascade controllers. The air fuel ratio uses a ratio controller. Note the limiters for the air and fuel controllers that ensure the fuel-air mixture is always lean.

Three important rules to remember are:

1. Cascaded controllers always use a differentiator Δ.
2. Feedforward controllers always use a summer Σ.
3. Ratio controllers always use a multiplier X.

The following is an example of the detailed information that was provided to the engineering department by a plant engineer in the field for a design change request.

TABLE 8-1 Signal Processing Symbols by MCAA

Function	Signal Processing Symbols
Summing	Σ or +
Averaging	Σ *n*
Difference	Δ or –
Proportional	*K* or *P*
Integral	∫ or *I*
Derivative	*d*/*dt* or *D*
Multiplying	×
Dividing	÷
Root extraction	√‾
Exponential	X^n
Nonlinear function	*f*(*x*)
Tri-state signal (raise, hold, lower)	↕
Integrate or totalize	*O*
High selecting	>
Low selecting	<
High limiting	≯
Low limiting	≮
Reverse proportional	–*K* or –*P*
Velocity limiting	*V* ≯
Bias	±
Time function	*f*(*t*)
Variable signal generator	*A*
Transfer	*T*
Signal monitor	*H*/, *H*/*L*, /*L*

Function		Signal Processing Symbols	
Logical signal generator		B	
Logical AND		AND	
Logical OR		OR	
Qualified Logical OR		>*n*	GTn
n – an integer		<*n*	LTn
		–*u*	EQn
Logical NOT		NOT	
Set memory			S, SO
Reset memory			R, RO
Pulse duration			PD
Pulse duration of the lesser time			LT
Time delay on initiation			DI or GT
Time delay on termination			DT
Input/output	Analog		A
	Digital		D
	Voltage		E
Signal converter	Frequency		F
	Hydraulic		H
	Current		I
Examples:	Electromagnetic or sonic		O
D/*A*	Pneumatic		P
I/*P*	Resistance		R

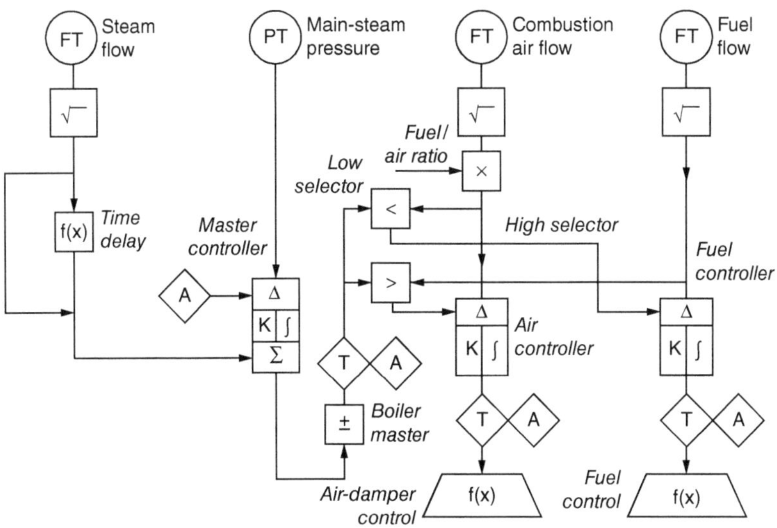

FIGURE 8-6 Logic diagrams.[3]

Problem

On a load rejection, or master fuel trip (MFT), the deaerator (DA) pressure is reduced below saturation causing the DA and boiler feed booster pump suctions to flash violently. The reason for the pressure reduction is that the boiler feed system is calling for water to replenish the drum. This, in turn, calls for more condensate to the DA, which condenses the steam blanket. On a load rejection, or MFT, steam supply to the DA is reduced or stopped completely.

Analysis

First Consideration

On a load rejection, condensate to the DA must be reduced or, in case of an MFT, stopped completely. The heat input to the DA, by the turbine extraction, has been reduced or stopped completely. Therefore, if the condensate is induced, it will absorb the heat in the DA, collapse the steam blanket, and drastically lower the DA pressure without lowering the storage tank water temperature.

The DA level has to be high enough so that the change in stored water volume in the DA equals the change in drum water volume ($\Delta V_{DA} = \Delta V_{drum}$). To accomplish this, the DA level setpoint (SP) is instantaneously reset to a lower level representing the volume change in the boiler drum. In other words, reduction of condensate flow to safe levels is done by reducing the level control setpoint (SP). The actual level did not change, therefore the controller receives a high signal and the condensate valve closes.

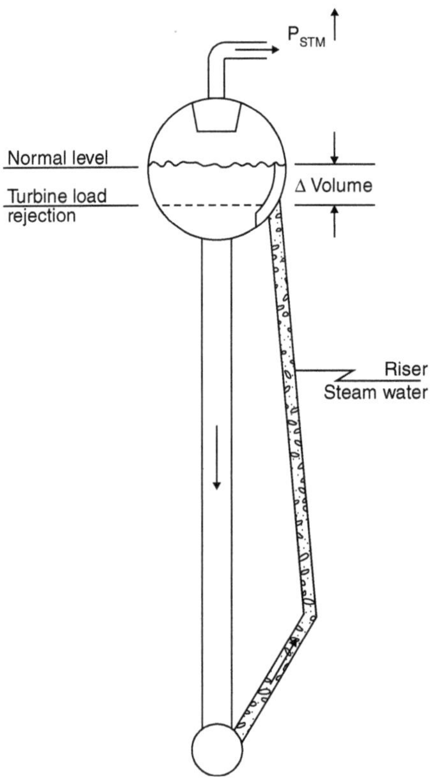

FIGURE 8-7 Drum level depression with sudden drop in steam flow.

Figures 8-7 and 8-8 depict what occurs. The sudden load change takes the DA pressure into the vapor dome of the Mollier diagram. Once there, the water flashes back to the saturation line. This process of flashing will continue until equilibrium is reestablished. Flashing occurs in the entire system, from the feed pump suction to the DA, endangering the pumps, DA internals, and personnel.

Second Consideration

As previously mentioned, because of the volume collapse in the boiler, the drum-level controller is calling for water from the DA. On an MFT, boiling stops, riser volume collapses, and drum level drops.

Firing rate determines boiling rate and, therefore, volume collapse. Thus, the change in drum level is proportional to load and ΔV_{drum} (Fig. 8-9). The change in volume is the water replenishment

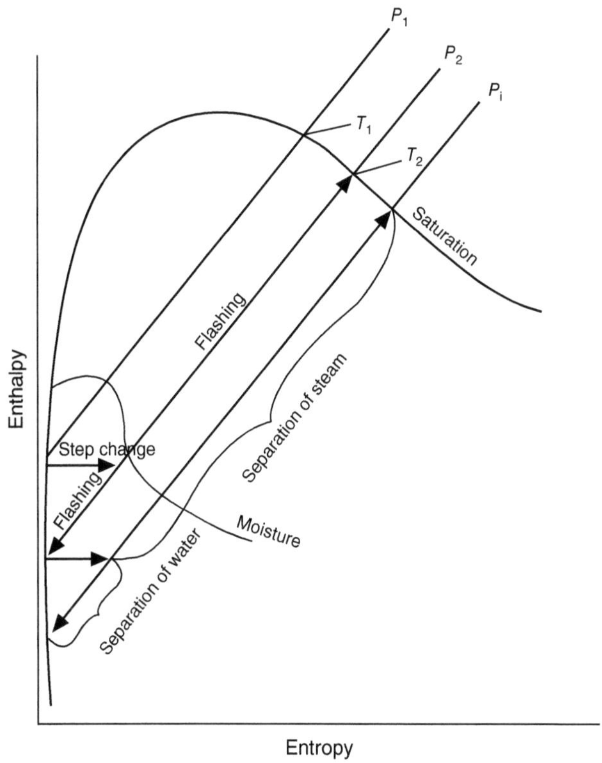

FIGURE 8-8 Mollier diagram representation of flashing.

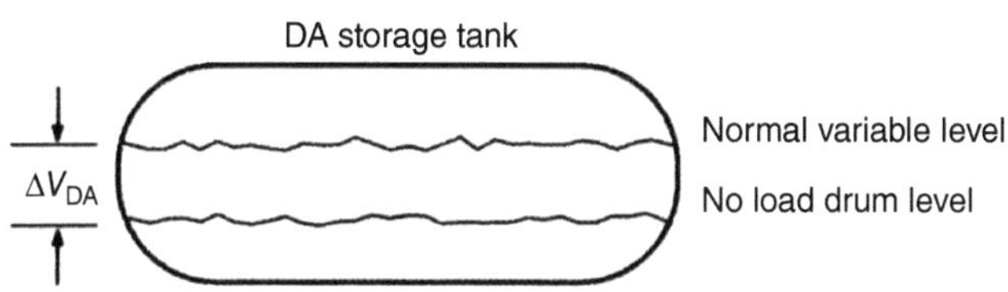

FIGURE 8-9 Level set point rise proportional to load.

required from the DA. The volume requirement will determine the DA level and SP change requirement. When water is taken from the DA storage tank, the DA makeup valves opens to replenish it. The makeup water sprays condense the steam and collapse the DA pressure; therefore, the pressure must be reestablished or flashing will occur.

Conclusion

The suggested solution is a two-point approach. One is to initiate startup pegging steam to restore pressure and the other is by variable SP. The SP will vary according to steam flow for the ΔV_{drum} requirement in equivalent DA differential level.

The startup DA level control (Fig. 8-10) is a *single-element controller* until steady feedwater flow is established to the drum. The system then switches to *three-element cascade, feedforward control*. The three-element controller, level setpoint is indexed by steam flow. The pegging steam valve is forced open by energizing a solenoid. The pressure setpoint is the same as it is for startup pegging steam.

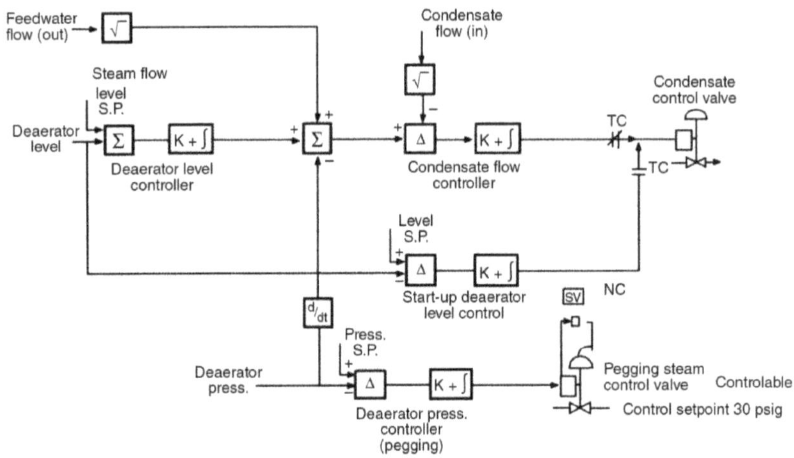

FIGURE 8-10 Deaerator level control logic.

PROPORTIONAL BAND, RESET, AND RATE

Within a controller, the following three functions are usually available. The functions that are used depend on the process requirements.

1. Proportional band (PB) is equal to the reciprocal of gain (K) which is:
   K = % output/% input = 100/PB
2. Integral ∫ is reset.
   ∫ = integral time, minutes (per repeat), reciprocal of rate
3. Derivative $d/d(t)$ is rate. Reciprocal of reset
   $d/d(t)$ = repeats/minute

Proportional band, in simple terms, matches input to output in a process, or visa versa. For example, if a storage tank outlet valve is opened, the tank level makeup valve will open to match the outlet flow.

Integral (reset) restores the tank to the setpoint level. In other words, a simple proportional band controller will match input to output but not restore the process to setpoint. As the makeup valve is opening, the level is dropping until inlet matches outlet flow. It requires reset to restore the process to setpoint (SP), which is the original level. Integral opens the inlet valve to allow more flow than the outlet valve until the level is restored to normal. Derivative is the rate at which the process is restored.

The difference between the process variable (PV) (level in the preceding example) and the SP is the error which the proportional, integral, and derivative function work to correct. These controllers are generally referred to as proportional-integral-derivative (PID) controllers.

CONTROL SYSTEMS TUNING

Rules for tuning controls were first developed and published by Ziegler and Nichols in the early 1940s.[4] The rules provided herewith were those same rules polished, presented, and described by David W. St. Clair in 1990.[5]

It was not until Ziegler & Nichols[1] analysis tools began to spread to the industrial world (about 1950) that the striking simplicity and technical validity of the rules began to be appreciated.

They presented two approaches—closed loop and open loop. The closed loop is used more frequently and is presented first.

Closed-Loop Technique

1. Change the reset time to its longest or highest setting (reset rate in repeats/minute to its lowest or shortest setting).
2. If used, change the derivative time (rate time) to zero or to its minimum.
3. Increase the gain (decrease the PB) until the loop maintains a small, sustained cycle. Note the period of the cycle, P_n, to be called the *natural period*. Implant this securely in your mind. Nothing so simple is more important.
4. Set the controller adjustments to:

$$K_c = K_{cu}/2$$

$$T_i = P_n$$

$$T_d = P_n/8$$

where

K_c = proportional gain %/% input
= 100/PB

K_{cu} = ultimate gain; that which produced the sustained cycle (u = ultimate)

T_i = reset or integral time, min (per repeat), reciprocal of reset rate, repeats/min

P_n = natural period, min, established in step 3

T_d = derivative time, min

These are not precisely the coefficients presented by Ziegler and Nichols, but they are close enough. St. Clair tended to reduce K_{cu} by 3 rather than 2, to avoid future problems, but he stated that this depends on the situation.

Open-Loop Technique

The other approach is open loop. Place the controller in manual, and when the process is sufficiently stable, make a small sustained change in the controller output, called a step change. The process response is expected to look something like Fig. 8-11.

Then set:

$$K_c = 1/RL$$

$$T_i = 4L$$

$$T_d = L/2$$

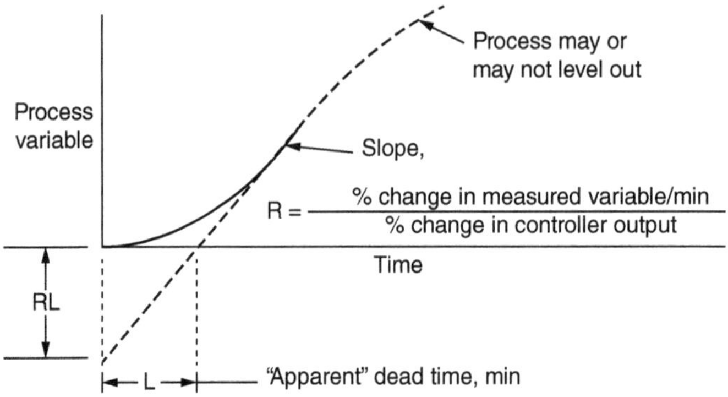

FIGURE 8-11 Typical open-loop step responses.

Note *L* in particular. Note also the *RL* combination that determines the gain. The coordinates, as labeled, do not yield the dimensions given for *R*. Normalize the slope calculations to give the dimensions given for *R*.

St. Clair stated that about the only time he used the open-loop approach was on slow processes, such as temperature and composition. On fast processes such as flow, pressure, and level, the closed-loop approach was much faster and more certain. Even when using the open-loop response for slow systems, he was far more interested in *L* than *R*—and may not even determine *R*.

On rare occasions, the slope *R* will continue to increase. That situation is not covered by these tuning rules. Also on rare occasions, the response may go in the opposite direction first. The boiler swell problem is a typically referenced example. That case also is excluded.

These simple rules are used in commercially available computer programs designed to automatically tune controllers.

DCR in 2003 took automation one step further by "patching in" engineers via the Internet to make DCS configuration changes and perform control tuning online. On the first project, the original design engineer for the DCS performed the system configuration changes for a project in the Dominican Republic while he was in Argentina. The I&C engineer who did the tuning for the same project was located in Colorado. However, to accomplish this, a good coordinator must be at the site. The time and money saved was substantial.

PROCESS FIELDBUS (FOUNDATION FIELDBUS) TECHNOLOGY

There are many process fieldbus protocols. One, for example, developed by Siemens is Profibus, which essentially uses digital highways to connect field devices, PLCs, and DCSs. Siemens is rapidly replacing copper wire 4-20 mA technology in new construction. With it, communication devices are replacing the traditional I&C technician's 4-20 mA signal generators. Because a digital highway unites all devices, there is no need to run individual wires from each device back to a PLC or DCS. For example, a workstation can be connected at any point in the system and control processes as if it were a control room function. Field devices can also perform functions that once were performed by the DCS. This reduces traffic on the highway. Motor control can also be integrated into the highway.

Field devices that use fieldbus technology can process information directly off the highway. For example, process information such as pressure and temperature needed for control valve positioning are accessible on the highway. There is no need to send it from the field transmitters to the DCS for processing and then send it back to a valve positioner.

For example, a flow transmitter can pick the information off the highway and process the data using its own function block. With analog 4-20 mA systems, this is only possible through the DCS. With their own function blocks, they can calculate fully compensated mass flow. The flow control valve positioner will, in turn, pick up the compensated signal.

WHO IS THE FIELDBUS FOUNDATION?[a]

Driven by their customers' needs, process control and manufacturing automation companies formed the Fieldbus Foundation to complete development of a single, open, international, and interoperable fieldbus. The Fieldbus Foundation is an independent, not-for-profit organization based on the following principles:

- *Fieldbus technology is an enabling technology; not a differentiating technology.*
- *Fieldbus technology is open and available to all parties.*
- *Fieldbus technology is based on the work of the International Electrotechnical Commission (IEC) and ISA (the International Society for Measurement and Control).*
- *Fieldbus Foundation members support and work with the international and national standards committees.*

New technology brings forth new problems. Breakers are addressed using dip switches; if you take a breaker from one cubicle to replace another breaker in a different cubicle—and don't readdress the breaker—you will operate the motor assigned to the wrong cubicle. For example, you moved the breaker A to the B cubicle; when you call for an A start, the B will start. In other words, if you replace a vacuum pump breaker with a cooling water pump breaker, and don't readdress it, when you initiate a start for the cooling water pump, the vacuum pump will start.

Connectors become the vital link to the highway; therefore, more than ever, the root cause of fieldbus-related problems is connector related not program or protocol related. The most common connector problem is loose pins in the harnesses for devices such as motor-operated valves and switchgear cubicles.

HART PROTOCOL

HART protocol allows digital communication upgrades using old 4-20 mA wiring. In other words, digital signals are superimposed on analog wiring. That is, the old wires are used like DSL (digital subscriber lines), where digital signals can be sent on the same line as the old analog signals. This allows substantial digital upgrading in existing plants by using the existing analog wiring. This is similar to the high-speed internet signals superimposed on operating phone lines.

REFERENCES

1. "Guide for Protection, Interlocking, and Control of Fossil-Fuel Unit-Connected Steam Stations," IEEE 502-1985.
2. Cerveny, R., *Weather's Greatest Mysteries Solved*, Prometheus Book, New York, 2009.
3. Baur, P. S., "Combustion Control and Burner Management," *Power*, September 1982.
4. Ziegler and Nichols, "Optimum Settings for Automatic Controllers," *Transactions of ASME*, November 1942.
5. St. Clair, D. W., *Controller Tuning and Control Loop Performance: A Primer*, January 1990.

[a]Compliments of the Fieldbus Foundation.

CHAPTER 9
COOLING TOWERS

A good general reference source for information for all different types of cooling towers is the Special Reports series published by *Power* magazine. Of course, the best detailed information can be obtained from your cooling tower instruction manual. The purpose here is to briefly describe different cooling tower designs, without the detailed elaboration provided by the manufacturer. This will serve as a basis for later evaluation of startup and testing requirements.

All wet-tower performance is based on evaporation. How the manufacturers design their towers to utilize evaporation for maximum performance is really not an important topic for startup. Each manufacturer will have its own design preference, and claims, regarding air–water distribution and so forth. The types of towers (dry, wet-dry, wet, forced, or natural draft) are site considerations. All considerations are economics and environmental related and are among the most important considerations during the conceptual design phase of a project.

The following terms are important to know:

Approach—the difference between the outlet water temperature and the air wet-bulb temperature. The approach at 100 percent humidity is zero, and no evaporation occurs. In other words, no cooling tower can cool water below the wet-bulb temperature of the air.

Cell—a term used to describe each section of a tower separated by partition walls.

Cold-water temperature—the outlet temperature of the tower basin water that returns to the steam condenser.

Counter flow—air that sweeps upward against the vertical fall of water.

Cross flow—air that sweeps horizontally through the vertical fall of water.

Drift—water droplets entrained in the air leaving the tower.

Drift eliminators—louvers located before the air outlet to capture water droplets, which escape the fill area, in order to decrease drift.

Fill—the material packing that breaks up water particles and increases the water surface area to enhance evaporation.

Inlet louvers—used to keep water from splashing out of the tower when fans are not running.

Range—the difference between the inlet (hot) and outlet (cold) water temperatures.

Prior to the 1920s, most of evaporative cooling for power plants was performed using spray ponds. All were replaced with cooling towers of one type or another starting in the 1920s. Spray-pond technology is still used for cooling tower blow down evaporation as well as for other wastewater disposals. Today, the type of cooling tower used depends on a site economic and environmental evaluation. The following are examples of typical cooling tower types from reference Bartz and Maulbetsch.[1]

Wet cooling systems

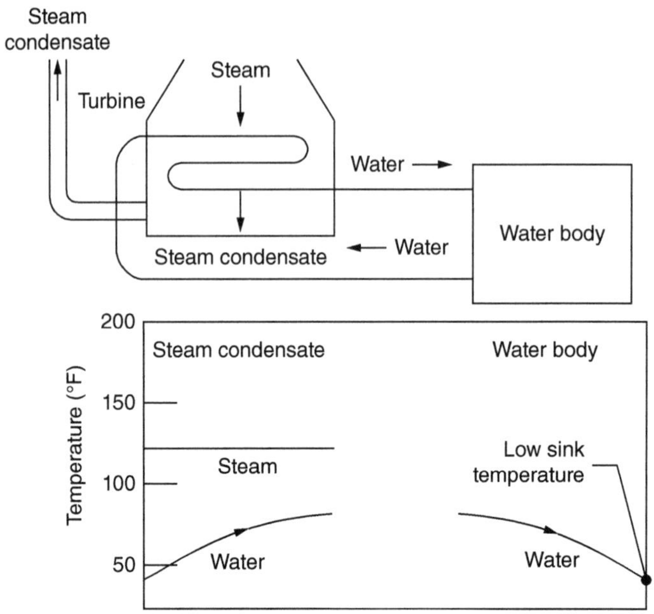

(*a*) Once-through cooling with natural water source.

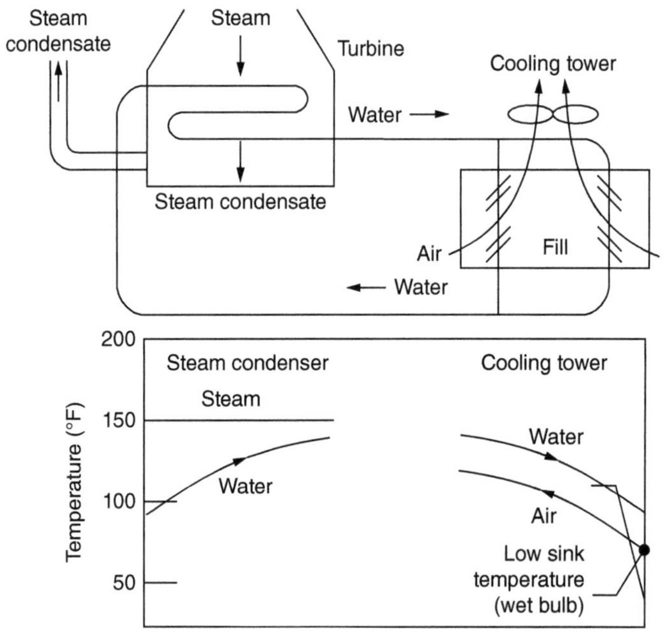

(*b*) Evaporative cooling tower.

Water-conserving cooling systems

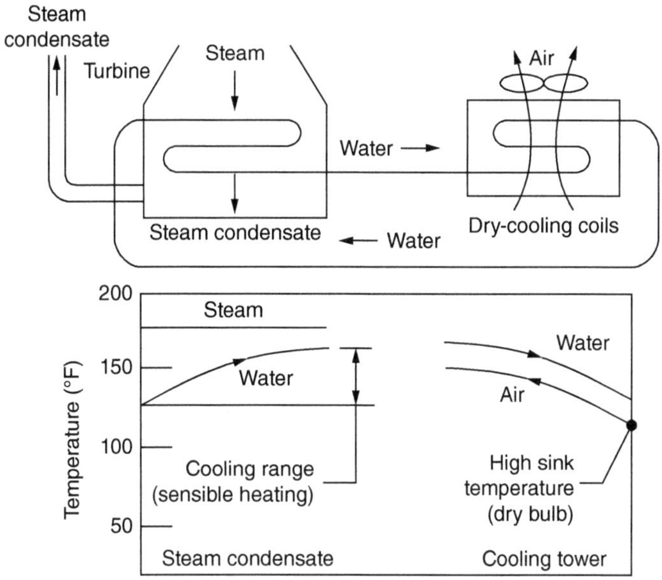

(*c*) Dry cooling.

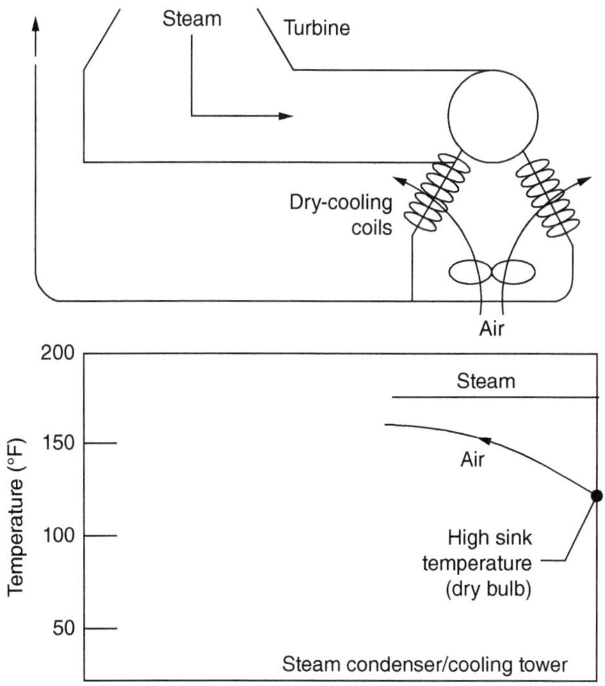

(*d*) Steam condensing.

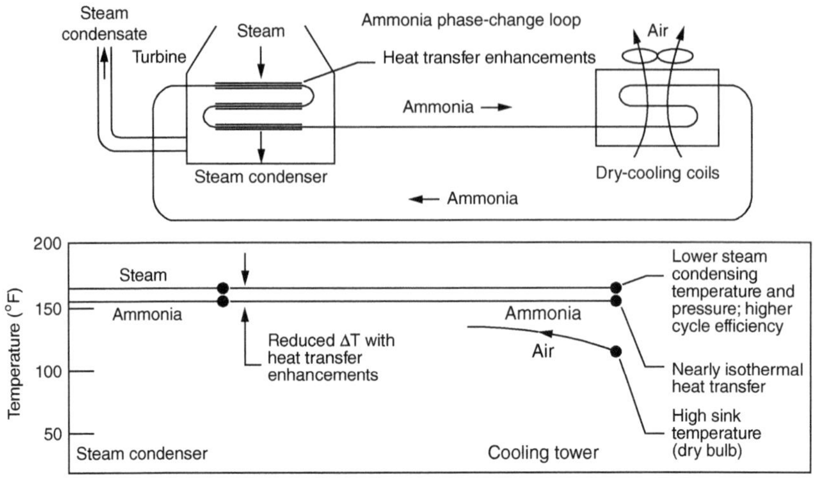

An advanced ammonia-based water-conserving dry-cooling system.

A 10-MW demonstration plant using an advanced ammonia dry cooling tower was built at the PG&E Kern Station near Bakersfield, California, in the early 1980s. A full report was published in 1987.[2]

WATER TREATMENT

Cycles of concentrations (CC) are the ratio of the dissolved solid concentration in the circulating water to the concentration in the makeup water. In order to keep the CCs down to reasonable levels, there must be a continuous rejection of circulating water. The rejection point is generally on the hot-water side for pressurized transfer and process purposes. The amount of rejection (blowdown) must take into account the cost of scale-inhibiting chemicals versus the cost of processing the blowdown water. Closed, circulating water system scale inhibitors are a major operating cost for power plants.

> It is standard practice to utilize the cooling tower evaporation capability to take water to as high a salinity as possible to minimize costs of any further concentration required. However, as salinity increases, the capital and O&M costs for several key steam-electric plant subsystems also increase. An optimum level for cooling-water salinity exists. It will vary with the type of plant, the plant configuration, and many site conditions. An evaluation of one situation indicates the optimum salinity region to be approximately 10,000 ppm based on the sidestream desalting process incorporating electrodialysis. However, the selection of the optimum sidestream desalter depends on feedwater [makeup water] salinity, unit evaporation pond cost and net annual evaporation rate. Regions will exist where ED, RO, or no desalter will prove to be the optimum.[3]

Figure 9-1 illustrates the different types of blowdown treatment for zero-discharge plants. All have an evaporation pond for the final effluent.

Microorganisms are also a major concern in closed evaporative cooling systems. Therefore, the water must be treated in order to prevent the organisms from growing. The bacteria, fungi, and algae form a slime that, if left unchecked, can overcome an open cooling tower where sunshine and warm water provide a life-supporting ecosystem.

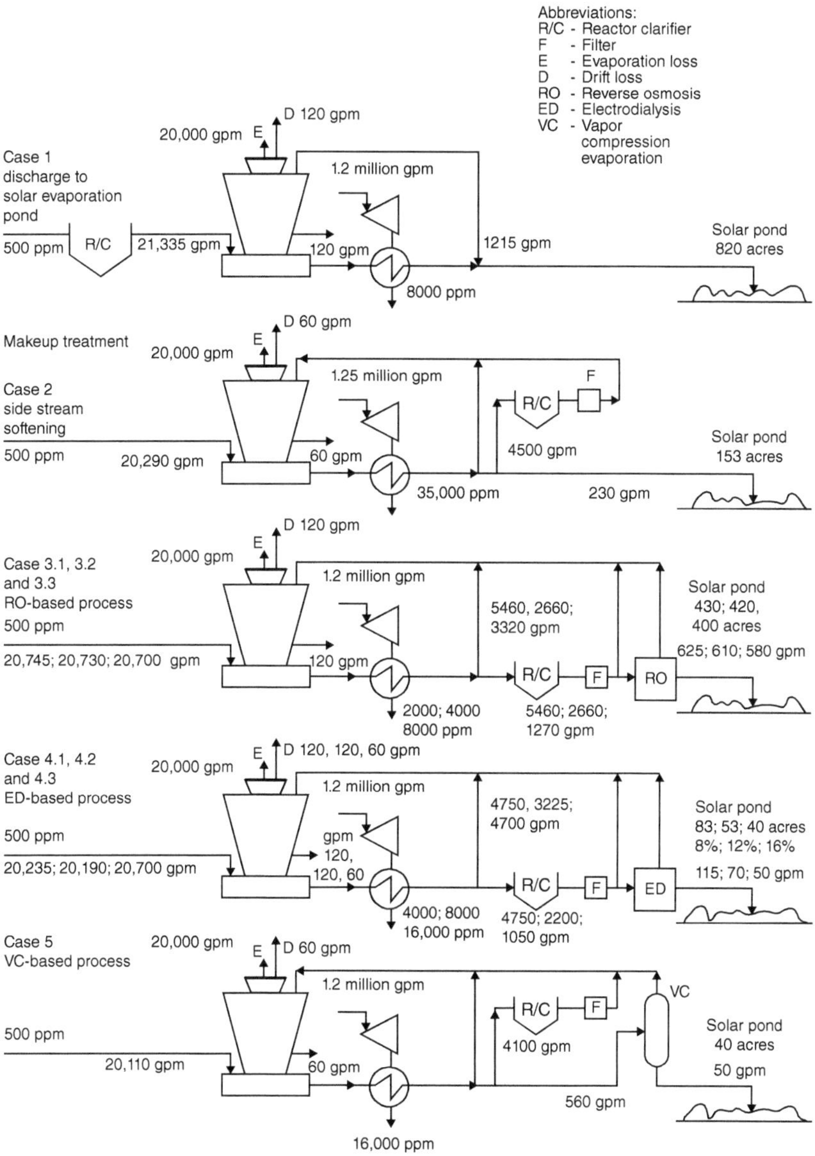

FIGURE 9-1 Solar pond load vs. size and blow down treatments.

Chlorine is the most inexpensive biocide but has a history of detrimental effects on wooden towers because it is an oxidizing agent and must be carefully controlled. Continuous chlorinating can cause detrimental environmental effects because it will be carried with the tower drift. Also, certain microorganisms can develop an immunity to it over a period of time. Because of the aforementioned, a well-planned biocide treatment program should be in place before the tower is trial operated. Indiscriminate, as well as inadequate, chlorination can cause problems later in the startup or commercial operation.

Although the most common, chlorine is not the only biocide that can be used. Other biocides can be used as well. Chemical suppliers can provide different non-oxidizing biocides as well. The following are excerpts from Smith and Muia.[4]

A laboratory technique, known as a *time-kill study*, can be used to choose the most effective microbiocide for a system. The test attempts to duplicate field conditions. It can be used to predict which microbiocide will be most effective in controlling the biological population.

The technique consists of adding varying concentrations of microbiocides to aliquots of a mixture of cooling water and biological slime deposits obtained from the system. Predetermined concentrations of each test microbiocide are added to separate aliquots of the water–slime mixture. Contact is maintained for an approximate period of time, usually 2 hr, and standard plate counts are then run on each sample. Counts are also run on recirculating water–slime mixture that contain no microbiocide. The overall effectiveness (percentage of kill) of each microbiocide is then determined by:

$$\text{Percentage of kill} = \left[\frac{(a\text{-}b)}{a}\right]100$$

where

a = control count
b = test sample count

A typical series of test results is shown in Table 9-1. Examination of the results indicates that microbiocide A is the best choice for this system. Feeding 25 mg/l of the compound into the system should significantly reduce the microbiological population. An important point to remember is that the percentage of kill in the laboratory will not necessarily be achieved under field conditions. Experience has shown that for a 95 to 100 percent kill obtained in the laboratory, a kill of approximately 60 percent will be obtained in the actual

TABLE 9-1 Typical Results from Microbiocide Time Kill Study

Control Plate Count (no biocide added) = 1,250,000 colonies/ml			
Microbiocide	Concentration Tested (mg/l)	Bacterial Plate Count after Contact Time (colonies/ml)	Bacterial Kill (%)
A	25	68,000	95
	50	16,400	99
	75	8000	99
B	25	750,000	40
	50	589,000	63
	75	620,000	60
C	25	630,000	49
	50	312,000	75
	75	101,000	91
D	25	390,000	89
	50	196,000	84
	75	155,000	88
E	25	260,000	79
	50	164,000	87
	75	60,000	95

system. Organisms are continuously added by incoming air and water, and the microbiocide is constantly being diluted by makeup water. Silt and slime in the system place an additional demand on the microbiocide.

It is not enough to feed and maintain a certain concentration of a microbiocide in the system and expect the system to remain trouble-free. Microorganisms, because of their ability to become acclimated to a treatment, can begin to multiply over a period of time, again developing slimes and deposits. Because alternating the microbiocide may not solve the problem, some method of monitoring the effectiveness of the microbiocide treatment program must be established.

Visual observations of specific problem areas of the cooling-water system will keep an operator informed of the extent and rate of slime buildup. The upper deck, sump bottom, and fill slats of the cooling tower must be routinely inspected, and temperature and pressure readings through the condenser must be taken regularly. Although visual observation and inspection can alert an operator to a biological problem, the bacteria may have already begun to form bacterial slimes.

Standard plate counts provide an excellent method of monitoring the bacteriological population. Plate counts will alert the operator to an increasing bacterial population before slimes form, thereby permitting corrective action to be initiated immediately after the biological buildup begins.

Data shown in Fig. 9-2 can be used to determine the most effective times for microbiocide addition. If the microorganism population in the cooling tower goes above point A, massive slime growths will build up in a short time—possibly forcing shutdown for cleaning. Therefore, the microbiocide should be added when the bacteria population approaches A.

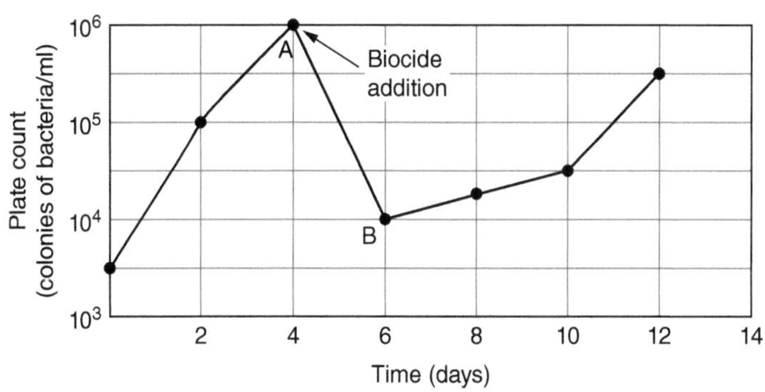

FIGURE 9-2 Bacteriological growth curve.

FANS

Cooling tower fans can be found up to 60 ft in diameter. They are relatively low speed and therefore can have resonance problems with the cooling tower structure. Excitation of the fan blade vibration can occur due to obstructions in the airflow path.[5] The obstructions are the fan gearbox supports, distribution piping, and so forth, that cause dead-space pulsation or turbulence that can excite blade vibration, which in turn can excite structural vibration.

The basic structural excitation frequency is commonly referred to as the bladepass frequency (BPF) and is calculated by the equation:

$$\text{BPF} = \text{No. of blades} \times \text{rpm}$$

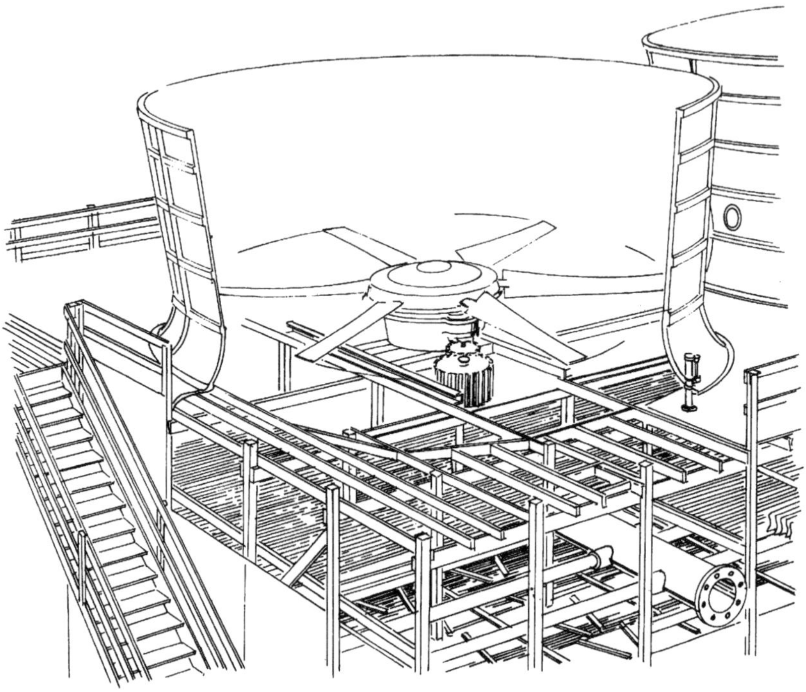

Cutaway of a typical cooling tower illustrating obstructions.[5]

Best results are obtained if this basic excitation frequency is kept at least three times as high as the fan-blade natural dynamic frequency.

$$f_d = [(f_s)^2 + K \times \text{rpm}]^{1/2}$$

where f_s and f_d are both expressed in cycles/minute and K is a constant, whose value is 1.5 for most fans. Determination of f_s, the static natural frequency can be determined by striking the blade and recording the first mode of vibration.[5]

From a startup point of view, if excessive vibration exists, the first thing that can be addressed are the obstructions in the flow path and reenforcement or added weight to the structure or to the blades. This is obvious because the fan speed and number of blades cannot be easily changed.

The easiest solution may be to disrupt the air-impulse excitation frequencies through removal or rearrangement of obstructions in the plenum. This will lower the harmonic constant of the tower, but can never remove all amplification potential.

A more positive method is to change the dynamic natural frequency of the fan and move it out of the critical range. This can be done either by changing the speed of the fan or by weighting the blades to change the natural frequency. In most cases, weighting of the blades proves to be the more economical solution to the problem.[5]

It should be noted that variable speed drives are not possible for use with cooling tower fans because of the resonance problems associated with them and the tower structure. If the fan speed were variable, there would be a high likelihood of the fan to pass through or operate at a critical point. If energy conservation is a question, such as in winter operation, a two-speed fan is usually the selection. If it is not operating at a critical at full speed, it is unlikely that it will at half speed. Criticals are multiples of fan speed such as 1, 2, 3, 4, etc.

VIBRATION PROTECTION[5]

Cooling fans up to 30 ft in diameter should have a system that can respond in a linear manner to preset vibration severity levels over a frequency range from 120 to 30,000 cycles/min. Provisions for a time delay, to avoid startup trip-out, are also important. These requirements are beyond the capability of most mechanical devices. Such devices set to trip on one frequency range may not respond to low-frequency vibration. It is also common to find that in order to prevent them from tripping out on startup, mechanical devices are set so high or are so remotely located as to be ineffective. Mechanical devices are also of no use in establishing monitoring systems.

Electronic vibration devices that monitor vibration levels based on velocity measurements are available to meet requirements of large-diameter cooling fans. However, a careful examination of their frequency response is required to assure protection at the large fan's lower operating speeds.

Figure 9-3 displays suggested vibration severity limits based on measurements of velocity vibration levels for fans up to 30 ft in diameter.

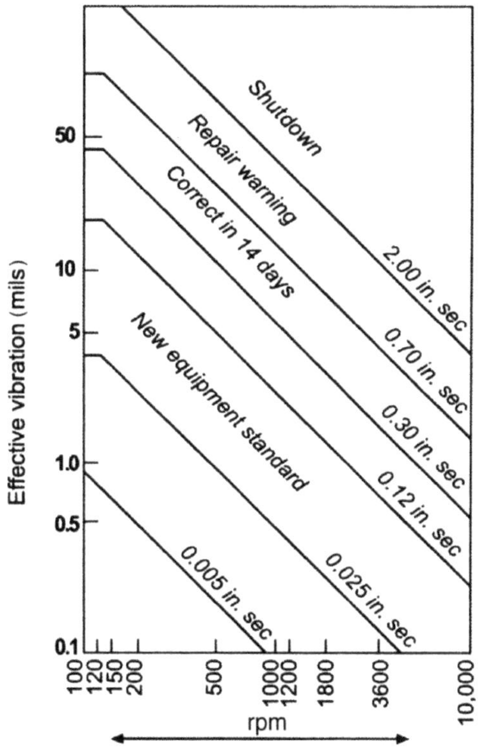

FIGURE 9-3 Vibration allowance curve.

Fans larger than 30 ft operate at speeds that are below the useful range of most velocity-limiting systems. These larger fans should have a system that reacts to displacement measurements to monitor fan condition. Systems based on proximity sensors are best suited for this application.

FIRE PROTECTION

All cooling towers should have a fire protection system. This is especially true of those that are constructed with combustible materials. The most common natural cause, as opposed to arson, of cooling tower fires is lightning strikes. Therefore, all towers should have lightning arrestors. Combustibles in the cooling water have also caused serious cooling tower fires.

Fires during construction are also a serious problem that should be addressed. A temporary fire-fighting system should be available until the permanent system is installed and operational. Towers with wooden decks and combustible fill can also be kept wet with temporary sprinkler systems until the permanent system is operational. If the tower fill is made of polystyrene, it presents a serious fire potential.

Access to the tower should be restricted during and after construction to minimize the question of vandalism or arson.

COOLING TOWER PRETRIAL OPERATION TESTING

If the tower is available for pretrial operational testing, the cooling tower fire protection system and the fans should be tested first. The fire protection system should be tested as soon as it is available, regardless of whether the tower itself is ready for operation. Some key points are as follows:

- The fire protection system trial operation has priority over all other tower startup activities.
- The tower should be checked for construction debris, scaffolding, tools, broken pipes, and damaged fill. The tower basin should be cleaned and ready for filling.
- The fan vibration protection system must be functional and the detection devices calibrated.
- Gearboxes must be filled with oil and the deck sight glasses checked.
- Motors should be run uncoupled for proper rotation, current, vibration, etc.
- The fans should be rotated by hand and the blade tip clearance checked between the blade and the velocity recovery stack (fan housing). The clearance is very important as far as fan efficiency losses are concerned.
- The fans should be run for vibration analysis and checked for proper blade pitch. Blade pitch is a function of motor current. If the fans are running at greater than full load current, a fan test should be performed. If the blade pitch has to be reduced (reducing air flow) and the air flow is normal, it is an indication the designer missed the anticipated losses in the fan system. Reducing airflow will reduce the overall tower efficiency. If the fans are running with low current, the pitch must be increased to increase air flow.
- Remember, fan loading is directly proportional to air density. Motor loading is based on the motor nameplate heat ratings. For winter operation (cold but dense air), the motor temperature will be low while the motor current is high. Therefore, if in the winter you are at the low-temperature rating of the insulation class, you can operate at a higher current. In other words, adjust the blade pitch for 1.15 × full-load current for winter operation; then as the air warms up in the summer, the current will drop to the full-load rating without having to adjust the blade pitch. See the discussion of motor insulation and NEMA class in Chap. 12.

- The tower water risers should be checked for installation of vacuum breakers. They can be mechanical or simply standpipes that will allow a water column collapse without collapsing the riser or distribution piping.
- If a water balance is to be performed, ensure that the risers have the proper "stopcock" valves installed for velocity traversing. The water balance checks that all risers, and their cells, have equal flow.

TRIAL OPERATION

Do not operate the tower until the circulating water treatment systems are ready for trial operation. When ready, place the circulating water system in service and balance the water flow to each cell. Chemicals for cooling towers are one of the most costly expendables for plant operation. Therefore, the system must be researched thoroughly to ascertain exactly what chemicals are needed and how much is absolutely necessary.

Towers that do not have inlet louvers can be a problem because water will splash out when the fans are not operating. This can cause severe icing problems in winter operation when some fans may be shut down. This becomes a balancing act, manipulating fan and bypass operation. Also, towers without division walls in the middle will allow wind to blow through, carrying the spray water with it. In the worst case, a louverless tower can be an environmental nightmare.

TESTING

For cooling tower performance testing use procedures and formulas from the cooling tower test codes and standards. Cooling Tower Institute code ATC 105[6] governs cooling tower testing as well as ASME PTC 23, Atmospheric Cooling-Water Equipment.[7] The CTI code is by far the most comprehensive of the two. The following are excerpts (tricks-of-the-trade) for a cooling tower performance test:

$W_L \approx W_{CW} \approx$ Circulating water flow through the condenser

$W_{C1} + W_{C2} \approx$ Sum of closed cooling water flows

$t_2 =$ Cold-water temperature

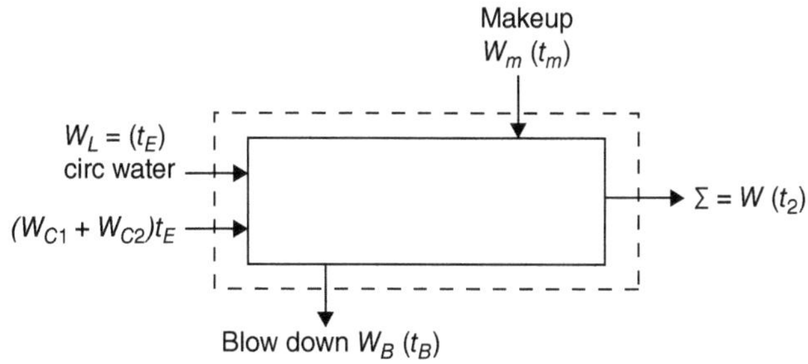

$$t_2 = \frac{W_{CW}t_E + (W_{C1} + W_{C2})t_E - W_m t_m + W_B t_B}{(W_{CW} + W_{C1} + W_{C2} - W_m + W_B)}$$

$$t_2 = 87.25°\text{F}$$

Note that the cold water temperature is derived from the weighted average of the total flow streams and is not the simple, single measurement of the cold water in the basin.

After solving for t_2, the range and approach can be found.

$$\text{Range} = t_{\text{HOT}} - t_{\text{COLD}} = t_R$$

$$\text{Approach} = t_{\text{COLD}} - t_{\text{WET BULB}} = t_{\text{app}}$$
$$t_H = 108.65$$
$$t_{\text{WB}} = 74.20$$
$$t_2 - t_{\text{WB}} = 87.25 - 74.20$$
$$t_{\text{app}} = 13.05$$

$$\text{Guarantee } t_{\text{app}} = 14$$
$$t_H - t_2 = 108.65 - 87.25$$
$$t_R = 21.40$$

The farther away the fans are from the switchgear, where motor data are measured, the greater the error in the actual motor operating conditions. Therefore, this loss must be accounted for.

Cable Data

Size	3-1/C	500 mcm
Length	3A	450 ft
	3B	570 ft
	3C	690 ft
	3D	820 ft
Total		2530 ft
Design HP		199

The total flow to the cooling tower is then adjusted for actual fan hp.

Line loss: $P_L = 3\,I^2 R = (\text{amps as read})^2\,(\text{3-phase cable resistance})$
$= 3(220.36)^2(0.0216)(2530/1000)$
$P_L = 7975.88$ watts

Measured power (4 motors): $P_M = (3)^{1/2}\,E\,I\cos\alpha$
$= (3)^{1/2}\,(455.5)(220.36)(0.89)(4)$
$P_M = 610.94$ kW

Power ratio: $(P_D/P_{\text{actual}})^{1/3} = 4(199)PF\,\eta/\eta(P_D - P_L)$
$= 4(199)(0.7457)/0.93(610.94)^{1/3}$
$R_P = 1.015$

Test water flow: $W = W_L + \Sigma W_C$
$= 64{,}052$ gpm

The total flow to the cooling tower is then adjusted for actual fan hp.

Corrected water flow for fan HP: $W = 64{,}052\ (1.015)$
$W = 65{,}013$ gpm
Design flow = 67,050
Predicted flow = 67,000
Corrected flow = 65,013
Ratio = 65,013/67,000 = 97%
67,050/0.97 = 69,250 gpm

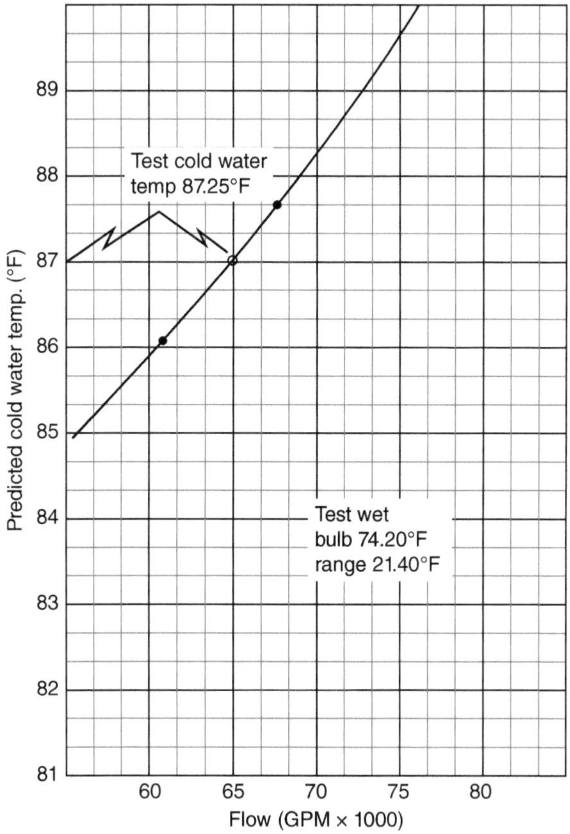

Using the manufacturer's curves, a new curve is drawn to determine the predicted design range versus actual range.

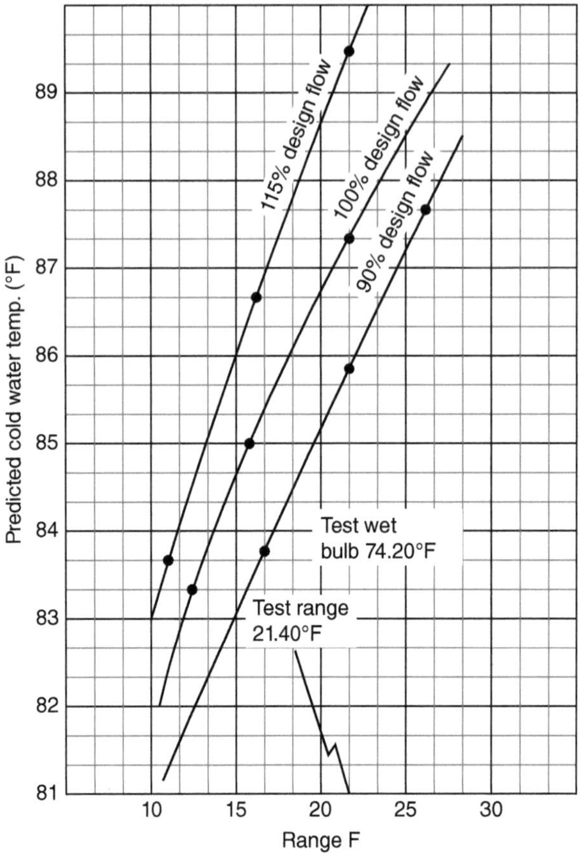

REFERENCES

1. Bartz, J. A. and Maulbetsch, J. S., "Are Dry-Cooled Power Plants a Feasible Alternative?" *Mechanical Engineering*, October 1981.
2. NTIS, PC A09/MF A01, Research Reports Center.
3. Westbrook, G. T., "Cooling Tower Salinity Optimization," *Power Engineering*, August 1970.
4. Smith, A. L. and Muia, R. A., "Controlling Micro-Organisms in Cooling Water System," *Plant Engineering*, June 1976.
5. Sylvester, J., "Low-Speed Dynamics Diagnose Large-Diameter Cooling Fan Failures," *Power*, February 1979.
6. "The Cooling Tower Institute Code," ATC 105.
7. "Atmospheric Cooling-Water Equipment," ASME PTC 23.

CHAPTER 10
CORROSION

This chapter covers common information regarding corrosion and some easy to follow prevention rules regarding materials and coating selection. The special report by Strauss and Puckorius[1] is a highly recommended document that should be a part of every plant engineers reference binder. The National Association of Corrosion Engineers (NACE) is the authority for corrosion prevention design.

DISSIMILAR METAL CORROSION

The most popular galvanic corrosion topic in the industry is dissimilar metal corrosion. It also has the most myth associated with it. First, to be classified as *dissimilar* metals, the metal must be different in respect to its position in the galvanic series. For example, most stainless and carbon steels are dissimilar metals, but connecting them either by welding or mechanically is perfectly acceptable. Welding stainless to carbon steel, of course, must be done using stainless filler metal. On the other hand, connecting brass to carbon steel can cause problems because of their positions in the galvanic series. This is only true if there is an electrolyte available. The electrolyte provides the galvanic current flow path. Secondly, the electrolyte must contain oxygen. An exception to the rule is bronze welding cast iron. The metals are positioned in the galvanic series in which dissimilar metal corrosion is anticipated. However, the cast iron normally is far larger than the bronze welding material. That is, corrosion also depends on how large the material is that is acting as the anode compared to one acting as the cathode. In this example, the anode is the cast iron, which is very large compared to the welding metal.

The anodic material is the sacrificial material and the cathodic is the protective. The easiest way to remember is by remembering that *sacrificial anodes are used in cathodic protection.* One example is zinc sacrificial anodes attached to steel exposed to seawater.

Ionization is used in water treatment to remove unwanted elements from water. In nature, ionization removes material that we don't want to lose. For example, it tries to restore the material man has taken from the earth back to its original state. We take iron oxide (ore) from the earth and make steel. Nature oxidizes it to its original state—iron oxide. The process requires electron flow and ionization. The most severe corrosion will occur when the dissimilar metals are immersed in an electrolyte that contains a continuously replenishing supply of oxygen. Remember, corrosion is an electrochemical process that requires a potential difference, current flow, and ion exchange.

For example, a brass shower head attached to a steel water pipe will cause deterioration of the steel and eventual failure. The water is an electrolyte with a continuously replenishing oxygen supply. This example is similar to a steel tubesheet and brass tubes in a condenser waterbox; the tubes will eventually cause a tubesheet failure if not cathodically protected.

On the other hand, it is common practice to install brass sprinkler heads on a steel pipe header for fire protection systems. If the system is a wet system, the oxygen in the static water will deplete, and oxidation will stop without causing corrosion damage. The electrolytic water supply is not continuously replenishing oxygen. If the sprinkler system is a dry system, there will not be an electrolyte. Also, note that on lube oil systems there may be brass fittings on steel pipe. The oil is neither a good electrolyte nor contains oxygen.

Having provided exceptions, now consider the rule. *When coupling brass to steel in a continuous service water process, dielectric couplings must be used to isolate the brass from the steel.* A backflow preventer for small-bore process lines is a massive brass apparatus when compared to small-bore piping. If coupled directly to a steel pipe, the pipe will rapidly deteriorate. Large-bore piping will general have cast steel backflow preventers.

GALVANIC SITES

Any site that can lead to a potential difference between two substances, when immersed in an electrolyte, can cause galvanic corrosion. The electrolyte is necessary for a current flow path.

As a matter of fact, the galvanic series is established by measuring the differential voltage (potential difference) between different metals. The electrolyte generally used for establishing the series is water, but in a practical sense, it can be any conductor containing oxygen such as soil. Buried pipes are submersed in an electrolyte.

Zero or little potential difference indicates similar metals. Large potential difference indicates dissimilar metals. The galvanic series for fresh and sea water are provided in Table 10-1. Figure 10-1 illustrates the potential differences for similar and dissimilar metals.

Deposits in stagnant areas such as crevices, or on a surface where boiling is taking place, can produce galvanic corrosion. Surface boiling can cause precipitation of dissolved solids. For example, flame impingement on the surface of boiler water walls can cause surface boiling, deposit buildup, and pitting. Crevices can produce deposit sites where galvanic action can take place. Stagnant areas can be idle location for deposits to accumulate. All of these can produce galvanic sites and corrosion.

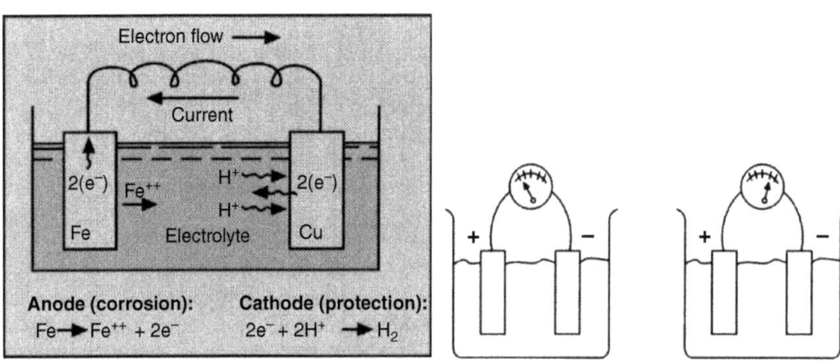

FIGURE 10-1 Similar metal and dissimilar metal potential difference.

TABLE 10-1 Galvanic Series for Fresh and Sea Water

Freshwater[1]	Seawater[2]
Magnesium	Magnesium
Magnesium alloys	Magnesium alloys
Zinc	
Galvanized steel or galvanized wrought iron	Zinc
	Galvanized steel or galvanized wrought iron
Aluminum 1100	
	Aluminum 52SH*
Aluminum 6053	Aluminum 4S
Alclad	Aluminum 3S
	Aluminum 2S
Cadmium	Aluminum 53S-T
	Alclad
Aluminum 2024	
	Cadmium
Mild steel	
Wrought iron	Aluminum A17A-T
Cast iron	Aluminum 17S-T
	Aluminum 24S-T
Type 410 stainless steel (active)	
Type 304 stainless steel (active)	Mild steel
Type 316 stainless steel (active)	Wrought iron
	Cast iron
Lead tin solder	
Lead	Ni-Resist
Tin	
	13% chromium stainless steel, Type 410 (active)
Muntz metal	
Manganese bronze	50–50 lead tin solder
Naval brass	
	18-8 stainless steel, Type 304 (active)
Nickel (active)	18-8, 3% mo stainless steel, Type 316 (active)
Nickel-chrome–iron alloy (active)	
	Lead
Yellow brass	Tin
Admiralty brass	
Aluminum bronze	Muntz metal
Red brass	Manganese bronze
Copper	Naval brass
Silicon bronze	
	Nickel (active)
Copper–nickel alloy	78% Ni–13.5% Cr–6% Fe (Inconel) (active)
G-bronze	Yellow brass
M-bronze	Admiralty brass
Silver solder	Aluminum bronze
Nickel (passive)	Red brass
Nickel–chrome–iron alloy (passive)	Copper
Monel	Silicon bronze
	5% Zn–20% NI, Bal. Cu (Ambrac)
Type 410 stainless steel (passive)	70% Cu–2% Zn–10% Sn (Composition G, bronze)
Titanium	88% Cu–3% Zn–6.5% Sn–1.5% Pb (Comp. M, bronze)
Type 304 stainless steel (passive)	Nickel (passive)
Type 316 stainless steel (passive)	78% Ni–13.5% Cr–6% Fe (Inconel) (passive)
Silver	70% Ni–30% Cu (Monel)
Graphite	18-8 Stainless steel, Type 301 (passive)
Gold	18-8, 3% Mo stainless steel, Type 316 (passive)
Platinum	

OXYGEN PITTING

Oxygen pitting is another galvanic corrosion mechanism.

Figures 10-2 and 10-3 are illustrations of galvanic sites occurring on similar metals and on the surface of single metal surface.

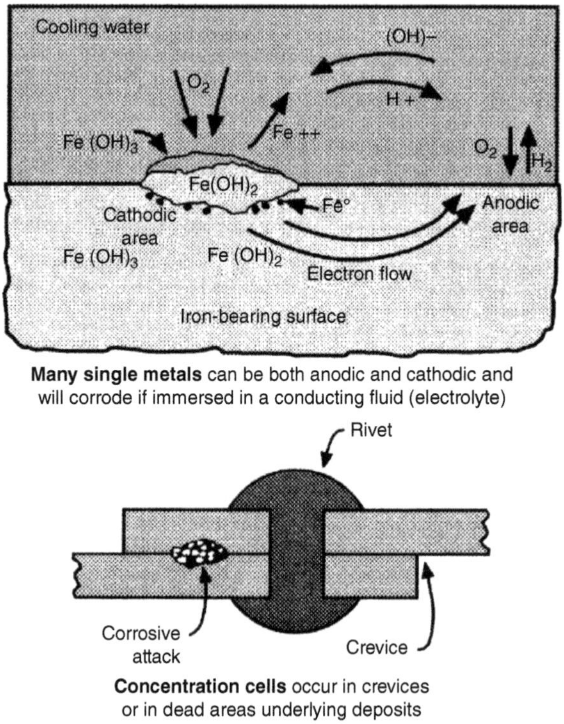

FIGURE 10-2 Examples of galvanic sites.

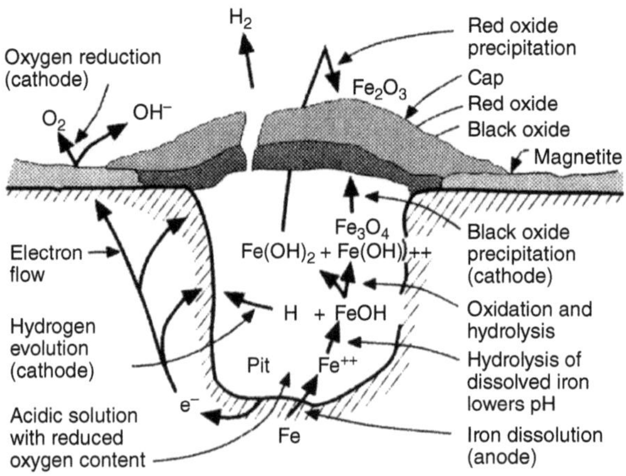

FIGURE 10-3 Cathode-anode reaction.

Pitting in this boiler feed pump at the interstage diaphragm were on the surface, but in the volutes small holes were discovered. The holes were the size of a ball point pen.

In the casing, the oxygen pitting almost penetrated the full thickness in very concentrated caverns. A pen could be inserted almost completely in some. The pump casing material was carbon steel and should have been specified as chrome steel.

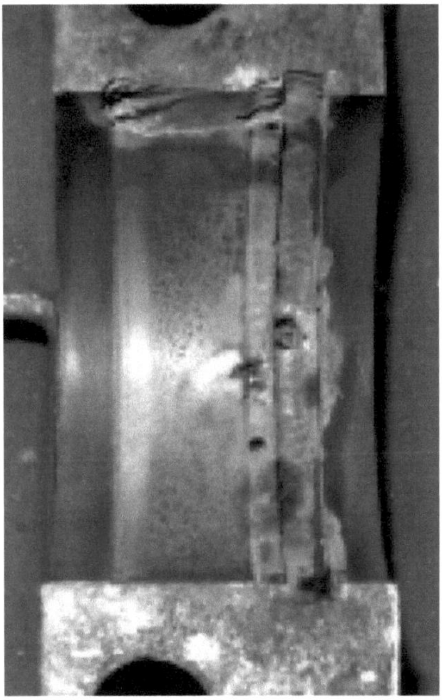

Chloride Pitting

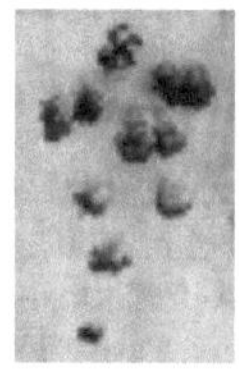

Chloride pitting of stainless steels is actually oxygen pitting made possible by the removal of the metal protective surface film by the chloride. The pitting has the same appearance as oxygen pitting of ferrous metals.

Stainless Steels

Corrosion of stainless steel in seawater is common because of the presence of chlorides, oxygen, and the electrolyte. The chlorides continually wash away the protective surface film on the steel, and pitting occurs.

If the majority of sailboat fittings are made from stainless steels, why then do they not corrode? The answer is simple—they are made from stainless steel that contains molybdenum. And, biological organisms are prevented from growing on them.

> It will be noted that the best performance under adverse conditions described may be expected from stainless which contains molybdenum in addition to chromium and nickel. Such compositions also represent the best choice of stainless steel for sea water service under more favorable conditions—as at high velocity—since there may occasionally be exposure at low velocity or other unfavorable conditions where the benefits of superior resistance will be needed.[2]

High velocity is relative. That is the velocity must be high enough to remove the stainless steel surface film to create an unfavorable condition. On the other hand low velocities that allow surface organism to grow are also unfavorable. Take for example, the velocity of water entering a condenser under vacuum between a tube and tube sheet. It the tube sheet is stainless steel and the tube titanium they will occupy a similar passive position on the galvanic scale. But if a leak between the two should occur the "high" velocity of the leak will wash away the protective film of the stainless steel. Once that happens the stainless becomes active and will occupy a position opposite the titanium position and will corrode.

In summary, if the protective passive film on the surface of stainless steel is washed away by any means the steel will rise in the galvanic series, become active, and will become subject to corrosion.

> Higher corrosion resistance is possible with Type 316, which has 2–3% molybdenum and extra nickel to stabilize the austenite, and Type 317, which has 3–4% molybdenum and slight increases in nickel and chromium.

Two so-called high-performance austenitic alloys are AL-6X and 254SMO. Each has over 6% molybdenum, and they are popular for seawater-cooled condensers.[4]

BACTERIAL INFESTATION

Iron Bacteria

Iron bacterial infestations are another phenomenon that can be difficult to eliminate. The bacteria feed on the iron oxide in service water and even potable water systems. It is easily identified because it turns the water bright red and leaves deposits in piping and pumps. The deposits, soft or hard, can eventually restrict or even plug strainers, pumps and piping systems. The most common remedy is chlorination shock treatment. If bacteria infests a deep well it will create a more serious problem, so much so, that abandoning the well may be cheaper than eliminating the bacteria. It has been observed in chlorinated water with normal basic pH. The iron oxide as well as bacterial excretions can apparently insulate the bacteria from biocides.

Anaerobic Bacteria

There are bacteria that can thrive in liquids that do not contain oxygen. They can cause considerable problems besides clogging pipelines, filters, and apparatus. One such is a *anaerobic bacterium* that thrives on fuel oil at the interface of water and oil in storage tanks. Therefore, water removal from fuel oil storage tanks should always be a routine procedure.

Sulfuric Acid–Producing Bacteria

There are some bacteria contained in river water silt that are *sulfate-reducing bacteria that generate sulfuric acid.* They can act very quickly. At two power stations located in the Caribbean close to rivers mouths, acid pitting perforated Type 316 stainless steel pipe within days. The pits were located under sediment at the bottom hemisphere of the pipes. The silt settled during startup interruptions in process flow. The only material found to resist this type of corrosion was nonmetallic. At one of the stations, the stainless steel pipe downstream of automatic cleaning filters did not suffer the damage that the upstream piping did. Unfortunately, the upstream piping was large bore >1000 ft long. Of course, the key is not to let the silt accumulate. Removal is possible in condenser waterboxes but difficult in pipelines unless purge water is introduced to flush the lines before stopping flow. It is highly recommended that condensers be flushed with fresh, clean water to remove silt and biological debris at the beginning of every outage. Also important is to ensure that when

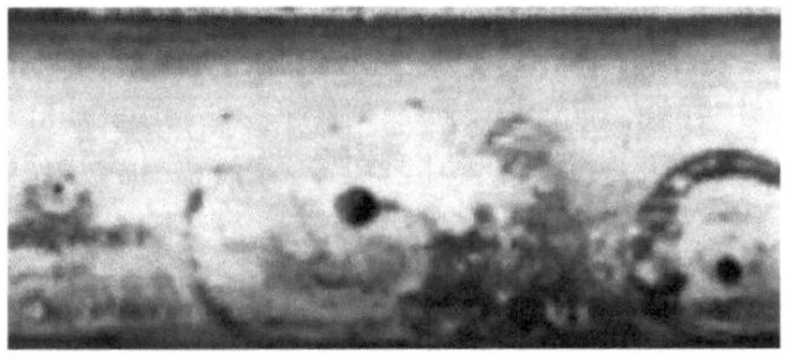

Bacteria caused pitting in Type 316 stainless steel pipe at a power station in Puerto Rico.

sending damage specimens to a lab, send a specimen with undisturbed deposits in the pipe. Even then, metallurgical labs with no field experience may clean the area affected and textbook-diagnose bacterial pitting as chloride pitting.

Several groups of bacteria produce acids and waste products that attack metal surfaces directly. One comprises sulfate-reducing bacteria. These organisms thrive in the absence of free oxygen and can convert water-soluble sulfur compounds to hydrogen sulfide. The sulfur is present naturally—usually as sulfate—or is introduced in treatment of cooling water with sulfuric acid. While methane and other gases may be produced, hydrogen sulfide is the most troublesome, since it is an acid that attacks mild steel, stainless steel, and copper alloys.

Desulfovibrio is the most well known among sulfate-reducing bacteria. Found in sea and fresh water, marine mud, and soil, it uses organic nutrients as food. In recirculating systems, this either comes from influent water or results from decomposition of other bacteria or algae. A favored area for growth is under existing deposits. It can be recognized by its distinctive corrosion pattern on mild steel—smooth, dish-shaped concentric rings, generally most prominent in the action of water flow—truly a "fingerprint" of corrosion.

The corrosive action of desulfovibrio can reach alarming proportions: It has been found to perforate steel condenser tubing in just six weeks. It is extremely aggressive, even in the presence of good corrosion inhibitors and in pH-controlled and chlorinated recirculating cooling water. The reaction with chromate and zinc salts forms reduced chromates and precipitated zinc salts, resulting in fouling deposits. Biocidal action on the bacteria is inhibited by two mechanisms: They are usually covered by slime, and the reducing environment of hydrogen sulfide reacts with chlorine to form harmless chloride salts.

Nitrifying bacteria are a second type. Emanating from the soil, they are corrosive organisms common to cooling-water systems. They react with ammonia to produce nitric acid, lowering pH markedly and resulting in general and local attack on steels and copper. The species is not affected by O_2 nor does it affect inhibitors like chromate or zinc.

Another group of corrosive bacteria is also acid-producing. Known as Thiobacillus, it can convert soluble sulfur compounds to sulfuric acid, and can also take hydrogen sulfide to sulfuric; oddly, the species is also a sulfide producer. The corrosive effect resulting from production of sulfuric acid by these bacteria is production of general attack wherever the organism contacts the metal.

One final group of bacteria meriting mention was discovered only recently. Called Legionnella Pneumophila, it is more familiarly known as Legionnaires' disease. While not destructive to cooling system components, it has been found extensively in all types of cooling towers in both industrial and air-conditioning systems. The bacteria are present in dust-laden air, and proliferate in cooling systems that are not treated adequately.[1]

CATHODIC PROTECTION

Condenser waterboxes and underground tanks and pipelines generally require cathodic protection. Some of these piping systems are circulating water lines, fuel supply lines, and fire protection piping. They are normally designed in accordance with NACE guidelines but may use another source specific to industry use. For example, condenser waterbox cathodic protection design guidelines can be found in an EPRI report by John Reason.[5] This report gives design guidelines for waterbox current densities similar to NACE guidelines for buried pipelines and structures.

The first survey that must be done for buried steel is the soil resistivity test. Current requirements for cathodic protection are determined by soil resistivity. The higher the resistivity, the lower the current density. They vary between 1 to 3 mA/ft^2 of buried, bare steel pipe per NACE guidelines. As a comparison, a seawater service carbon steel waterbox with dissimilar metal tubes and tubesheet can vary between 50 and 200 mA/ft^2 depending on the salinity of the water and tube and sheet material per EPRI guidelines.

Engineers will calculate the total surface area of the metal, then multiply by the coating failure rate. For example, if the listed expected failure rate for a certain coating is 0.5 percent, the total surface area is multiplied by 0.005 to obtain the possible exposed surface area.

Galvanic (sacrificial) anodes are generally only practical for buried structures, pipe, and so forth in low-resistivity soil and in waterboxes in freshwater service. The current requirements and service life of high-resistivity soil and saltwater service waterboxes may make sacrificial anode protection unrealistic.

For waterboxes, sacrificial anodes are attached directly to the waterbox using mounts cast into the anode material. For buried pipelines, the anodes are buried adjacent to the pipe to be protected. The condition of the waterbox anode is visually checked periodically. The condition of the buried anode is checked by reading millivoltage at a test box provided at the surface. The box contains a cable connection to a test electrode and one to the metal being protected.

The structure-to-earth potential is most widely used to determine the need for corrosion protection. The copper–copper sulfate half-cell is used by many pipeline companies as a reference electrode in making the potential measurement. In some cases, if the potential of the pipe to such an electrode, remote from the structure, is less negative than –0.85 V (less negative is synonymous with less anodic), cathodic protection is applied until the desired value is obtained. In other cases, iron rods are used as reference electrodes and a structure potential difference of –0.3 V is required for a safe condition. In still other cases, a pipe is considered safe from corrosion only if forced drainage is applied until the potential of the structure is depressed 0.3 V below its original value.[2]

Regardless of the method for determining the adequacy of protection or the deterioration of the anodes, this criterion must be spelled out clearly by the engineer of record. It is the plant engineer's responsibility to accumulate this criterion and establish the signature information for maintenance.

EQUIPMENT LAY-UP

Introduction

The following is an edited version of a procedure written for the lay-up of a 600-MW combined-cycle unit. It is an example of good lay-up practices. Of course, this should serve only as a guide because all equipment suppliers should be consulted before lay-up of their particular equipment is performed. Also, the procedures should be written by experienced professionals.

The primary lay-up procedures as recommended here are for an intermediate time period of longer than 30 days, but not exceeding 3 months. This primary recommendation takes into account the fast return to service requirement.

Should the lay-up extend beyond 3 months, a change in the procedure is recommended for the major components consisting of the heat recovery steam generators (HRSGs) and the turbine generator. This supplement is included and identified at the end of this short lay-up procedure.

The main objectives in properly storing power plant equipment during an extended outage are protection against corrosion, contamination, subfreezing temperature, and physical damage.

The potential severity of exposure must be assessed carefully to determine the degree of protection required. For suitable protection against corrosion, water and/or oxygen must be eliminated.

It is preferable to store plant equipment in warm, dry, ventilated, indoor locations whenever possible. Even equipment designed for outdoor use is susceptible to deterioration during prolonged idle periods when exposed to rain, dust, insects, and condensation. Note that condensation due to temperature change is more likely to contaminate the equipment when idle than when in service.

Caution

Equipment stored outside should never be covered with plastic tarps without a heating source to prevent moisture from condensing inside and on equipment. Simple light bulbs, with a drop light for protection, are sufficient. Plastic is a semi-permeable membrane. Moisture will pass through it (due to osmotic pressure) during the daily temperature changes from night to day. The moisture from the saturated atmosphere will condense during these cycles, continually wetting and drying internal, as well as external, metal surfaces under the tarp.

The key factors responsible for corrosion during lay-up are moisture and oxygen. Corrosion prevention is achieved when either moisture or air are eliminated from the system. There are two methods to control these substances: wet and dry lay-up. *Wet lay-up* is based on the elimination of oxygen from the system, while *dry lay-up* is based on the elimination of moisture.

Considering the requirement to restore the plant to service in a relatively short time, a wet lay-up of most of the steam generating and condensate equipment is best suited. Complete removal of all moisture required for dry lay-up is very difficult to achieve, especially in steam generators. Periodic operation of equipment and systems will ensure their protection and availability for restart.

Other equipment such as main turbines and condensers, main steamlines, and feedwater heater shell sides are protected by dry lay-up, yet remain ready to return to service in a reasonable time. Still other equipment, of necessity, must remain idle. Application of protective coatings is an effective means of preserving auxiliary equipment in storage. Correctly selected and properly applied coatings provide effective protection, ease of application and removal, and maximum coverage at low cost. Table 10-2 provides a list of preservatives frequently used for layup. Recommended procedures for protection and maintenance of each piece (or group) of plant equipment during the storage periods are described herewith.

TABLE 10-2 Preservatives List

Service	Product
Light grease for valve stems and threads to prevent rusting	Alvania No. 1
Light oil to fill idle crank cases and gearboxes	Ensis Engine Oil 30W
Heavy coating for shafts, sleeves, bolting (must be removed with solvent)	Ensis Oil L
Light oil for bearings, governor parts	Ensis Oil S
Corrosion inhibitor for use in turbine lube oil	VSI-27 Code 65249
Vapor barrier (Ronco Chemical Co.)	Ferro-Gard-50
Moisture detectors	Desiccant Color Pellets

The scope of this material is intended to cover as much as is practical for proper storage. Field conditions may indicate specific problems requiring special consideration. It is strongly recommended that a log be kept of pertinent data relating to each piece of equipment laid up as subsequently outlined and an inspection schedule be adhered to and, if necessary, altered in accordance with findings.

Heat Recovery Steam Generators (HRSGs)

Gas Side Lay-Up

1. Cover the stack to keep heat in and moisture out.
2. Heating and circulating water in the boiler (forced circulation boilers) will transfer heat to the gas passages, raising dewpoint to prevent moisture condensation.
3. Add radiant heaters in the boiler passages if transferred heat is insufficient to maintain temperature higher than dewpoint or if primary source of transferred heat is not available.
4. The boiler tube end-turn boxes should be heated by inserting electric strip heaters into the passages through available capped openings. This will prevent any differential of temperature occurring between the main gas passage and the sealed end turn boxes.

Water Side Lay-Up

1. Steam drum metal surfaces should be protected during wet lay-up by blanketing the vapor space of the drum with nitrogen. Purge all air with nitrogen. Maintain a 1.5-psig minimum cap. The water level in the drum should be held at normal level.
2. Deaerated boiler feedwater, chemically treated with morphaline to adjust the pH to 9.5–10.0 and with hydrazine to scavenge oxygen, should be circulated through the economizers and evaporators. Hydrazine residuals should be maintained at 200 ppm.

> **Caution**
>
> More damage has been done to HRSG outside fin tube surfaces due to neglect during storage than to internal surfaces. The tube fins are very thin and can deteriorate rapidly if not protected. HRSGs must be protected from the outside elements and radiant heat applied to the surfaces to prevent external condensation.

Steam Side Lay-Up

1. Superheaters and main steamlines are protected by nitrogen blanketing. When nitrogen is injected into the HRSG steam drums, air is purged out through piping low-point drains. Nitrogen pressure should be set and kept about 1.5 psig minimum to prevent air from being introduced into the system.
2. Maintain steam pipe traps in good working order to drain any condensate formed as a result of steam venting into the superheaters and main steamlines.

Sootblower Lay-Up

Sootblowers contain both a rotating assembly and a traveling assembly. The lances may remain in the retracted position.

1. Chain drives, rotating and traveling, should be coated with Ensis Oil L.
2. Worm gears of the traveling carriage should be coated with Ensis Fluid S.
3. Gearbox reservoirs should be filled to normal operating level with Ensis Engine Oil 30W.
4. Exercise the blowers once a month with the removable hand crank to ensure there is a protective coating on the gear drives.

Scavenger Air Blower Lay-Up

Heat recovery steam generator sootblowing systems must contain a scavenger air blower mounted in an insulated protective enclosure.

1. Install a heat lamp inside the insulated protective housing of each unit blower. This will prevent moisture from accumulating on the motor and blower surfaces.
2. Turn the blower shaft by hand once a month.
3. Coat the motor and blower shafts with Ensis Oil L.

Boiler Circulating Pump (BWCP) Lay-Up

1. If the pumps for the HRSGs are housed in an insulated enclosure, they do not need further protection.
2. These pumps will be operated periodically; therefore, normal operating maintenance must be followed. Operation of these pumps is required to heat the gas passages of the HRSGs to a temperature above the dewpoint. This is accomplished by circulating treated water, heated by cal-rod heaters, through the economizer and evaporative sections of the HRSGs.
3. Install temperature-monitoring equipment in the gas passages, and operate the BWCPs when this temperature falls to a value approaching dewpoint.

Steam Turbine and Condenser Lay-Up

Steam Turbine Lay-Up

1. The steam turbine including shafting, blading, and casing, will remain in a dry condition.
   a. The turbine building should always be maintained at a temperature not less than 50°F. This will help prevent freezing and condensation forming on equipment due to radical temperature changes.
   b. Lubricate turbine pedestals as normally required.
   c. Coat exposed portions of turbine shafts with Ensis Oil L. Apply protective coating of Ensis Fluid S to governor end parts. Coat internal portions of shaft and governor parts not splashed with oil with Ensis Oil L.
   d. The turbine should be placed on turning gear for about 30 min in every 24-hr period. Continuous operation of the turbine on turning gear is not recommended because this could result in excessive bearing wear due to the very thin lubrication film obtained at such a low shaft speed.
2. Circulate warm dehumidified air through the main condensers, through the low-pressure (LP) turbines, into the crossover pipe, and into the high-pressure (HP) turbine. See the Condenser Lay-Up section for additional information. Another stream of warm air should be introduced through the gland steam leak-off to the packing boxes of the HP turbine. Connect a 1500-W blower dryer with ductwork to the gland steam piping at the removable dresser coupling on the vented side of the gland exhauster. The leakoff lines running from the low pressure (LP) turbine packing boxes will be closed off. Drain all steam and waterlines thoroughly—this includes exhaust hood spray lines, as well as extraction steam and shell drains. Drain valves should be closed after draining. Drains that allow a visual check of water drainage should be checked weekly, or as determined by experience. The bearing oil system should be kept in operation at all times during shutdown. This practice will prevent formation of rust and scale in the supply and return lines. It also will prevent rusting of journals and control surfaces by keeping them

bathed in oil. Add Shell VSI-27 Code 65249 to oil as per Shell Oil Company instructions for vapor protection in drain lines, lube oil tank, and pedestals.

a. Oil temperature should be maintained between 50 and 90°F. The AC lube and seal oil pump may generate sufficient heat for that purpose. If not, the reservoir cal-rod heaters are to be used.

b. The oil tank vapor extractor should be kept in operation to eject hydrogen vapors from the oil system.

3. The electrohydraulic control (EHC) system should be placed in service once a month and exercised. The control valves should be operated to ensure that valve gear and servo-mechanisms are in good working order.
4. Take a lube oil sample every 3 months and submit it to a laboratory for testing. If sample shows moisture or sludge, cycle oil through the oil purifier.

Main Condenser Lay-Up

1. For the shell side of the main surface condenser, it is recommended that dry lay-up be used. This is based on the removal of moisture from the system.
2. To accomplish this dry lay-up:
   a. Completely drain the condenser and the hot wells.
   b. Follow by blanketing the system with dry warm air.
3. The dry warm air for this blanketing is generated by portable 3-kW radiant heaters placed at four locations in the dry-air flow path and adding a dehumidifier in the system.
4. The manhole covers located at the bottom of the hotwells are to be replaced with Plexiglas covers, modified to accept a 4- to 6-in-diameter flex hose. One flex hose is to be connected from one hotwell manhole cover to the discharge side of a 150-cfm dehumidifier. Dry air blown into the hotwell travels through the condensers and both LP turbines and is vented at the other hotwell manhole cover through another flex hose into the suction of the dehumidifier. Sensitive Magnehelix gauges may be used to measure ΔP across the system and dehumidifier. Use thermometers to check dewpoint.
5. Add treated desiccant pellets for moisture detection, and inspect weekly at each Plexiglas hotwell cover.

Main Condenser Water Side Lay-Up

1. Localized deposits of sludge create oxygen-deficient zones, which may disrupt the protective oxide film on the condenser tubes and cause pitting. For long-term lay-up, draining the waterboxes and hydraulically flushing the tubes is recommended. Inspection during cleaning will determine the extent of flushing. Lay-up dry.
2. Clean all surfaces of waterboxes and tubesheets. If necessary to inhibit corrosion, coat needed surfaces with Debecote.

Generator Lay-Up Moisture Protection

1. Check generator moisture drains weekly.
2. Monitor seal oil system and check gas purity.
3. Make insulation checks of generator bearings weekly.
4. Make insulation checks of rotor and stator weekly.
5. Check ground detection devices periodically.

Cooling Gas

If hydrogen cooling gas is to be maintained in the generator during the shutdown, proceed as follows:

1. The seal oil system will remain in continuous service. It operates as part of the bearing oil system for the turbine, which is its backup.
2. The cold gas temperature by resistance temperature device (RTD) measurement must be maintained between 15 and 20°C.
3. Hydrogen coolers should be cleaned and laid-up dry.

Balance of Plant Lay-Up

Condensate System

1. Isolate the condensate pumps and condenser hotwells from the condensate flow path as follows:
   a. Install a jumper between the existing pipe flange on the deaerator drain to the condensate storage tank and the suction line to the condensate transfer pump.
   b. Open the bypass line at the boiler fill station to connect the discharge of the condensate transfer pump to the inlet of the gland steam exhauster.
   c. Close both suction and discharge valves of both condensate pumps.
2. Treated lay-up water should be circulated from the deaerator storage tank through the condensate transfer pump, gland seal condenser, and feedwater heater and back again to the deaerator.

Boiler Feed System

1. Fill the boiler feed piping with treated lay-up water from the deaerator storage tank, through both the boiler feed pumps and up to the economizer inlet valves.
2. Chemicals to stabilize the lay-up water should be added at the suction side of the boiler feed pump, which can then be used to recirculate the contents of the deaerator storage tank. Use the boiler feed pump recirculation lines.
3. After initial blending, the treated water should be sampled for pH and hydrazine weekly.

Boiler Circulating System

1. Lay-up water, treated with morphaline to adjust the pH to 9.5–10.0 and 200 ppm of hydrazine to scavenge the oxygen, should be circulated through the economizers, steam drums, and evaporators. See Wet Lay-Up of HRSGs.
2. Protect the steam drum metal surfaces during wet lay-up by blanketing the vapor space of the drum with pressurized nitrogen. Steamlines and extraction lines are also protected with nitrogen.

Circulating Water System

This system provides the cooling water for the auxiliary equipment consisting of the lube oil cooler, hydrogen cooler, and condenser vacuum pumps.

1. Drain and refill the cooling tower basin with fresh service water.

2. Increase the pH of the recirculating water to 8.8–9.2 by adding dilute liquid caustic.
3. System to be operated for closed cooling-water system cooling.

Heat Exchangers

Feedwater Heaters

1. Treated water will circulate through the tube side of these heaters.
2. Provide a nitrogen blanket for the shell side of the heaters. This will maintain an oxygen free atmosphere for the extraction lines as well as the steam side of the heaters. Maintain a 1.5-psig minimum cap.

Deaerator and Storage Tank

1. Treated water should be recirculated through the boiler feed pumps back to the deaerator and storage tank. Circulate monthly to prevent chemical stratification in the system.
2. Maintain a nitrogen pressure of 1.5 psig minimum on the deaerator vapor space and steam supply line.

Gland Steam Condenser

1. Treated water will circulate through the tube side of this exchanger from the condensate flow.
2. The shell side will be kept dry by warm air being admitted on the vented side of the heater and flowing through the heater into the packing boxes of the high-pressure turbine.

Auxiliary Coolers
This equipment consists of the following:

1. Generator hydrogen coolers.
2. Bearing cooling water coolers.
3. Turbine lube oil coolers.
4. Condenser vacuum pump coolers.
5. Boiler feed pump lube oil coolers.
6. HRSG circulating pumps water-cooled bearing and seal casings.
7. Turbine hydraulic power unit coolers.

These coolers, all of which will remain in service as needed, require normal protection and maintenance.

Pumps

Boiler Feed Pumps

1. Isolate and fill the boiler feed pumps with treated condensate and Ferro-Gard 50-G.
2. Rotate alternately once a week to prevent any formation of chemical dead spots or accumulation of oxides in the pump internals.

Condensate Pumps

1. Isolate the pumps by closing the suction and discharge valves.
2. Completely drain the pumps, then flood with a 6-to-1 mixture of condensate and Ferro-Gard 50G and close vents. Rotate pumps alternately once per week.
3. If motor is not removed, it should remain coupled to the pump. The coupling should be lubricated per manufacturers instructions. Outside surfaces of coupling shall be coated with Ensis Oil L.
4. All exposed shafts should be coated with Ensis OiL L.
5. Oil-lubricated bearings shall be flooded to coat all internals with Ensis Engine Oil 30W.

Condensate Transfer Pump

During lay-up, this pump will serve to circulate condensate from the deaerator storage tank through the heat exchangers and back to storage. This pump will require normal protection, maintenance, and lubrication.

Condenser Vacuum Pumps

The two rotary, two-stage, water-cooled and -sealed, package-skid-mounted pumps are located on the ground floor of the turbine-generator building. This building will be heated and maintained at a minimum of 50°F at all times. Flood the pump with a 6-to-1 condensate and Ferro-Gard 50G mixture and close vents. Rotate weekly by hand.

Auxiliary Pumps

They shall be inspected prior to operation and winding insulation resistance checked. This section applies to the following:

1. Service water pumps.
2. Equipment cooling-water pumps.
3. Bearing cooling-water pumps.
4. Sump pumps.
   a. Strainer backwash sump.
   b. Cooling tower blowdown sump.
   c. Turbine house blowdown sump.
   d. HRSG blowdown sump.
   e. Auxiliary boiler fuel oil tank sump.
   f. Pond effluent.
   g. Lube oil area sump.
   h. Circulating water pump house sump.

These pumps require normal protection, maintenance, and lubrication. They shall be operated as required.

Circulating Water Pumps

1. These pumps take their suction from the circulating intake structure at the cooling water basin and circulate the water through the condensers before discharging into the cooling tower distribution headers. It is recommended that these pumps be laid up and stored in place.
2. They should be pulled out, cleaned of all sediment and debris, painted with bitumastic or epoxy, and reinstalled.
3. All exposed shafts should be coated with Ensis Oil L.
4. Rotate pumps by hand weekly.

Miscellaneous Plant Equipment

Condensate and Demineralized Water Storage Tanks

Because these are carbon steel tanks, lined with epoxy and vented to the atmosphere, there is no reason for nitrogen blanketing to limit the corrosiveness of the water stored during lay-up.

Instruments and Controls Lay-Up

1. Instruments located within the turbine building do not require additional protection because the building will be maintained at 50°F minimum at all times.
2. The HRSG instrumentation is located within heated enclosures provided for each unit. Because of the instrument locations, no further protection is recommended.
3. The instrument air compressors should remain in service, supplying air to prevent moisture from accumulating in the lines. Vent lines biweekly.
4. The condition of the air dryers should be carefully checked to ensure no moisture carryover.

Auxiliary Boiler

The auxiliary boiler system consists of the auxiliary boiler, the feed pumps, electric superheaters, and associated valves and piping.

1. This system should be protected from corrosion by wet lay-up, as per the manufacturer's instructions.
2. Increase the boiler water pH, and introduce nitrogen into the drum vapor space and superheaters. Maintain a 1.5-psig-minimum cap.

Valves

Due to the multitude and variety of valves used, only recommendations of a general nature can be made.

1. Moving parts of valves such as the shaft or stem should be covered with a light coating of Ensis Oil L.
2. All threaded valve stems should be coated with Alvania No. 1.
3. Boiler safety valves should be coated with Ensis Oil L on finished surfaces.

Extended Time Lay-Up

Operation of the gas turbine simple cycle with poor closure of dampers isolating the heat recovery boilers is very detrimental to the boiler, steamlines, and eventually, the steam turbine. This will occur with intermittent buildup of vapor in the steam drum and steamlines, causing corrosion and scaling in these areas. This can eventually lead to exfoliation of boiler and steamlines with subsequent damage to steam turbine blading.

The primary lay-up procedures, as previously recommended, are for an intermediate time period of 30 days but not longer than 90 days. If the lay-up extends beyond 3 months, or if vapor pressure buildup occurs in HRSGs, then changes in the lay-up procedures are recommended.

HRSG—Gas Side Lay-Up

1. Blanking plates, or some other positive closure, shall be designed and installed after the isolation damper and in front of the close-off damper so as to seal off the gas passage completely.

2. Sufficient electric strip or radiant heaters are to be located within the furnace so as to maintain the temperature above the dewpoint.
3. Do not operate the BWCPs continuously. Operate periodically to promote mixing of the chemicals in the boiler water side.
4. Make and install lightweight cap for stack.

Steam Turbine Lay-Up

The turbine bearing oil system need not be kept in continuous operation. The following procedures shall then apply:

1. Clean the lube oil tank and overhaul the lube oil pumps. Add Shell VSI-27 Code 65249 to the lube oil for vapor barrier protection of drain piping.
2. Circulate oil through the system for 4 hr every 2 weeks. Alternate pumps being used. Check tank for water accumulation, and check lube oil quality once a month. Place turbine-generator on turning gear for 2 hr.
3. The oil tank vapor extractor should be shut down. Rotate it by hand once a week.

Generator Lay-Up

1. The generator should be purged of all hydrogen gas. Use normal purging procedures.
2. After purging, fill the generator with air. Heat the air as necessary by placing electric heaters in the bathtub below the machine. Maintain a temperature above the dewpoint.
3. Operate the seal oil system as outlined previously for the turbine oil system.
4. Place a large room dehumidifier in the generator. Operate continuously with drain to outside (Table 10-2).

REFERENCES

1. Strauss, S. D., and Puckorius, P. R., "Special Report, Cooling-Water Treatment for Control of Scaling, Fouling, Corrosion," *Power*, June 1984.
2. Uhlig, H. H., *Corrosion Handbook*, sponsored by The Electrochemical Society, John Wiley & Sons, Inc., 1948.
3. Cotton, I. S., "Optimize Oxygen Control in Your Boiler-Feed System," *Power*, October 1980.
4. "New Stainless Steels Prolong Life of Key Plant Components," *Power*, April 1984.
5. Reason, J., "How to Design Condenser Cathodic-Protection Systems," *Power*, June 1984, from the Electric Power Research Institute, "Current Cathodic Protection Practice in Steam Surface Condensers," CS-2961, project 1689-3, May 1983.

CHAPTER 11

DIESELS AND GAS TURBINES

Diesels and gas turbines are combined in this chapter because large, central-station diesels and gas turbines share some common commissioning requirements. Where steam thermal electric stations require many systems and milestones—on the average 100 to complete (see App. P)—the diesel and gas turbine do not. Those common requirements require attention to the following six essential systems:

1. Fire protection system.
2. Lube oil system.
3. Cooling water system.
4. Fuel system.
5. Inlet air system.
6. Control system.

Tracking progress on these six systems will help ensure the project is delivered on schedule.

FIRE PROTECTION

Neither diesel nor gas turbine should be trial-operated without the fire protection systems in service, making that system critical for fuel delivery.

Diesel lube oil pits should be diked for each engine so that a fire at one engine does not spread throughout the entire engine gallery. If the dikes were not part of the design, build temporary ones for startup. Diesel engines will have leaking fuel as well as lube oil.

CO_2 testing is left to the last days before first firing because of the lock-out-tag-out restriction for entering compartments, so all preparations have to be made far ahead of that date. The compartments must be light and smoke bomb tested for leaks prior to concentration testing. Gas turbine CO_2 cylinder heads should be provided with enough spares to ensure that the CO_2 system testing does not enter critical path. Cylinder discharge heads are provided with reset keys (grenade pins), which invariably get lost prior to CO_2 concentration testing. Call the vendor for replacement keys. The compartments that contain combustibles must have combustible detectors. Remember, turbines fired on gas will not likely need protection against fires caused by fuel leaks because they will simply blow up if there are leaks. Flooding the compartments with CO_2 after gas is detected will take the compartment atmosphere out of the combustible range to prevent explosions. The main concern for actual fires is from liquid combustibles such as lube oil and fuel oil.

LUBE OIL

Large, central-station diesels consume enough lube oil that they never require oil changes. Because they use and consume so much oil, their lube oil system piping is usually large diameter. Therefore, the system should be cleaned mechanically. Also, the velocity required for adequately flushing such large-bore systems is not achievable. The critical path comes when the lube oil is flushed through the engine.

Gas turbine lube oil systems will require flushing as a primary means of cleaning. But, like the diesel, as much piping as possible should be cleaned manually, that is, disassemble the piping and shop clean it. (Refer to Chap. 6.)

With both diesel and gas turbines, the lube oil system cleaning can become critical path quicker than in steam stations because there are not a lot of systems left to give up scheduler's "float."

COOLING WATER

Both diesel and simple-cycle gas turbines share similar cooling-water systems—usually dry cooling towers (fin-fan coolers). The cooling water for both are closed cooling-water systems. Make sure the coolers or towers have proper air circulation to function properly. Make sure they are installed in accordance with the manufacturer's instructions the first day you arrive on site. Coolers mounted too close to each other, or on the ground, will ensure air recirculation and deficient water cooling. Ensure there is a sufficient surplus of glycol or other corrosion inhibitor warehoused on site.

FUEL SYSTEM

Central-station diesel fuel oil systems will normally be heavy residual systems. These systems will require a dedicated effort to commission. There are many centrifuges and other fuel-filtering and -conditioning equipment that will require preparation and troubleshooting. Flushing of these systems is imperative to avoid critical path delays. Gas turbines that fire residual fuels require the same fuel treatment.

Gas turbines can fire heavy residual fuel, but the overall domestic industry tendency is to fire gas. This means that the fuel gas receiving, metering, and conditioning skids must be checked out and leak tested prior to receiving gas. Don't forget gas monitoring, fire detection, and prevention systems, including pipeline cathodic protection and lightning arresting systems.

Gas turbine combustion compartment flexible hoses must be cleaned prior to commissioning. Do not use air "burst blows" for cleaning flex hoses. Sonic shock waves at the discharge flange can unravel the inner spiral lining. They are best cleaned with small pigs designed for that purpose. Purchase them from a "pig" supplier. Do not field fabricate pigs. Use soft pigs with a cleaning solvent such as Blue Gold (vaporless and nontoxic). You will find this method to be much faster than air blowing. This is the preferred method, by some, for cleaning all gas turbine compartment air and gas piping.

INLET AIR

Inlet air filters should have sufficient spare parts to ensure that small items such as filter cleaning solenoids (puffers) do not delay commissioning. Loading filter and evaporative cooling media should be scheduled ahead. Check with the media supplier and ensure the media can remain dry until the inlet air system is needed. Some evaporative cooler media will deteriorate if not kept wet.

CONTROLS

Diesels will have fewer than 100 control loops to check, frame gas turbines will have hundreds, and solid fuel plants will have thousands. It is important to list all the loops and track all cable and conduit installation as well as loop checking progress. The best source for tracking is the project's "Device Book." This book is important for any kind of plant startup checkout but governs all checkout and commissioning progress for diesels and gas turbines.

Owners and project managers gauge commissioning progress primarily by tracking progress of the aforementioned seven systems.

DIESEL ENGINES

Introduction

A good general reference source for information for all the different types of diesels and their accessories is "Oil and Gas Engines," a special report published by *Power* magazine.[1] Of course, the best *detailed* information can be obtained from your engine instruction manual. The DEMA (Diesel Engine Manufacturers Association) standards govern diesel engine operation.

Low-speed diesels are two-stroke cycle engines and are not suited for rapid start, emergency operations. They are preferred for central-station use and can be found to be 50 MW or larger in a single unit, with efficiencies of ~48 percent. They also can use residual fuels. The four-stroke cycle engines are used in all large, power-generating plants for emergency and black starting. Because of their low inertia flywheels and generator rotors, they have maximum acceleration potential required for emergency starts.

High-Speed Diesels

Fuel System

The fuel system discussed here is a typical diesel system provided as an external fuel supply from an external day tank for the engine. Because the engine fuel pump is a positive displacement pump, the fuel not being consumed by the engine is recirculated back to the fuel tank. Because of the recirculation, the heat generated by the pump must be dissipated, or the tank will heat up and eventually explode. Therefore, a fuel oil cooler is placed in this cycle to cool the fuel on the return cycle to the tank. A fuel shutoff valve with a "kill switch" (emergency stop) must also be installed in this system.

It is not good practice to construct the fuel system with threaded connections because of the potential for leakage. Also, if there is a siphon loop higher than the top elevation of the tank, a means of priming the system should be provided. Besides the obvious reason for avoiding leakage, leaks can cause a loss of prime when the engine is in "standby" mode. The tank suction pipe should have a foot valve to hold the prime. On the engine side, there will be a hand-operated fuel-filter priming pump. The system must be primed every time the filters are changed, or the air will compress during startup attempts, preventing the ejectors from operating.

Engine Cooling-Water System

The typical engine cooling-water system is provided as an external heat-exchange system. The water is normally a 30 percent glycol mixture. This system cools the engine cooling system and fuel oil coolers. As such, it must be interlocked to trip the engine on high temperature. The cooling-water pumps are sometimes part of the engine accessories. The normal heat exchanger used is a fan-cooled radiator-type cooler and is located outside on the shaded side of the room. The fans are usually interlocked to start as the temperature of the water rises. The number of fans running depends on the temperature, which is directly related to the ambient air temperature.

The engine room is usually cooled by ventilation with the flow from the (shaded) side of the building out the sunny side. The system inlet louvers and discharge fans open and start automatically on an engine start.

General Engine Lubrication[2]

Base selection of suitable oil on:

- Engine rating.
- Engine speed.
- Numbers of cylinders.
- Oil filter system.
- Fuel specifications.
- Operating load factor.

These characteristics are more important than lube oil cost. Engine manufacturers will not recommend any particular brand of oil but will, instead, specify viscosity and detergent level. Knowledgeable operators submit their selection to the engine manufacturer for approval or comments.

For viscosity, the recommendation on heavy-duty, high-output diesel or dual-fuel engines can be a value of 1000 SSU (Saybolt Seconds Universal) at 100°F and 75 to 80 SSU at 210°F. These oils should contain oxidation inhibitors.

For detergents, high-compression (11:1) spark-ignition gas engines may require a minimum detergency level equal to MIL 2104A standard. Recently, ashless detergents have been chosen for gas engines. High-output, medium-speed diesel engines may run on a Series 3 oil, which has higher detergency level than given in MIL 2104A.

The oil supplier can be requested to analyze used crankcase oil samples periodically, preferably weekly.

Engine Controls

The engine control cabinet is separate from the engine control unit (ECU), which is mounted on the engine itself. The cabinet should always be located in a separate room or building for safety and comfort. Engine crankcase explosions can cause severe injuries or death to anyone in the engine room. Access to the engine room should be only for engine inspection and not control panel operation.

The engine control cabinet interfaces the ECU and the plant and engine external systems' controls via a PLC. The manufacturer usually provides the cabinet and the contractor with the interconnecting wiring.

Engine Design Considerations[3]

Investigative studies of generators, exciters, and voltage regulators can maximize reliability of the diesel generator set. Specifications should include the following:

- First-try start capability.
- 10 s to rated speed and voltage from emergency start.

- Full-load acceptance in 30 s.
- Voltage recovery to 90 percent minimum-rated voltage in 1 s.
- Frequency drop not over 12 percent, with recovery to 105 percent rated frequency in 10 s after motor-load application.

Rapid automatic start acceleration to full speed in a suitable time, with load acceptable in a predetermined sequence, is vital for critical standby power. Factors affecting start and load acceptance include time lag in the air-start procedure, cranking time, inertia of engine, flywheel, and generator.

Load acceptance depends on the combustion air availability and response of the engine governor. An electric governor reduces overspeed and recovery time to rated speed. Hydraulic governors only sense speed, with a time of 0.5 s from no load to full load. Electric governors, on the other hand, sense both speed and load and can change rack position of the injection pumps on transient load before any change in frequency. Thus, full travel is reduced to within one-fifth of a second. Governor selection is usually determined by the engine manufacturer who is responsible for the qualifications.

Rapid starting is a function of engine, flywheel, and generator inertia. Distribution of the mass between the flywheel and generator depends on electrical requirements. Some manufacturers choose to retain conventional flywheels and omit stator shift on the premise that generator maintenance for standby service is less demanding than for continuous duty. Because the engine and generator-rotor inertia resist acceleration during startup, a low inertia appears desirable. However, electrical characteristics and mechanical design determine minimum inertia. Large inertia reduces transients during load change at the expense of longer acceleration period, whereas small inertia may cause high overspeed on load rejection.

Comment

An overspeeding diesel is a *run-away* diesel. It is usually caused by the induction of an alternative fuel into the air intake. The source could be its own crankcase oil, from a leaking turbocharger or chemical fumes around the engine. However, it can also be caused by slow governor response and load rejection of an engine-generator set with low flywheel and generator rotor inertial forces that retard acceleration. It can cause crankshaft breakage and engine casing damage.

Keeping an eye on critical temperatures and pressures by keeping an hourly log is normal for continuous operation. However, greater dependency is placed on safety for critical standby power. Troubleshooting shutdown and alarm annunciators should be provided for:

- High bearing temperature.
- Low oil pressure.
- Low turbo lube oil pressure.
- Low crankcase oil level.
- Oil heater malfunction.
- Low water pressure.
- High or low standpipe level.
- Shutting down of circulating pump.
- Jacket heater malfunction.

- Low starting air pressure.
- High or low fuel oil pressure.
- High oil-filter or strainer differential.
- High or low fuel oil day-tank level.
- High main bearing temperature.
- Engine vibration.
- Thrust bearing wear and failure of turbocharger.
- High crankcase pressure.

Keep extended periods of light-load and/or no-load operation to an absolute minimum for reliable operation.

Utilities have always stressed the need for overload capability for a period well in excess of commercial or municipal installations. These, in accordance with DEMA standards, are limited to 10 percent overload for a 2-hr period. However, a 25 percent overload for a period of 120 hr was demonstrated by turbocharged diesels for peaking service. Also, electrical insulation and generator construction should be able to withstand a 105°C temperature rise for an ambient temperature of 40°C.

Engine Failure Prevention[4]

Diesel engines (four-stroke cycle, emergency, black start) and compressors are entering service today in power blocks ranging as high as 16,000 hp (~12 MW) and brake-mean-effective pressure ratings up to 250 for 20-cylinder engines. These large units are proving to be subject to more crankshaft failures than their older, smaller brothers. This has resulted in increasing interest in the causes of these failures and in the precautions that must be taken to prevent them. Remember that replacement cost of a broken crankshaft can cost about 15 percent of the price of a new reciprocating engine.

Figure 11-1 shows a typical bending break caused by misalignment. Here, the break crosses the crank web starting in the web fillet radius adjacent to the crankpin. Subsequently, the break weakens the web and causes random fracture in the main journal.

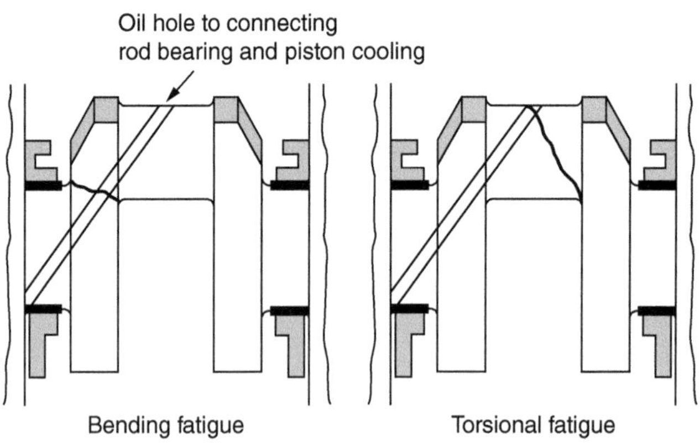

Typical bending failure due to misalignment is shown at left. On the right is torsional fatigue failure as a result of operating at or near critical speed

FIGURE 11-1 Crankshaft fatigue failure types.

A torsional-vibration break (right side of Fig. 11-1), is generally across the nodal point of the shaft elastic system. Starting in a fillet, the break advances at 45° across the crankpin nearest the engine flywheel. Such brakes result from operating at or near a critical speed of the crankshaft system's natural vibration frequency.

Installation Checks[5]

Temperature effects on the foundation are well documented. Excessive engine heat can distort the foundation, causing misalignment and bearing damage with subsequent crankshaft failure. Manufacturers have set up crankshaft deflection limits for each type and size of unit. The reading is taken by a dial indicator, or strain gauge, between webs at a distance from the shaft centerline prescribed by the manufacturer.

The crank web check should be made in the hot condition. Visual check of the base is also necessary to ensure that the main journal saddles are properly aligned.

Base alignment of a cold engine produces a slightly bowed condition, which becomes perfect horizontal alignment when the engine and foundation block are in the hot-operation condition. Under hot conditions, deflection readings of the individual crank throws are then taken. Experienced operators allow 0.002 to 0.003-in sag at center of base (for a V-16 engine) to compensate for thermal expansion. Final hot readings should be under 1 mil or the manufacturer's suggested limits.

Bowing distortion—or humping of the foundation block, leading to crankshaft misalignment—may be due to uneven shrinking or curing of certain types of concrete. This process may continue over an extended period of time before permanent distortion occurs. Generally, bowing of the top surface results in about a 5:1 ratio in crank deflection between webs of the crankshaft.

Temperature-differential problems are reduced if the engine is mounted on steel rails. The rails provide insulating air space and allow shimming of individual bearing supports in the base structure. Flat bars, 2 to 4 in thick, distribute engine weight to grout. Chocks—short steel plates, with their faces ground parallel and about 1 in thick—may be interposed between rails and engine base. Steel shims between rails and chocks give proper alignment of the engine crankshaft.

Check the grout because another major cause of crankshaft failure is deteriorating grout. If left unchecked, this can become a serious problem after an engine enters service. Damaged bearings, and ultimately a broken crankshaft, will result unless the unit is promptly regrouted. Grout deterioration can be caused by:

- Loose foundation bolts due to insufficient torque.
- Using the wrong grouting material. Certain cement and iron mixes not impervious to oil will disintegrate if cement particles become saturated from leaking oil. This can cause the foundation base to separate from its support and be distorted enough to disturb the initial alignment. Modern practice is to use special, oil-resistant grouts (normally resin epoxies) having high shear and temperature strengths.

Hot bearings caused by misalignment can cause crankcase explosions. A hot bearing can generate an oil mist within an air–oil combustion range and also the ignition energy to ignite it. The primary explosion ruptures the crankcase, expelling a hot-oil mist into the engine room where it mixes with fresh air and ignites, causing a secondary explosion. To prevent these explosions from happening, the cause must be addressed, that is, improper engine alignment.

High-Speed Diesel Package Pretrial Operation

Read the manufacturer's instruction manual thoroughly. The first check is of the engine placement itself:

- Check the engine grout. Grouting is not cosmetic—it is the structural support for the engine. Leveling shims are not designed for structural support. Anchor bolts are just that—they do not support the engine. They are designed to anchor the engine in place. Make sure the grout has reached its maximum strength before trial operating the machine.

- Check the torque of the anchor bolt nuts.
- Check the engine air intake for desiccant bags and loose material.
- Check that the exhaust is clear and free of debris.
- Remove shipping bars holding the flywheel in place.
- Check for other shipping hold-downs.
- Check the lube oil specification, and fill the crankcase sump.
- Check starting battery charge if a motor starter is used.
- Check that the control cabinet has been energized and all indicators function properly. Check that the PLC is energized and functions properly. Check that the battery charger is functional.
- Check that the lube oil and cylinder jacket heater is operational. Check that the oil temperature is between 100 and 120°F and the jacket water temperature is 120°F or in accordance with the manufacturer's instruction manual. Make certain that the circulating water through the heater is flowing. If the heater is hot, and the piping isn't, there is a valve closed somewhere.
- "Shoot," with a signal generator, the annunciator and trip input/output loops to and from the engine-generator set and ensure all function properly on the PLC and panel display.
- Prime the fuel system.

The next step is to check out the auxiliary systems:

- Check the engine room ventilation system operation.
- Check the closed cooling water system water level and operation.
- Check the heat exchanger operation.
- Check the air compressors and compressed air system if an air start system is incorporated.

High-Speed Diesel Package Trial Operation

All exits from the engine room should be open and the room cleaned of all construction materials and debris. The initial cranking of the engine should be with the ECU fuel control cable disconnected to prevent firing on the first roll. Follow the manufacturer's instructions for initial operation. On initial roll, watch for excessive movement or vibration; fuel, oil, or water leaks; and any unusually noises.

- Fire the engine up to speed, and immediately check the emergency stop to ensure it functions properly.
- Refire the engine up to no-load, full speed and monitor engine operation. Check generator voltage and temperatures. Recheck the emergency stop, and ensure the engine stops without overspeeding.
- Refire the engine and perform the generator electrical and synchronization checks.
- Load rejection tests should be performed at 25, 50, 75, and 100 percent to ensure that the engine does not overspeed. This has to be done before the emergency start and load tests are conducted.
- Conduct emergency start tests.

GAS TURBINES

The ASME International Gas Turbine Institute disseminates information and provides education regarding gas turbine development around the world.

Thermodynamically, the gas turbine is a Brayton cycle machine, whereas the steam turbine is a Rankine cycle machine. Both cycles combined constitute the combined cycle. "Closed cycle" is an erroneous term often used to describe the combined cycle.

Aeroderivatives gas turbines such as the GE LM2500 (25 MW) and LM6000 (40 MW), designed for mobile operation, are lightweight and therefore ideal for use in war ships. They are also excellent choices for peaking and small cogeneration plants that have frequent load swings. They are self-contained and have internal antifriction bearings, and the auxiliaries are attached to the engine exterior.

Note the different sections of aeroderivative in Fig. 11-2. There are two turbine sections in the aeroderivatives: one driving the compressor and the other driving a generator. The jet exhaust transition piece has been replaced by a turbine inlet transition piece and power drive turbine.

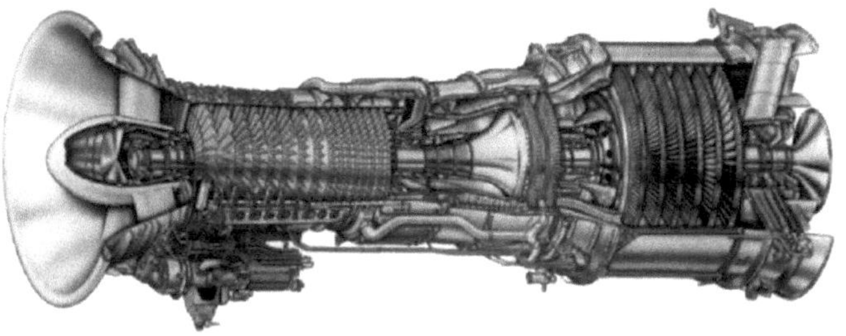

FIGURE 11-2 GE LN2500 gas turbine.

Industrial gas turbines are heavy machines suited better for base-load operation. They were referred to as "frame" machines because they were built on stationary, framed platforms. The platforms support both the stator and the rotor assembly-bearing pedestals as well as all auxiliaries. Today, there are frame machines that operate at capacities >400 MW. These machines use plain bearings.

High-Temperature Corrosion

High-temperature corrosion is referred to in the industry as "green rot" because of the green patina look of the deposits in the damaged areas. High-temperature corrosion is caused primarily from sodium and vanadium in the fuel or sodium that is airborne. It can be controlled by fuel washing and the use of fuel additives. Fuel washing removes sodium from fuel that has seawater contamination and magnesium additives inhibit vanadium corrosion. All additives add to the ash deposits in the turbine sections; therefore, frequent cleaning is required.

> When the gas turbine is operated intermittently as in marine, locomotive, or in peaking service, the deposit breaks off on shutdown or on startup after shutdown, so that it comes to an equilibrium thickness. With reasonable stall margin in the compressor, the unit can be operated in this manner more or less indefinitely without any special equipment for cleaning the turbine. If the gas turbine, however, is operated continuously without shutting down, deposits continue to form at a more or less constant rate, and as the turbine inlet pressure

builds up to the compressor stall pressure, it is necessary to clean with some solid material such as nutshells, rice, or refinery catalyst, or to shutdown and wash. Thus, if continuous operation is required, a means of cleaning under load must be made available.[5]

Conclusions from an EPRI report[5] were as follows:

1. Hot-gas-path corrosion, when identified, was attributed to improper operation of the fuel treatment equipment or to inlet air impurities.
2. The main difficulty with low-grade fuels is in the deposits that build up in the turbine from the inhibitors used to prevent corrosion by vanadium. The units must be water-washed at intervals, and this may significantly reduce the unit availability.
3. Daily (or more frequent) testing of the fuel is necessary to verify that the fuel treatment is adequate.

It is recommended that the decision to use low-grade oils in gas turbines be based upon economic considerations that include the additional equipment and the operational changes required as indicated in the survey information.

Explosion Protection

Gas turbines need to be purged after an ignition failure trip before a refire is attempted. This is especially true in combined-cycle installations where HRSGs are in the gas path. Large quantities of fuel can accumulate in a simple cycle silencer or in a combined-cycle HRSG gas path, which can be extremely dangerous. The purge cycle is the same as for fired boilers. The purge cycle must displace 5 volumes of air, purge for 5 min, or purge in accordance with the manufacturer's instructions, whichever is the greatest. First, remember the purge cycle is performed at the gas turbine (GT) purge speed and may require 10 to 15 min to purge a GT and HRSG combination adequately. Secondly, it is essential to take combustible analyzer readings of the gas path before attempting a refiring sequence after a misfire.

This photo was taken after a gas turbine HRSG transition explosion. The explosion occurred after a refire was attempted without proper purging.

REFERENCES

1. "Oil and Gas Engines," *Power*, September 1980.
2. Kauffmann, W. M., "Follow Oil-Change Guidelines to Prevent Diesel Catastrophe," *Power*, June 1983.
3. Kauffmann, W. M., "How to Design Your Stand-by Diesel Generator Unit for Maximum Reliability," *Power*, December 1979.
4. Kauffmann, W. M., "Prevent Engine-Crankshaft Failures," *Power*, September 1980.
5. "Worldwide Survey of Current Experience Burning Residual and Crude Oils in Gas Turbines," AF-1243, TDS-78-833, Final Report, Electrical Power Research Institute (EPRI), December 1979.

CHAPTER 12
ELECTRICAL

An excellent tool as a field reference for recent graduates, as well as the experienced electrical startup engineers, is the "Red Book," IEEE Standard 141, IEEE Recommended Practice for Electric Power Distribution for Industrial Plants;[1] the "Green Book," Recommended Practice for Grounding of Industrial and Commercial Power Systems;[2] and the "Buff Book," Recommended Practice for Protection and Coordination of Industrial and Commercial Power Systems.[3] See the IEEE color book series.

In addition, ABB offers the Silent Sentinels publication,[4] "Applied Protective Relaying," which is considered by many to be the "bible" for electrical protection. Silent Sentinel publications from 1960 to 1989 have been formatted on CD and are available free. Readers can e-mail tyler.j.kerner@us.abb.com for a free version. Or, browse for "abb silent sentinels" on the Internet. The following discussion and schematic illustrate motor protection explanations found in the Silent Sentinel publications:

> *4.0 Protection of Motors Rated 1500 HP and Above (Fig. 12-1)*
>
> *4.1 Thermal overload protection is provided by temperature relays, Device 49; the type DT-3 or the MPR relay. The type DT-3 relay or the MPR relay monitors stator temperature by means of one or more RTD's and is preferred where blocked ventilation or high ambient conditions may exist. The MPR optionally monitors 6 winding RTD's, 4 bearing RTD's and 1 load case RTD. Also available is the PRO*STAR which provides superior stator and rotor thermal models.*
>
> *4.2 Locked rotor protection is provided by an overcurrent relay Device 51. This is completely adequate if the allowable locked rotor time exceeds the acceleration time. If acceleration time exceeds locked rotor time, then this protection must be obtained from a zero speed switch operating in conjunction with Device 51, or a distance relay, type KD-10—Device 21, operating in conjunction with a timing relay, Device 2.*
>
> *4.3 Principle fault detection for a large motor is provide by the differential relay, Device 87, using either of two schemes. The preferred differential protection, due to better sensitivity and lower cost, consists of three IT instantaneous current relays or one three-phase type 50D relay energized from three flux-balance-current transformers, type BYZ (or equivalent) for IT, or GS type sensors and type GRD relay mounted on the motor terminals. The alternate differential scheme utilizes a total of six current transformers and three type CA or type 87M differential relays, Device 87. The latter method requires the current transformer ratio to be based on the full load current rating of the motor. Although this considerably lessens the fault sensitivity, it has the advantage that the cable circuit to the motor is included within the differential zone. When the IT, 50D, or GRD relay differential scheme, Device 87, is employed, two Device 50's, and Device 50G, are essential for cable protection. These relays are also recommended as back-up protection with the type CA or 87M relay scheme. Alternately, the MPR relay supplies the 50 and 50G functions. The MPR relay can also be equipped with the high speed phase differential function, Device 87.[1]*

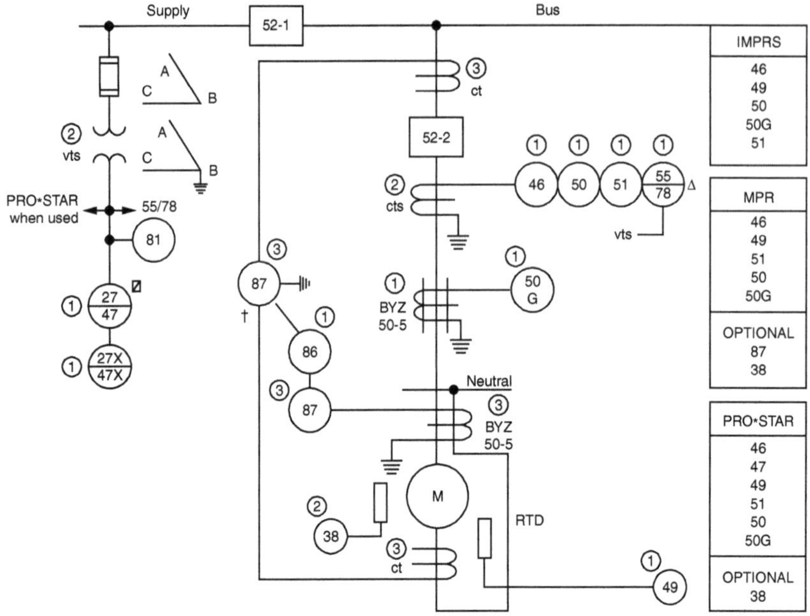

Note: (a) Phase ct's should be approximately 150% of full load.
(b) The MPR, PRO•STAR and IMPRS are microprocessor relays containing several functions.
† Alternate to BYZ scheme. BYZ scheme is preferred.
▨ Must apply CVQ or 60Q when device 46 is omitted.
Δ Applied on brushless synchrous motors.
(1) Represents the quantity of devices required.

FIGURE 12-1 Protection of motors rated 1500 HP and above.

SCHEMATIC READING TUTORIAL

Recent graduates of all disciplines will have all the fundamental engineering skills for their introduction to the startup profession, with the possible exception of the basic skills for reading industrial electrical diagrams. The following is a brief tutorial on common electrical drawings for recent graduates as well as for nonelectrical engineers and technicians.

Figure 12-2 illustrates a basic logic diagram for a simple pump arrangement.

The first skill is to understand elementary and control wiring diagram, motor starter, and circuit breaker terminology. The control wiring diagram of an AC circuit (Fig. 12-4) illustrates enough information to allow tracing wires and troubleshooting. The elementary diagram (Fig. 12-5) illustrates the same thing but in its simplest form. It should be mentioned that motor starters (See 42 of App. G) are breakers for low-horsepower, low-voltage service. They are used for frequent opening and closing of a circuit. Circuit breakers (See 52 of App. G) are breakers used for medium- to high-horsepower, medium-voltage service.

The pump in Fig. 12-3 is operated locally and has "interlocks" for a level permissive and a level trip. The high-high level is an alarm located on the local annunciator panel. The B, U, O, R, C, and W refer to wire markings. They could have been numbers, or color-coded, or both. All motor starters are *energized to close and de-energized to open.* All drawings show contacts in the "shelf," or de-energized, state. They can be thought of as coil-operated switches so they can be operated automatically or from a remote location.

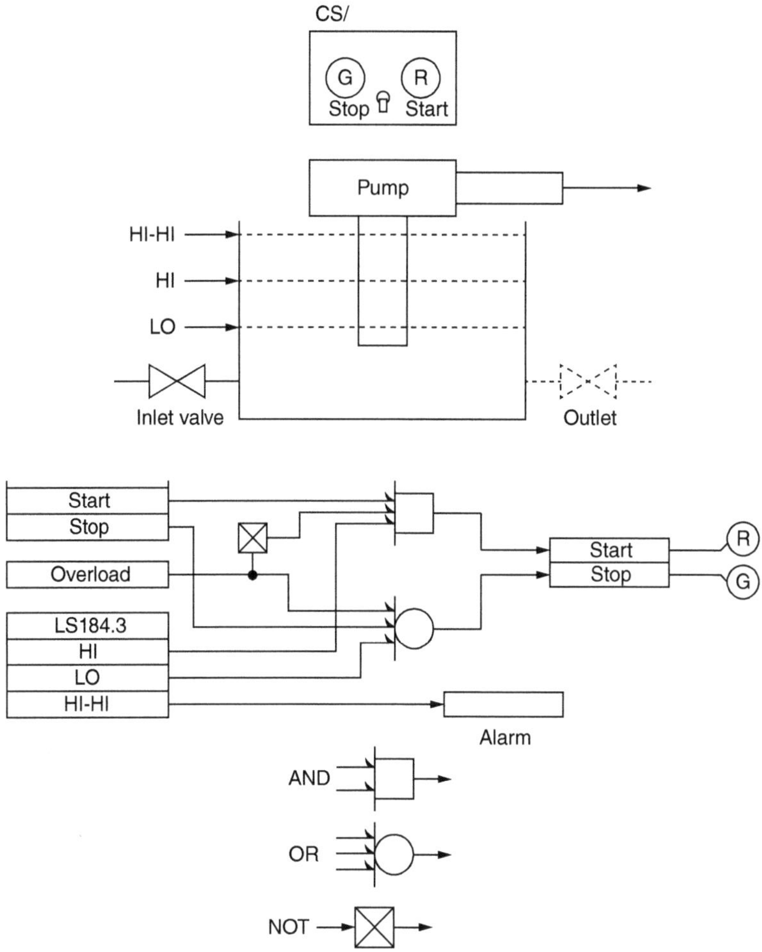

FIGURE 12-2 Typical simple pump installation with logic diagram.

Notice that the motor leads in Fig. 12-4 are labeled T_1, T_2, and T_3. The load or line leads are Ls. When it is said in the field or specified in the startup procedures to lift the "T" leads for run simulation, these are the leads that they are referring to.

> **Note**
>
> Pulling the fuses will de-energize the control transformer and prevent breaker operation. The 42 contacts and 49 thermal overloads are an integral part of the "contactor." The thermal overloads (heaters) can be changed to match the overload protection requirements of the motor. Heating strips in the contactor buckle when overheated, from over current, tripping the contactor. Therefore, they are referred to as heaters.

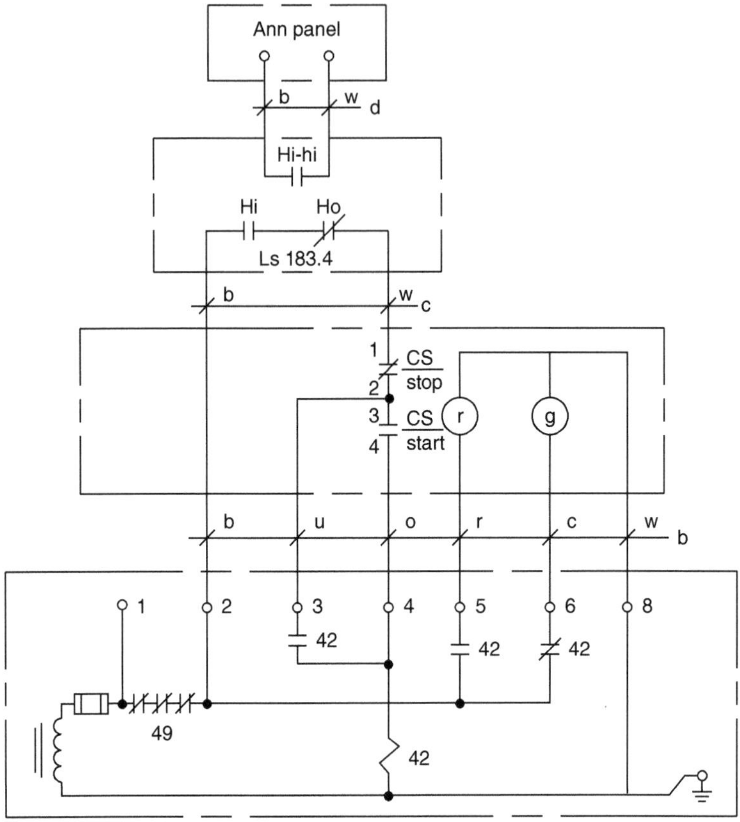

FIGURE 12-3 Control wiring diagram (CWD) with AC control.

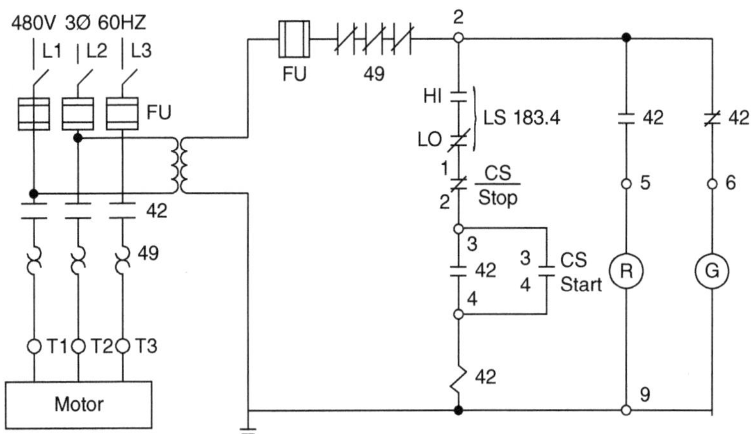

FIGURE 12-4 Elementary diagram with AC control.

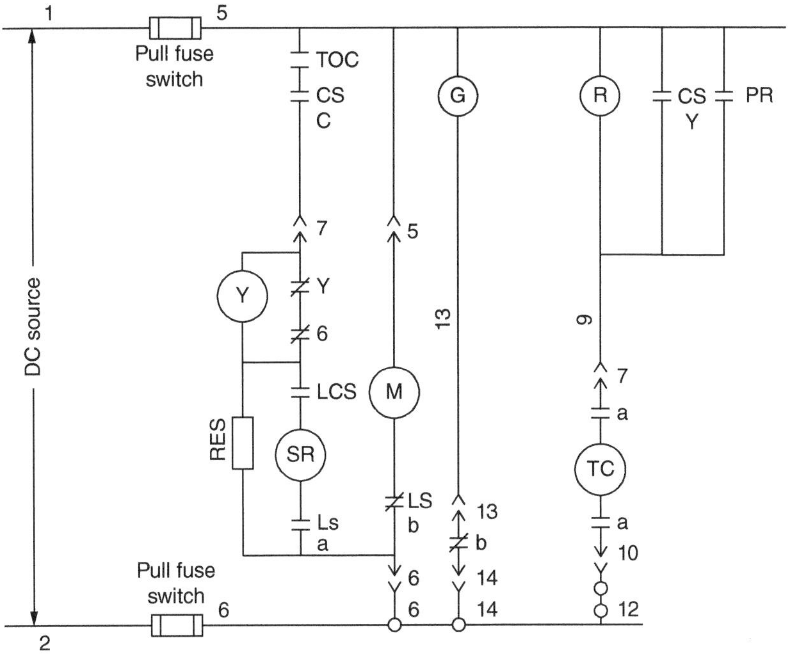

FIGURE 12-5 CWD with DC control.

Assuming a breaker tripped because the heaters were undersized, startup technicians have been seen holding the contactors in with a screwdriver, which is an example of a very bad assumption and a potentially lethal mistake if a solid fault had actually occurred.

Electrical logic and elementary diagrams are rarely provided unless specifically requested by the owner in the contract documents. Engineers work exclusively with control wiring diagrams (CWDs) and their associated connection and interconnecting wiring diagrams.

To add system interlocks to the breaker diagram, the architect-engineer's designer will extend the right side of the motor schematic. He or she will then change the symbols in the schematic to match his or her own design convention, The "legend" for the drawings will be on the first page of a drawing set and must be reviewed and understood in order to understand the drawings completely. No two companies use the same format.

If a PLC is used, only the final devices will be shown on the drawing (e.g., the process contacts "PR" in Fig. 12-5). Separate PLC logic will be rotated counter-clockwise 90° so the document can be scrolled as "ladder logic."

Earlier (before PLCs), the circles for the trip coil (TC) and start relay (SR) were replaced with a typical relay coil symbol as in Fig. 12-6. Today, the diagrams may resemble PLC ladder logic for conformity. All devices and contacts contain the main device designation 52. For example, the auxiliary switch contacts a, b, and c are 52_a, 52_b and 52_c. They are sometimes called MOCs, or mechanically operated contacts. A linkage to the breaker open-and-close mechanism operates the auxiliary contacts. The Y is replaced by a 52Y. The Y is a symbol for the anti-pump logic that prevents the breaker from reclosing if it trips while the start button is still depressed or the momentary start switch is closed. The SR is replaced with an X, for the spring release coil, and is 52X. Limit switches (LS) and

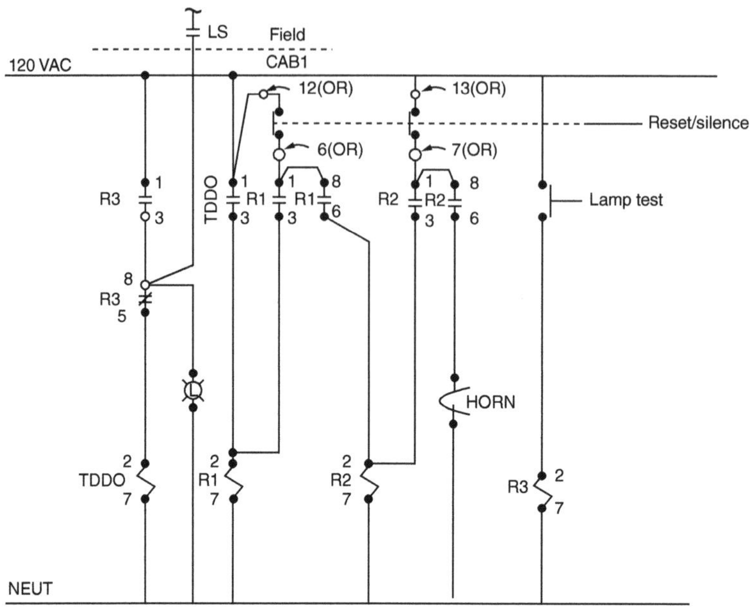

FIGURE 12-6 Alarm CWD with silence and reset.

interlock switches (IS) that are operated by the breaker position are 52LS and 52IS, respectively, and so forth, for all the devices. The symbol TOC stands for truck-operated contacts that make or break when the breaker is inserted or withdrawn from the cubicle. See App. G for further descriptions of devices and their operation.

Figure 12-7 is a sequence diagram for the schematic in Fig. 12-8 taken from a GE instruction manual.

All drawings are read for DC from + to – and AC from "hot" to neutral as in Figs. 12-5 and 12-6. The Fig. 12-6 diagram can be rotated counterclockwise and the traditional relay coil symbols replaced with "circles." Although it would remain "relay" logic, it would conform to the "ladder" logic depiction. Note the time-delay drop-out relay (TDDO). This relay is necessary in order for the "reset/silence" circuit to drop out when the alarm silence pushbutton is depressed. Also note that if the level switch (LS) were a pressure switch, a time-delay pickup (TDPU) relay could be used in that circuit to avoid nuisance alarms.

The pushbutton used in the preceding CWD has "stackable" contacts so that additional contacts can be added. The number of additional contacts used is determined by the contact springs. There will be a point when the button cannot be depressed because too many springs have been added. Therefore, if additional contacts are contemplated, a relay configuration should be substituted. In this case, a separate rung would be added that would have R4, and the pushbutton contacts would be replaced with the relay contacts.

Some other important terminologies are as follows: A relay that is used to power another relay is called an "interposing" relay. Contacts that are not powered by the same voltage source as the relay coil are called "dry" contacts. For example, a 24-V thermostat closes and powers a relay coil with 120-V or 240-V contacts to start an air conditioner. The air conditioner contacts are called dry contacts because they do not have the same voltage source as the coil.

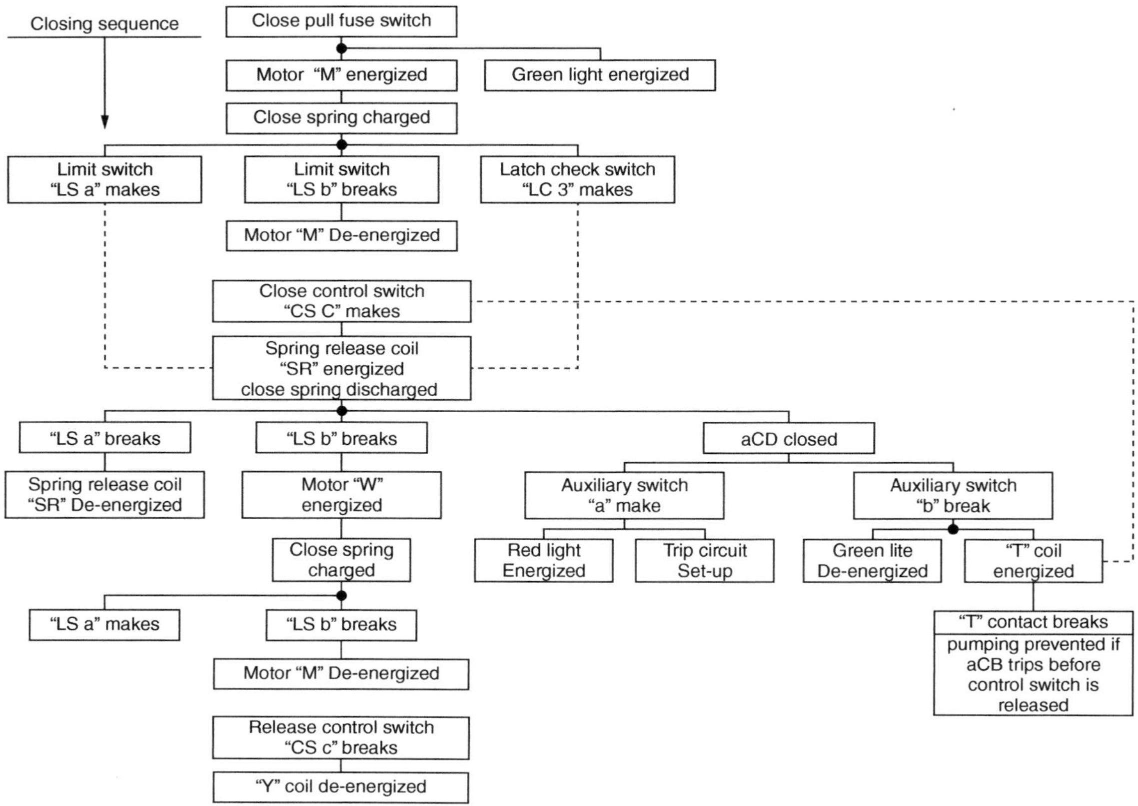

FIGURE 12-7 Sequence diagram.

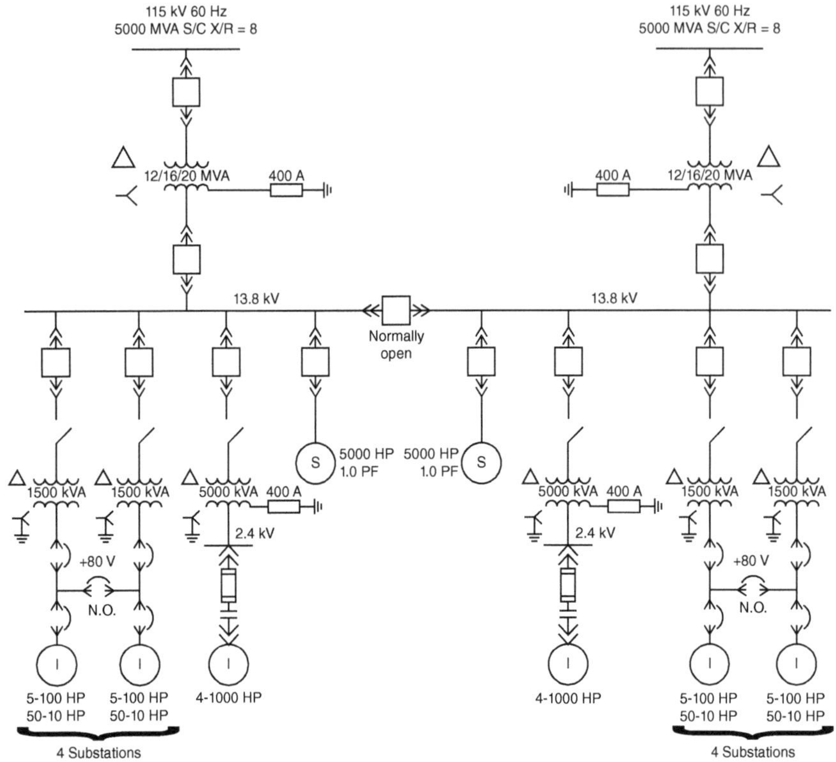

FIGURE 12-8 Preliminary one-line diagram.

Relays are self-contained devices that contain both coils and contacts. Although they are depicted separately on the schematics, they are physically in the same container. Ice cube relays are so called because of their small, clear plastic containers.

SINGLE-LINE OR ONE-LINE DRAWINGS

For design purposes, the one-line diagram is used in conjunction with the plant physical layout for system protection and fault current analysis. It is termed "single line" because it shows only one of the three power phases. The one-line drawing is the basis for most startup electrical testing. From it, the circuits and devices are yellow-lined complete as each is tested by startup.

> **Note**
>
> High voltage is 15,000 volts and above. Medium voltage is 601 to 15,000 volts. Low voltage is 600 volts and below.

A preliminary diagram is shown in Fig. 12-8 with the major current carrying devices. Figure 12-9 illustrates a developed one-line diagram in which the protective devices have been added.[3]

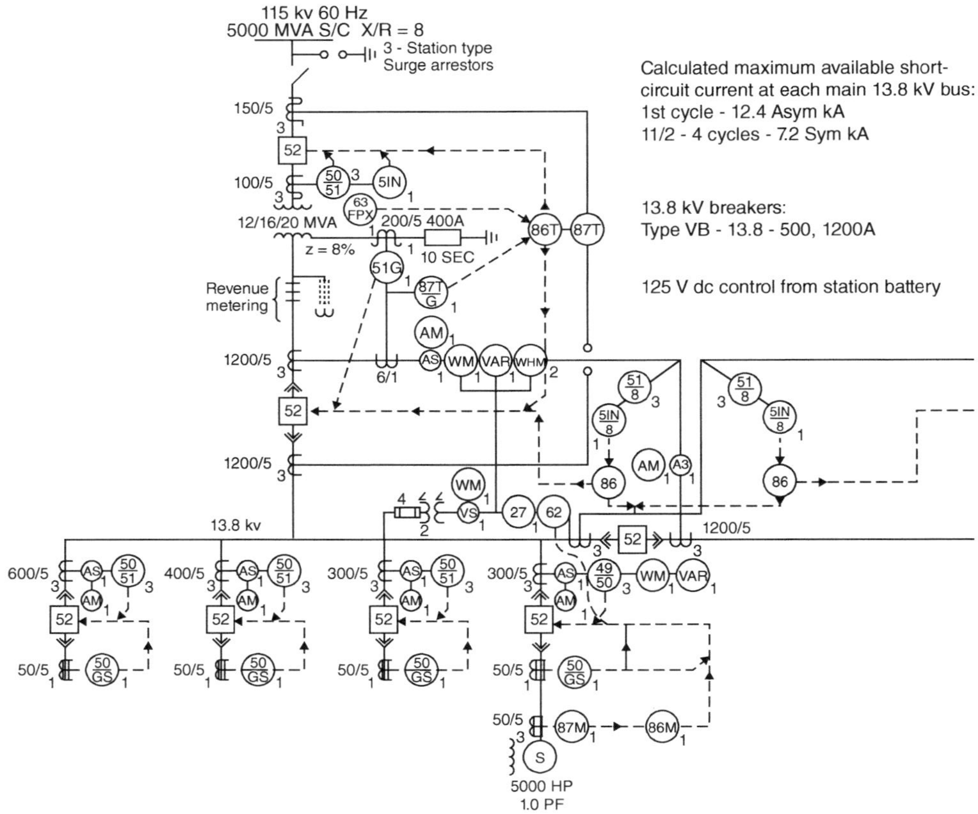

FIGURE 12-9 Fully developed one-line diagram.

All device numbers conform to ANSI/IEEE C37-2, Standard Electrical Power System Device Function Numbers.[4] A list is attached as App. G.

The preliminary single-line drawings in Fig. 12-10 illustrate three different types of electrical systems used in the industry: radial, loop, and ring bus.

The following items should be shown on all one-line diagrams:

(1) Power sources including voltages and available short-circuit currents

(2) Size, type, ampacities, and number of all conductors in kilovolt-ampere rating

(3) Capacities, voltages, impedance, connections, and grounding methods of transformers

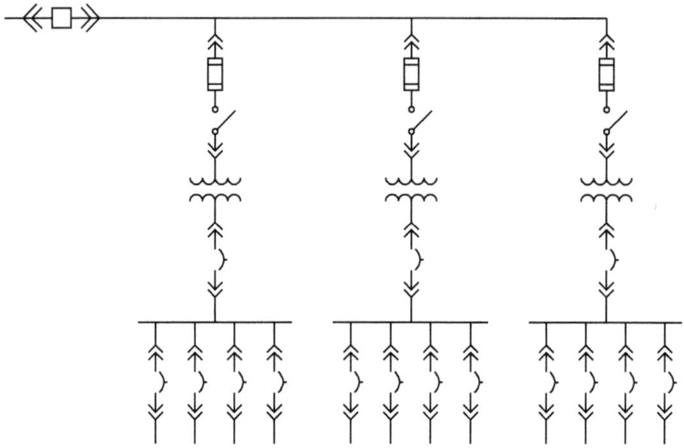

FIGURE 12-10*a* Expanded radial system.

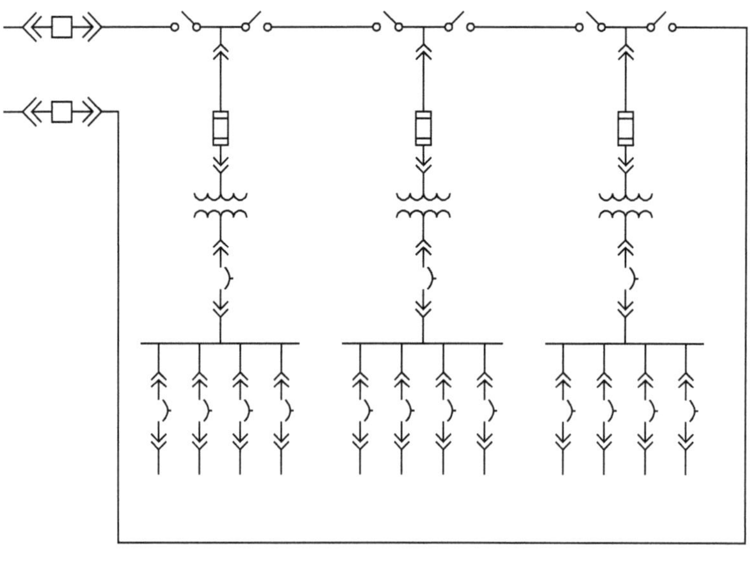

FIGURE 12-10*b* Primary loop system.

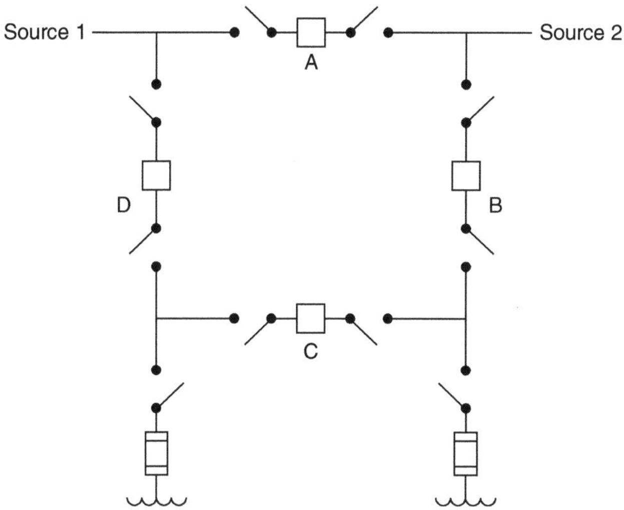

FIGURE 12-10c Ring bus system.

(4) Identification and quantity of protective devices (relays, fuses, and circuit breakers)
(5) Instrument transformer ratios
(6) Type and location of surge arrestors and capacitors
(7) Identification of all loads
(8) Identification of any other distribution system equipment
(9) Type of relays and relay settings on utility supply lines for both phase and ground faults

The one-line diagram should show proposed future additions, and the effect of such additions should be a part of the original system planning.

The actual drawing should be kept as simple as possible. It is a schematic diagram and need not show geographical relationship. Duplication should be avoided.[1]

ELECTRICAL CONSTRUCTION TERMS AND PRACTICES

There are many terms and practices used in electrical design and construction that will be discussed in startup meetings. The startup engineer should be familiar with these terms and practices.

Cable and Raceway Systems

Cable and raceway systems should be installed in accordance with the engineering drawings and the National Electrical Code, NFPA 70. Although the code exempts electric utilities from the code provisions, variance will not occur without engineering documentation.

Raceways—any system that carries conductors. For example, conduit, cable bus, cable ducts and cable trays are all raceways.

Conduit—metal tubing used as an electrical raceway. Electrical metallic tubing (EMT) is never used in power plant construction. It is thin-wall tubing and cannot support the

TABLE 12-1 NEC large conduit size and support requirements.[5]

Conduit Size (in)	Maximum Distance Between Rigid Metal Conduit Supports (ft)
1	12
1¼–1½	14
2–2½	16
3 and larger	20

load of an operator if stepped on. Therefore, Rigid conduit is used. It is heavy wall tubing and can withstand the rigors of an industrial environment. Conduit ½ and ¾ in wide must be supported every 10 ft and at least 3 ft from a junction box in accordance with the National Electrical Code (NEC). Larger conduit is supported in accordance with Table 12-1. The standard length of conduit is 10 ft versus 20 ft for pipe.

Liquid tight—flexible metal conduit with a plastic covering used to connect conduit to apparatus that may have movement. It cannot be longer than 3 ft from an entrance.

Strut systems—the metal framing, channels, hardware, and fittings used to support cable raceways. Channel is commonly referred to by the manufacturer's name, such as Unistrut, Superstrut, etc. Where industry standards and good workmanship are rigorously enforced, framing and channel are never cut and welded in lieu of using the proper attaching, splicing hardware, and fittings provided by the manufacturer of the channel. However, this is rarely enforced in developing countries and frequently disregarded in developed ones. The NEC does not spell out the use of either method.

Conduit bodies—fittings used for pullboxes and 90° bends. They are sometimes referred to as "Cond-O-Lets" or incorrectly called LBs. A type LB conduit body is illustrated in App. V. A way to remember the differences is an LB is a fitting with "linear" inlet and a "bottom" outlet. Type LL has a linear inlet and left outlet with the inlet facing away from the observer. Type LR is the opposite. Type C is a straight inlet-outlet body. The NEC limits the total combined equivalent quarter bends to 360° between access fittings. Conduit bodies are most often used to satisfy this requirement.

Entrance caps or boxes—used for aerial service entrances or motor cable entrance.

Explosion proof—refers to enclosures. If an internal explosion were to occur, the explosion would be contained and would not cause an ignition source outside.

Conduit seals—seals to prevent flammable or explosive mixtures from traveling from a hazardous space to a nonhazardous one.

Intrinsically safe—refers to a device that in normal operation, if faulted, will not produce sufficient energy to result in the ignition of a flammable or explosive environment.

Wires and cables—specified by type in accordance with their use. Wires larger than 4/O (0000) are normally called *cables*. The NEC appendices list all wire/cable types and their physical characteristics and ratings. Distributors will have a list of applications for all wires and cables.

Types TW and THW have a smooth, glossy surface that makes them easier to pull through conduit. They are preferred by construction engineers for low-voltage applications. Types THHN, RHH, and XHHW have higher ratings but are harder to pull. There are many more types that are selected by the engineer for their rating and physical exposure classifications.

Armored cable is commonly used for direct burial. Consideration for all direct burial cable should take into account ground settlement. If settlement occurs, the cable must not pull free of the entrance box or from terminations. This can happen in areas where the water table rises close to the surface during seasonal changes. Where the plant is located by a river or on a delta where the water table is close to the surface, cables must be installed in duct banks supported by piling. There is no way to prevent direct burial cable from rising and sinking with the rise and fall of the water table in such locations.

Note

Excavation had started before this photo was taken. If cable pullout occurs, it usually happens during startup, making it a commissioning problem.

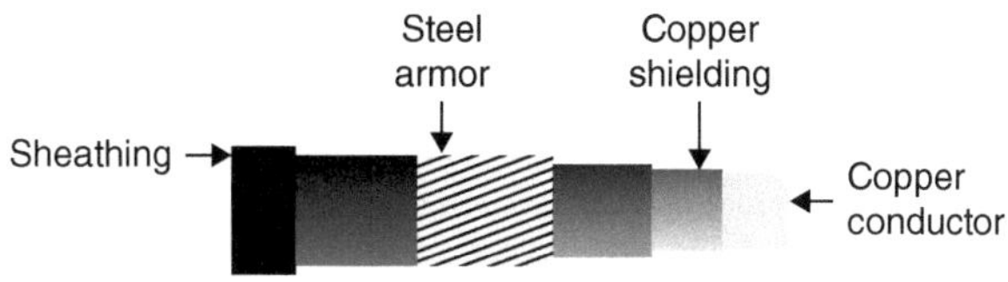

Typical XLPE armored cable

Direct burial should not be a design consideration if future cable and wire replacement is a maintenance consideration. This is especially true of control wiring where it would be impossible to re-pull a single damaged conductor.

Terminations—for medium- and high-voltage cables, must be installed so that electrical stress deterioration of the insulation does not occur. If the shield is abruptly cut,

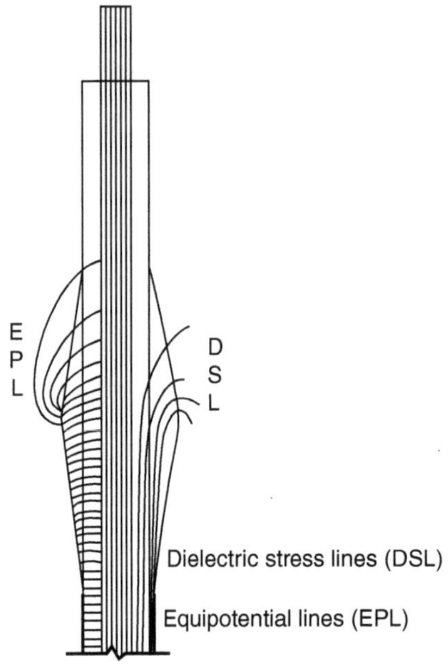

FIGURE 12-11 Stress profiles resembling magnetic field profiles.

there will be a sharp concentration of electrical stress that can destroy the insulation. A damaged stress cone will gradually change the stress from a normal equipotential profile to an insulation (dialectic) stress profile as illustrated in Fig. 12-11. Note that the dielectric line are cut short for this illustration to save space. The profile is represented as lines only for illustration purposes. Actually the profiles are similar to magnetic fields which are also commonly illustrated with lines.

Corona Phenomenon

The partial discharge problem can be described as follows:

The technology of testing and maintaining solid dielectrics has been greatly complicated by the long-term effects of partial discharge (also know as corona). As voltages have been pushed higher, and extruded plastic dielectrics have been developed to take the place of oil-impregnated paper insulation, partial discharge and subsequent treeing have added a new dimension to dielectric problems.

The problem is that it's very difficult to manufacture a solid dielectric that's totally uniform. Cable manufacturers go to extensive quality-control extremes to ensure that the plastic they extrude onto is pure, and is as free of defects and cavities as possible. But even if they achieve a perfect product, there is still the possibility of external mechanical damage causing a cavity in the conductors, or the semi conducting layer that surrounds them.

Because of the lower permittivity of the air in such a void, a high-voltage stress develops across it. As a result, air and gas within the void ionizes, and a gaseous discharge is created within the void. High-speed ions striking the walls of the void cause further erosion, which enlarges and extends the void. Ozone generated in the entrapped air adds further chemical attack on the void walls. This is the basic mechanism behind the formation of trees in solid insulation. These electromechanical trees grow slowly through insulation over a period of days or years, until they reach a point at which a complete discharge path is created, and the insulation breaks down catastrophically.

The cause of the tree growth in a particular situation are hard to pin down, though it is known to be accentuated by the presence of moisture and physical damage. It is virtually impossible to measure the amount of treeing in a length of installed cable and to predict its remaining useful life.

Partial discharge is essentially a gaseous phenomenon and is normally associated with high-voltage conductors suspended in free air, where it is often referred to as corona. The mechanism in this case is ionization of the molecules of air close to the conductor in regions of high electrical stress. These ions move away from the conductor into areas of lower electrical stress. The result is the familiar buzzing sound heard around transmission lines.

Corona is most likely to occur at a sharp point or irregularity on the high-voltage conductor, because the voltage stress is greater at points where the conductor surface is sharply curved. At night, light cause by corona discharge often can be seen around pots on a high-voltage system.

In gas-insulated equipment, corona cannot be allowed. Because of the nature of the gaseous dielectric used and the short clearances, corona would inevitably escalate into complete breakdown. Much of the success in designing gas-insulated equipment is in the design of conductors with smoothly curve surfaces, to avoid points of stress concentration. Quality control

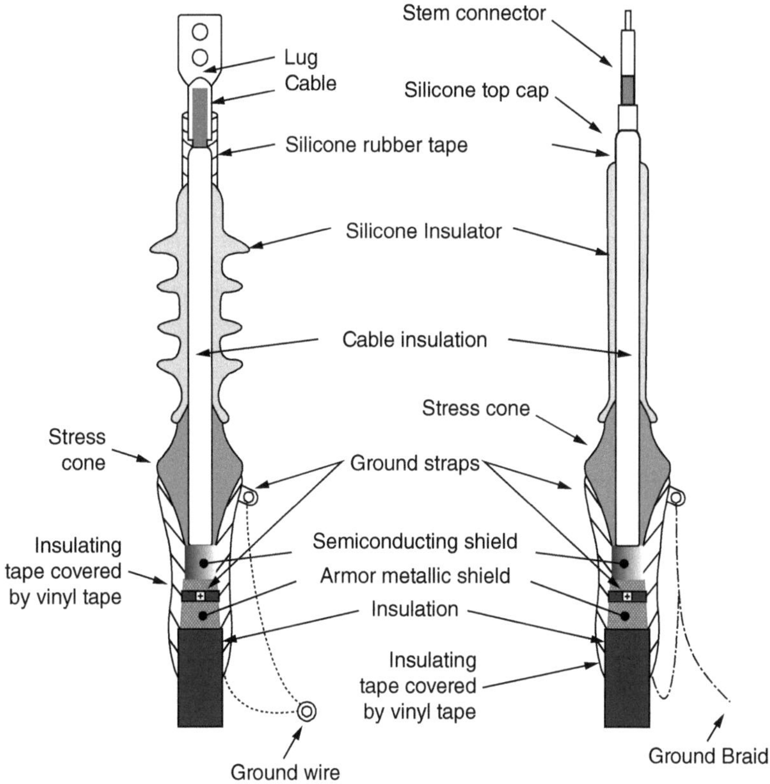

is also vital to ensure that no manufacturing burrs are left on machined parts, or metal debris left in the gas-enclosing chamber.[6]

Pot heads—poured porcelain type terminators for Class 2, nonpressurized moisture proof terminations.

Cabinets

Cabinets should be set on steel rails embedded into a "housekeeping" platform. They are set on rails to ensure that they remain level in order to prevent distortion. Concrete surface is rarely perfectly level after curing.

The NEC, in accordance with voltage and exposure, specifies clearances between cabinets and other components. Here is an example of a properly mounted cabinet with an improperly routed pipe run.

If an electrician were working in this cabinet, sitting on a bucket, leaning backward, his or her body could complete a ground-fault circuit via this pipe run. The NEC requirements are written to guarantee safety, not comfort.

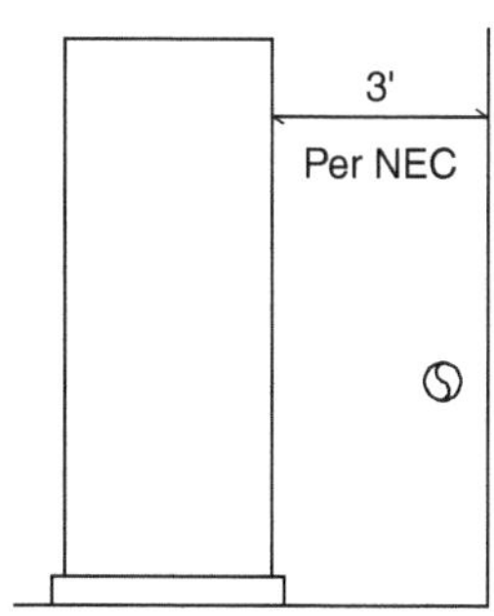

FIVE IMPORTANT BASIC ELECTRICAL STARTUP POINTS

Continuity

Continuity checks are the first electrical checks that must be made by construction to ensure that all wiring is terminated according to wiring drawings. This does not check drawing errors, but it will provide substantial aid in troubleshooting control circuits during operational tests by confirming that the wires are terminated at the point designated on the drawings. Note that a continuity check does not stop with the continuity verification from the control room to the breaker. Also, make sure that power cables terminate at the proper motor. Be cautious! Designations penned on a motor by craft or plant personnel as aids should always be considered incorrect; they must always be verified and corrected. This may seem remedial, but too often when one team checks the control wires and another the power cables, it invites a mishap at the time of initial energization.

Meggering Prior to Energization

Meggering of power circuits is another essential test prior to trial operation. It is used to determine circuit resistance to ground. This test will determine cable or winding damage, improper splicing, or the presence of moisture prior to energization. It uses current leakage to check insulation. A voltage is applied with a mega-ohmmeter to test the insulation resistance. This test should not be confused with hi-pot (high-potential) tests, which are used to determine insulation-resistance breakdown. Capacitance tests also check insulation-resistance breakdown without the potential damage possible with the high voltages used in hi-potting. Hi-potting and capacitance tests are usually limited to large cables, generators, and transformers and serve to determine before-and-after conditions related to shipment, installation, and service life. The latter will be an important part of the component history. It is essential that the resistance test data sheets be available prior to energization to ensure the tests have been completed.

Check Motor Protective Devices

Relays used in motor circuit protection must be set prior to system energization. These relays are not factory set because relay time-current curves must be matched with motor curves. To do this, the relay curves are matched using tap settings with the motor-starting curves by the engineer. These settings are given to the startup engineer for relay calibration. Microprocessor protection also requires the engineer's data. However, no calibration is necessary.

When hiring a third-party testing company to test motor protection and calibrate relays, the contract must be clear as far as what the calibration requirements really are.

Careful consideration must also be given to motor starting time. There is a big difference between starting a pump and a fan. A large fan will require a much longer time (as high as 10 times) to overcome inertial forces accelerating to operating speed than a typical large pump. The difference is compounded when utilizing "soft-start" technology, which lowers the starting voltage and amperage, thereby extending the starting time. Therefore, the startup engineer must develop a plan and implementation strategy beforehand, which will be the basis for actual trial operational field tests.

The question of whether or not a "black-start" generator set has the capacity to start a vital motor places another element in the equation. Black-start capability can affect the starting time and line voltage, which are additional factors considered using soft starters.

Check Control Circuits

After all preliminary checks are completed, it is time to test the control circuits. This test must check the design logic, the final drawing accuracy, and the field installation. The control circuit checkout must be accomplished prior to energizing power circuits. Testing for proper operation prior to system initial operation can be accomplished in some installations by installing jumpers across the breaker test-position switch terminals and installing mechanical devices on the breaker auxiliary open/close–position switch to simulate actual energized operation. On modern installations, test options are provided in which the auxiliary relays are connected via a test "umbilical cord" for that purpose. Motor starters can be placed in a similar mode by removing the "heaters." The best method, if the sequence allows, is simply to disconnect the power, T leads, in the breaker. These methods all allow the motor control circuit to be energized and operated prior to energizing the power circuit. If the leads are disconnected, it also ensures that the circuit will operate properly with the breaker in the "racked-in" position.

Because no power circuit will be energized, the control logic can be tested for proper operation at a relaxed pace. On complex circuits, it is imperative that a procedure be used so that no interlock is overlooked. The most important thing to remember is that the motor control portion of the startup procedure is separated from actual power energization. This greatly reduces the risks of energizing a power circuit without adequate motor control protection. It also simplifies the checkout procedure by separating the control circuit checkout from initial operation checkout procedures. This permits engineers to direct their full attention during initial operations toward the operation of the process equipment.

Check Motors for Rotation and Run-In

The final electrical check prior to initial operation is the motor rotational check. A 2-hr run is recommended so that the motor can stabilize at its operating temperature. After it has reached a stable temperature, baseline reference data can be taken for motor maintenance history. After the motor has reached operating temperature, the high-reading resistance temperature device (RTD) should be determined. The high-reading RTD then should be used for annunciation and tripping.

If the breaker trips instantaneously while closing during the first trial operation, the first thing to check is the instrument transformers' nameplate data and ensure that all transformers, for all three phases, are the same. Needless to say, that is the first thing that should have been checked during pretrial operational testing. Regardless, instrument problems seem to remain the principal cause of circuit malfunction during startup—way ahead of actual faults.

ELECTRICAL TESTS

An excellent resource for electrical testing for the electrical startup engineers, as well as startup management is Gill's *Electrical Equipment Testing & Maintenance*.[7]

The following standards should be available to the startup engineer:

- IEEE 43-2000, "Recommended Practice for Testing Insulation Resistance of Rotating Machinery."
- IEEE 62-1995, "Guide for Diagnostic Field Testing of Electric Power Apparatus—Part I: Oil Filled Power Transformers, Regulators, and Reactors."
- IEEE Standard 48-1996, "IEEE Standard Test Procedures and Requirements for Alternating Current Cable Terminations 2.5kV Through 765 kV."

- IEEE Standard 400-2001, "IEEE Guide for Field Testing and Evaluation of the Insulation of Shielded Power Cable Systems."
- IEEE P400.1, "IEEE Guide for Making High Direct-Voltage Tests on Power Cable Systems in the Field."

GROUNDING

In a generating station, there are four grounding systems tied to the ground grid: (1) lightning; (2) station service power; (3) equipment, defined in IEEE Standard 665; and (4) instrumentation and control system grounding, defined in IEEE Standard 1050.

Soil-Resistivity Tests

Ground-grid design must adhere to the applicable codes and standards such as IEEE 80, Guide for Safety in AC Substation Grounding. In order to design the grid, the engineer first must have all pertinent soil-resistivity information. This information should have been acquired by hiring an experienced firm that provides reputable test data to be used for the entire plant grounding system. If necessary, these tests will be performed to account for seasonal changes in soil conditions, not just a "one-shot" test in the summer.

If there is a question about the adequacy of the grounding grid due to failure of the ground potential tests, it is recommended the Wenner, four-point method of testing the ground resistivity be performed to compare design test data with actual site conditions. The test is described in detail in IEEE 81, Guide for Measuring Earth Resistivity, Ground Impedance and Earth Surface Potentials of a Ground System.

1. Place small electrodes in a straight line at intervals a. The depth should not exceed $0.1a$. The spacing should be equal in a straight line. For example, place electrodes at 20, 40, and 80 ms.
2. Pass a current through the outer two electrodes, and measure the voltage between the inner two electrodes.
3. Calculate the resistivity as follows:

$$\rho = 2\pi a R$$

where $R = V/I$

Grounding System Resistance Tests

The fall of potential method, or three-point test method, is the fundamental method for measuring the resistance to earth of a grounding system. Ground-resistance testers are commercially available with kits including leads and electrodes. The tester will be supplied with an instruction manual describing in detail how to perform the ground-resistance test.

The fall of potential or three-point test, is a simple test using two test rods (Y and Z) driven into the ground to a depth of 9 in. They are placed in a straight line from an electrode (X) of the grounding grid to be tested. Rod Z is driven, preferably, 10 times the diagonal length of the grid. Drive probe Y in the ground near X and every 10 percent of the distance to the outer probe Z in a straight line. Follow the instructions provided with the tester. Plot resistance versus distance. Ground resistance of 0.05Ω to 0.5 Ω should be expected.

INSULATION (DIELECTRICS)

How do dialectics fail: There are several mechanisms that can cause dialectics to break down. Under normal conditions, dielectrics insulate because, unlike metals, they have very few free electrons available to conduct current. If the voltage stress across the dielectric is increased beyond a certain point, additional electrons are torn from atoms by collisions with the fast-moving free electrons. These additional electrons move from negative to positive potential, forming an increased current through the dielectric. Heat generated by this current aids in the release of more electrons, resulting in the cascading effect of breakdown.

When an arc is struck through a liquid dielectric such as mineral oil, gas is released by decomposition of the oil. Ionization of this gas helps to maintain the arc. At the same time, solid particles may line up under the influence of the electric field, creating a path along which an arc may be struck through the liquid. This is one reason why liquid dielectrics must be regularly tested and maintained.

An arc struck through, or over the surface of, a solid insulator leaves a permanent scar on the surface, or punches a pinhole right through the material. Unlike liquid or gaseous dielectrics, the damage in not automatically repaired when the arc ceases. However, the effect may not be as immediately catastrophic as it is in a gas. A moderate inoffensive arc may continue for some time through the pinhole in, say, a motor-winding insulation, without causing any immediate further damage. But it's just about inevitable that at some time, usually when a switching voltage surge enters the motor, this minor fault will escalate into a major short circuit, possibly causing considerable damage to the motor. That's why sensitive equipment that can detect minor faults is an essential part of electrical protection.[11]

All major plant wiring, cabling and electrical apparatuses will have some sort of dielectric material that will require testing. Two examples of dielectric material to be tested are cable insulation and transformer oil. The presence of acetylene in transformer oil samples is a sign of arcing damage.

Insulation Resistance Tests

This is a DC nondestructive test to check insulation resistance called "meggering." The term *megger* is a registered trademark of AVO International, formerly the James G. Biddle Company, but it is used throughout the industry regardless of manufacturer. The tester reads millions of ohms resistance or megaohms. Generally any reading less than 1 megaohms/1000 V (check your specs) indicates problem insulation. This is normally done on the cable "reel" before installation and then after installation. After installation, cables >4000 V are insulation proof-tested using a DC high-potential (hi-pot) tester. This test is performed regardless of whether the Megger tests passed or failed. *Occasionally, a Megger test will fail, yet when testing at a higher voltage, using a more exacting procedure the test will pass. This may be due to surface leakage.* (See use of the "guard" tap on the Megger in the following setup procedure.) The hi-pot test will be conducted for cables using 80 percent of the factory final-test voltage. Maintenance tests use 65 percent of the factory test voltage. Therefore, the cable specification must be checked and the supplier advised of the test voltage to be applied so as not to void the cable warranty.

All motors should be meggered prior to being energized. If the motor fails the test, it should be heated to drive out moisture and retested. A rule-of-thumb for motor and generator hi-pot test voltage is two times the rated voltage plus 1000. Therefore, if a motor rated voltage is 4160 V, the hi-pot voltage will be 9320 V. Table 12-2 shows Megger test voltages commonly used. Temperature correction base for rotating equipment is 40°C.

Wire, cable, and transformer Megger tests must be corrected from ambient temperature to 20°C. During commissioning, the exact values are not as important as the trend over

TABLE 12-2 Commonly Used Megger Test Voltages[12]

System Voltage	Megaohms	System Voltage (kV)	Megaohms
216	1.2	34	35
480	1.5	69	70
2400	3.4	138	239
4160	5.15	230	231
7200	8.2	345	346
13800	14.8	500	501

time for maintenance. Therefore, if the correction during startup is not made, at least the ambient temperature should be recorded on the test data sheet for future trend evaluation.

A thorough, initial maintenance test will record the resistance for at least five temperatures of the same device or cable. The results should be a straight line on semi-log paper because the relationship is exponential.

The value of the resistance reading is not as important as the trend. If plotted on semi-log paper, the readings break radically from the straight-line trend, and insulation problems are likely. But generally, any reading below 1 megaohm is a sign of insulation breakdown.

For each temperature, a Megger reading was taken for the same equipment that yields a correction curve for that equipment, such as a cable or motor winding. See example Fig. 12-12. The corrected curve falls off gradually with time. Any sudden drop will indicate impending insulation breakdown. See Fig. 12-13.

Voltage Rating of Cables	Mega-ohmmeter Voltage
Up to 300 V	500 V
300 to 600 V	500 to 1,000 V
2,400 to 5,000 V	2,500 to 5000 V
Above 5,000 V	5000 V

The test setup for Megger and hi-pot testing is fairly simple. See the cable test connections in Fig. 12-14. The "guard" tap on the tester is normally not used during startup. It is

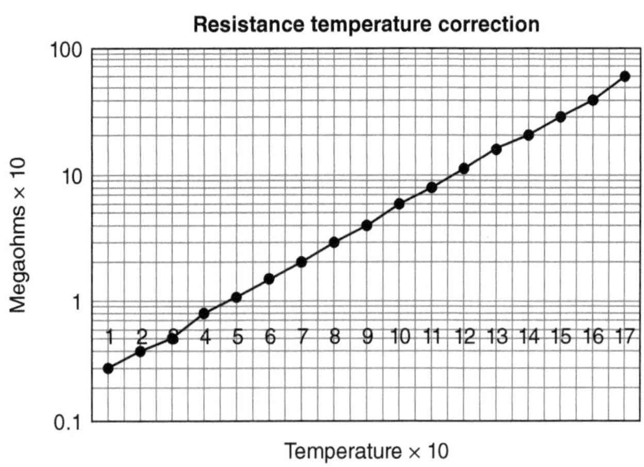

FIGURE 12-12 Resistance megaohms versus temperature.

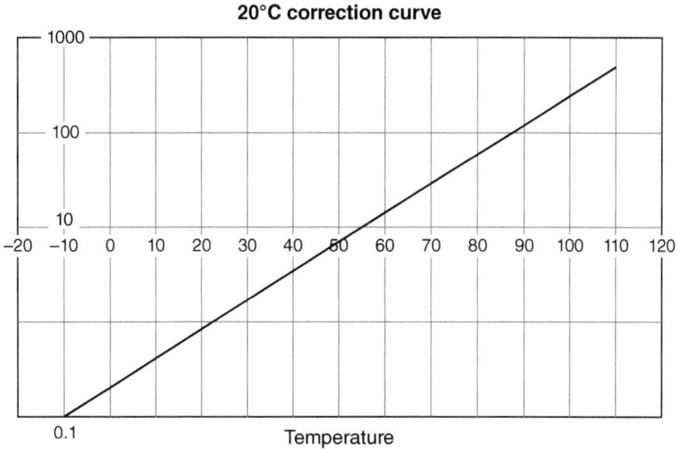

(a)

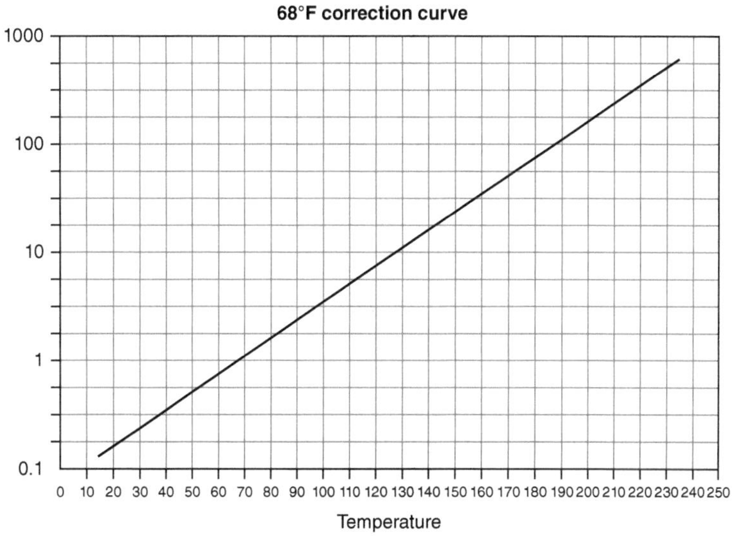

(b)

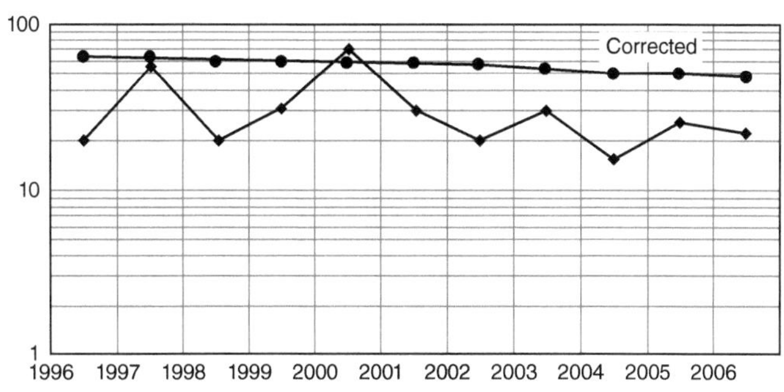

(c)

FIGURE 12-13 Megaohm vs temperature correction.

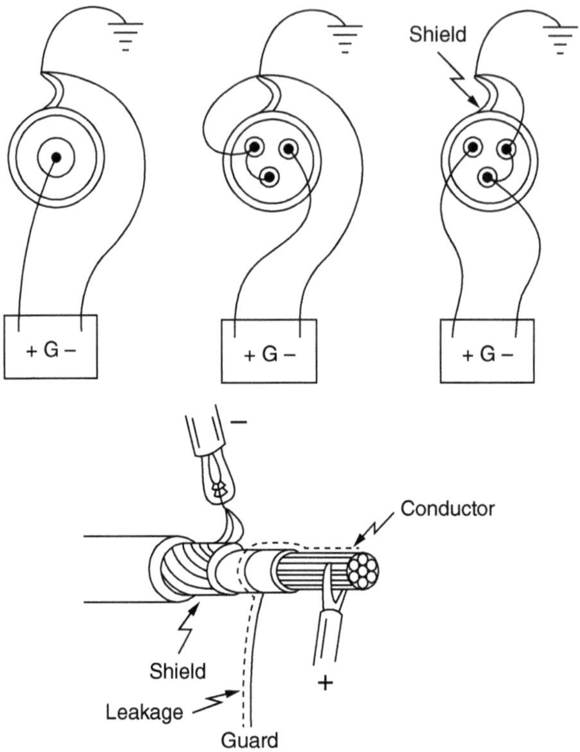

FIGURE 12-14 Megger connections.

used if the cable end insulation surface could have been contaminated such that a leakage current occurred from the conductor over the insulation to the neutral connection. To use the guard, twist a copper wire around the insulation and connect it to the guard lead. This will eliminate any surface leakage that could adversely affect the readings. Tester taps L and E are equivalent to L and N or plus and minus for the + and – leads.

PROOF-TESTING

> **Caution**
>
> Proof-testing should not be conducted without consulting the manufacturer. If that is not possible, then proof-testing should be limited to AC high-potential tests. In some cases, DC high-potential tests have been known to damage otherwise acceptable cable.

The most common proof-test is the high potential test (hi-pot) for startups. For both startup and maintenance, the most common hi-pot is the "step voltage test." The test voltage is applied in equal percentage steps and held for 60 s at each step up to the maximum voltage. At each step, a milliamp reading is taken. Usually, the number of steps is not less than eight. At the maximum voltage, milliamp readings are taken at 30 s and every minute thereafter

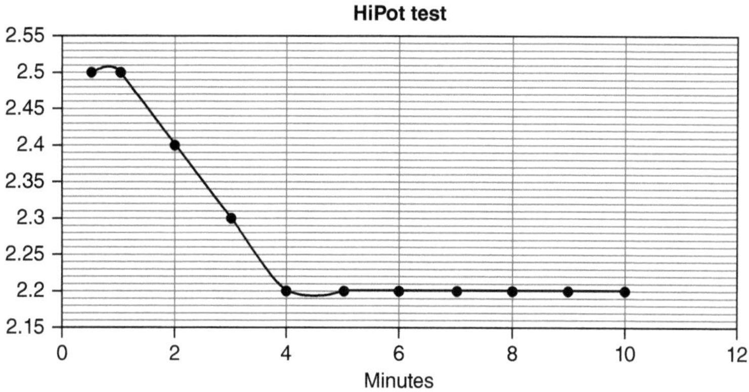

FIGURE 12-15 High-Pot test results.

for at least 6 min. A curve is plotted for leakage current versus time. The current leakage represents the insulation resistance. If there is no increase in current and the current steadily decreases, the cable is good.

If plotted on normal graph paper, the drop can be magnified by modifying the *Y*-coordinate scale. A continuous rise indicates insulation breakdown. The initial high point is the result of charging current not leakage current (Fig. 12-14).

The step and decay are performed together. The step test takes the voltage from zero to the final voltage: for example, 2, 4, 6, 8, 10, 12, and 14 kV. At each step, a milliamp reading is taken. At 14 kV, the milliamp decay is read at 30 s and the at 1, 2, 3, 4, 5 and 6 mins.

The charging current generates the initial current rise that occurs after the test voltage is applied. This current will drop off as the conductor insulation becomes charged to the applied voltage. After that, the current will remain constant or will bleed off very slowly. If the current rises, it indicates current flow through a failure in the insulation. Ohm's law applies ($V = IR$). If measuring resistance, it initially will be low; then it will increase as the insulation is charged. If the resistance drops, it indicates current flow through a failure in the insulation.

When proof-testing, always be prepared to replace the cable being tested. The test is used to test for incipient cable breakdown. If the cable breaks down, it must be replaced. If replacement will cause an unscheduled outage or significant intolerable delay, then the test should be rescheduled until the proper preparations can be made. See Table 12-3. This is a table of DC hi-pot voltage values for ICEA cables from 1968 and later.[1]

TABLE 12-3 Hi-Pot Values

Insulation Type	Insulation Level %	5 1	kV 2	15 1	kV 2	25 1	kV 2	35 1	kV 2
Elastomeric: Butyl and	100	25	19	55	4149	80	60	–	–
oil base	133	25	19	65		–	–	–	–
Elastomeric: EPR	100	25	19	55	4149	80	60	100	75
	133	25	19	65		100	75	–	–
Polyethylene, including	100	25	19	55	4149	80	60	100	75
cross-linked polyethylene	133	25	19	65		100	75	–	–

Note: Colum 1—installation tests made after installation but before service; column 2—maintenance tests, made after cable has been in service.

These test values are lower than for pre-1968 cables because the insulation is thinner. Hence, the AC test voltage is lower. The DC test voltage is specified as three times the AC test voltage, so it is also lower than for older cables.[1]

Cable Fault Locating

A hi-pot test can be applied to underground cables to locate faults. The "thumper" uses much higher voltages than the normal hi-pot because the intent is to cause a failure that is audible. The test is referred to as a "thumper" because of the muffled sound the fault makes when the high voltage discharges under ground. Generally, it will take many voltage discharges as people walk the underground right-of-way listening for the thumping noise. There are other, commercially available fault detectors that can be used, or should be used, before the thumper test is used.

Plant Component Testing

All electrical equipment must be tested before it is energized for the first time. If construction has claimed that a test was already completed without documentation, then it must be repeated by startup. This rule should not be broken unless proper testing documentation can be provided. Factory testing is not a substitute for field-testing. *Shipping, handling, storage, and installation can all have damaged, loosened, or changed calibration and settings of electrical apparatus.*

Wire and Cable

All wires and cables must be identified and marked in some way by construction. They are then "rung out" to determine if the right wire or cable is landed in the correct location. The method used is by earphones, a bell, or a continuity tester. The positive lead of the test device is clipped to a cable to be identified, and the common side is connected to ground. At the other end, a lead connected to the same ground system is touched to the cable; if there is continuity, the ear phone will click, the bell will ring, or the tester will go from infinity to 5Ω or less. Construction will also Megger test all power wires and cables >400 V and hi-pot all medium- and high-voltage cables to ensure they were not damaged during installation.

MOTOR STARTER AND SWITCHGEAR TESTING

Grounding for Starters and Breakers

The grounding system for starters and breakers must be a solid, low-resistance one. The ground should go directly to the grounding grid, not via structural steel. A poor ground is worse than no ground because it gives the appearance that the system is safe when it really is not. Therefore, the first test to perform before beginning any other tests is the ground-resistance test. All grounds from the equipment to the ground "grid" should have <5Ω resistance.

Instrument Transformers

The general consensus is that the secondary neutrals for these devices should be connected to ground at the entrance to the relay room rather than at the device. See ANSI/IEEE C57.13.3 [B2][9] for additional information.

The nut is loose on this grounding bus interconnection between cubicles. This may not have been noticed without a micro-ohm or ground-resistance test of the bus.

Motor Starters

All starters and cubical contents are checked for cleanliness, damage, tightness of connections, and proper operation. All interlock functions are tested by triggering field devices or by signal generation with the T leads lifted or the breaker in a "full" test position. The breaker contacts are micro-ohm (low-resistance) tested to ensure good contact and historical signature. New contacts should have in the neighborhood of 2 μΩ resistance with 200 μΩ acceptable. They are then overload trip tested by primary injection. *Personnel from a reputable electrical testing company should conduct primary injection tests. The test equipment and/or procedures exceed the experience of the typical electrical commissioning engineer.* Primary injection tests the overload devices. Early discovery of breaker problems can reduce delays during crucial startup operations.

Switchgear

Switchgear and bus bars should not be DC hi-pot tested because the sharp edge in the breaker and bus connection will produce corona and tracking, which can damage the installation. A micro-ohm resistance (Ductor) test of breaker contacts and bus connections will determine contact and bus integrity. Micro-ohm testing of the contacts on an annual basis will trend deterioration.

PROTECTION DEVICES

Relays

All electromechanical relays must be tested and calibrated. The switchgear should be tested for trip timing via the protective device.

- Long-time pickup.
- Long-time delay.
- Short-time pickup.
- Short-time delay.
- Instantaneous pickup.
- Instantaneous delay.

These devices should also be checked only by personnel from a reputable electrical testing company. The test equipment and/or procedures exceed the experience of the typical electrical commissioning engineer.

Electromechanical relays have operating characteristics that apply only to that particular relay. For example, the coil that drives the operating mechanism, or disc, moves proportional to the current applied to the coil. (The disc is similar to late-model utility revenue meters.) No two relays have exactly the same characteristics. Therefore, each relay has to be calibrated for the pickup current and the speed at which the disc will reach the trip point. The pickup point is the point at which the disc starts to move. The time the disc takes to rotate and close the contacts is the time-delay setting. Each relay class will have a curve associated with it that matches the pickup/time characteristics of the coil–disc operation. The relay curve must be matched with the operational curve of the equipment that it is protecting, such as a motor.

Each relay can have more than one curve to select from. Each curve is associated with a relay "tap"; by changing a tap setting, the appropriate time-current curve is selected.

Figure 12-16 is a typical field data sheet. The curves represented as dashed lines are the characteristic curves for only one relay model. One is a relay model selected by the engineer for a pump motor and the other for a fan motor. The curves represented by the solid lines are the starting characteristic for the motor. If the full-load current, starting time, and current are known, the "across-the-line" operation curve can be drawn with the relay curve overlaid on the graph. The in-rush is the locked (blocked) rotor current, and the 1.15 is the motor service factor current rating.

The paper used is semi-log in order to represent the millisecond readings and the large integer readings on one graph.

Microprocessors

Microprocessors have revolutionized protection. Electromechanical devices require special skills to set, whereas simply inputting the engineer's settings into an interface module can set microprocessor devices. An 80-ms trip setting is only an input into a menu cell for a microprocessor device. For a relay, set the time dial and time the closure until the desired results are obtained. Another advantage is that microprocessor settings do not drift with time or vibration.

NEMA MG-1, Section III Large machines, Part 20 Large Machines-Induction Machines, Subsection Test and Performance 20.8 Temperature Rise, sets the standards used in the following example of commissioning this large vertical turbine pump motor. Applicable sections of MG-1 should be consulted for different motors and updates. However, all data required should be supplied by the motor manufacturer's specifications included in their instruction manuals.

Figure 12-16 is a typical field data sheet used for the initial run of a large motor using a soft starter. The rectangle on the current-time curve represents the protection blockout for initial in-rush during motor starting. The soft-starter settings must be changed if the motor cannot start within the overall plant design criteria. For example, if the motor in-rush trips a black plant generator-set because of line undervoltage due to overload droop, the start time, initial voltage, or both, must be changed.

The insulation class (F) for the motor in Table 12-4 allows it to run at its service factor at an ambient temperature far greater than 40°C.

Air-cooled motors have standard NEMA ambient temperature ratings of 32 to 104°F (0–40°C) and 50 to 104°F (10–40°C) for water-cooled motors. For service in ambient conditions other than the standard [>104°F (40°C)], the motor must be derated by operating it at a lower load. NEMA sets the insulation class that governs the operating temperature.

NEMA describes two temperature measurement methods: one by thermo detector and the other by RTD. The total temperature allowed is the rise + standard.

The box indicates the setting field in a microprocessor protective device. No curves or calibration are necessary. Pickup and time are verified, but no calibration is required.

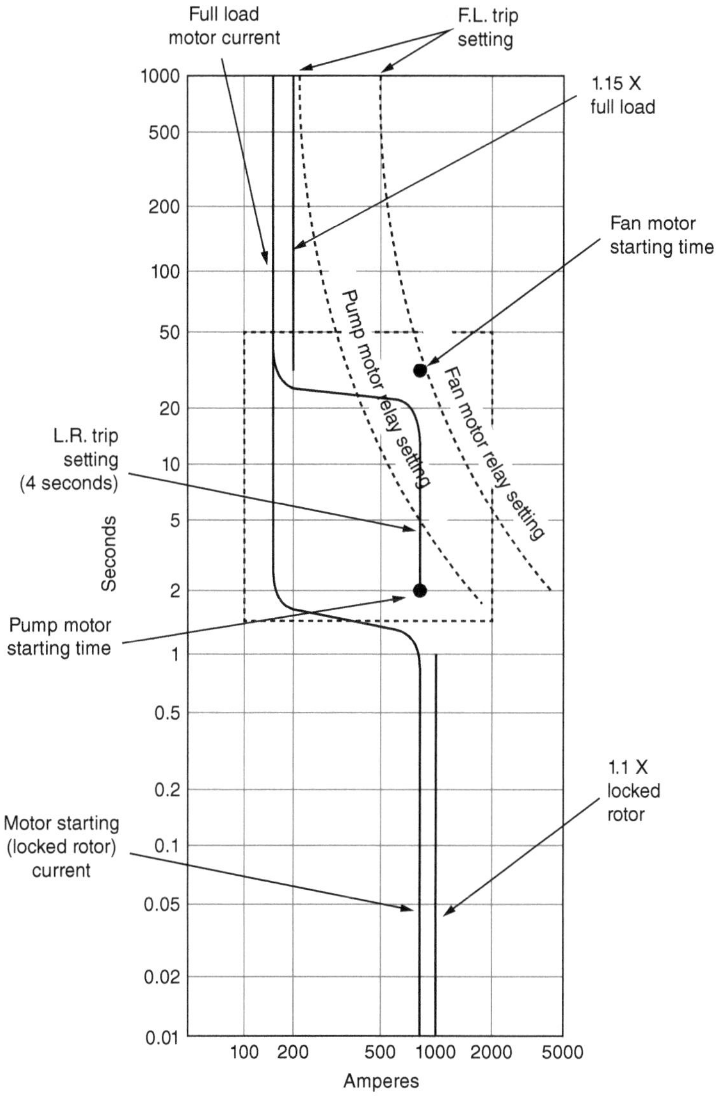

FIGURE 12-16 Time-current motor start curves.

TABLE 12-4 Data from NEMA MG-1.

Insulation Class	Ambient Standard	Measured Rise	RTD Rise
A	40	40	65
B	40	65	90
F	40	90	115
H	40	115	140

Microprocessor settings can also be placed into groups. For example, group 1 could be the "actual data" and group 2 could be the "factory default" data. Or group 1 could be "soft-start" data and group 2 could be "across-the-line" data.

Motor Protection

Application of motor protection is the extent of the startup engineers' expectancy with regard to verifying the match between the apparatus and protective device. For motors, there are five characteristic categories that have to be determined in order to input settings into a protective device, aside from voltage and frequency. The data are found on the motor nameplate and the engineer's or manufacturer's specification data sheets.

Another "heat"-related motor classification is the service factor. If motors are specified with a service factor higher than unity (1.0), such as 1.15, it will allow the motor to run hotter for limited periods due to the higher load capability. A service factor of 1.15 will give a 10°C tolerance. This should be taken into consideration when evaluating the insulation class. As can be readily seen, the service factor and insulation class are related. Consider special lubricants if the ambient temperatures exceed 120°F when considering the motor temperature rise. A startup engineer must be conscience of the motor characteristics. For example, in cold weather, insulation service factor may not be an issue, but if there are severe climate changes, this can become a serious concern.

Switchgear

All connections in the switchgear and bus must be checked for tightness. Typical torque values for different size bolts can be found in App. S, but it is highly recommended that the values provided in the manufacturer's instruction manual be used. Fastener torque values vary by type, grade, and manufacturer. The breaker contacts and the bus connections are then micro-ohm (low-resistance) tested to ensure good contact and historical signature. Contacts should be tested using 100A-DC minimum for best results. The expected contact resistance is between 200 and 250 $\mu\Omega$. Bus resistance should be $<20\ \mu\Omega$.

It should be noted that vacuum-bottle contacts are rapidly replacing air contacts because of significant reduction in capital and maintenance costs. The vacuum-bottle contacts require only maintenance checks and recording of the external contact wear indicator readings. Air at 40°C is dielectric, but at 4000°C it is a conductor; therefore, an arc will be produced and sustained when breaking contacts in air until the voltage dissipates or the arc is extinguished. Vacuum bottles can be expected to last for more than 1 million operations, or the life of the plant, because no arcing occurs across the contacts.

The breakers should be test tripped by operating all protective devices. All interlock functions are tested by triggering field devices or by signal generation. This usually has to be done with more than one breaker at a time. The starting and tripping of boiler fans, using a typical example, will require both the A and B fan breakers, for both the induced draft (ID) and forced draft (FD) fans, to be in the test position in order to check all interlock functions. Large, central stations may have as many as eight pair of ID and FD fans.

Soft-start motors will have factory settings for time-sequence starting. It is important that the startup engineer determine whether the sequential starting overriding criterion is to minimize across the line in-rush or to minimize hydraulic surge and shock-wave propagation in the process system. If the main design criteria are for acceleration and deceleration control, to protect the process system, then the sequential setting may be different (longer time settings) than for in-rush control.

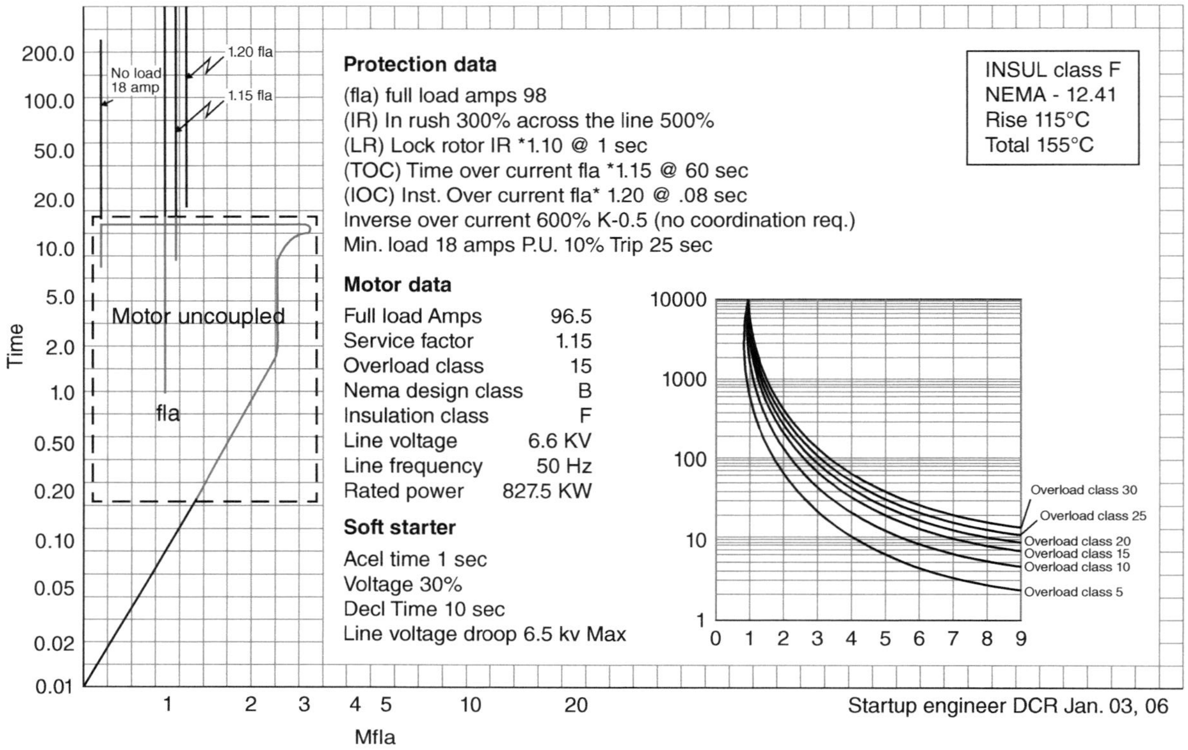
Protection data
(fla) full load amps 98
(IR) In rush 300% across the line 500%
(LR) Lock rotor IR *1.10 @ 1 sec
(TOC) Time over current fla *1.15 @ 60 sec
(IOC) Inst. Over current fla* 1.20 @ .08 sec
Inverse over current 600% K-0.5 (no coordination req.)
Min. load 18 amps P.U. 10% Trip 25 sec
Motor data
Full load Amps 96.5
Service factor 1.15
Overload class 15
Nema design class B
Insulation class F
Line voltage 6.6 KV
Line frequency 50 Hz
Rated power 827.5 KW
Soft starter
Acel time 1 sec
Voltage 30%
Decl Time 10 sec
Line voltage droop 6.5 kv Max
INSUL class F
NEMA - 12.41
Rise 115°C
Total 155°C
No load 18 amp
1.20 fla
1.15 fla
Motor uncoupled
fla
Time
200.0
100.0
50.0
20.0
10.0
5.0
2.0
1.0
0.50
0.20
0.10
0.05
0.02
0.01
1
2
3
4
5
10
20
Mfla
10000
1000
100
10
1
0 1 2 3 4 5 6 7 8 9
Overload class 30
Overload class 25
Overload class 20
Overload class 15
Overload class 10
Overload class 5
Startup engineer DCR Jan. 03, 06

INSTRUMENT TRANSFORMER

Current transformers come in many different types: wound, bar primary, window, bushing, double-secondary, split-core, air-core, tripping, and auxiliary. The window type, often referred to as a "doughnut," is the most commonly used for motor protection. The primary is the actual, current-carrying conductor that extends through the transformer window. The standard secondary rating is 5 A. Primary injection for field-turns ratio testing is not practical. When used in conjunction with electromechanical devices, the secondary circuits are tested, if transformer terminals are provided, to ensure the circuit/device impedance or "burden" (circuit resistance plus device reactance) does not introduce an error in the calibrated relay pickup.

This diagram illustrates, via a single-line diagram, proposed test zones.[12]

An important safety precaution is that the transformer secondary terminal must be shorted to ground to maintain the coil discharge, prior to disconnecting the secondary circuit. Instrument transformer nameplate data must be checked for conformity with engineering documentation to ensure that all three phases have the same transformer ratios.

Potential transformers should be turns-ratio-tested and the circuits checked by secondary injection. These transformers are best tested in the factory and shipped with certification. If they are not, they can be tested in the field. Reasonable accessibility is usually available. The standard secondary voltage is 120 V. Verification of polarity should be battery tested. Instrument transformer nameplate data must be checked for conformity with engineering documentation to ensure, foremost, that they have the same transformer turns ratios.

Power Transformers

Power transformers must be thoroughly inspected prior to installation. If impact recorders were installed for shipping, they must be checked prior to unloading. The startup engineer should verify impact recorder information on the receiving and inspection report. Large transformers will be shipped without oil. Before oil is added, the transformer internals must be inspected for loose components such as tap changers, baffles, and bracing. The tanks should have a nitrogen blanket that must be purged with dry air for inspection. After inspecting the transformer, the external components such as radiators, fans, bushings, and instrumentation need to be installed.

The startup engineer should check all accessories for proper installation. All local devices for alarm and trip should be checked. These will include temperature, pressure, and level instrumentation. The cooling pumps should be operated and checked. After all protective devices are verified, the transformer should be insulation tested in accordance with the manufacturer's recommendations.

Transformer Protection

Transformers may be subjected to three main damaging conditions: overload, short circuit, and over voltages. Persisting overloads result in heat rise, deterioration of transformer insulation, and later failures on moderate over voltages.

Short circuits phase-to-phase or phase-to-ground result in transformer winding insulation failures. These faults must be cleared quickly to avoid the danger of fire, transformer mechanical damage, and connected system instability.

Over voltages resulting from lightning switching surges or ground faults are usually of a transient nature. However, these over voltages may subject transformers to dangerous voltage stresses unless protective means are provided to prevent the application of these stresses to the windings.

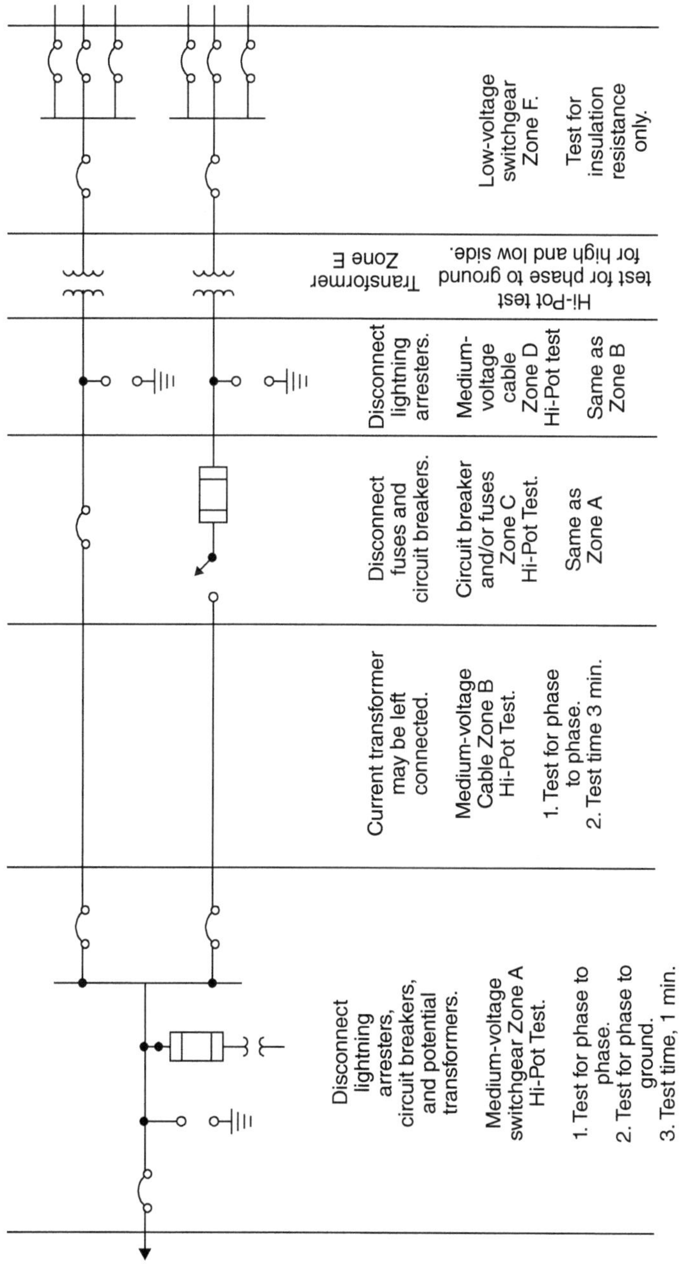
Disconnect lightning arresters, circuit breakers, and potential transformers.
Medium-voltage switchgear Zone A Hi-Pot Test.
1. Test for phase to phase.
2. Test for phase to ground.
3. Test time, 1 min.
Current transformer may be left connected.
Medium-voltage Cable Zone B Hi-Pot Test.
1. Test for phase to phase.
2. Test time 3 min.
Disconnect fuses and circuit breakers.
Circuit breaker and/or fuses Zone C Hi-Pot Test.
Same as Zone A
Disconnect lightning arresters.
Medium-voltage cable Zone D Hi-Pot test
Same as Zone B
Transformer Zone E
Hi-Pot test test for phase to ground for high and low side.
Low-voltage switchgear Zone F.
Test for insulation resistance only.

Overload Protection

Overload protection equipment includes: liquid level gauges, oil temperature indicators, thermal relays, hotspot temperature equipment, vacuum (negative pressure) gages, positive pressure inert gas protection equipment, and forced air cooling.

Liquid level gages provide high-level oil indication, which can occur in sustained overloads. It also provides low-level indication, alerting the operator to a possible leak, which may cause winding flashover. The liquid level gage usually sounds an alarm and is standard on all transformers.

Oil temperature indicators operate at the top of the transformer and sound an alarm at high oil temperatures. This also is a standard transformer device.

Short-Circuit Protection

The mechanical pressure relief device is the only short-circuit protection device which is standard on a transformer. It is mounted on the transformer cover and relieves excess internal pressures. Alarm contacts are furnished at no additional cost. This device can be considered a last resort for overload protection.

Breakers in the transformer primary with overload relays provide short-circuit and overload protection. The relays are available with time delay and instantaneous settings.

The instantaneous relay of the breaker in the primary circuit of a single transformer, with an instantaneous trip secondary breaker, should be set a little above the asymmetrical short-circuit current that the breaker will see during a three-phase fault close to the transformer secondary. This is usually high enough to prevent relay operation on transformer magnetizing inrush and will coordinate with other protective devices.

If the transformer primary feeder breaker trips when the circuit is energized during installation tests and there is no fault, the instantaneous relay current setting should be increased just enough to prevent relay operation on magnetizing inrush current.

The minimum setting of the instantaneous relay on a feeder supplying several transformers will usually be dictated by the need for accommodating their total magnetizing inrush current. This will be about 10 to 15 times the normal current. Such settings would be high enough to prevent tripping the main feeder breaker with a solid secondary fault on one of the transformers and full load of the others.[11]

The Buchholz relay is a gas detector that is mounted in the transformer conservator line to the main tank. If gas is formed because of an internal fault, the relay will trip an alarm and capture a sample.

These lightning strike counters are connected to the lightning arrestor grounds.

Transformer Testing

Power transformer testing is, in most cases, best contracted to a third party because the testing will take startup personnel from routine tasks in order to be dedicated to the transformers.

The **AC high-potential test** is recommended for acceptance testing because it subjects the transformer to the same internal stresses encountered during normal operation. The test

TABLE 12-5 Transformer Power Factor Acceptance Values[12]

Test[11]	% PF Good	Marginal	Investigate
H winding	0.5 or less	0.5 to 1.0	Above 1.0
X winding	0.5 or less	0.5 to 1.0	Above 1.0
Interwinding	0.5 or less	0.5 to 1.0	Above 1.0
Bushing UST	Below 1.0	—	—
Oil-main-tank, tap changer, switch compartment	0.5 or less	Above 0.5	Replace if above 2

Note: "H" is the standard representation for the primary transformer windings and "X" represents the secondaries.

should be limited to 75 percent of the factory test. For maintenance testing the transformer should be tested for only 3 min at its rated voltage.[12]

A **power factor test** can also be performed. Power factor is represented as:

$$PF = EI \cos \theta / EI$$
$$= I_R / I_C$$

This is because, with the secondary side open, the no load resistance is limited to the dielectric current and capacitive current. Companies such as Doble Engineering will provide detailed testing procedures with their test sets.

The Buchholz relay is a gas detector that is mounted in the transformer conservator line to the main tank. If gas is formed because of an internal fault the relay will trip an alarm and capture a sample.

Table 12-5 illustrates general acceptance values for power factor testing.

A transformer turns ratio (TTR) test is also used to determine open or short circuits. Ratios should not deviate more than 0.5 percent of the nameplate defined by ANSI 57.12 section 6.1.[12] See IEEE C57.12, IEEE Guide for Installation of Oil-Immerse EHV Transformers 345 kv and Above.

Oil dielectic tests are a very important part of the plant maintenance history. The tests are preformed using ASTM standards. Five samples are tested and averaged.

GENERATORS

Generators must have heat applied well in advance of preliminary testing in order to drive off moisture that entered during site installation and assembly. Hydrogen-cooled generators will need a pressurized air test. The equation used for air leakage testing is illustrated here. All generators will require a stator winding insulation resistance test, high-potential or power-factor test, and a no-load test.

Generator Air Leakage Test Calculation

$$L_A = 238\ V/H\ [(P_i + Pb_i)/(273 + T_i) - (P_f + Pb_f)/(273 + T_f)]$$

where:

V = generator volume

H = hours of testing

P_i = initial generator pressure, in-Hg
P_f = final genator pressure, in-Hg
Pb_i = initial barometric pressure, in-Hg
Pb_f = final barometric pressure, in-Hg
T_i = initial generator air temperature
T_f = final generator air temperature

The insulation resistance test should be used to determine the dryness of the generator prior to proof-testing. High-potential or power-factor tests should be conducted in accordance with the manufacturer's instructions. Prior to load testing and synchronization, a no-load test is conducted by running the generator to speed with the leads disconnected in order to check voltage buildup.

BATTERIES

In many countries, battery rooms are common household utilities. They are usually located in kitchens, under-the-stair space, or any place an inverter and battery bank can be installed. The ventilation for the hydrogen generated in batteries is negligible compared with the normal occupancy ventilation requirements. This is true for most spaces where batteries for uninterruptible power supplies (UPS) are located. They are commonly referred to as "the inverter," even though the inverter is only one part of the UPS system. Floors generally can be rinsed if electrolyte is spilled or leaked; so elaborate acid-proof floor and wall coverings are seldom a consideration. Exploding batteries are not a consideration. But dropping batteries during handling can happen, and the resultant spill-splash can be hazardous to the eyes or health. Quick access to running water is usually available but never a prior consideration by owners. Safety is rarely considered, mainly due to ignorance.

A UPS system for residential use is sized for refrigeration, communications, television, and lighting. The UPS consists of a battery, a battery set or battery bank, a battery charger, and an inverter. The charger/inverter, as the name implies, either is providing AC power to charge the batteries or taking power from the batteries to provide AC power for utilities. In the first case, the unit is rectifying AC to produce DC power; in the second, it is inverting the DC to produce AC power. It is as simple as that.

The quality is the only marked difference between the previously mentioned home or office "inverter" and an industrial UPS system. The size dictates the engineering requirements in an industrial environment, which are, in turn, governed by the applicable codes and standards. For example:

- IEEE 450, "Recommended Practice for Maintenance, Testing, and Replacement of Vented Lead-Acid Batteries for Stationary Applications."
- IEEE 484, "Recommend Practice for Installation Design and Installation of Vented Lead-Acid Batteries for Stationary Applications."

If the codes and standards are followed, the guideline will fall neatly within commonsense considerations regarding environmental, health, and safety factors. If the ventilation is not adequate for human occupancy, it will generally not be adequate for the batteries, charger, and inverter. If required, using the IEEE standards for hydrogen concentrations and the battery supplier hydrogen release information, the adequacy of the space ventilation is easily determined and checked against the mechanical engineer's space ventilation data. However, human occupancy and equipment (charger/inverter) ventilation demands are much greater than hydrogen ventilation requirements. Battery hydrogen disperses rapidly and requires little ventilation to prevent accumulation. Providing a hose bib for rinsing the battery room floor and an eye wash/shower station are commonsense considerations.

Testing

Battery and connections—Each battery voltage should be measured. The voltage should be within 0.05 V of the manufacturer's rating. After connecting each battery, including interconnections, it should be resistance tested. Check the resistance by applying a load across the battery from the adjacent battery terminals in order to include the battery connection resistance. The resistance is the battery Δ voltage/amperage reading. The values should not be greater than 5μΩ or an individual variance not greater than 10 percent. These, as well as the overall "bank" resistance, are used for maintenance and, therefore, should be recorded graphically for trend observation.

Battery bank load test—The batteries should be load tested by disconnecting the battery charger and applying the design load to ensure the batteries can maintain the emergency equipment in service for the specified duration. That usually is dictated by the heaviest load, which will normally be turbine DC lube oil pumps. The load test must include, however, "all" DC loads, including the vital AC inverter.

Inverter test—The normal AC supply should be interrupted to ensure that the inverter picks up the load as an un-interruptible supply. Ensure the normal power is picked up and the inverter dropped out on restoration of power. Ensure that the voltage, as well as frequency, are within specifications.

Charger test—With the batteries disconnected, check the charger output voltage and ensure that it is within specifications.

> **Note**
>
> All UPS systems, local as well as remote, must have central control room, charger failure, and low-voltage alarms.

DISCUSSIONS

Power Factor

In a simple, DC electrical circuit, the power available is the voltage times the amperage—that is, $E = IR$. The only thing opposing current flow is the system resistance. However, in utility AC systems, it is not this simple. In the utility system, reactive power also opposes true power. In utility power generation, $E = IR \times$ Power factor.

Active power is power that flows in one direction from the point of generation, such as from a power plant to the distribution system. The direction of flow is always one way to the user. Reactive power, on the other hand, flows back and forth in either direction and is retained in the system.

There are two components of reactive power: *inductive reactance* and *capacitive reactance*. Examples of inductance circuits that produce reactive power are generators, motors, and transformers. Magnetism is required to induce current in these machines. Voltage, in this case, must be raised or magnetism excited before current will flow. In other words, *current lags voltage in an inductive circuit.* Examples of systems with capacitance are systems with capacitors or synchronous motors. Also, underground cables and large distribution systems produce a capacitance. The capacitance exists between the conductor and ground; therefore, they produce a capacitive reactance. The underground cable insulation and the aerial cable air separation between the conductor and ground form a capacitance. Introducing a current to a capacitor causes a voltage rise. Therefore, *current leads voltage in a capacitance circuit.*

Figure 12-17 illustrates the effects of reactance on capacity.[12] Note the phase angle diagrams for resistance and reactive power.

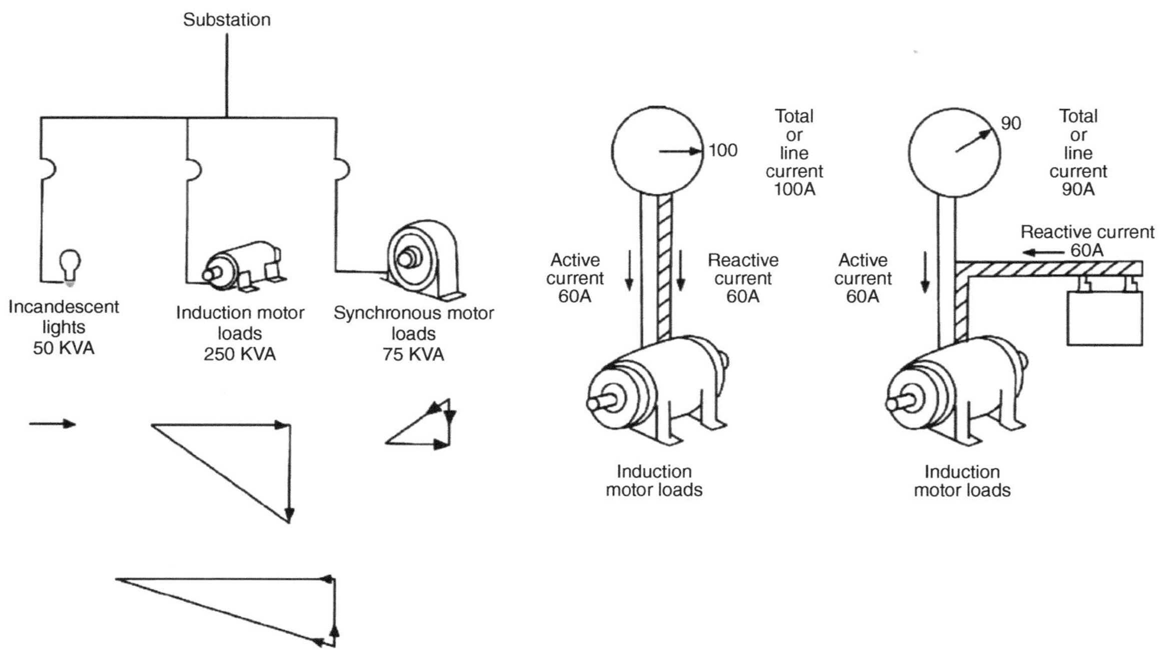

FIGURE 12-17 Effects of reactance on capacity.

Either *lagging* (inductive reactance) or *leading* (capacitive reactance) opposes efficiency. Balancing the reactive loads to produce a unity power factor will increase productive power generation. A unity power factor (1.0) will indicate 100 percent performance or total active power. Because active power is the only power available for useful work, it is usually the only basis for revenue. Therefore, generators want to produce as much active power (volt-amps, VA) and as little reactive power (volt-amps reactive, VAR) as possible.

For power generation, the reactive power is a result of excitation of the generator field. The excitation creates a magnetic field from which active power is generated. The more the field is excited, the more reactive power is produced, creating a "lagging" power factor. Utility generators will be designed for a 0.85 lagging power factor. This will offset the capacitance reactance of the large distribution grid. These machines are more expensive than higher power-factor machines because the reactive power generated by increased excitation causes higher temperatures. Therefore, the generators require more core copper windings and better cooling methods.

Independent power producers are concerned with producing only active power and, therefore, purchase generators with higher power factors such as 0.95. They operate as close to unity as possible. But there is a degree of caution because generators have certain limitations wherein continuous operation outside the parameters would result in generator damage.

The specific loading and power-factor limitations are dependent on each machine design and are not universal. Therefore, the curve for one machine should never be used for another. Use only the capability curves supplied by the manufacturer for each machine. In the reactive capability curve, load versus power-factor capabilities is illustrated. The limitations are based on field, armature, and localized heating of the stator core.

Greatly underexciting the generator to a point where the rotor angle increases to a level such that the driving torque on the generator shaft becomes higher than the synchronizing torque developed by the generator windings must be avoided. That is, underexciting to a low-leading power factor could cause a generator pole slip. This means that the generator accelerates out of synchronism for an instant in time. This causes massive current flow, high vibration, and substantial overheating. The typical power factor where this could be experienced is 0.75 leading or less, at approximately 50% load. In small-grid systems, this could also "pull" the system voltage down to unacceptable levels due to extreme leading MVAR being imported by the generator. These characteristics vary considerably from generators of different designs and applications. MVAR has both over- and underexcitation limitation setpoints calibrated that match the generator characteristics to prevent these conditions from happening.

Ferro-Resonance

Ferro-resonance is, simply stated, series resonance. *Ferro* refers to the iron core of a power transformer. In mechanical terms, *resonance* refers to that point where a disturbance frequency in a system equals the natural frequency of that system. This allows the disturbance to accelerate beyond the intended design limitations of the system.

Resonance can occur in any AC series circuit with inductive (L) and capacitive (C) reactance. Reactance is dependent on frequency. If frequency increases, for example, inductive reactance decreases and inductive reactance increases. The frequency at which the two are equal and 180° out of phase can be said to be the natural frequency. That is, if the power frequency exactly equals the natural frequency of an LC circuit, it is said to be in resonance. At that point the capacitance and inductance components are exchanging energy in increasing oscillations. Without circuit resistance, the oscillations will vary rapidly and drive toward infinity. In simple terms, series resonance occurs when $L = C$ in a series circuit.

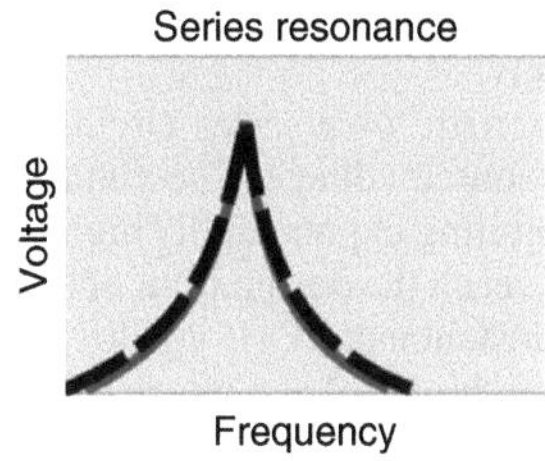

$$Z = (R^2 + X^2)^{1/2}, \text{ where } X = X_L - X_C$$

or

$$I = E/R, \text{ when } X_L - X_C = 0$$

In the preceding diagram, the voltage is rising toward infinity. It might be said that this was purely theoretical and a good topic for classroom discussion or laboratory experimentation.

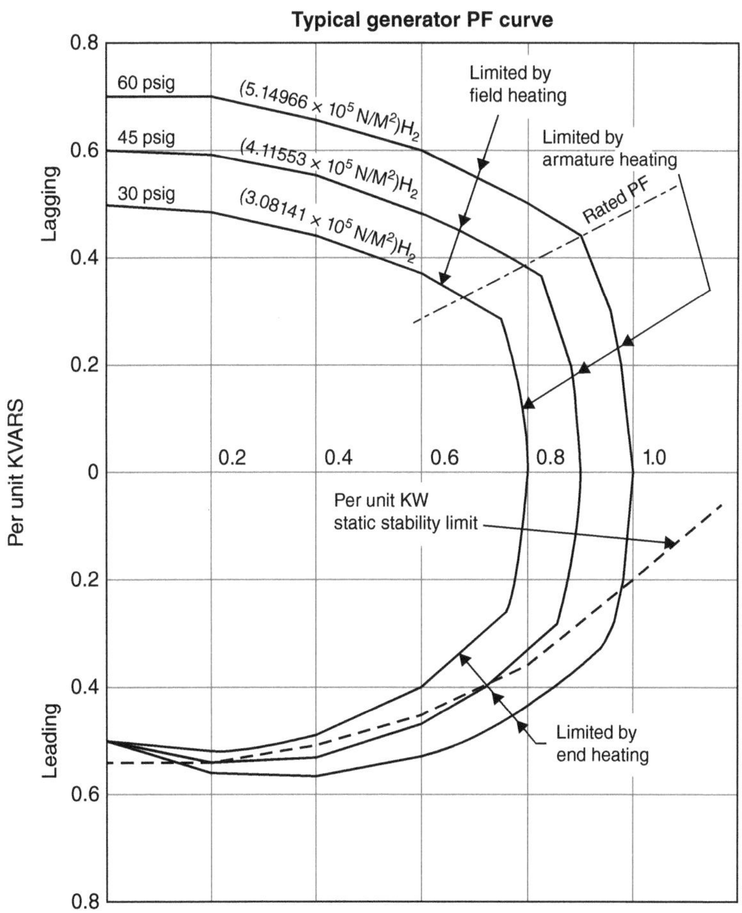

Someone who has had the experience of closing in on a high-voltage power circuit that was in resonance will know differently.

The energy storage characteristic of a series circuit explains its capacity to achieve voltages much higher than the source voltage. The voltage measurement will resemble an oscillating sinusoidal wave exhibiting exponentially higher peaks at each cycle. Within two to three cycles, the voltage will reach the design limit of the weakest link in the system. In a grounded system, it will seem instantaneous. (If you lift the ground, it will take longer and can be observed with an ordinary high-voltage meter.) The weak link can be either a surge protector, potential transformer, or service transformer connected to the primary line. The weaker of them will explode.

One of the most common sources for ferro-resonance in power systems is found in the relationship between buried cable and power transformers. The buried cable provides the capacitance reactance and the transformer the inductance reactance. Both aerial and buried cable are a source of capacitance reactance. However, the buried cable makes a much better capacitor and, therefore, can be equal and opposite in strength to the inductance of a power transformer in a relatively short distance.

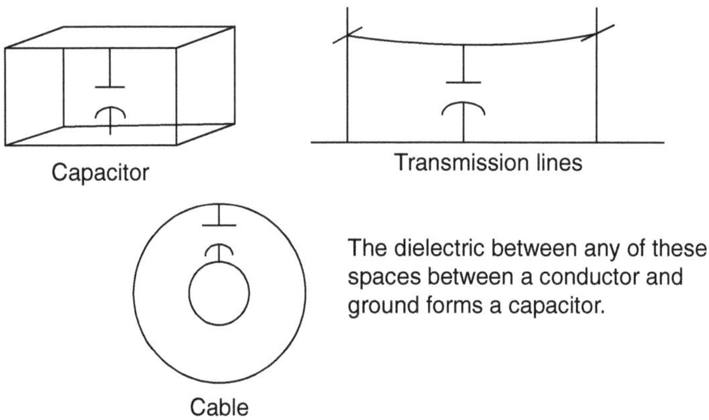

System design can contribute to this phenomenon as well as mitigate it. First, if the main transformer connected to a system (e.g., transmission line) is a wye-wye (Y-Y) transformer, the resonance will more likely form destructive oscillations. If, however, the oscillations are dampened by a delta winding, as in a Y-Δ transformer, the oscillations can be controlled. Delta tertiary transformer windings, or zigzag transformers with a secondary delta, can also be used to control the "third harmonic" effects.

In one case, a 5-m: underground transmission line between a hydro power station and substation was experiencing dangerous ferro-resonance problems. The power plant main transformer was a Y-Y. To remedy the problem, startup engineers installed three pole-top transformers at the substation. Then, they connected the pole top primaries to the transmission lines in a grounded Y configuration. The three pole-top secondaries were "busted out" into a delta. The delta secondary served as a harmonic trap and eliminated the ferro-resonance problem. The no-load current flow in the floating delta was 20 A.

Switchyards and Substations

Switchyard and substation construction varies with aesthetics and constructability. Constructability is mainly material and erection costs. For example:

- Lattice structures are light and can be flown into remote regions. But assembly in the field is more tedious unless the structures are preassembled.
- Beam structures are lower in cost than lattice structures and are assembled in the field rapidly. They, of course, are not recommended for flying into remote regions.
- Tube structures are more attractive, difficult to modify in the field, and the most costly.

None of the designs are selected solely from an electrical engineering point of view.

From a startup point of view, the engineer must check grounding and clearances.

Remember, at 4000°C air is a conductor. If air switches are opened under load, air will become a conductor, and the arc will not immediately be broken.

Partial-load break switches are designed for breaking residual currents such as residual-magnetizing current of a transformer or line-charging current. But they will maintain the conduction path through air if the switch is opened under load. Breaking residual currents will produce impressive arcing.

Switch versus structural clearances must be checked to ensure that in the open position, the clearances are maintained. Basic structures must meet electrical needs and real estate limitations but must also meet NEMA specified clearances.

Clearances for "strain bus" (cable bus) must have allowances for sway and sag. All buses will sag, including rigid bus, to some extent, which is normally acceptable as long as the clearances are maintained. Some cables may even sway in the absence of wind. This is due to forces caused by line current that alternately repulse and attract the conductors if the forces are in resonance with the natural frequency of the conductors. The forces during a fault that force the conductor apart can be severe and must be used in the design and selection of insulators and structures. There are many types of antivibration devices available to eliminate vibration and sway.

Note

A rule of thumb for bus sag is 1/150th of the span where two supports are used and 1/200th of the span where three or more are used.

All lightning arrestors, structures, fences, gates, and operating devices in a switchyard must be "hard" grounded directly to a grounding grid. All operating devices and control panels should have steel grating to stand on that are grounded directly to the grounding grid. Coarse gravel is used as a dielectric for the ground surface within the switchyard.

Most switchyards with oil circuit breakers and oil-filled transformers have containments around this equipment in the event of a spill. If a fire suppression system is also installed, the containment must take into account the deluge water so that the suppression system does not spread a fire rather than suppress it within a confined space. The containment must be designed to discharge rainwater and account for snow and ice accumulation. These considerations normally utilize oily water separators to discharge water and capture oil. Therefore, commissioning of this equipment must be considered in the startup plan.

An uninterruptible power supply will be necessary for operation of the circuit breakers. Large switchyards may have a control room for the UPS and electrical protection equipment. Emergency lighting will be required as well as normal illumination for the yard. The emergency lighting should consider a black plant startup. The UPS should have low voltage and loss-of-charger alarms.

Electrical Protection

The "made simple" way to describe how protection devices operate is to recall how electromechanical devices function.

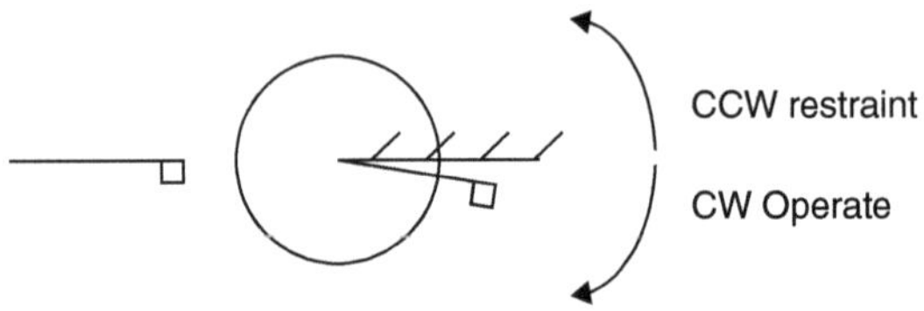

Instrument transformer polarity

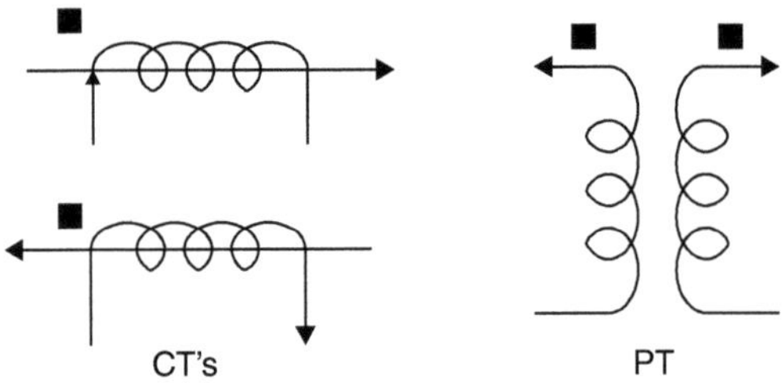

Directional Relays
There are two inputs: voltage for restraint and current for pickup. To get operating torque, current and voltage need to be going in the same direction. [Note: A quick test for correct directional operation is to reverse the voltage from the potential transformer (PT) so it is going in the same direction as current. If it is, the relay will pickup.] To get restraint, *E* and *I* source will have opposite instant direction. If a fault occurs, voltage drops below that which is caused by normal load swings. When that happens, restraint is lost and instantaneous overcurrent pickup occurs. In the accompanying sketch, the current flowing through the upper-left coil produces current out of the secondary coil. Current flowing out of the primary, bottom coil, produces current flowing into the secondary coil.

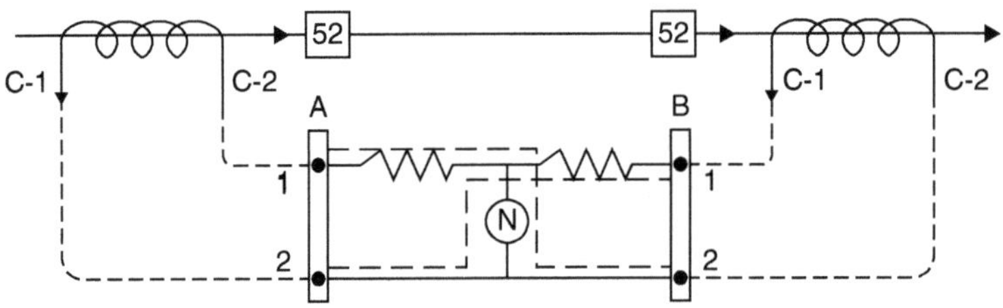

Differential Relays
An equal current in paths C-1 and C-2 from B_1–A_1 and A_2–B_2 produces a null current at N, which is the operating coil of the relay.

A fault between the current transformers (CTs) will reverse the current in one of the CTs, causing a current at the null point or operating coil. A fault outside the CTs will have no effect.

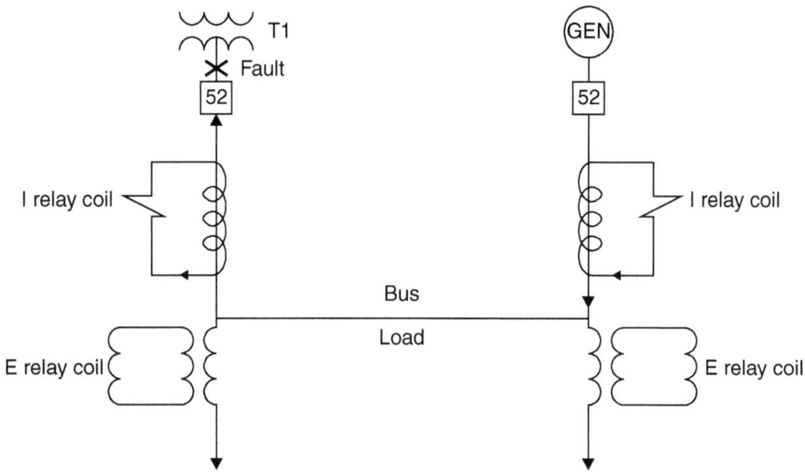

T52-1 will trip, therefore, the generator will not feed the fault; but it will keep feeding the load.

Distance Relays

Distance relays come in many types: impedance, reactance, offset distance, and mho. The mho units, can be represented on an R-Y diagram as illustrated in the accompanying figure.

Ohmic relays are relays actuated by two electric quantities, current and voltage, and the phase angle between them. The operating range of these relays is "ohmic reach."

Any value of voltage, current, and phase angle present in the relay's operating coils that results in a vector impedance inside the ohmic circle, produces relay contact closing torque or power in an inductive circuit. These devices cannot be set until generation is high enough to establish the true impedance of the transmission system.

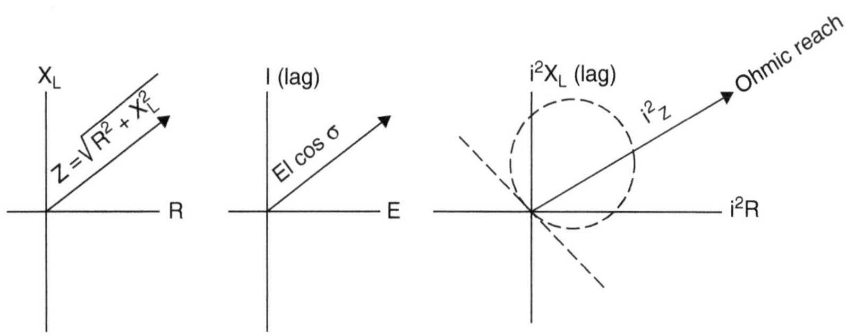

GROUNDING RESISTORS

Grounding resistors are often used to limit ground-fault current. Their purpose is to protect circuit interrupters from damage or insulation breakdown such as with oil circuit breakers (OCBs). They are rated in amperes-seconds. Both current and time ratings are crucial. The resistor bank must withstand a fault current long enough for the interrupter to clear the fault.

REFERENCES

1. "Applied Protective Relaying," Westinghouse Electric Corporation, Relay-Instrument Division, Newark, New Jersey, 1982.
2. General Electric Guide, *GE Medium Voltage Switchgear*, GET 6600F.
3. "Applied Protective Relaying," Westinghouse Electric Corporation, Relay-Instrument Division, Newark, N.J. adopted by ANSI/IEEE as Standard Device Numbers, Standard C37.2-2008.
4. National Electric Code, NFPA 70.
5. Reason, J., "Why Dielectric Failure is the Major Cause of Electrical System Breakdown," *Power*, July 1981.
6. Gill, A. S., *Electrical Equipment Testing & Maintenance*, Upper Saddle River, N.J.: Prentice-Hall, 1982.
7. ANSI/IEEE C57.13.3.
8. NEMA 12.41.
9. Lazar, I., "Transformer Protection and Relaying in Industrial Power Plants," *Power Engineering*, May 1979.
10. ANSI 57.12.
11. ANSI/NEMA MG-1, American National Standard, Motors and Generators, National Electrical Manufacturers Association, Arlington, Virginia.
12. IEEE Standard 141, Recommended Practice for Electric Power Distribution for Industrial Plants.

CHAPTER 13
FANS

The plant engineer should be familiar with the type of fan he or she is commissioning, its characteristics and behavior, and the relationship among the control components. A good general reference source for information on different types of fans is AMCA[a] 201, Fan Applications Manual. The best *detailed* information can be obtained from your fan instruction manual. The objective of this discussion is to briefly describe different fan designs, without the detailed elaboration provided by AMCA. This will serve as a basis for later evaluation of startup and testing requirements. Fan performance testing is standardized in ASME PTC-11-1984 and AMCA 203-90.

Table 13-1 defines some important engineering terms related to fans.

Figure 13-1 illustrates fan performance curves for different blade designs. The radial tip (RT) fan, which uses a backward-inclining (BI), forward-curved (FC) blade, is the centrifugal fan most often used in boiler forced draft (FD) and induced draft (ID) designs. The reason for using RT fans for FD and ID designs is that they have a stable operating range from shutoff to full capacity. This is unlike the backward-inclined blade, which has a notable unstable region that has to be passed to reach full capacity. Even the backward-curved design has an unstable region—but not pronounced enough to prevent operating from zero to 100 percent. The axial fan has a definite, pronounced unstable region similar to axial compressors.

Axial fan[1]

Centrifugal fan

[a]Air Movement and Control Association International.

TABLE 13-1 Engineering Terms Related to Fans

Term	Definition
SP—static pressure	Pressure (head) exerted on the surface of a containment, such as breaching with no velocity impingement, at the inlet and outlet of the fan. An atmospheric inlet cancels out in the calculations; the SP is then simply the outlet pressure.
Velocity pressure	Pressure exerted on the surface of an object, perpendicular to the flow stream, by the air or gas velocity. Velocity pressure is not measured unless the fan is rated in TP rather than SP. It is small compared to SP and can be calculated using the inlet and outlet areas and volumetric flow.
TP—total pressure	Sum of the static and velocity pressures at the inlet and outlet of the fan. It is the same for pumps where it is referred to as total dynamic head (TDH).
Capacity	Volumetric flow in cubic feet per minute (CFM).
Standard air	Standard used by manufacturers as the reference air conditions for a fan. If the fan curves are using standard air conditions, then fan performance must be corrected to site conditions. Standard air is given as 70°F at sea level (29.92 in Hg) and a density of 0.075 lb/ft^3.
Water gauge	A water manometer used to determine the inlet and outlet pressures. The inlet may be negative with respect to atmospheric pressure, therefore subtraction of the negative number will yield a positive addition to the discharge pressure, e.g., $SP = SP_{outlet} - SP_{inlet}$; $5 - (-1) = 6$.
Air horsepower	Theoretical horsepower of the fan at 100% efficiency.
Brake horsepower	Actual horsepower required to drive the fan, which includes all inefficiencies.

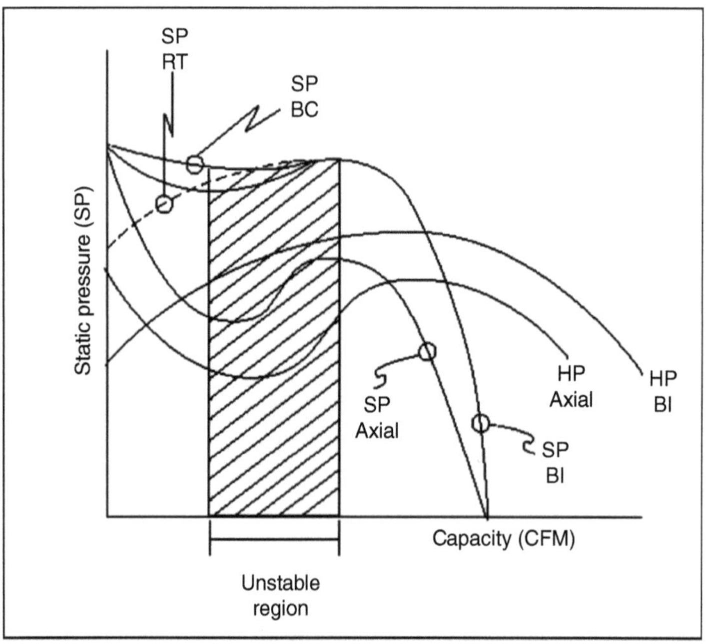

FIGURE 13-1 Fan performance curves for different blade designs.

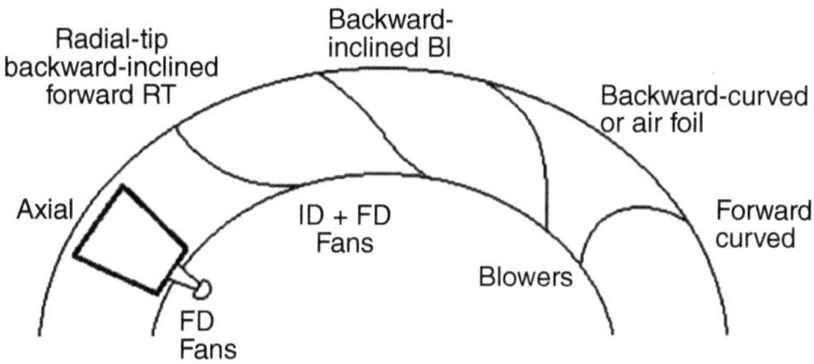

FIGURE 13-1 (*Continued*)

Generally, these unstable regions can be passed fairly easily when operating only one fan. If the region of instability is pronounced, as with an axial fan, it may not be possible to pass this region with a second fan if the first is at full load. The reason is that the fan will be held in the region of instability by the static pressure of the first fan. The full-load backpressure of the first fan will prevent flow of the second fan when the full-load and partial-load static pressures match. Therefore, the first fan flow must be reduced to allow the fans to come up together, each sharing the load. That means reducing unit load.

Understanding the fan curves and how to determine static or total pressure is important. Generally, the capacity "flow" is determined during air quality control and performance testing at full load just prior to commercial operation. With accurate flow and pressures, the curves can be verified. However, during initial trial operation, only accurate "pressure readings" are sufficient to determine or assess operational problems. Fan capacity tests can be conducted, if necessary, after the airflow is determined for the air quality control tests.

Remember, inlet pressure control via inlet guide vanes alters the fan characteristics by changing the design inlet ambient pressure. Also, fan horsepower and efficiency will improve because the air is being guided into the inlet. Therefore, the fan curve will not

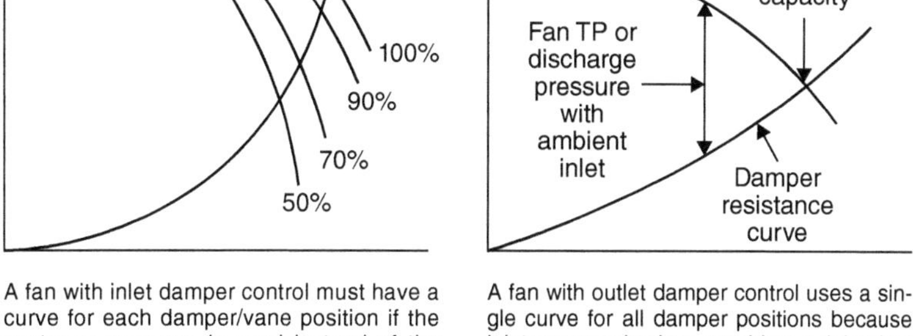

A fan with inlet damper control must have a curve for each damper/vane position if the upstream pressure is used instead of the downstream (actual fan inlet pressure).

A fan with outlet damper control uses a single curve for all damper positions because inlet pressure is always ambient or inlet duct pressure.

be accurate until full load is reached, where true ambient design conditions are satisfied and guide-vane alteration of airflow ceases. For example, the inlet pressure will follow a system resistance curve with a family of performance curves, one for each percentage point of inlet damper position. This characteristic is the same for a variable speed fan. For the outlet damper–controlled fan, the performance will follow a single-fan performance curve. For startup purposes the inlet pressure can be measured on the downstream side of the inlet dampers, if it can be done without endangering the fan. In this way, the measurement will put the differential pressure on the performance curve similar to the outlet damper–controlled fan.

Curves A, B,C, etc. (Fig. 13-2) are characteristic ratio curves, that is the ratio of damper resistance to the total system resistance.[2] Note that they all have similar flow characteristics of valves.

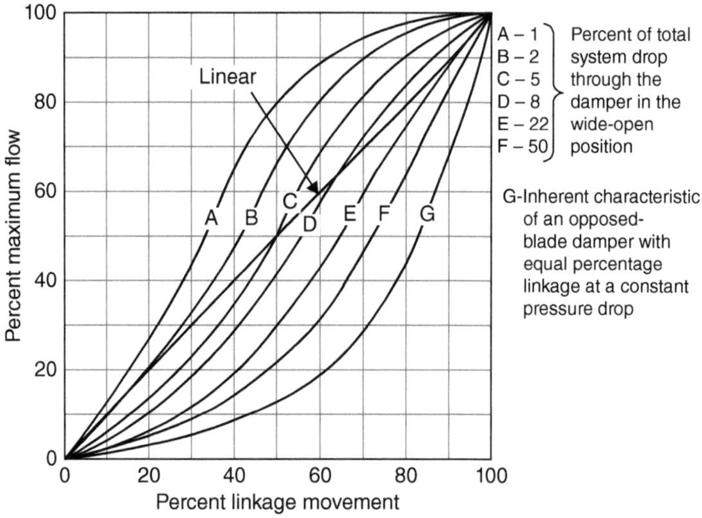

FIGURE 13-2 Flow vs. damper position curves.

The louvre and opposed-blade louvre are illustrated in Figs. 13-3a and 13-3b.[3] The opposed-blade louvre is generally selected to improve the downstream flow characteristic. The double dampers in Fig. 13-4 are sealed with an outside air blower. They are used for user-safe internal maintenance operations.

Also, the capacities, pressures, and horsepower must be adjusted for temperature. The following formula ignores density or specific volume change, "v"—as in $Pv = RT$ of the perfect gas laws—but is sufficiently accurate for startup purposes. However, if the startup engineer is dealing with large temperature and/or density differences then the perfect gas law must be used. For example, if dealing with used equipment that has been relocated to site at a different altitude, the density must be taken into account.

$$P_{\text{test}} = P_{\text{design}}\,[(T_{\text{design}} + T_{\text{absolute}})/(T_{\text{test}} + T_{\text{absolute}})]$$

$$Pv_{(\text{test})} = Pv_{(\text{design})}\,[(T_{\text{design}} + T_{\text{absolute}})/(T_{\text{test}} + T_{\text{absolute}})]$$

Note: $R = 55$; the gas constant for flue gas cancels in the calculation.

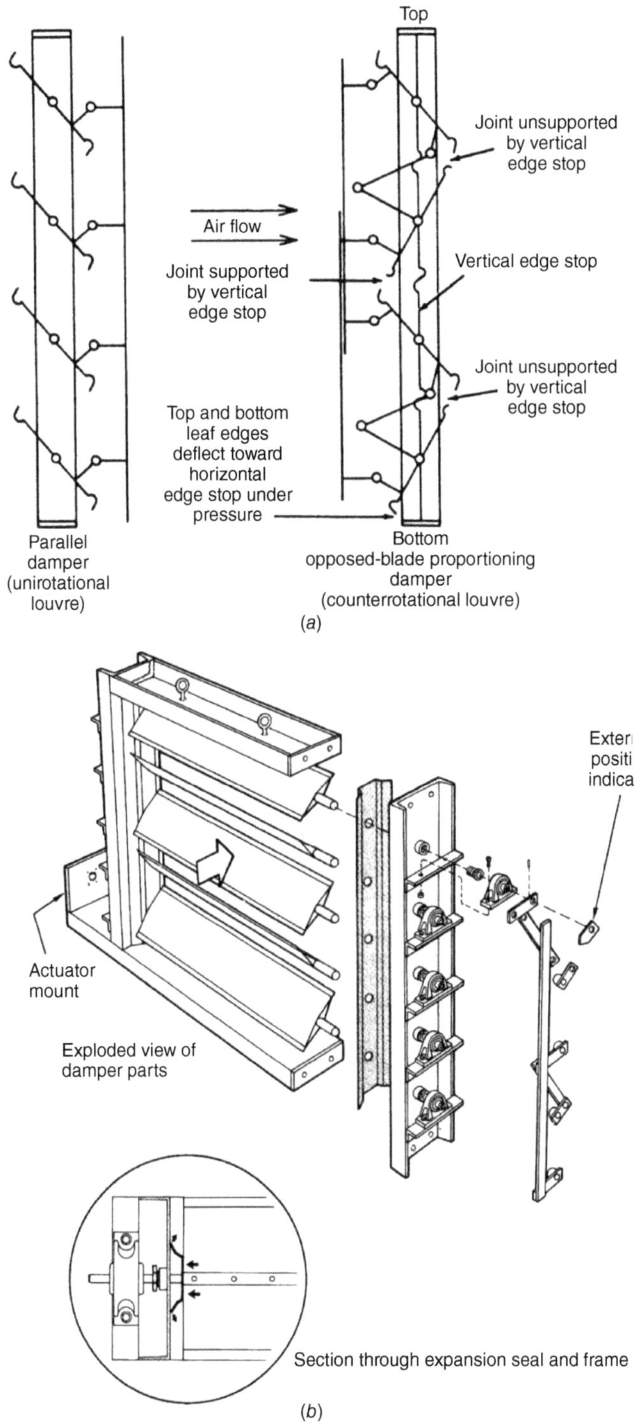

FIGURE 13-3 Unirotational and counterrotational louvres.

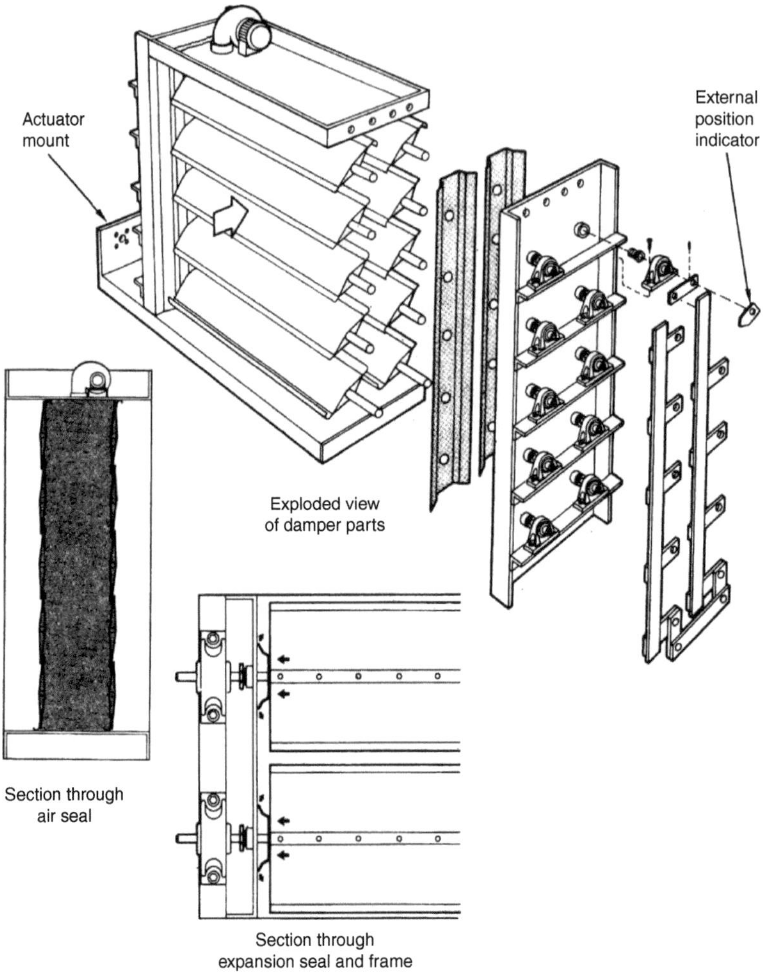

FIGURE 13-4 Air sealed damper.

Generally speaking, the plant engineer will not have to make calculations regarding the change of fan head and capacity. Nevertheless, it is important to know that the relationships determine how a change in impeller diameter will change capacity, head, and horsepower. There are four basic dimensionless equations used for that purpose: the three previously mentioned parameters and the fourth for specific speed. All are the same for evaluating pumps as well.

1. Q/ND^3
2. H/N^2D^2
3. hp/N^3D^5
4. $N_S = N \times Q^{1/2}/(H \times 0.75)$

where Q = flow
H = head
N = speed
D = diameter
ρ = density

Specific speed will be given by the manufacturer at the design point. As shown, it varies when any of the variables change. Also note that the other equations simplify by cancellation of equal variables. For example, if we wish to know what effect varying the fan speed will have on capacity (without changing impeller diameter), the term D^3 cancels out in the algebraic equation.

$$Q_1/N_1D^3 = Q_2/N_2D^3$$

$$Q_2 = Q_1(N_2/N_1)$$

PRETRIAL OPERATIONAL CHECKS FAN INSPECTION

The first thing that must be done to ensure there are no startup time delays is to inspect the fan for suction seal clearance and the breaching for temporary internal bracing used for shipping and construction erection. Usually, ductwork and breaching are braced with structural steel or pipe to keep them from twisting or collapsing. The temporary bracing can be identified by its attachment. If it is attached by tack welding, then it is temporary. Of course, checking the design drawings will confirm if it is temporary or permanent. Seals must be tight enough to prevent circulation from the outlet into the inlet, decreasing fan capacity, but they must not rub. Verify the clearance against the fan instruction manual.

Lube Oil and Cooling Water Systems

The lube oil and cooling water systems require flushing before the fans can be run. Therefore, it is prudent to start the activities early in case they do not clean to acceptable standards prior to the scheduled trial operation date. If the bearings require grease, lift the caps and ensure they were properly packed with grease at the factory and that they have not been sabotaged.

Dampers and Guide Vanes

All dampers and inlet guide vanes must be checked for proper operation. Check the linkage bearings for proper grease application. The drives should be disconnected and placed at the 50 percent position prior to preoperational checks. Ensure that they are stroking in the right direction. If the drive is tested at the limits, zero or 100 percent, and it travels in the wrong direction before the limits are verified, it can be damaged. After verifying the travel direction, verify the stop limit switch settings while uncoupled.

Repeat this process with the dampers or vanes coupled to the drive. Check that the drive is connected with the shortest distance between the drive arm and the damper linkage. It will be the fixed end of the duct for hot-gas service. Hot ducts must be anchored so they can expand up or away from the anchor point. For a hot duct supported by a base, the bottom is fixed and the top expands upward. Note that the duct expansion will change the stroke.

If the duct expands upward, the linkage should be connected at the bottom. The effects of thermal expansion must be minimized. Verify the lever arm will move the damper full open and close. Check the internal clearances of the blades.

Note

If an electric drive trips on overload, especially for units that have been in service for a long period of time, check the bearings. The most common cause for drive overload is damaged or seized bearings due to the lack of proper lubrication and maintenance.

TRIAL OPERATION

The most frequently encountered problems commissioning fans are duct and fan vibration. Fan vibration is always measured at the fan and motor bearings. The most frequent cause of fan vibration is fan wheel imbalance followed by inadequate pedestal design. If the vibration is the highest at the outboard motor bearing and diminishes moving toward the outboard fan bearing, the motor pedestal design should be examined. If the vibration is highest at the fan bearings and diminishes moving toward the outboard motor bearing, the fan should be balanced. Refer to Chap. 21.

The fan manufacturer supplies the architect-engineer with the mass requirements for the mounting pedestal design. But the mass is only one factor in pedestal design. For example, the pedestal must also be proportioned properly. If it meets the mass requirements but is very tall compared to a short, wide-based pedestal, it may encounter resonance vibration problems. If, after fan balancing and bearing inspection, the motor bearing vibration is still high, increasing inboard to outboard, then the pedestal needs to be modified. The cheapest and quickest field modification is to add concrete. This does not require an engineering vibration analysis.

The most frequent cause of duct vibration is inadequate surface stiffening. The most common root cause of this problem is designing the duct solely for structural strength based on internal pressure, without full consideration for vibration. The cheapest and quickest field modification is to add structural stiffeners to the sheet metal. This also does not require an engineering vibration analysis. However, if the root cause is ductwork design configuration rather than structural flimsiness, it usually is not feasible to modify it due to other equipment and structural interferences. Therefore, modification of the duct configuration should only be considered when other problems such as high motor load and low capacity restrictions are also present. Even then, internal flow modification devices should be considered first before modifying the ductwork itself.

Hot-gas fans, such as boiler-induced draft fans, must be monitored for motor overload during low-unit capacity, cold-air startups. For example, a typical ID fan motor will reach full-load current at a minimum boiler purge (approximately 25 percent volumetric flow). The reason for this is the horsepower is determined by mass flow. Therefore, cold-air testing can only be performed with the boiler forced draft fans at minimum load. Boiler purging is based on boiler volume, not air mass flow.

REFERENCES

1. Wagner, C. E. and Johnston, K. H., "Axial Mechanical Draft Fans for Energy Conservation," ASME 74WA/Pwr-12.
2. Liptak, B. G., and Venczel, K., *Instrument Engineers' Handbook*, Chilton Book Company, Radnor, Pennsylvania, USA, 1980.
3. Affatato, A., "Control and Isolation Dampers for Power Plants," *Power Engineering*, April 1981.

CHAPTER 14
FIRE PROTECTION

A plant engineer's nightmare begins when he or she attempts to extinguish a minor fire only to find the extinguisher will not discharge. The engineer then retrieves another and finds it won't discharge either. Running out of extinguishers, he or she unreels a hose station only to find there is no water. On to another hose station only to find it does not have a hose but does have a water supply. By this time, the fire is out of control. The engineer then has to go back and get the hose from the other station and connect it to the station with water. This is, too often, a true and reoccurring startup story.

After commissioning the raw water system, the first mechanical system that must be commissioned and turned over to operations is the fire-fighting system. Commissioning must also ensure that the operations group is trained in plant fire fighting by public firefighters and the public firefighters are familiarized with the plant and the plant fire-fighting systems. If the operations group is not ready to accept that responsibility, then the commissioning team must. A fully automatic fire protection system does not automatically pass casualty risk from the commissioning team to the operations group. It cannot be overstated that fire fighting personnel protective clothing and fire brigade training are essential elements of fire protection. Most fires are extinguished before they become out of control, or trigger automatic system activation, because manual equipment is functional and plant personnel are trained.

GENERAL PRACTICE

Fire protection must be integrated into plant systems in order to function properly. Consideration must be given for plant security and heating, ventilating, and air conditioning (HVAC) as prime examples. Building architecture that provides multiple-exit fire-rated doors, windows, and walls for permanently occupied spaces such as control rooms is a good example. Service and fire water supply integration is commonplace. HVAC duct fire dampers and smoke exhaust systems are a common integration in all nuclear plants by regulation. Electrical and HVAC systems must be automatically deenergized on activation of smoke or fire detection. Do not deluge a transformer or generator that is energized. Ventilation duct fire dampers are UL rated at zero airflow and will not close if there is airflow in the duct unless they have been modified to do so. Wall penetrations for raceways must be sealed with fire retardant against air, smoke, and fire infiltration.

Containments must be designed such that when filled with firewater, they don't overflow burning combustibles into unprotected areas. Containments for combustibles must be provided with oily water separators that will drain water spills, rain, or deluge water from the containment as well as separate and dispose of combustibles. Manual shutoff is as important as automatic start. Too much water can cause as many problems as too little. Containments, such as condenser pits under steam turbines, serve to reduce a potentially uncontrollable, three-dimensional fire to a controllable, two-dimensional one. Foam systems must be considered for all potential three-dimensional fires—such as tank farms, multiple-unit engine-room oil pits, and containments for multiple vessels or tanks—where a fire can spread from one to another.

A sprinkler system activated by smoke detectors may not extinguish a smoldering electrical fire within an electrical enclosure. Also, personnel may not be able to enter an electrical room due to toxic smoke, and it may be hours before a continuous thermal sensor (CTS) fire wire detects the fire. If the fire spreads, and is hot enough, the CTS can activate a sprinkler system in electrical equipment rooms after the fire has broken out of the cabinet.

Smoke detection, smoke exhaust fans, and fire damper systems integrated into electrical switchgear or DCS rooms are far more practical for fighting cubicle fires than automatic sprinkler systems. After smoke is detected in rooms equipped with smoke exhaust fans, the smoke can be exhausted, the room safely entered, the source of the fire located, and the fire extinguished with portable CO_2 cylinders before the smoldering cable or equipment burst into flames. If automation is desired, cabinet smoke detectors can be used to activate dispersion of fire-extinguishing gas within the cabinets.

On the other hand, cable fires in cable spreading rooms, vaults, or raceways are best extinguished with water because not only does water extinguish the flame, but it also cools the area so it will not reignite.

> Carbon dioxide is ineffective after a fire has spread 1 ft laterally and two electronic console shelves in height; there on water must be used to extinguish the fire.[1]

Therefore, if a fire is not extinguished by CO_2, water must be applied.

Note

Portable extinguishers must be regularly maintained and available in order to extinguish fires before they activate automatic systems.

HALON

In the 1980s, scientific research indicated that Halon was an ozone-depleting agent, and the Montreal Protocol of 1987 required a phaseout of new production. In 1989, the Halon Alternative Research Corporation, a nonprofit trade association, was formed to promote the development of Halon alternatives. Therefore, control rooms, computer rooms, and other occupied spaces are generally protected by fire suppression gases other than Halon. No occupied space is protected with oxygen scavengers such as carbon dioxide.

PUMPS AND SURGE TANKS

Jockey pumps hold fire system pressure just below the low-pressure start of the main fire pumps and are low capacity. The jockey pump helps minimize surge when the main pumps start. If the system has leaks, the jockey pump will cycle on and off. No attempt should be made to install bladderless accumulators to compensate for system leakage. The water in the accumulator will absorb the air in a short time and go solid. If an air system is used for continuous air replenishment, it will require a special air compressor with a pressure rating exceeding the station's air system pressure.

Usually, the main pump is electric motor driven. The backup should be diesel-driven in case of an electrical blackout.

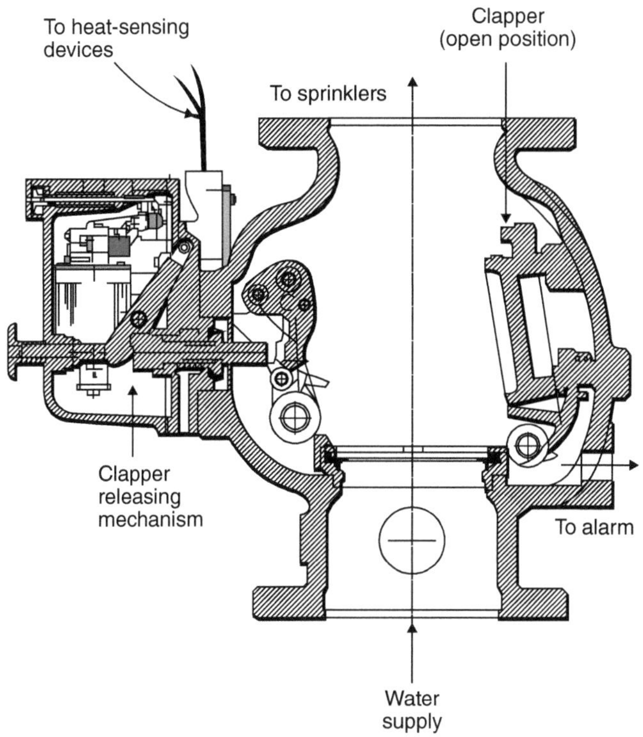

Deluge valve

PIPING SYSTEMS

Fire protection piping is a perfect example of a dynamically unbalanced system. Because all branches are held at deadhead (zero flow) until a control valve or hydrant is opened, the sudden discharge produces shock waves in the piping, causing unbalanced thrusts. Therefore, the piping must be restrained above ground with thrust restraints and below ground with thrust blocks. See Chap. 17 for thrust problems and solutions. The thrust restraint requirements generally surpass seismic restraint requirements.

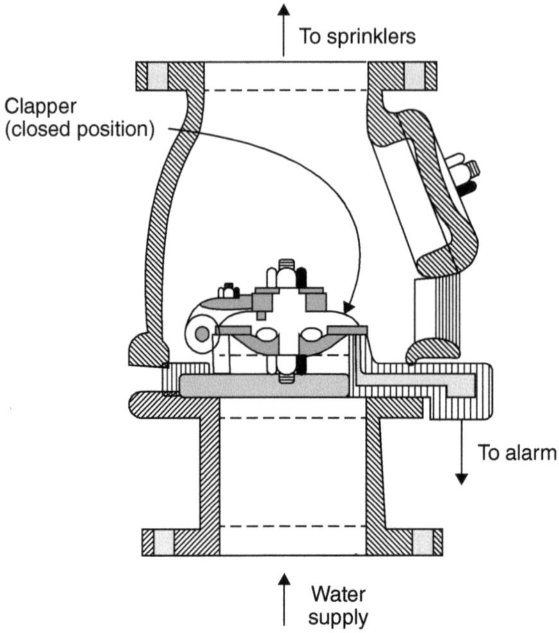

Sprinkler system valve

> **Note**
>
> Fire systems are a rare exemption to the rule regarding differential metal or galvanic corrosion. Because the systems are stagnant for long periods of time, a replenishing oxygen supply does not exist. Although the system contains the conductor (water) for galvanic current, it does not contain a replenishing oxygen supply and, therefore, will not corrode the steel pipe where brass valves or sprinkler heads are attached. Also, because fire system piping is stagnant, surge tanks without bladders cannot be used because the water will absorb the air buffer and the tank will go solid. If it is replenished to reestablish the level, the surge tank becomes an oxygen-replenishing device.

VALVES

Fire control valves are connected to the piping with mechanical couplings, sometimes generically referred to as "Dresser" couplings, so the valves can be removed quickly. The control valve gates or flappers are held closed by either water pressure in wet systems or air pressure in dry systems. The release of the upstream pressure allows the gate to open, releasing the supply water. This can be done manually for a dry deluge system or by the melting of fusible links in sprinkler heads for sprinkler systems. Once the gate or flapper is released, it opens and triggers an alarm. Figures 14-1 and 14-2 are examples of deluge and sprinkler valves.

FIGURE 14-1 Deluge valve.

FIGURE 14-2 Sprinkler valve.

DRY VERSUS WET

Dry systems are used for deluge fire fighting. When a fire detection device detects a fire, the deluge valve is tripped and all sprinklers are flooded simultaneously. Dry systems are also used in areas cold enough to cause freezing or if there is significant concern for water leaks.

A wet system is used when individual sprinkler operation is desired. Therefore, the wet system uses fuses to activate individual sprinkler heads.

If the pressure on the wet and the dry side of a dry system are equal, the system is flooded and must be purged and reset to the dry condition. The dry-side air pressure holds the supply valve flapper closed. Because the flapper air-side surface area is larger than the water port, the pressure required to hold it closed is less than the water pressure.

Foam systems must be used to fight liquid fires such as burning fuel, oil, or chemicals. Containments for storage of such liquids normally have foam fire-suppression systems. The foam forms a barrier between the liquid and the oxygen supply. The notable exception is transformer oil fires, where deluge systems are used.

COAL SILOS

Coal silos or bunkers must be provided with a means of injecting CO_2 or other oxygen scavenger from the feeder deck. Directing a water stream or throwing dry ice into a burning silo from the top will cause a dust explosion, which can rupture the silo, blow a coal dust cloud to atmosphere, and cause a potentially fatal secondary explosion. The CO_2 type of oxygen-scavenging injection system must be installed prior to the first delivery of coal. It cannot be an afterthought when hot spots appear on the sides of the bunker.

> **Note**
>
> Tripper decks and crusher buildings must be kept clean of coal spillage, and the dust collection systems must keep the air clean of any airborne coal dust. Fires and explosions in these areas can be far more devastating than in silos. Coal spills that cover conveyor or motor bearings are especially dangerous. The coal can cause bearing damage, which in turn can become the ignition source for a dust explosion.

COAL MILLS

> **Note**
>
> Coal's reaction to storage, transport, and preparation is not consistent and varies from region to region. Some coals, say from Columbia, are very "friendly." They are not highly susceptible to auto ignition or explosion. On the other hand, coals from the Powder River Basin (PRB) in the western United States are dangerously unstable, and special consideration must be given for their handling.

Coal-mill explosions can be as dangerous as silo explosions. Usually, mill explosions rupture the primary air breaching, releasing a cloud of coal dust and causing a secondary explosion potentially more lethal than the primary explosion. The primary source for these explosions is hot coal in dead spaces in the mill or breaching. Usually, mill fires or explosions do not occur on shutdowns. Therefore, inerting a mill while shutting down is not generally necessary unless it is shut down because of a fire.

Steam injection can heat coal to the point that the coal will begin to self-generate heat up to the ignition temperature. A shutdown long enough for a bed deposit to generate a spontaneous combustion explosion is more likely if it were moistened and preheated with a steam purge. Furthermore, steam inerting can make matters worse by wetting the coal in the mill and making a restart difficult. Remember, "an explosive $CO\text{-}H_2$ mixture may also result from the steam hot coke reaction."[3]

On the other hand, when a fresh supply of oxygen is introduced during the restart of a mill full of coal, it may explode if it is not inerted immediately prior. On a restart, steam purge will at least eliminate the oxygen required for an explosion. Also, steam inerting on a restart does not allow time for the coal to generate enough heat to self-ignite.

The best mill-fire (or coal-smolder) detectors used to determine the need for mill inerting or fire extinguishing are mill CO monitors. If CO is detected, the mill should be emptied into the boiler, in service, or shut down and inerted until CO is no longer detected. Also, it is extremely important to keep the mill outlet temperatures below the coal self-heat generation temperature.

AIR PREHEATERS

Fire detection in air preheaters is by far the most difficult aspect of fire protection in power plants because of the small initial ignition point compared to the vast surface area of the heater and the high speed of the rotor. As a consequence, a fire is usually large and out of control before it is detected by fire detectors or plant operators. The fire is usually started by deposits and can quickly reach metal temperature combustion, which is around 1300°F. (Refer to Chap. 5 for root causes for deposits and fires in air preheaters.) By this time, deluge systems are of little use, and the fire must be fought with hose stations. Once the metal reaches a temperature that supports combustion, the carbon within the metal will burn to rust. Air preheaters contain ample amounts of metal combustibles with the vast amount of heat-exchange plates in the bundle sections. Once ignited, these fires are difficult to extinguish. One such fire at the Itabo Station, in the Dominican Republic in 2003, took an extraordinary and courageous effort on the part of station operators and maintenance personnel to extinguish. They had to use fog spray nozzles to get close enough to the heater in order to remove access covers. By this time, the fire was blow torching out from expansion joints, threatening the surrounding structures. Once the covers were removed, the team used multiple hose stations to spray water into the heater gooseneck fashion. It took 6 hr to extinguish. Fortunately, the heater did not seize and stop, which would have made it impossible to cool the rotor and extinguish the fires as the elements passed the doorways. If the rotor had stopped, and the fire had not been extinguished, the entire station would have been threatened.

An infrared, or equal, detection system must be used to detect deposit hot spots in air preheaters. The detection system must have "bottle up" compensation in order to allow the cold-end temperature to equal the hot-end temperature without tripping an alarm when the boiler is bottled up. The deposit temperature of self-ignition is in the range of 600 to 800°F. This is below the metal combustion temperature; therefore, the fires can be controlled using a deluge system if detected early. The deluge system must be sized at least twice that required for water washing. Once the metal combustion temperature (1300°F) is reached, the deluge system will be ineffectual.

Note

Gas temperature monitoring and operator view ports are useless devices for early air preheater fire detection.

SOLID FUEL STORAGE

The traditional method for fire protection in coal storage areas is compaction. Compaction prevents or minimizes air infiltration. Remember that the type and source, plus moisture and oxygen content, determine a coal's self-heat generation temperature, which can be as low as the ambient temperature. The piles are normally compacted in 1-ft layers. Woodpiles are difficult to compact and are generally piled high to maximize rain runoff to prevent water infiltration. In the case of wood, water provides more oxygen and hydrogen for wood decomposition, self-heat generation, and, subsequently, fires. Dry ice is the only proven way to extinguish woodpile fires. A hole is made near a hotspot and a front-end loader drops dry ice into the hole and covers the hole. The dry ice cools hot wood and scavenges the oxygen in the pile, extinguishing the fire.

CONVEYORS

Sprinkler control valves for conveyor systems must be contained in heated spaces for freeze protection. The sprinkler system must be a dry-pipe system. Tunnels must have air locks to prevent natural draft from carrying a fire from one structure to another.

FIRE BARRIERS

Fire barriers must be installed in ducts, building penetrations, between large transformers, and between multiple diesel engine lube oil pits so that a fire can be contained where it started.

REFERENCES

1. *Manual for Industrial Safety and Fire Protection*, Sandia Laboratories, Albuquerque, NM, 1971.
2. Foote, H. M., "Automatic Sprinkler Systems," *Plant Engineering*, October 13, 1977.
3. Benson, Maloney, and Diaz-Tous, "Defend Your Mills Against Fire, Water, and Mechanical Failure," *Power*, May 1982.

CHAPTER 15

FUELS AND COMBUSTION

FOSSIL FUELS

All hydrocarbon fuels come from the same source, whether coal, petroleum, or gas. Decaying vegetal matter accumulating in marshes and bogs was covered hundreds of millions of years ago with the earth's upheavals and rising and receding oceans during cyclic ice ages. Under extreme pressures and temperatures, the different fossil fuels were formed from this decaying vegetal matter.

It is reasonable to assume that the rate of consumption of these fuels will eventually equal maximum possible extraction. That is, all the world's reserves will be under production. At that point of peak production, no more growth in the fossil energy field will be possible. After reaching this threshold, energy efficiency will have to increase in direct proportion to the depletion of reserves. Because the majority of petroleum production is utilized in the transportation industry, primarily in the automotive industry, it is reasonable to assume alternate means of transportation will be in place before alternate forms of energy. In the power generation industry, a complete shift to natural gas is occurring, for new construction projects, due to improved extraction methods using shale fracturing and horizontal drilling. Development still continues in the field of fusion power production, which, if successful, could replace all fossil-fuel power production. Hydropower will reach a similar threshold, but renewable energy sources such as solar and wind will not have a significant impact. From 1890 until 1920 in the United States, electrical generation rose from practically zero to 40 billion kWhr. After 1920, electrical power generation rose to more than 3,800 billion kWhr in 2006. Considering that power production, as well as mechanized land travel, was virtually zero just before the turn of the 20th century, this is remarkable.

HYDROCARBONS (COAL)

The extent of the coalification of decaying vegetal matter determines its ranking as a fuel (Table 15-1). It is ranked as follows:

Peat (years)
Lignite
Subbituminous
Semianthracite
Anthracite
Graphite (billions of years)

TABLE 15-1 Classification of Coals by Rank after ASME-ASTM Method[1]

Rank	"Mineral-matter-free" basis			Additional requisites physical properties
	"Dry"		"Moist" minimum heating value	
	Minimum % Fixed Carbon	Maximum % volatile matter		
I. Anthracite:				
1. Meta-anthracite	98	2		
2. Anthracite	92	8		
3. Semianthracite	86	14		Nonagglomerated
II. Bituminous:				
1. Low volatile	78	22		
2. Medium volatile	69	31		
3. High volatile *A*	Less than 69		14,000	
4. High volatile *B*	Less than 69		13,000	
5. High volatile *C*	Less than 69		11,000	a. Agglomerating, nonweathering b. Agglomerating, weathering c. Nonagglomerating, nonweathering
III. Subbituminous:				
1. Subbituminous *A*	Less than 69		11,000	Weathering, nonagglomerating
2. Subbituminous *B*	Less than 69		9500	
3. Subbituminous *C*	Less than 69		8300	
IV. Lignite:				
1. Lignite	Less than 69		Less than 8300	Consolidated
2. Brown coal	Less than 69[a]		Less than 8300	Unconsolidated

[a] Coals having more than 69 percent fixed carbon shall be classified according to fixed-carbon content regardless of heating value.

Peat is not actually coal—and is not ranked by the American Society for Testing and Materials (ASTM), but is a product of the first stage of coalification. Peat has a high, inherent moisture content and is on the low end of the solid fuel scale. Solid fuels with high moisture content such as peat and garbage with high putrescibles such as fruit and vegetable (green waste) will not support combustion unless dried. They require more energy to dry, in order to support combustion, than can be converted to useful energy. The main reason for this is the moisture is inherent, not free. These products must be pre-dried using some form of external heat in order to be used as fuels; otherwise, they will not support combustion.

Coal can be centrifuged to eliminate most of the free moisture, and all the remaining free moisture can be removed by air drying. This is done for coal-slurry, pipeline conveying systems. Wood-burning boilers also can use flue-gas drying to remove free moisture from the fuel.

Although graphite is the most advanced in the coalification process, it is not actually ranked as a fuel. It lacks the properties required as a fuel because of its high-ignition temperature requirement, and it is far more useful for other purposes.

From the Table 15-2 averages, it can be observed that the heating values for anthracite coal are not necessarily higher than those of bituminous. Also, there is no real definitive

TABLE 15-2 Typical Composition of Solid Fuels[2]

Fuel	Proximate Analysis (%)				Ultimate Analysis (%)						Heating Value (Btu/lb)	
	Moisture	Volatile Matter	Fixed Carbon	Ash	Carbon	Hydrogen	Oxygen	Nitrogen	Sulfur	Ash	As Reported	Dry Basis
Wood												
Oak, dry					50.2	6.0	43.3	0.1		0.4		8320
Pine, dry					52.6	7.0	40.1			0.3		9610
Peat	56.8	26.0	11.2	6.0	23.1	9.6	59.6	1.3	0.4	6.0	3600	8330
Lignite	34.8	28.2	30.8	6.2	42.4	6.7	43.3	0.7	0.7	6.2	7210	11,000
Bituminous coal:												
Low, vol, Pennsylvania	3.3	20.5	70.0	6.2	80.7	4.9	5.3	1.1	1.8	6.2	14,310	14,800
Low, vol, Arkansa	3.4	16.2	71.8	8.6	79.6	4.3	4.8	1.7	1.0	8.6	13,700	14,180
Med, vol, Virginia	3.1	21.8	67.9	7.2	80.1	5.0	5.2	1.5	1.0	7.2	14,030	14,480
Med, vol, Alabama	2.8	23.8	62.4	11.0	76.9	4.8	5.1	1.5	0.7	11.0	13,450	13,870
High, vol, A, Kentucky	3.1	35.0	58.9	3.0	79.2	5.7	10.0	1.5	0.6	3.0	14,290	14,750
High, vol, B, Ohio	8.2	36.1	48.7	7.0	68.4	5.6	16.4	1.4	1.2	7.0	12,160	13,250
High, vol, C, Illinois	12.1	40.2	39.1	8.6	62.8	5.9	17.4	1.0	4.3	8.6	11,480	13,050
Subbituminous, Colorado	19.6	30.5	45.9	4.0	58.8	6.0	29.6	1.3	0.3	4.0	10,130	12,600
Anthracite coal, Pennsylvania	2.3	3.1	87.7	6.9	86.7	2.2	2–9	0.8	0.5	6.9	13,540	13,860
Semianthracite, Pennsylvania	3.0	8.4	78.9	9.7	80.2	3.6	4.7	1.1	0.7	9.7	13,450	13,870
Semianthracite, Virginia	3.1	10.6	66.7	19.6	70.5	3.5	5.0	0.8	0–6	19.6	11,850	12,230
Bagasse, dry					43.2	6.0	48.0			2.8		7,980
Coke, by-product	0.8	1.4	87.1	10.7	85.0	0.8	1.2	1.3	1.0	10.7	12,690	12,790
Coke breeze	12.0	4.2	65.8	18.0	66.8	2.5	11.2	0.9	0.6	18.0	10,200	11,590
Charcoal	12.0	1.9	83.1	3.0	84.0	2.3	10.7			3.0	11,300	12,850

dividing line between the different ranks of coal. For example, a Kansas low-grade subbituminous may be classified as lignite in Texas.

For the purpose of analyzing combustion heating value, moisture and ash are the most important for solid fuels. In an analysis of types of fuel, moisture is given as inherent moisture, not surface moisture. Surface, or free, moisture, depends entirely on surface area. For example, 1 kilo of large pieces of coal has a low surface area while the same kilo of crushed coal has a high surface area and coal silt or pulverized coal has a very high surface area.

The purchase and storage of solid fuels must take into consideration fineness because of surface area and the ability of fuel to retain moisture. The finer the fuel, the greater its ability to retain moisture. Grindability must also match that which the mills are designed for. If a Hardgrove index of 50 is used for a mill performance curve and the mill is designed for a low-carbon-rank bituminous and a high-carbon-rank anthracite is used, the mill performance will suffer substantially. Typically, 50 is used for peak mill output for bituminous coal. Grindability can be thought of as the reciprocal of hardness. Coal grindability is classified by the Hardgrove index. Anthracite, the hardest coal with the highest fixed carbon content, has an index of 40 to 45. Bituminous has an index of 45 to 60, and subbituminous, the softer coal with the lowest fixed carbon content, has an index of 60 to 90. In other words, anthracite is the hardest and has the lowest grindability.

Grading of coal should not be confused with ranking as described earlier. Grading of coal is standardized by ASTM. Grading is based on five physical characteristics or properties: size, heating value, ash content, ash-softening temperature, and sulfur content. Therefore, a grade specification might look like the following: 2–4 in, 140-A4-0-F20-S 0.7. The heating value is expressed in the nearest hundreds, for example, 14,000 Btu/lb is expressed as 140. The rank and grade might be (60–90) 2–4 in, 140-A4-0-F20-S0.7.

All solid fuels have a self–heat-generating temperature that is dependent on the fuel properties, moisture content, and oxygen. The process is called *exothermic oxidation*. After coal (or other solid fuel) is crushed, the surface area is increased dramatically, exposing it to greater amounts of moisture and oxygen. Some coal will start to self-generate heat at ambient temperature and will eventually reach a spontaneous combustion ignition temperature.

Wood fuels react in a similar manner as oily rags. But spontaneous combustion will not occur unless the self-heating temperature is reached. In some parts of the United States, a sawdust pile will decompose into a black liquor waste, within a 10-yr period, and disappear into the watertable without ever generating enough heat to catch fire. In other parts of the country, the same pile may self-ignite in a matter of days. Too much water (or too little oxygen) will inhibit spontaneous combustion. Unfortunately, too much water means that the fuel must be dried in order to burn. Therefore, adding water can be self-defeating. (Refer to Chap. 14 for other problems in power stations associated with solid fuel fires and explosions.)

ANALYSIS OF SOLID FUEL

Analysis of coal or coke should be conducted in accordance with ASME PTC 3.2, Solid Fuels.[2] Analysis of wood and agricultural waste products should also be conducted using the sampling guidelines set forth in PTC 3.2. Analysis of fuels is conducted first at the mine, then at the place of usage, and finally where it is burned. The definitions are "run-of-the-mine," "as-received," and "as-fired."

The mine owner is required to ship a grade within certain guidelines and therefore must provide a mine site analysis. The consumer must determine that the fuel purchased is within contractual guidelines and performs the as-received analysis. Degradation in shipment and storage require this analysis for contractual reasons. The as-fired analysis must be performed in order to determine fuel consumption. First, the boiler efficiency is determined by the losses method using ASME PTC 4.1, Steam Generating Units.[3] From the boiler efficiency, the fuel consumption can be determined using the input–output method. In other words, once the efficiency is known, the fuel consumption can be determined. Fuel feeder flow measurements for solid fuels are not considered accurate enough for efficiency calculations.

Figure 15-1 is an example of boiler efficiency that was determined by the losses method. The fuel consumption was then adjusted by iteration until the efficiency in the input–output

REID GARDNER F-W spec coal CALCULATION SHEET 1
REV 07-sep-09 06:38 AM

SYMBOL		DESCRIPTION		UNIT	
		HEAT LOSSES		Btu/lb A.F. Fuel	%
LUC		Unburned carbon in refuse	=	2,18	0,027
LG'		Heat in dry flue gas	=	325,40	4
Lmf		Moisture in "as fired" fuel & limestone	=	318,05	3,94
LH		Moisture from burning of Hydrogen	=	348,28	4,281
LmA		Moisture in the air	=	4,18	0,051
LCO		Formation of Carbon Monoxide	=	0,00	0
LB		Surface radiation and convection	=	19,50	0,24
Ld		Sensible heat in flue dust	=	41,93	0,515
LmL		Moisture in limestone(included in fuel)	=		0
LCA		Calcination of limestone(included in Cb)	=		0
Lp'		Heat in bottom ash	=	0,00	0
LSB		Heat in sootblowing steam	=	0,00	0
LBD		Heat in blowdown	=	0,00	0
Lp		Heat in ash coolers	=		0
		MANUFACTURE'S MARGIN			1,5
	L	= Btu / lb A.F. fuel	=	1.059,51	14,6
i H*f*	*Hf*	Chemical heat in fuel to be obtained			
	Hf	= Btu / lb A.F. fuel	=	8.160,00	

	HEAT CREDITS PER POUND		Btu Credit/ lb A.F. Fuel
BAe	Entering air	=	-9.118.401
mAe	Moisture entering with the inlet air	=	-68.262
fe	Measured fuel rate	=	372.885,00
B	heat credits/lb	=	-24,64
Be	Heat credits/hr	=	-9.186.663

Air Preheater Air Leakage	=	0,0%
Boiler Efficiency (By Losses) Eff1 = 100 - L/H*f*+B x 100	=	85,48 %
Losses Eff1= 85,48 Capacity Factor	=	0,85
In/Out Eff2= 85,48 Corrected Net Output	=	275,00 MW
Wfe= 372.885 Lbs/Hr AF T/YR	=	1.388.251
SFC = 5,05 Lbs/KW (AF) BDT/YR	=	967.194
SFC = 3,52 Lbs/KW (BD) LST/YR	=	0
NPHR = 22.129 Btu/KW		

FIGURE 15-1 Solid fuel consumption calculation.

equation equaled the efficiency in the losses equation. From the as-fired fuel consumption and unit capacity factor, the annual fuel consumption and net plant heat rate (NPHR) for as-fired and bone-dry fuel can be determined. It should be pointed out that capacity factor (the ratio of the actual power production to the rated power production) can be greatly affected by fuel moisture. In pulverized-coal units, the resident time of the fuel in a mill can be greatly affected by the moisture of the coal. If moisture is above specification, the boiler capacity will be reduced because of mill capacity reduction associated with mill residency time.

Biomass plants can have much more serious problems regarding fuel analysis than coal plants. The same guidelines should be followed for each, however, taking into account the different fuel-receiving methods. For example, at several wood-fired plants in California during the 1980s, the fuel costs were millions of dollars more than estimated for the first year of operation. The reasons for such discrepancies included inadequate fuel specifications in the fuel-purchasing agreements, inadequate fuel-receiving systems, and inadequate boiler-efficiency testing.

There are always two things to remember when estimating the annual fuel budget: (1) you must receive the exact fuel you pay for, and (2) you must account for all the fuel received and burned.

The fuel receiving system conceptual design must incorporate fuel unloading, weighing, and sampling into one system at one location. The system must be designed such that the owner's operators can easily perform all operations and analyses to make an exact accounting of fuel received.

The fuel-purchase agreements must incorporate sufficient specifications to prevent fuel suppliers from mixing fuel grades. For example, fuel must be purchased by grade, which must be dependent on percentage of fines, ash, and moisture. Limits must be placed on acceptable ash content. The limits must include free, as well as inherent, moisture and ash. The sampling procedure must be structured to do so. You don't want to pay for dirt scooped up in the supplier's fuelyard or water accumulated, supposedly in delivery.

The acceptance tests for the plant must use testing procedures that will become an integral part of future plant operations. These tests must provide the accounting department with the most accurate fuel utilization figures possible. Also, a belt scale should be incorporated with an automated conveyor grab sample procedure to give an accurate, continuous accounting of fuel delivery.

FUEL UNLOADING AND SAMPLING

For biomass fuel, the laboratory should be an integral part of the truck unloading building with all the necessary lab equipment immediately available. The laboratory equipment for all fuels must be properly specified for the application. The operators should be trained in both sampling techniques and sample analysis. Further, they should be schooled in the importance of their work, the use and meaning of uncertainties, and the revenue consequences of error. Fulfilling these requirements for this type of operation, it will not be necessary to employ chemistry lab technicians because the procedures will be routine and repetitive.

Biomass as-received ash and moisture content should be determined using laboratory methods by the unloading operators. Because the operator takes the sample, the storage and drying part of the procedure will take far less effort and time than packaging and delivery to another building or third-party lab. Biomass samples should not be held at the unloading point more than 3 hrs before they are analyzed.

The fuel unloading facilities must be an integral part of the fuel-sampling system. This means truck scales must be an integral part of truck dumpers. Ship or train unloading conveyors must incorporate scales. Biomass truck dumper shacks must contain the fuel laboratory.

The best biomass fuel unloading facilities will incorporate a truck dumper that raises the entire truck—tractor and trailer. The dumper arms should be designed so they are totally supported by weight sensors that permit the dumper to be a dumper and scale combination. This arrangement allows the operator to be in control of all fuel received at one location. The dumper design must also consider that a representative sample must be taken automatically for the operator.

The greatest inaccuracies in fuel sampling can be introduced at the fuel suppliers' facilities. Mixing different grades of fuel does this. An example is if the fuel supplier mixes high-moisture-content fines with a relatively coarse-grade fuel. This can be done easily by wood suppliers by loading layers of chips, sawdust, and then chips again. The sawdust layer contains large quantities of water used to cool saws. The weight of the sawdust is substantially higher than the chips. However, the volume of chips is much greater and has the most influence on the fuel analysis. Furthermore, if the operator takes a sample off the back of a truck, where the supplier loaded the best fuel, the as-fired versus as-received fuel accounting will differ substantially in the long run. If the prevailing price for sawdust is $10/ton and $40/ton for chips, imagine the difference between not only cost, but also the quantity of fuel purchased. In a year's time, this could account for millions of dollars between as-received and as-fired accountability. It is very important that the as-received fuel be analyzed in such a manner so as to accurately determine its true quality. This should be used to determine the bulk rate paid to suppliers. The analysis should consist of moisture and ash content.

These inaccuracies can be controlled by six important considerations:

1. The owner's operator takes the sample.
2. The system is designed such that samples taken are representative of the entire load.
3. The system is designed such that the unloading operator can observe the full load and verify its content.
4. The owner has developed adequate fuel quality classification criteria for procurement.
5. Fuels of different grades are delivered separately. This is especially true with biomass fuel. Sawdust, bark, or bark dust should not be delivered with chips.
6. Procedures for sampling, sample reduction, and analysis must follow established standards. These procedures should follow the sampling and sample reduction procedures described in detail in PTC 3.1[4] regardless of whether they are for coal, coke, or biomass.

The larger the sample, the more accurate the analysis will be. However, only crucibles of fuel will be analyzed in the laboratory. A sample reduction procedure must be followed to reduce the gross sample to a laboratory sample representative of the whole. The number and size of the samples must follow the table from PTC 3.2[2] for performance testing. The PTC 3.2 method is based on ash content. The accuracy using this method will be within ±10 percent for the ash content. If performance test accuracy is required, the code stipulates that to increase the ash analysis accuracy to ±5 percent, the increments stipulated must increase four times. For an accuracy of ±3.3 percent, the increments must be increased nine times. An increment is a sweeping motion through a fuel stream from a conveyor or a digging motion into a load. For as-received testing, the procedure should be modified to reduce the size of the gross sample, but the same testing equipment should be used.

Once the sample increments have been collected, they must be crushed to the size stipulated in the code. After they are crushed, they must be passed through a large-hole riffle to thoroughly mix them. They are then crushed again and passed through a medium-size riffle. Finally, they are crushed so as they will pass through a small-riffle screen. At each step, the sample is reduced in weight until no less than a 15-lb sample remains. By this time, the sample should be truly representative of the gross lot within the accuracies stipulated.

For samples of biomass fuel, the increment crushing and sample reduction procedure must obviously be altered. Nonetheless, the more closely the procedure follows the code, the more accurate the results will be. If sawdust, or other fines, are mixed with wood chips the chip size must be reduced so that both can be mixed to give an accurate representative sample—the reason being the chip moisture will consist mainly of inherent moisture of approximately 20 percent, while the fines could be as high as 50 percent moisture consisting mostly of free moisture. Commercial grinders are available that can reduce the chips without adding heat that will drive off moisture, reducing the accuracy of the analysis. One suggestion is to freeze the increments with liquid nitrogen, then crush them.

FUEL-WEIGHING SYSTEMS

Automatic fuel-weighing systems can be used to determine the weight of fuel on a conveyor within ±0.25 percent. Given the weight and belt speed, the delivery rate can be determined on a continuous basis. But, the true delivery rate is far from accurate. However, the weight includes moisture and assumes a constant density. This means the grade and moisture content must be determined to establish the true fuel feed or delivery rate, which is the bone-dry feed rate for a particular grade.

CONTINUOUS MOISTURE MONITORS

Generally speaking, continuous moisture monitors are not used except to determine the moisture of a substance of homogeneous density and uniform moisture content such as grain or corn. The reason is that the calibration of such instruments is affected by characteristics like density and hydrogen content. For example, if the density is not homogeneous, then there is no set calibration. This is especially true for neutron and capacitance instruments.

Infrared systems measure fuel moisture over the surface of the belt, which reflects back to a sensor. This method does not measure the moisture through the depth of the belt material, and therefore, it is not considered accurate. Nuclear devices give accurate moisture, but they have to be recalibrated for each sample density and hydrogen content. They are only good for lab analysis and only for homogenous fuels. Capacitance moisture detectors are also impossible to calibrate for variable density and particle size.

Instruments that measure surface moisture are much more inaccurate than those that measure a cross-section sample. The method generally used for cross-section measurement is to slipstream a sample through a metering chute. This is accomplished by use of a sample screw conveyor that collects the sample and returns it to the feed stream again. Nonetheless, their accuracy for solid-fuel systems is questionable.

LABORATORY MOISTURE DETERMINATION

Once a laboratory sample is obtained by the reduction method, the laboratory tests can begin. The first laboratory test is to determine the moisture content. The procedure should define percent moisture of the volatiles driven off at ≈105°C. A large convection oven specified especially for fuel drying will be needed for sample storage and drying.

Coal

Wet appearing coal is first air dried at room temperature or in heating ovens 10°C to 15°C above room temperature. Drying should not continue after the weight change of the sample is not more than 0.10 percent per hour because further drying may cause oxidation. The air-dried coal, or coal appearing dry, should then be crushed to pass an 840-μm (No. 20) sieve. The sample is then dried in a convection oven between 104°C and 110°C for 1 hr. The ASTM (D 271-48) procedure is outlined in detail in PTC 3.2.

Biomass

Biomass samples should be ground to sawdust composition and dried in the same oven used for coal sample drying. "The drying time for wood saw dust should be 16 hours at 100–104°C because one hour was found to be inadequate."[5]

Volatile Matter

Once a dry sample is available, it is heated to 950°C for 7 min for coal in special ovens for this purpose. Wood should be heated for no more than two minutes according to Mingle and Boubel.[5] The samples are heated until the blue luminous flame of distillation and consumption disappears.

Ash Test

Dry coal is gradually heated to 500°C within 1 hr and then to 750°C within 2 hr. The sample is heated to constant weight. Mingle and Boubel[5] found that repeated heating and weighing required by the code for coal was not required for wood. They determined that the samples were completely ashed within 30 min at 750°C.

Proximate Analysis

The moisture, volatile matter, and ash content tests yield what is termed the *proximate analysis* of the fuel, that is, ash, volatile matter, and carbon.

Ultimate Analysis

The ultimate analysis is conducted by burning a sample in a Heraeus-type furnace in which the constituents are absorbed and weighed to determine content percentage. The constituents are carbon, nitrogen, hydrogen, sulfur, and oxygen by difference.

CALORIFIC VALUE

The heating value of the fuel is determined by firing a fuel sample in a *bomb calorimeter*. The higher heating value (HHV) is the gross heating value. The lower heating value (LHV) is the net value and assumes that the latent heat of evaporation for moisture in the fuel is not recovered in the by-products of combustion; that is, the flue gas temperature exits the unit above the condensation temperature. ASME PTC 4.1, Steam Generating Units,[3] uses the HHV for performance testing, and the losses due to moisture in the combustion air and fuel are accounted for in the calculations.

The HHV for solid fuels can be calculated if the ultimate analysis is known using Dulong's formula:[1]

$$HHV = 14{,}540C + 60{,}958\,(H - O/8) + 4050S$$

where

C = fraction of carbon by weight
H = fraction of hydrogen by weight
O = fraction of oxygen by weight
S = fraction of sulfur by weight

Kent's Handbook for Mechanical Engineers' provides two formulas: Dulong's, $HHV = 145.4C + 620(H - O/8) + 41S$, and Vondracek's, $(160.5 - 0.112C_1)C + 486(H - O/10) + 45S$, where C_1 is the percent carbon of the ash and moisture-free basis. The Vondracek formula can be used for both solid and liquid fuels.

PETROLEUM

In the latter part of the 19th century, specifically the 1890s, petroleum refining was nothing more than a distillation process. Crude oil was being distilled to produce kerosene for lamp oil, which was used as a substitute for whale oil. Crude at that time was rated as sweet or sour according to its sulfur content. Oil with a high sulfur content, when burned in kerosene lamps, had a bad or sour smell—therefore, the term *sweet* or *sour crude* was coined. A use for other distillates was not yet developed, and they were discarded into the environment.

Most people of the period were familiar with the fermentation and distillation process for making alcohol. Therefore, for the oil pioneers of the day, distillation of kerosene, which was the major product, was probably no mystery. The other fractions from distillation were discharged to the environment similar to the moonshiner's dregs. For exploration, they simply drilled wells where there was oil seeping out of the ground, or they drilled where someone had the misfortune of striking oil trying to find water.

The fermentation and distillation process illustrated in Fig. 15.2 is what is required to make ethanol from farm produce such as corn. The distillation process requires a heat

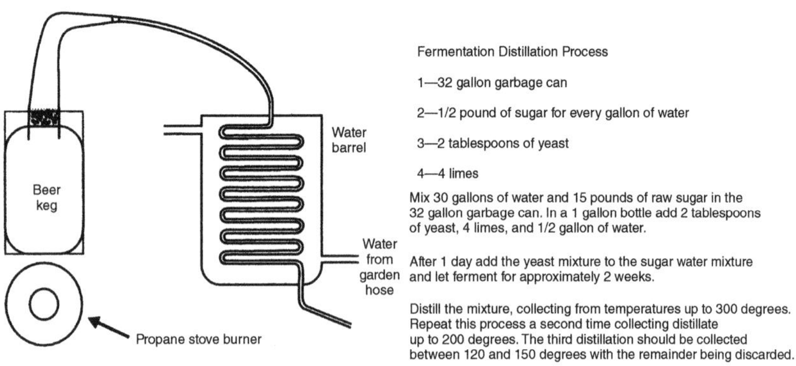

FIGURE 15-2 Typical home made still.[7]

source to generate the steam from the fermented batch. In other words, fuel is burned to make fuel.

> **Note**
>
> The past 100 years have included the second industrial revolution, the technological revolution, travel to outer space, and major political revolutions. These developments are unparalleled in thousands of years of human history. All this flurry of activity and development pushed the development of fossil fuel and atomic energy. This giant leap forward in industrial and technological development can be easily followed, browsing the Internet for the history of petroleum and gasoline. Any engineer with a curiosity and desire to understand recent modern history should take the time to do so. If we're addicted to anything, it is to science and technology—not oil. Oil has just supported our quest for the betterment of mankind through the furtherance of science and technology.

An excellent book for plant engineering professionals interested in the basics of the petroleum refining industry is *Petroleum Refining for the Non-Technical Person*, by William L. Leffler.[8]

In the power industry, diesel and residual oil are the primary *petroleum products* (liquid fuels) used to fire boilers, gas turbines, and diesel engines. In central stations, No. 2 fuel oil is primarily used for lightoff and warmup, after which residual fuels are fired. The ASTM standards for petroleum products set the standard for testing liquid fuels. The ASTM methods and terminology provided in the following section can be found in detail in the cited ASTM standards.

HEATING (CALORIFIC) VALUES

The heating value is determined by burning a fuel sample in the presence of an oxygen atmosphere under pressure. The heat release is absorbed by water in a jacket surrounding the "bomb" calorimeter. The temperature rise of the water determines the heat release. The water released in the combustion process is condensed, and the latent heat of evaporation is

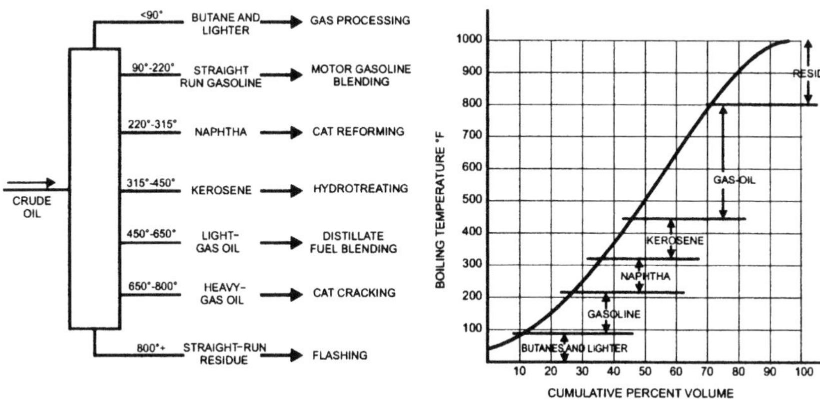

FIGURE 15-3 Distilling crude and product distribution.

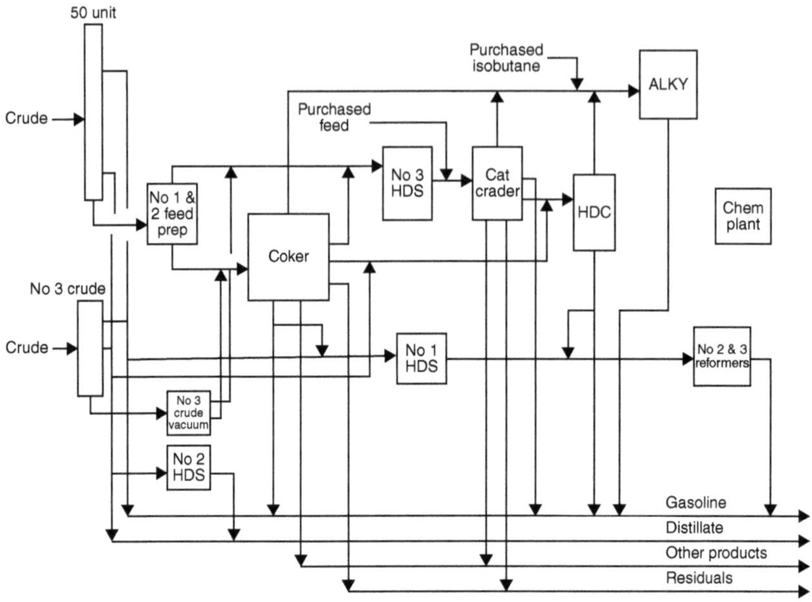

FIGURE 15-4 Typical simplified refinery flow diagram.

given off to the water jacket. For field use, the HHV for all fuels can also be determined by calculation. For performance testing, the calorimeter test is always used.

The lower (or net) heating value is a fictitious value that would be obtained if the water vapor in the products of combustion could be cooled to 77°F (25°C) and still remain completely in the vapor state. In other words, the lower heating value is equal to the higher heating value minus the latent heat of evaporation of the water formed.[1]

Precise measurement of the HHV for liquid fuels can be determined by either of the following standards: ASTM D240-87, Standard Test Method for Heat of Combustion of Liquid Hydrocarbon Fuels by Bomb Calorimeter,[9] or ASTM D2382-88, Standard Test Method for Heat of Combustion of Hydrocarbon Fuels by Bomb Calorimeter (High-Precision Method).[10]

Approximations for heating value of a liquid fuel can be determined by calculation if the specific gravity, or API, is known, using the following equations:[1]

$$\text{HHV} = 22{,}820 - 3780d^2 \text{ Btu/lb}$$

where

$$d = \text{specific gravity } 60/60°\text{F}$$

For example:

$$d = 0.68$$
$$\text{HHV} = 22{,}320 - 3780(0.4624) = 20{,}572$$

Or

$$\text{HHV} = 18{,}650 + 40(\text{API} - 10) = \text{Btu/lb for fuel oil}$$
$$\text{Degree API} = (141.5/\text{spgr } 60/60) - 131.5$$

$$= (141.5/0.68) - 131.5$$
$$\text{API} = 76.59$$
$$\text{HHV} = 18{,}650 + 40(76.59 - 10) = 21{,}313$$

Note the difference between the results using formula (1) and formula (2).

Note

For distillate fuel oils and residual fuel oils having gravities higher than 18° API, a close approximation of the gross heating value is Btu per lb = 18,250 + (40 × API). For residual fuels having gravities less than 18° API, Btu per lb = 17,690 + (58 × API). A correction should be made if water and sulfur are present in appreciable amounts. See Reference 12.

Note

The heating values given, such as 18,250; 18,650; and 17,690; that are corrected by specific gravity for particular fuels are reference values. The heating value for the specific fuel grade must be known in order to correctly use the correction for specific gravity. If the fuel-grade heating value is 19,638 Btu/lb for mid-continent distillate or 19,178 Btu/lb for Venezuelan cracked distillate, none of the preceding formulas will yield an accurate correction for gravity. Therefore, it is recommended that you use the heating value for the specific grade that is burning be substituted for the heating values in the formulas provided.

The heating values for gaseous fuel can be calculated if the constituents of the gas are known. For most fuel gases, the heating value is given in Table 15-2.[1]

The following sample calculation, from Fuels and Combustion, Smith and Stinson, for a manufactured gas mixture is given using molecular weights and heating values from the Table 15-3. This could also apply to well gas with different constituents. Also, ASTM D3588-91 Standard Practice for Calculating Heat Value, Compressibility Factor, ASTM 1945-81 Standard Method for Analysis of Natural Gas by Gas Chromatography and ASTM 1826-88 Standard Test Method for Calorific (Heating) Value of Gases in Natural Gas Range by Continuous Recording Calorimeter can be used to determine gas properties.

Example Compute the heating value, in Btu/ft^3, for the following gas mixture at 60°F and 14.7 psia:

$$CO_2 = 4.6\%$$
$$N_2 = 7.0\%$$
$$H_2 = 49.7\%$$
$$C_2H_6 = 2.5\%$$
$$O_2 = 0.4\%$$
$$CO = 24.9\%$$
$$CH_4 = 10.9\%$$

TABLE 15-3 Heats of Reaction[16]

Name	Formula	Molecular Weight	Δh^a evaporation Btu/lb	$-\Delta h$ combustion,^{b} or heating value, Btu/lb Higher	Lower	Δh formation^c Btu/lb mole
Methane	CH_4	16.04		23,861	21,502	–32,179
Ethane	C_2H_6	30.07		22,304	20,416	–36,401
Propane	C_3H_8	44.09	147.1	21,646	19,929	–44,647
n-Butane	C_4H_{10}	58.12	155.8	21,293	19,665	–53,626
Isobutane	C_4H_{10}	58.12	141.4	21,242	19,614	–56,576
n-Pentane	C_5H_{12}	72.15	157.5	20,914	19,340	–74,400
n-Hexane	C_6H_{14}	86.17	157.4	20,771	19,233	–85,480
n-Heptane	C_7H_{16}	100.20	156.8	20,668	19,157	–96,470
n-Octane	C_8H_{18}	114.22	156.1	20,591	19,100	–107,500
2,2,4-Trimethylpentane	C_8H_{18}	114.22	132.2	20,556	19,005	–111,500
n-Nonane	C_9H_{20}	128.25	155.7	20,531	19,056	
n-Decane	$C_{10}H_{22}$	142.28	155	20,483	19,020	
n-Tetradecane	$C_{14}H_{30}$	198.38	154	20,358	18,927	
n-Hexadecane	$C_{16}H_{34}$	226.43	154	20,318	18,808	
n-Pentatriacontane	$C_{35}H_{72}$	492.93		20,350	19,032	
Ethylene	C_3H_6	28.05		20,276	20,276	22,478
Propylene	C_2H_6	42.08		21,032	19,683	8,776
Isobutene	C_4H_8	56.10	157.7	20,716	19,367	–6,017
Octene	C_8H_{16}	112.21	158.0	20,349	19,000	
Cyclopentane	C_5H_{10}	70.13	174.7	20,175	18,825	
Cyclohexane	C_6H_{12}	84.16	168.8	20,026	18,676	
Benzene	C_6H_6	78.11	186.3	17,986	17,259	21,079
Toluene	C_7H_8	92.13	177.3	18,245	17,424	5,157
o-Xylene	C_8H_{10}	106.16	175.9	18,438	17,547	10,510
Methyl alcohol	C_2H_6	32.0	502	9,770	8,644	
Ethyl alcohol	C_2H_5OH	46.0	396	12,780	11,604	
Propyl alcohol	C_3H_7OH	60.0	295	14,500	13,300	
Butyl alochol	C_4H_9OH	74.1	254	15,500	14,284	
Acetylene	C_2H_2	26.04		21,460	20,734	97,485
Hydrogen	H_2	2.016		60,958	51,571	0
Carbon (solid, graphite)	C	12.01	To CO_2	14,087		0
Carbon (coke)	C	12.01	To CO_2	14,540		0
Carbon (coke)	C	12.01	To CO	4,440		0
Water, liquid	H_2O	18.016				–122,890
Water, vapor	H_2O	18.016				–103,968
Carbon monoxide	CO	28.01	To CO_2	43,43.6		
Carbon dioxide	CO_2	44.01	147.1			–169,182

[a] Latent heat of evaporation of fuel at constant pressure, 77°F.
[b] $-\Delta h$ combustion, or –heating value at constant pressure, 77°F, each compound in its normal state at 77°F. (For example, CH_4 is a gas at 77°F, so heating value is for gaseous fuel. C_5H_{12} is ordinarily a liquid at 77°F, so the heating value for it is for liquid fuel.)
[c] Δh formation, or heat of formation of compound, from its elements at constant pressure, 77°F.

TABLE 15-4 Data for the Example

Component	Moles	Heat of Reaction (Btu/lb)	Heat Released (Btu)
CO	0.249	4343.6×28.01	30,294
H_2	0.497	$60,958 \times 2.016$	61,076
CH_4	0.109	$23,861 \times 16.04$	41,718
C_2H_6	0.025	$22,304 \times 30.07$	16,767
Total			149,855

Equation of state:

$$V = RT/P$$

$$R = (1545/\text{molecular wt})\ \text{ft/°R}$$

$$T = 460 + 60 = 520°\text{R}$$

$$P = 14.7 \times 144 = 2117\ \text{psf}$$

$$V = (1545 \times 520)/2117 = 379\ \text{ft}^3$$

$$\text{HHV} = 149,855/379 = 395\ \text{Btu/ft}^3$$

COMBUSTION

Combustion takes place in three phases for solid fuels: distillation, oxidation, and reduction. The combustion of fuels in a perfect world should take place in the boiler furnace. The convection pass of a boiler should be the dark path where steam is heated by hot gases. In the worst-case scenario, combustion occurs in both the furnace and the convection pass. In the worst-case scenario example, wood is fired in a stoker-type boiler where, theoretically, it will burn as a bed of fuel following the simplification of the three-phase process:

After surface distillation of a solid particle of fuel the following reactions occur:

$$\text{Oxidation: } C + O_2 \rightarrow CO_2$$

$$\text{Reduction: } C + CO_2 \rightarrow 2CO$$

$$H_2O \rightarrow OH + H$$

$$OH + H_2 \rightarrow H_2O + H$$

$$CH + O_2 \rightarrow CO_2 + H_2O$$

In Fig. 15-6, the ideal combustion bed is illustrated. Figure 15-7 illustrates what actually occurs. The ideal bed is one that can only be maintained with coal fires. Figure 15-5 illustrates what happens when firing wood. As the wood dries and distillation takes place, the wood particles' surface burns continually, reducing its size. Eventually, the weight of the particle reaches a weight equal to the carrying capacity of air that is equal to the primary air velocity. It is then carried upward, suspended in the secondary air volume into the convection pass and precipitator. Looking into the convection pass, a transparent

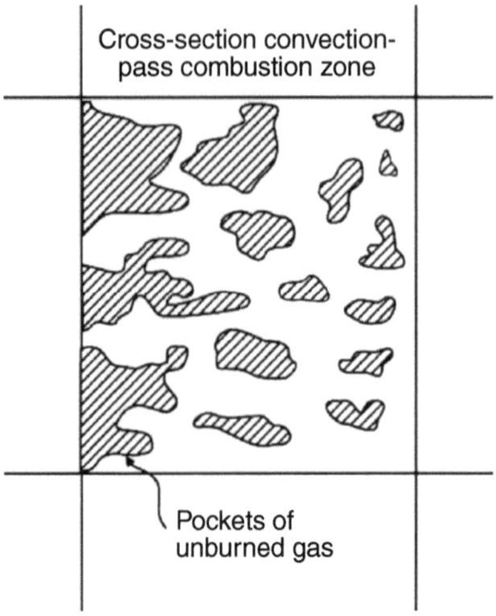

FIGURE 15-5 Flame pockets in convection pass.

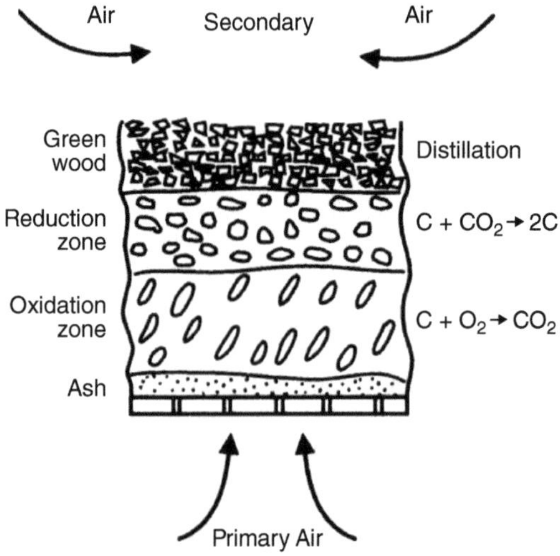

FIGURE 15-6 Ideal stoker bed combustion.

luminance can be observed with flames from pockets of burning gas as illustrated in Fig. 15-5. The fine particles of char can only be observed when they are collected with the ash at various ash collection points.

The degree of char, or carbon, carryover can be determined by the carbon content of the ash. This carryover is directly dependent on combustion efficiency, which is in turn dependent to a great extent on furnace residency time.

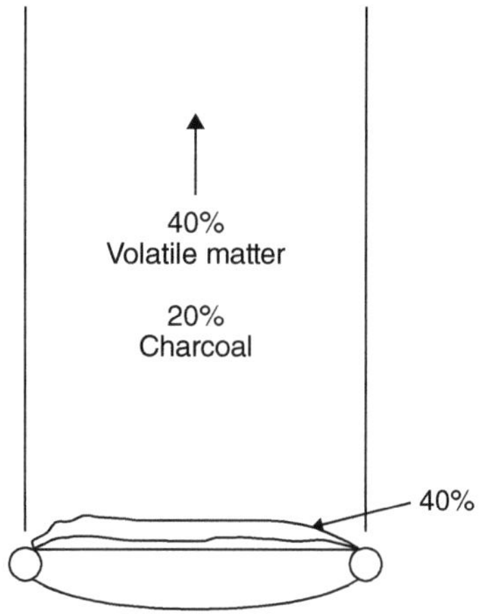

FIGURE 15-7 Wood burning stoker combustion.

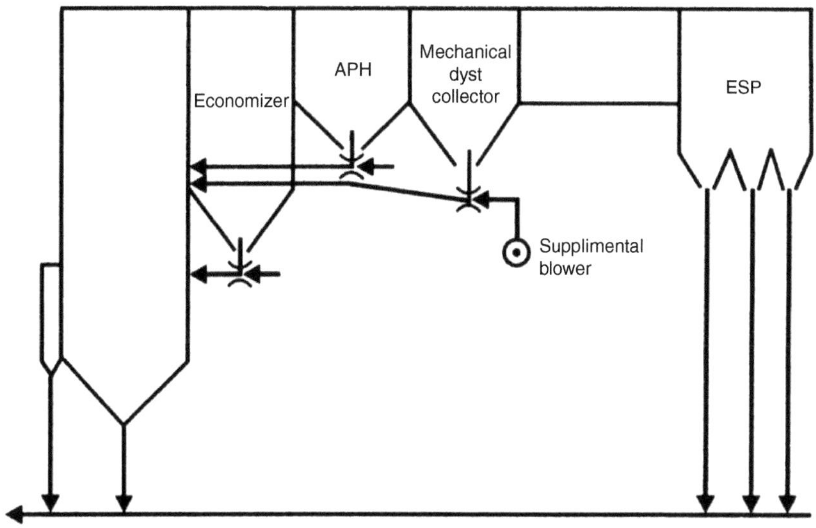

FIGURE 15-8 Experimental char re-injection system.

In the mid-1980s, startup engineers from DCR Engineers wrestled with the problem by attempting to re-inject the char back into the furnace. This was done by using supplemental blowers with venturis, as illustrated in Fig. 15-8. The major problem was that the pressure in the furnaces was higher than at the collection points. The reason was, the closer the induced draft fan was to the collection point, the higher the negative pressure. Later, a major U.S. boiler manufacturer combined the collectors into one large, single-cyclone dust collector. The success of the new concept was based on the re-injection point, where they designed a loop seal that prevented furnace pressure from blowing the particles back into suspension in the gas stream. They also effectively increased the residency time of the fuel so that no combustion took place in the convection pass. At that time the term *circulating fluidized bed* (CFB) was coined.

The aforementioned example serves to illustrate the importance of ash analysis during commissioning of any solid-fueled boiler.

COMBUSTION ANALYSIS

Two of the most important parameters in the combustion process analysis are O_2 and CO. Continuous monitoring is usually provided for monitoring oxygen but not always for carbon monoxide. However, carbon monoxide is the better of the two for monitoring combustion efficiency. Oxygen is usually used as the control parameter for excess air. The commissioning engineer should be thoroughly familiar with the type of monitors being used. A common misconception was that Orsat analyzers measured O_2 on a wet basis. The Orsat apparatus gas samples were saturated with water at all times. But because the sample was kept at ambient temperature, if there was a decrease in sample volume, the water volume decreased proportionally. Therefore, the water had no effect on the analysis. Both in-situ and extraction (with dryer) monitors provide O_2 readings on a dry basis. Engineers use the amount of O_2 theoretically required to burn 100 lb of fuel and then multiply by a safety factor "excess air." The theoretical air used in these stoichiometric calculations do not include moisture; therefore, the flue gas analysis should not. The following example is often provided to illustrate the difference between the wet and dry readings:

$$\text{Wet } N_2\ (90\%) + H_2O\ (5\%) + O_2\ (5\%) \rightarrow O_2 \approx 5\%$$

$$\text{Dry } N_2\ (90\%) + O_2\ (10\%) \rightarrow O_2 \approx 10\%$$

Although it would appear that the wet basis is the correct reading, that is not the case. Furthermore, too much excess air raises the APH outlet temperature and overloads the ID fans and air quality equipment. For example, if the manufacturer's boiler specification requires 3 percent flue gas oxygen and the unit is run with 5 percent for safety, all other parameters will increase, such as exit temperature and gas volume. As the volume increases, so does the gas velocity in the furnace and convection pass. The result is less residency time and less heat transfer, which causes the exit-gas temperature to rise and boiler efficiency to go down.

Not too long ago, the only analysis used was the Orsat, which was not continuous. The operators had to watch other parameters such as flame appearance, stack opacity, and exit temperature. During commissioning today, this is still the case because the automated equipment has not been proven in a reliability run.

Make sure that tramp air isn't leaking into the flues upstream of the monitors. Tramp air can cause O_2 trim to reduce the stoichiometric air at the burners into an explosion range. Usually, engineers will run the excess air up during low-load trial operation because the monitors and controls have not been proven. But, as full load is approached, all these parameters—visual as well as analytical—become important. Also, take into consideration that extractive-type monitors can be dangerous in that the sample may not be analyzed for several minutes, during which time airflow remains constant, even in load swings that can take the boiler into an explosive range.

Monitor CO during startup so combustion trends can be observed. Usually, 200 ppm of CO or less is optimum, but a feel for the unit must be developed. Constant burner and airflow adjustments are required to obtain optimum conditions.

SOLID-FUEL HANDLING

Too often, fuel handling in the fuelyard is not supervised by experienced engineers totally dedicated to that purpose. The reason for this is that fuelyard duty during commissioning is quite often left to operations personnel or to a junior startup engineer assigned the task. To ensure the proper fuel is delivered to the fuel bunkers, experienced commissioning engineers must supervise fuel handling in the fuelyard. They also must train operations fuel handlers how to stack and recover fuel in a manner such that the best fuel is delivered first. During startup, the best and driest fuel must be used first because the unit is undergoing trial operation. Initial trial operation is not the time to be testing the unit to see if it can handle the upper limits of the fuel specification. For example, the driest fuel must be selected for initial trial operation to maximize stability so the controls can be tuned. This may mean constructing a temporary fuel shelter if storm conditions prevail before first fires. The cost of a shelter is far less than the cost associated with weeks of delays trying to handle wet fuel during initial trial operation. A long discussion on how to cope with plugged chutes, feeders, and mills can be defined, but the root cause for plugging is always directly related to wet fuel. Eliminate it and eliminate the problem. During startup, always be cautious of foreign materials entering the conveyor systems and bunkers. Large concrete chunks with embedded rebar can play havoc with conveyors and feeder belts. Construction is generally not finished at this time, and construction debris is usually available. Close monitoring must be performed to prevent it from being thrown into the fuel-conveying system. Also, the presence of noncombustibles such as log-landing debris and coalyard mud cannot be introduced into the fuel feed.

MILL TESTING

The most important mill performance test is the mill outlet *fineness test*. Mill performance is dependent on coal grindability and moisture. If the coal is wet, more primary air and more resident time in the mill are required to dry it to the point that it is light enough to be carried out of the mill into the furnace. The problem can make setting the mill classifiers difficult. If the classifiers are set wider than normal to increase unit load (in essence compensating for wet coal), the carbon content of the ash will increase dramatically. If, for example, the classifiers are set at 75 percent, and to achieve the optimum coal fineness with spec coal the setting is 65 percent, high carbon in ash will be a direct indication that they are indeed set too wide. Airflow was increased to carry more fuel into the furnace, but it didn't burn completely. As illustrated with the stoker example, the particles of coal are too large to burn in the furnace. The furnace is not tall enough to provide adequate residence time to burn fuel with a fineness larger than specified. Therefore, burning occurs throughout the convection pass, and finally, unburned carbon falls out in the ash collectors.

WINDBOX PRESSURE

The burner windbox pressure must be adjusted and set higher than the furnace pressure. This must be done with a temporary manometer. The manometer must be clearly visible from the burner front at all times during first fires and trial operation. If the pressure in the windbox should become negative, in relation to the furnace pressure, combustibles can be drawn into the windbox and explode. Many startup engineers have died as a result.

FUEL-FIRING PRECAUTIONS

Liquid Fuels

Before first fires, the gun extensions into the furnace must be set to the proper distance inside the furnace. That means entering the furnace to check the proper gun distance setting. Once set, they must be marked at the burner front as a reference point. If any adjustments are required to establish the proper flame, that reference point is used.

If atomizing steam is used, the system must be checked before firing to ensure that the lines are heated all the way to the guns. Also, check that a bypass and drain system is provided at the guns to ensure that superheated steam reaches the burners. If there is no bypass and adequate drain system, the steam will condense and blow into the furnace. If condensate, rather than superheated steam, is injected while attempting to burn fuel, it will blow the fuel with water into the furnace, preventing ignition. Most causes for loss of flame detection are inadequate atomization, not improper flame detector calibration or orientation. Too often, multiple attempts are made to establish fires while making adjustments to the detectors, all the while raw fuel is being collected in the deep center of the air preheater (APH) baskets.

Note

The fuel cannot be seen in the APH because it blows deep into the baskets. Therefore, a visual inspection usually does not reveal the seriousness of the situation. Refer to Chap. 14. Once the steam is finally heated, atomization begins and fires are established. Now the APH will be ready for ignition.

Coking

When burning heavy fuel, once the oil is preheated to design temperature, atomization in the burner tip must be switched from steam to air or mechanical atomization. If the preheated oil continues to be atomized with steam, the steam will overheat the oil, and it will "crack" into fractions, including petcoke, that will form on the burner tip and collect in the APH. This will be noticeable by a radical fluctuation in flue gas CO. Petcoke, if ignited in the APH, will cause a serious fire hazard. Refer to Chap. 14.

Gas Fuels

Note

Heaters must be fired and tested before liquefied petroleum (propane) gas systems can be placed in service. Also if the fuel flow exceeds the heater capacity, the entire fuel line—from the heater to the burner front—will freeze tripping the boiler.

The auto ignition temperature of methane is 1076°F, and corresponding adiabatic compression pressure from the perfect gas law $Pv = RT$ is 2121 psi. Therefore, diesel ignition of natural gas is impractical. Natural gas ignition in an internal combustion engine can only be achieved by spark. An exception is a diesel engine, which can use a seed feed of diesel fuel or spark ignition. It cannot diesel-ignite natural gas. Also, natural gas has narrow explosive or flammability limits. Below ~5 percent, it is too lean and above ~15 percent, it is too rich, even in presence of an ignition source. Browse Wikipedia and/or The Engineering ToolBox for LEL (lower explosive limit) of natural gas for a more precise explanation.

Natural Gas Analysis

The following is an analysis of natural gas used for performance testing. See App. S for practical use of the analysis.

Constituent Analysis

	percent	C	H
CO_2	0.20		
N_2	0.26		
CH_4	97.70	0.977	3.908
C_2H_6	1.66	0.033	0.099
C_3H_8	0.15	0.005	0.012
C_4H_8	0.02	0.001	0.002
C_4H_{10}	0.01	0.000	0.001
	100	1.016	4.022

Combustion Gas Analysis (Products)

	Percent
CO_2	2.96
O_2	15.42
CO	0
N_2	81.62

Stoichiometric Reaction

$$C_{1.016}H_{4.022} + bO_2 + 3.76\,bN_2 \rightarrow aCO_2 + bO_2 + 3.76bN_2 + eH_2O$$

$$a = 1.016$$

$$e = 4.022/2 = 2.011$$

$$b = 1.016 + 2.011/2$$

$$= 2.022$$

Actual Reaction

$$aC_{1.016}H_{4.022} + bO_2 + 3.76\,bN_2 \rightarrow 2.96CO_2 + 15.42O_2 + 81.62N_2 + eH_2O$$

$$a = 2.96/1.016 = 2.91$$

$$b = 81.62/3.76 = 21.71$$

$$e = (2.91)(4.022)/2 = 5.86$$

Furnace Vibration Burning Natural Gas

One serious problem that has been encountered in small, gas-fired boilers is severe furnace vibration. In one unit in 1984, the vibration when warmup gas was fired was so violent it appeared that furnace structural damage might occur. The immediate startup solution was to modify the secondary airflow by closing one of 10 registers. That, in theory, took combustion out of resonance with the furnace. Therefore, setting the registers at different positions was the final solution. This appeared to be similar to piping systems in which the pulsations caused by a pump impeller passing the discharge diffuser came into resonance with the piping system's natural frequency. This causes the piping system to sway back and forth or vibrate violently.

An investigation of other boilers by Frantisek L. Eisinger and Robert E. Sullivan was presented in an ASME paper entitled, "Elimination of Thermoacoustic Furnace Vibration in a Gas-Fired Package Boiler: A Case Study."[13] An abstract is presented here. The full report can be downloaded from the Internet.

> A severe furnace vibration was developed in a gas-fired package boiler during an initial operation. The vibration occurred at higher loads and was characterized by an acoustic standing wave residing inside the furnace in the front-to-rear direction along a burner axis. Theoretical predictions indicated that the thermoacoustic behavior of the burner/furnace system operated in an unstable range explaining the underlying cause of the vibration problem. For the elimination of the vibration problem, a two-step solution was envisioned: (1) The mode of fuel injection was modified in order to affect the flame's oscillatory behavior and its interaction with the acoustic waves in the enclosed spaces of the burner conduits and the furnace. This step was successful in mitigating the vibration in a still thermoacoustically unstable system. (2) A full elimination of the vibration may be achieved by making the system thermoacoustically stable by modifying the cold air portion of the system. Both the step 1 solution, which suppressed the thermoacoustic oscillation, and the step 2 solution of making the system thermoacoustially stable will allow the boiler to operate without vibration at all loads.[13]

This paper gives a detailed description of the issues involved. They briefly characterized the phenomenon as follows:

> The phenomenon of acoustic vibration or pressure oscillations in a gas caused by a constant heat source is characterized as a Sondhauss tube or Rijke tube phenomenon based on the Boundary conditions of the respective tubes and also the mode of heat input.

Refer to Chap. 21 for a explanation of the same problem (furnace vibration caused by resonance).

Furnace Convection Pass-Tube Vibration

A similar solution to avoid Karman vortex shedding resonance can be designed into furnace-tube layout. That is, the tubes are installed in irregular patterns, or baffles are installed by startup, to break up the shedding vortices so they do not come into resonance with the tubing natural frequency and cause vibration. The root cause of vortex shedding problems is that the laminar flow of the fluid enables the pulsation to transmit uninhibited. Creating turbulence eliminates the problem. Refer to Chap. 21 for further discussions regarding these and other furnace-related vibration problems.

Fuel Contaminants

One of the worst contaminants in fuels is sodium. From a purely lay nonanalytical point of view, sodium chloride [with a melting temperature of 97.8°C (208°F) and boiling temperature of 883°C (1621°F)] will be in the liquid phase throughout the convection pass and binds ash to the superheater tubes. Elemental sodium sublimates at 2330°F. Sodium chloride can be inherent with stockyard manure and lignite, or it can be introduced with free moisture from saltwater during shipment. Salt can also be introduced to fuel in tankers that use saltwater for ballast. The elimination of sodium chloride in liquid fuel is a fairly straightforward ion-exchange process. But, removing sodium from solid fuels is much more complicated because the ion exchange requires that the fuel's moisture content be elevated. Many studies of this problem have been conducted. For example, the ASME Research Committee on Boiler Fouling gives degree of fouling for different levels of contaminants.[14] "The Impact of Ash Deposition on Coal-Fired Plants," written by Jim Williamson and Fraser Wigley,[15] gives predictability indices so one can be prepared for fouling when it occurs.

The key for successful commissioning is to determine what contaminants are in the fuel before firing the boiler. The most important step is to have an analysis made of all the ash constituents, not just the ultimate analysis. Knowing what to expect and what additives to be used, plus being prepared, are the principles used to minimize problems during commissioning.

REFERENCES

1. Smith, M. L., and Stinson, K. W., *Fuel and Combustion*, McGraw-Hill, New York, 1952.
2. ASME PTC 3.2.
3. ASME PTC 4.1.
4. ASME PTC 3.1.
5. Mingle, J. G., and Boubel, R. W., "Proximate Fuel Analysis of Some Western and Bark," *Wood Science* Vol. 1, July 1968.
6. *Kent's Mechanical Engineers' Handbook*, McGraw-Hill, New York, 1952.
7. Power Barge Antilles.
8. Leffler, W. L., "Petroleum Refinery for the Non-Technical Person," PennWell, 1985.
9. ASTM D240-87.
10. ASTM D2382-88.
11. Equation 1 is from Misc. Pub. 97, 1927, National Bureau of Standards.
12. "Significance of Tests of Petroleum Products," report prepared by ASTM Committee D-2 on Petroleum Products and Lubricants. See also Reference 6, *Kent's Mechanical Engineers' Handbook.*
13. Eisenger, F. L., and Sullivan, R. E., "Elimination of Thermostatic Furnace Vibration in a Gas-Field Package Boiler: A Case Study," *Journal of Pressure Vessel Technology*, ASME, June 2009.
14. ASME Research Committee: A Boiler Fouling.
15. Williamson, J. and Wigley, F., Taylor and Francis, Washington DC, US, 2009.
16. Selected Values of Properties of Hydrogen, National Bureau Standards Circluar C461, API Research Project 44.

CHAPTER 16

FEEDWATER HEATERS

The Heat Exchange Institute is the governing authority on condensers and heat exchangers (www.heatexchange.org). The following definitions are from HEI Tech Sheet 127:[1]

Pressure Classification

Low-pressure heater: A heater located (with regard to feedwater flow) between the condensate pump and either the boiler feed pump or, if present, an intermediate-pressure (booster) pump. It normally extracts steam from the low-pressure turbine.

High-pressure heater: A heater located downstream of the boiler feed pump. Typically, the tube-side design pressure is at least 1500 psig, and the steam source is the high-pressure turbine.

Intermediate-pressure heater: A heater located between the booster pump and the boiler feed pump. Usually, the tube-side pressure is within 200 to 300 psi of the low-pressure heaters, and the steam is extracted from an intermediate-pressure turbine.

Orientation

Horizontal: Most heaters are of this configuration. These are the most stable in regard to level control, although they occupy more floor space. Disassembly is by means of either shell or bundle removal. Most are floor mounted, although some are mounted in the condenser exhaust neck.

Vertical, channel down: Although these conserve floor space, the amount of control area available for liquid-level fluctuation is less. Disassembly is by shell removal. Installation and removal may be more difficult than for horizontal heaters.

Vertical, channel up: These are the least frequently used. Disassembly is by means of bundle removal. If a subcooling zone is present, it must extend the full length of the bundle, because the water must enter the bottom and exit at the top end of the heater.

Zones

Zones are separate areas within the shell in a feedwater heater.

Condensing zone: All feedwater heaters have this zone. All of the steam is condensed in this area, and any remaining noncondensable gases must be removed. A large percentage of the energy added by the heater occurs here.

Subcooling zone (optional): The condensed steam enters this zone at the saturation temperature and is cooled by convective heat transfer from the incoming feedwater.

Desuperheating zone (optional): The incoming steam enters this zone, giving up most of its superheat to the feedwater exiting from the heater.

CLOSED FEEDWATER HEATERS

In Fig. 16-1, the level in the condensing section is below the tube bundle. Also, the condensate enters the subcooler section through a bottom inlet and exits from a top outlet. Therefore, the centerline of the condensate outlet nozzle is not necessarily the centerline of the normal water level. The centerline of the normal water level is labeled on a plaque attached to the shell. Control of the closed feedwater heater level is relatively simple because of the slow response of the process. Load changes of the turbine do not create radical upsets that happen in open heaters and boilers.

Both the tube side and shell side of the heater will have relief valves. The tube side is protected against water expansion causing overpressure. If the tube-side waterbox inlet and outlet valves are closed, steam can continue to heat the water, causing the isolated section to overpressurize. The shell side is protected against overpressure from a condensate or feedwater system tube failure.

The turbine must be protected against water induction coming from the feedwater heaters. The water induction source may be associated with faulty level control or a tube leak. An extraction steamline water induction block valve must be installed in the line. This valve is separate from the steam non-return valve (NRV), which protects the turbine from overspeed.[2]

Also, turbines must be protected against overspeed caused by the release of a heater's stored energy. For example, after a turbine trip, flashing condensate (release of the heater's stored energy) can reverse flow back into the turbine, causing an overspeed condition.

OPEN FEEDWATER HEATERS (DEAERATORS)

Oxygen pitting is probably the most destructive form of corrosion. The reason is that it "eats" deep into the mass of a metal rather than the surface. It tends to localize creating large, deep craters with sharp edges or even caverns that can penetrate completely through the metal in a short period of time. The mechanics of the reaction are galvanic, in which a small area lacking oxygen becomes anodic while the surrounding area with excess oxygen becomes cathodic. The result is a very concentrated loss of metal at the anode. It is especially damaging as the temperature rises, increasing the rate of reaction.

> To remove oxygen from water, theoretically all one has to do is elevate its temperature to the saturation point. The reason for this is that oxygen will not dissolve in water at saturation. For water at ambient pressure the saturation temperature is 100°C. However, this will only liberate about 90% of the gas. The rest must be liberated by dilution. That is a gas, in this case steam, is passed through the water in order to liberate the remaining 10%. The mechanism actually reduces the partial pressure of the water by diluting it with steam; this can be explained by Dalton's and Henry's gas laws. For every condensate and feedwater system there exists an optimum point in the cycle to combine both methods. That point has to be before the water temperature is high enough to accelerate corrosion.[3]

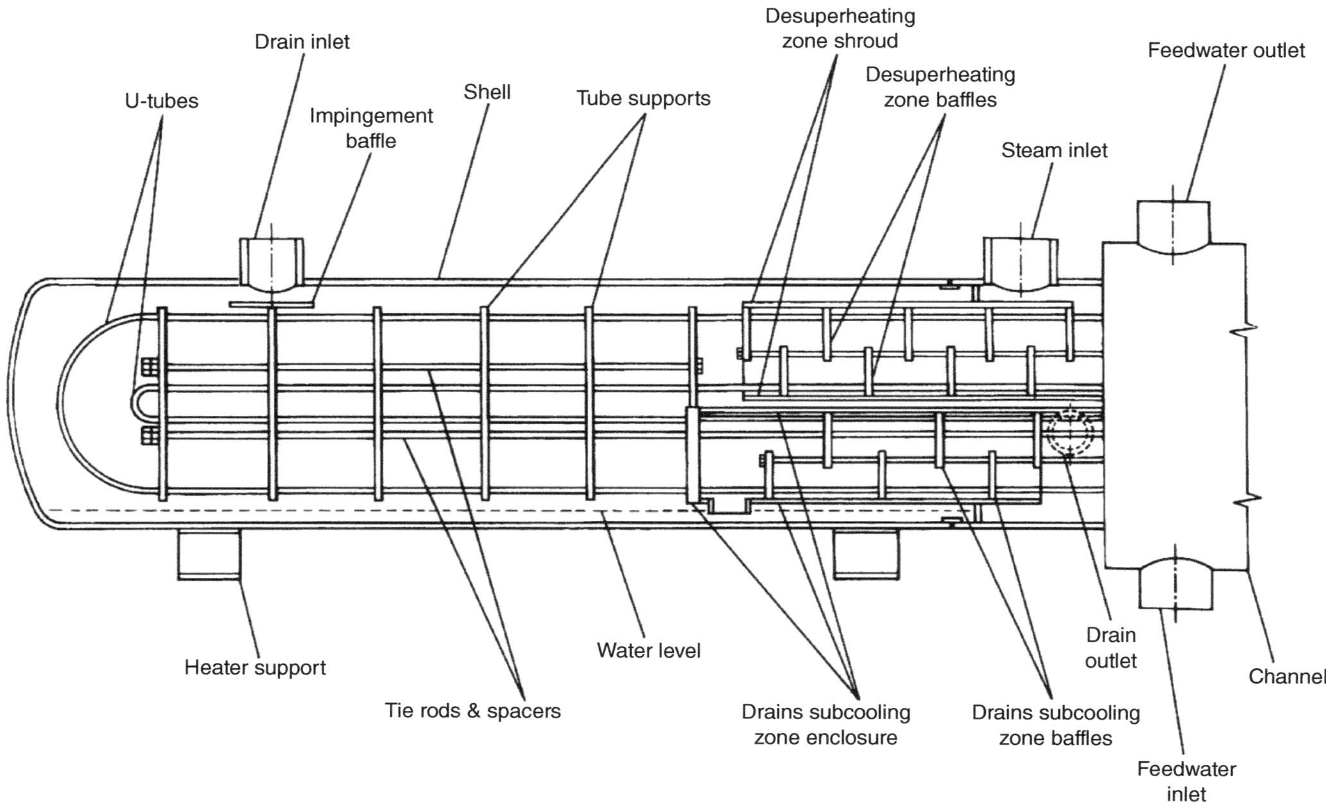

FIGURE 16-1 Horizontal feedwater heater.

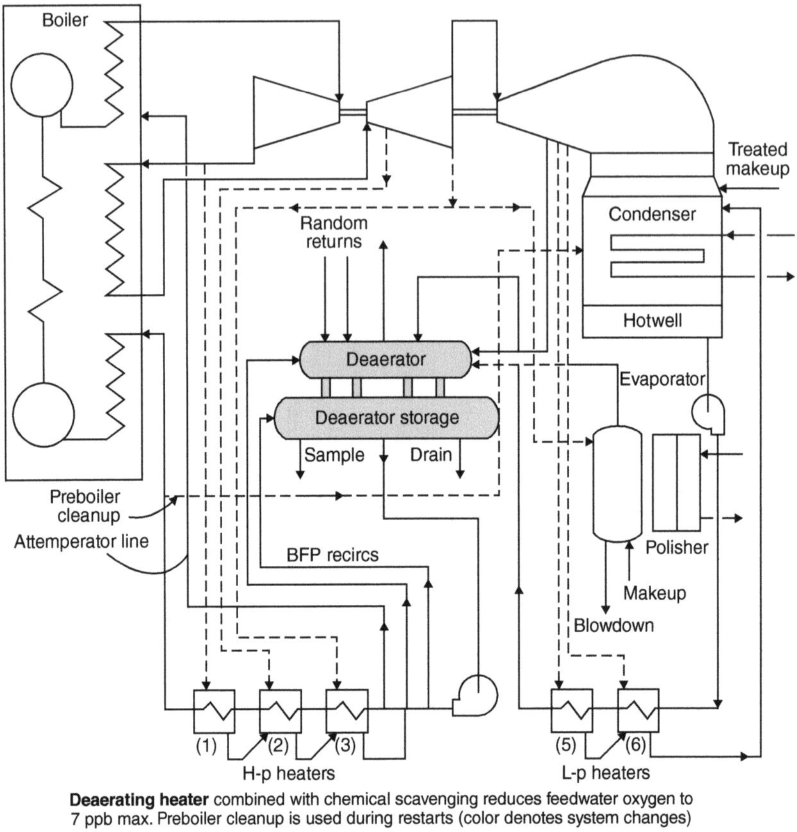

FIGURE 16-2 Deaerator and feedwater heater train.[3]

The source of most of the oxygen is air leakage into the condenser and oxygen in the makeup water. Most of the air can be ejected from the condenser because the water is at saturation as it condenses. The air is removed in the condenser air-removal section. However, below the surface, water is subcooled, due to the water "head," and can absorb oxygen that leaks in below the surface.

The remaining oxygen must be removed in a deaerator (DA). There are two basic types of deaerators: atomizing and tray. In the atomizing type, atomized water is sprayed into a high-velocity stream of steam, which further breaks up the water droplets, diluting the water with steam. In the tray type, inlet water is sprayed over trays, where it is reduced to thin layers. High-velocity steam is admitted in a cross-flow that breaks up the water particles as they drop from one tray layer to the next. In both types, steam is vented from the deaerator to eject the gases, which will include some carbon dioxide, ammonia, and oxygen. The usual guaranteed deaeration is < 7 ppb dissolved oxygen.

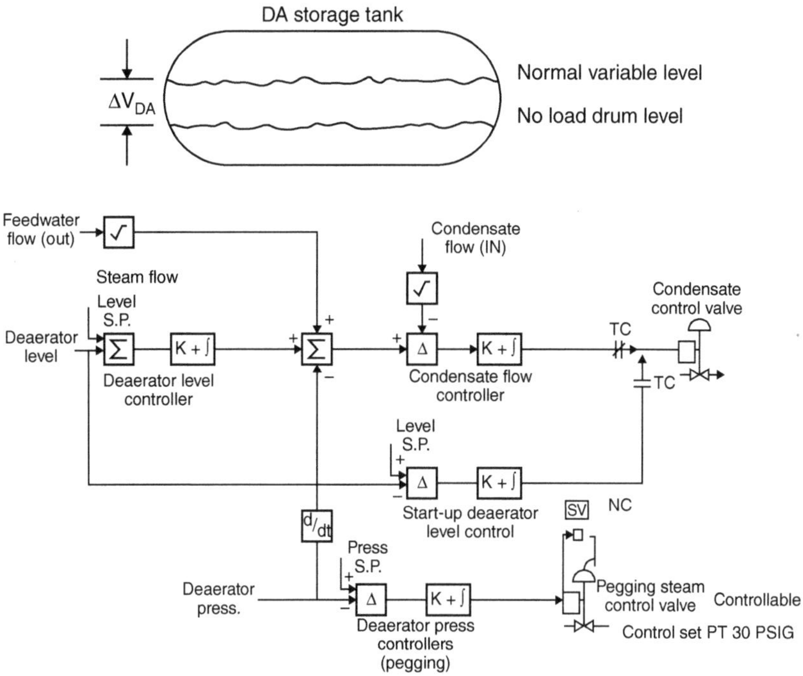

FIGURE 16-3 Typical DA level control with flash protection.

DA Flashing

A DA is an open feedwater heater. That is, the heat exchange is accomplished by mixing the water directly with steam. This poses some problems for startup. During unit load transients, the DA pressure can drop well below the saturation point of the water in the DA storage tank. When this occurs, the water will flash to steam. The flashing can be very violent and can damage the trays and nozzles in the deaerator section. It can also damage boiler feed or boiler feed booster pumps.

Because of the flashing phenomena, it is important to review the DA control scheme. DA flash protection must be incorporated into the control configuration. No means of mechanical control such as vortex eliminators, etc., can suppress flashing. See Chap. 8 for a detailed discussion on DA flashing. See Fig. 16-3 for an example of a DA flash protection control system design using a vaiable level setpoint

PRETRIAL OPERATIONAL CHECKS

- The deaerator section of the DA must be checked to ensure that the hardware has been installed, such as trays and nozzles. They are sometimes shipped separate from the vessel.
- Check that the storage tank has been cleaned.

- Check that the storage tank outlet to the boiler feed pumps has a coarse mesh screen. The screen is needed to prevent trays or other hardware from entering in the event of flash damage.
- Because the DA has a large amount of stored energy, second only to the boiler, the extraction steam supply line must have a power assist non-return valve. The valve is necessary to prevent this energy from forcing steam back through the turbine, causing an overspeed. The operation of the NRV must be checked by actual initiation of a turbine trip.
- The high-high level trip must be checked to ensure that the power-operated water induction prevention gate valve closes and the turbine side drains open.
- Enssure pegging steam is available.
- Ensure that the oxygen scavenger injection system is available for chemical deaeration during startup.
- Ensure that the level control system has temperature compensation. The differential between an ambient-level column and the actual storage tank level can be as high as 12 in in a large, high-pressure unit. Refer to the Chap. 5 drum level compensation discussion for further elaboration.

TRIAL OPERATION

The DA level control and the startup pressure control (pegging steam) are the principal element for startup attention. However, the turbine water induction prevention devices must be tested. They are very important in the event level control fails. ANSI/ASME PDP-1[3] and TDP-2[4] govern the design and operation of these devices. A copy of the latest revision of these standards must be available in order to instruct the operation personnel on recommended safe operation. The standards call for testing of all the devices once a month.

It should be noted that NRVs are not good water-induction-prevention valves. The reason is that the water rises in the lines with no differential pressure against the valve. Furthermore, the water can cause the large swing gates to distort, allowing the water to rise unrestricted.

If water enters the turbine while operating at rated speed or while carrying load, do not trip the unit if the turbine vibration and differential expansion are satisfactory and there are no other signs of distress. Search out and isolate the source of water immediately. The primary cause of damage is casing distortion created by the cold water. Removing the heat from the turbine will only aggravate the problem.

HEATER TROUBLESHOOTING

If you think a heater performance is suspect, check the water level. If the shell water is covering more tubes than designed, its performance will be negatively affected. This can be manifested in the downstream heater, not necessarily in the heater suspected.

If the piping configuration and water level have been checked and everything appears as designed, the quickest way to find out the real source of the problem is to perform a heat balance. For example, if the DA performance is in question because the condensate outlet flow is below normal, restricting unit full-load operation, the problem could be the upstream heater drain flow and not the DA itself. A heat-balance evaluation would determine the real source of the problem.

Compare the test data against the plant design heat balance. All parameters are known, with the exception of steam and drain flows. Steam and drain flows will indicate whether there is a problem. The heaters may have to be followed upstream until all flows go into the DA. Note in the sample calculation, the No. 1 heater drain had a flow element in the line. For troubleshooting purposes, it is acceptable to check the indicated flow against the calculated flow. However, when performing a full-system heat balance, either the condensate flow or feedwater flow measurement should be selected as the basis for the calculation—not both. Likewise, drain flows must be determined by heat balance. Figure 16-4 is an example of a DA mass flow balance.

The most common cause for deficient heater drain flow is high condensate level, which causes a loss of steam condensing surface area. (Make sure there is no tube leak.) The second is incorrect piping.

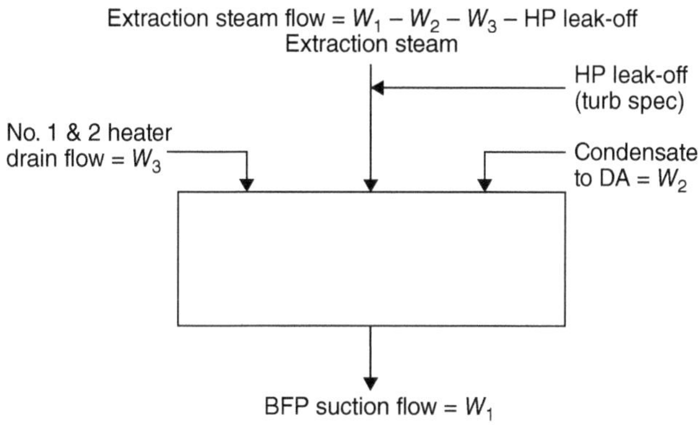

FIGURE 16-4 Example of a DA mass flow balance.

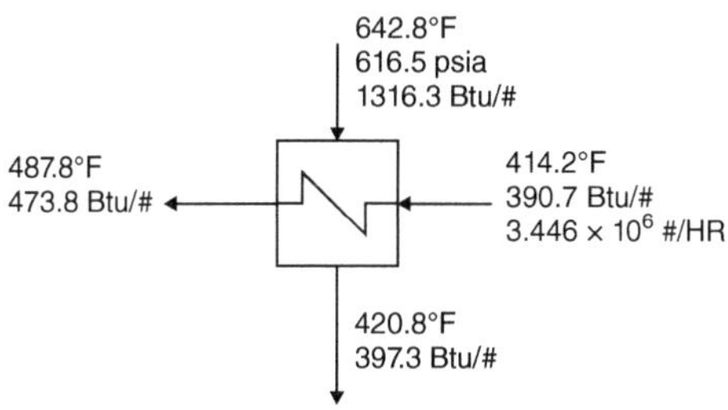

$$W_{STM} = \frac{W_{FW}\,\Delta h_{FW}}{\Delta h_{STM}}$$

$$= \frac{3.446 \times 10^6\,(473.8 - 390.7)}{1316.3 - 397.3}$$

$$= 311{,}602 \text{ \#/HR} \Leftarrow$$

$$W_{STM} = 303{,}700 \text{ \#/HR} \Leftarrow$$ See flow element calculation sheet

FIGURE 16-5 Sample of heat balance calculation of the first upstream heater from the DA.

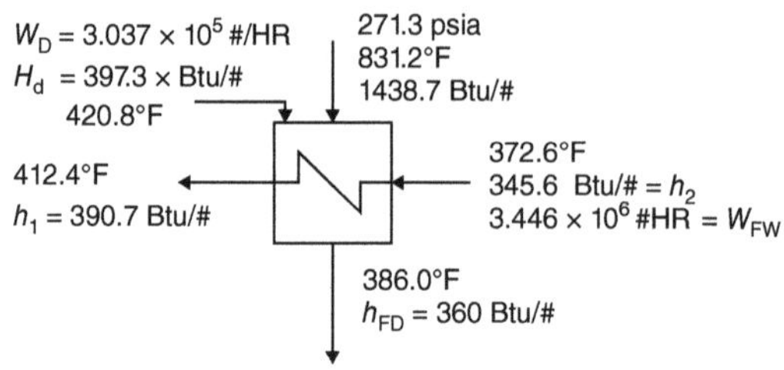

$$(W_{ESTM} + W_D)\, h_{TD} = W_D h_D + W_{ESTM}\, h_S - W_{FW}\,(h_1 - h_2)$$

$$W_{ESTM} + W_D = \frac{W_D h_D + W_{ESTM}\, h_S - W_{FW}\,\Delta h}{h_{TD}}$$

$$W_{ESTM} + W_{ESTM}\,\frac{h_S}{h_{TD}} = \frac{W_D h_D - W_{FW}\,\Delta h}{h_{TD}} - W_D$$

$$W_{ESTM} = \left[\frac{W_D h_D - W_{FW}\,\Delta h}{h_{TD}} - W_D\right] / \, 1 - \frac{h_S}{h_{TD}}$$

$$= \left[\frac{3.037 \times 10^5\,(393.3) - 3.446 \times 10^6\,(390.7 - 345.6)}{360} - 3.037 \times 10^5\right] \Bigg/ - \frac{1438.7}{360}$$

$$= 134.699\ ^{\#}/\text{HR} \Leftarrow$$

FIGURE 16-5 *(Continued)*

REFERENCES

1. Tech Sheets are periodically reviewed by the HEI and are updated. Visit www.heatexchange.org for the latest version.
2. "Recommended Practices for the Prevention of Water Damage to Steam Turbines Used for Electric Power Generation," ASME TDP-1-1998.
3. "Fossil Fired Plants," ANSI/ASME PDP-1.
4. "Nuclear Plants," ANSI/ASME TDP-2.

CHAPTER 17
PIPING

PIPING AND INSTRUMENT DIAGRAMS

Piping and instrumentation diagrams (P&IDs) are the most frequently used documents in industrial plants. They provide basic guidelines for operations and maintenance and must be understood by all disciplines. They are derived from simple process flow diagrams (PFDs) that are used to illustrate how systems function. However, they are much more detailed than the PFDs.

Many companies have standardized their P&IDs using ISA-5.1-1984 (R1992), Instrumentation Symbols and Identification.[1] The first sheet of a set of P&IDs will be the legend for that set of drawings. Most legends follow the International Society of Automation (ISA)[a] standard. Some companies may deviate from the standard, so it is always advisable to review the legend before reviewing the P&IDs.

P&IDs are dimensionless drawings and do not show where devises, components, or piping are located within a facility. Nor do they show the length of piping. However, most diagrams show pipe and valve sizes, which are part of the pipe identification number on the drawing. If you are unfamiliar with P&IDs and/or the facility in which you are working, it will be difficult at first to locate small devices. A simple rule to follow when trying to locate instruments in the field is to first locate the largest piece of equipment (such as a tank, pressure vessel, or pump) that the piping is attached to. Then, follow the piping from that piece of equipment to the instrument you are looking for. After you become familiar with the piping systems, it will be easier to follow the piping to locate devices.

DEVICE TAGGING

All instrumentation and control devices in the field should have identification tags. The tag numbering system will generally correspond with the ISA standard. The device on the diagram will have a designation in both the upper and lower hemispheres of the circle, representing the tag. The upper designation is the device type. For example, a flow transmitter can have a FT designation. The first letter is the process variable. The second is the device type. A flow control valve can be FCV. The second letter is the modifier "control" and the third letter designation, if used, is an additional modifier. In this case, it is "valve." Some engineering firms use three rows within the tag "circle," which is a variation of the ISA standard. In the

[a]Original named as Instrument Society of America (1945); changed to Instrumentation Systems and Automation Society in 2000 and International Society of Automation in 2008.

example in Fig. 17-1, flow element #1 is designated for system 100. Sample designations are listed in Table 17-1. Examples of tag designation types are listed in Fig. 17-2. Note that a solid line through the center of the circle means the instrument is visible to the operator. A dashed line means it is not visible.

FIGURE 17-1 Tag designation.

TABLE 17-1 Sample Designations

A	Analysis	A	Alarm
D	Density	C	Control
E	Electricity	E	Element (flow, temperature, etc.)
F	Flow	H	High
H	Hand	I	Indicator
J	Power	G	Glass (gauge)
K	Time	L	Low
L	Level	S	Switch
M	Moisture	T	Transmitter
P	Pressure	V	Valve
Q	Quality	W	Well (thermal)
R	Radiation		
S	Speed, frequency		
T	Temperature		
V	Viscosity		
W	Weight		
Y	State (e.g., energized/deenergized)		
Z	Position (e.g., position switch)		

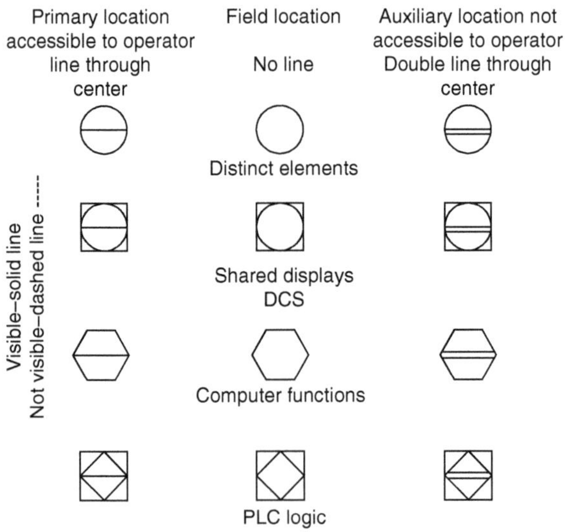

FIGURE 17-2 Device symbol types.

Signals that are transmitted from one device to another are categorized by type and designated by drafting lines similar to the example in Fig. 17-3.

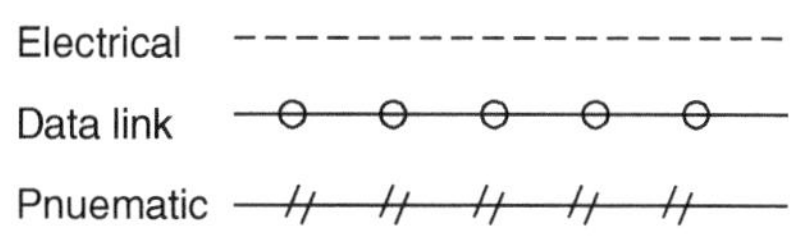

FIGURE 17-3 Line types.

GENERAL ARRANGEMENT DRAWINGS

General arrangement (GA) drawings are used to locate large equipment in a plant. Orientation of the equipment is always based on a "plant north" reference. Plant north may not be magnetic north, but it is usually within 15°. The most visible objects in a power plant are the stack and boiler. Establish the orientation using them as reference. Multiple pieces of equipment are labeled A, B, C, and so on, from north to south, west to east, or in accordance with the operating company's standards. For multiple-unit stations, make sure the right unit is being used.

Sometimes for large facilities, the engineer provides an area index. The index is a plan layout that will provide an identifier for areas such as turbine building and crusher building. The index will be the key for locating the plan and elevation equipment location drawings.

First-Time Exercise Take out a set of P&IDs and study the legend while reviewing the systems. Once familiar with the P&IDs, get the general arrangement (GA) drawings and locate a large piece of equipment. Find plant north and go out and study the physical arrangement of the equipment with regards to the GAs and P&IDs. If you take construction drawing into the field, never read them by turning north to orient yourself. That is the first sign that you are a "greenhorn." Discipline yourself to read them in any orientation.

The following ASME codes apply to piping systems as American National Standards (ANSI) and are used for application and installation of piping.

B31.1, "Power Piping."[2] (This standard is also used for district heating systems. Note also that the International District Heating Association provides a *District Heating Handbook*.)

B31.3, "Process Piping."[3]

B31.4, "Pipeline Transportation Systems for Liquid Hydrocarbons and Other Liquids."[4]

B31.5, "Refrigeration Piping."[5]

B31.7, "Nuclear Piping" was withdrawn in 1971 and thereafter, nuclear piping has been covered by ASME B&PV Code, Section III, Subsections NA, NB, NC, and ND.[6]

B31.8, "Gas Transmission and Distribution Piping Systems."[7]

B31.9, "Building Services Piping."[8]

B31.11, "Slurry Piping Systems."[9]

The following codes and standards provide pipe specifications such as outside and inside diameters, wall thickness, weight per foot, and standard test pressures: ASME/ANSI B36.10, "Welded and Seamless Wrought Steel Pipe"[10] and ASME/ANSI B36.19, "Stainless Steel Pipe."[11] These data can be found in supplier catalogs.

ASME/ANSI B16 Series, "Standards of Pipes and Fittings,"[12] provides dimensional data as well as weights and tolerances for pipe fittings. These are also the dimensional data provided in supplier catalogs.

ASTM Standard specifications are used by designers to specify pipe for engineering and procurement documents, e.g., ASTM A335/A335M-95a, "Standard Specification for Seamless Ferritic Alloy-Steel Pipe for High-Temperature Service."[13] A list is provided in App. E.

Title 49 CFR Parts 190–199, "Department of Transportation, Pipeline Safety Regulations,"[14] cover pipelines for transport of hazardous liquids and carbon dioxide and should be used in conjunction with B31.4.[4]

Manufacturers Standardization Society of the Valve and Fitting Industry, Inc. (MSS) provides SP 58, "Standard Practice for Pipe Hangers and Supports, Design and Manufacture,"[15] and SP 69, "Pipe Hangers and Supports-Selection and Application."[16] A full list of their publications is listed in App. D.

American Water Works Association (AWWA) provides numerous standards, such as ANSI/AWWA C950, "Standard for Fiberglass Pressure Pipe,"[17] AWWA M9, "Manual of Water Supply Practices,"[18] the standard "Ductile Iron Pipe & Fittings for Water and Other Liquid,"[19] and many more.

The Ductile Iron Pipe Research Association (DIRPA) provides piping publications such as "Thrust Restraint Design for Ductile Iron Pipe."[20]

SPECIAL TERMS

Specification number—number used by ASTM for pipe specifications such as A53, A106, or A335.

Schedule—thickness of the pipe (e.g., Standard, 40, 80, and 160).

Note

The diameter of pipe up to 12 in is the inside diameter; for 14 in and larger, the diameter specified is the outside diameter. Also, Standard pipe has the same wall thickness as Schedule 40 up to 10 in. For 12 in and larger, Schedule 40 is thicker. The standard length for pipe is 20 ft.

Class—class schedule delineator such as Standard (Std), Extra Strong (XS), and Double Extra Strong (XXS). This is not to be confused with "class fitting" as in the strength class of valves and pipe fittings, e.g., Class 150, 300, 600, 900 flange, etc. It is linked to pipe schedule. Class and type are listed in the ASTM specifications.

Grade—designation used to separate grades of pipe of the same material specification, delineating different maximum allowable stress values by grading different temperature and pressure ratings. (Grade can be either numeric or alphabetic.) Grade P91, or simply Grade 91, refers to stainless-steel seamless pipe containing 9 percent chrome and 1 percent molybdenum for A335 pipe. As of this writing there are 18 grades for specification A335 pipe. Browse "P91 Industry Alert"[21] for urgent information regarding safe installation and maintenance of P91 pipe.

Note

Table 17-2 shows that the maximum allowable stress values are the same for A53 and A106 pipe up to 650°. After that, they drop, which means the maximum allowable pressure drops also. That is, the maximum allowable operating pressure is temperature dependent below –20° and above 650°.

TABLE 17-2 Maximum Allowable Stress Levels

Spec No.	Grade-Specified Minimum Tensile (Kpsi)	Maximum Allowable Stress Values for Metal Temperatures (°F) not Exceeding:			
		–20 to 650	700	750	800
A53	48	12.0	11.7	10.7	9.0
	60	15.0	14.4	13.0	10.8
A106	48	12.0	11.7	10.7	9.0
	60	15.0	14.4	13.0	10.8
	70	17.5	16.6	14.8	12.0

Nominal composition—basic chemical composition without percentile.

P numbers—numbers assigned by Section IX, "Welding and Brazing Qualifications" of the ASME Boiler and Pressure Vessel Code.[22] This number is used to reduce the amount of welding procedure qualifications by grouping metals of similar base metal characteristics together. These groups are the P numbers 1, 2, 3, 4, 5, etc., and should not be confused with the pipe grade P. That is, they are all numerical. When a welder refers to P5 pipe, for example, he or she may be referring to a welding (material) group, not the pipe specification or grade.

Seamless pipe—pipe formed by forcing a billet over a mandrel versus welded seam pipe that is formed by rolling and welding steel plate. Standard seamless pipe is manufactured up to 26 in after which welded pipe is used.

Water hammer—term used for explosive expansion of condensate in steam lines, explosive flashing of water to steam in waterlines, and shock waves created in waterlines by the sudden opening or closing of valves. All types are potentially dangerous.

In cross-country pipelines, if a pump station loses power or a valve is suddenly closed, the sudden loss of pressure in the line will send a pressure shock wave down the line.

Column separation will occur behind the wave, causing a vacuum cavity to form in the line. When the pressure drops below saturation, the water will boil at ambient temperature, filling the cavity with steam. When the shock wave hits an obstacle in the line, such as a nonreturn valve, it will be reflected back up the line. When the reflected shock wave hits an obstacle (such as the closed valve) within a large vapor cavity created by the first wave, it will fill that cavity explosively. This can also happen when a valve is opened too quickly, creating a pressure wave upstream of the valve. Another form of water hammer is the sudden opening of a fire hose nozzle, which sends a shock wave back up the line as a reaction force shock (pressure) wave that strikes all obstructions and pipe elbows.

The initial warmup for steam transport lines can generate large quantities of condensate. If the warming steam is admitted faster than the condensate can be drained, the condensed water will flash back to steam and hammer the line because the condensing causes a vacuum to form, which is relieved by the water flashing back to steam. The flashing will continue until pressure–temperature equilibrium is established or the condensate is drained. Usually once initiated, serious damage will occur before the flashing subsides. This can happen during warmup of long geothermal or cogeneration steam transport pipelines as well as turbine warm up. Condensate and steam differential temperatures can also allow differential metal temperatures between the top and bottom of steam headers, causing permanent distortion.

A sudden loss of pressure in a deaerator can cause flashing in the storage tank as well as in the pump suction lines. The flashing will continue until the pressure is restored or the pressure and temperature return to saturation. Because of the stored energy in deaerators, this can be a very serious problem. For further information refer to Chap. 16.

Steam hammer—term used to describe the hammering effect of a sudden release of steam into a system or its sudden isolation. One example is the sudden opening of a boiler steam isolation valve with high drum pressure but low superheat flow, allowing a sudden release of low temperature steam into a main steam header, at lower pressure, but high superheat temperature. The example here is a two-boiler operation with a single steam turbine. The resultant release of high-density steam will cause a steam hammer all the way through the piping system and turbine. The hammer can have long-lasting detrimental effects on the turbine. This can happen by simply cracking a boiler isolation gate valve under pressure.

PIPE THRUST

Pipe thrust is caused by an unbalance in velocity and pressure within a piping system. This rarely occurs in process piping systems, with the exception of abnormal conditions described previously for water and steam hammering. Unbalanced pressure and velocity forces required to initiate line movement during steady-state flow conditions are usually not sufficient unless there is a change in pipe diameter. These forces are easily balanced by the designer with the piping layout. Needless to say, these systems must be designed to prevent transient surges and hammer. They must incorporate, for example, valve speed limits and proper condensate draining. Two exceptions under normal operating conditions are the sudden release of steam or water to atmosphere and system harmonics—for example, the opening of safety or relief valves and fire protection nozzles generates sudden unbalanced velocity and pressures within the system.

The sudden opening of fire protection spray valves and nozzles, coupled with a system network that has changes in pipe diameter, require that buried networks have restraints at all directional changes.

Because the average process system is balanced, resonance can cause a piping system to start swaying. For example, if pulsation caused by a pump impeller passing the discharge is in resonance with the piping natural frequency, the pulsations will cause the pipe to start swinging. This normally occurs at low flows; therefore, increasing the minimum flow will take the system out of resonance. If pipe seismic sway bracing is installed, this phenomenon may never be noticed.

The following sample problem is given for analysis of unbalanced forces within a piping system:

> A temporary pipeline runs parallel to a river to a 90° elbow. It then runs normal to the river 50 ft and discharges into the river. It is proposed to lay the pipe on the ground unrestrained. The pipe is used to discharge high-velocity flushing water from a process piping system. It is believed by the plant manager that thrust caused by the discharge will whip the pipe and endanger plant personnel. The startup engineer must show that the thrust at the elbow is less than the restraining force by which the pipe is held in place by dead weight and ground resistance. If the restraining force is not greater than the thrust, the line will require restraints.

Given

$$Q = \text{flow rate} = 2000 \text{ gpm} = 2.2 \text{ ft}^3/\text{s}$$

$$\upsilon = \text{specific volume} = 0.016 \text{ ft}^3/\text{lb}$$

$$P_2 = 0, \text{ at pipe discharge}$$

Pipe is 6 in Schedule 40

$$A = \text{pipe transverse area}$$

$$= 28.89 \text{ in}^2 = 0.2 \text{ ft}^2$$

$$\delta = 90°$$

$$\cos 90° = 1$$

$$\sin 90° = 0$$

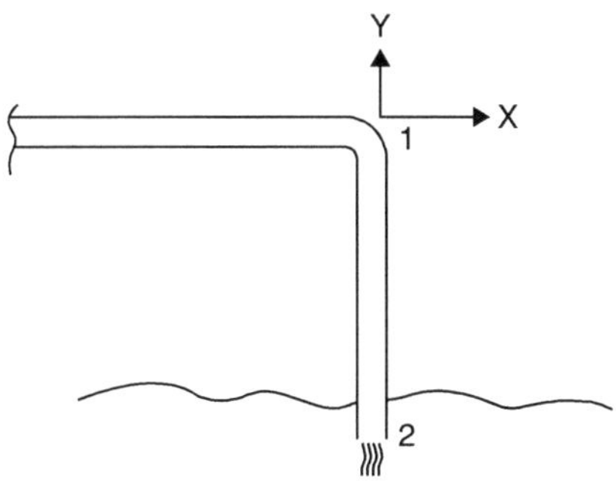

Find Pipe Thrust

v_2 = velocity at discharge

v_1 = velocity at elbow

H_L = head loss between discharge and elbow

P_1 = pressure at elbow

P_2 = pressure at discharge = 0

$v_2 = v_1 = Q/A = 2.2/0.2 = 11$ ft/s

$f = 0.016$, friction factor for clean commercial steel and wrought iron pipe

$$H_L = P_1 = fLv^2/D_{2g} = 0.016 \times 50 \times 11^2/0.2 \times 32.2 \times 2 = 7.5 \text{ ft} = 3.3 \text{ psi} = 475 \text{ lb/ft}^2$$

Solution

Momentum

$$F_x = A/\upsilon g\,(v_1 \cos \delta - v_2 \sin \delta)^2$$
$$= [0.2/(0.016 \times 32.2)]\,(11 \times 1 - 0)^2 = 47$$
$$= 47$$
$$F_y = A/\upsilon g\,(v_1 \sin \delta - v_2 \cos \delta)^2 = -47$$
$$F_M = [47^2 + (-47)^2]^{1/2} = 66$$

Pressure

$$P_x = H_L = 475 \times 0.2 \times 1 = 95$$
$$P_y = -95$$
$$F_P = (F_x^{\,2} + F_y^{\,2})^{1/2}$$
$$= [95^2 + (-95)^2]^{1/2} = 134$$
$$F = 66 + 134$$

F = 200 lbf

Find Restraining Force

(DIPRA standard, Thrust Restraint Design for Ductile Iron Pipe 5th ed.), Edition 2002,[20] in conjunction with ANSI/ANNA C150/A21.50[23] were used as guidelines for this solution.

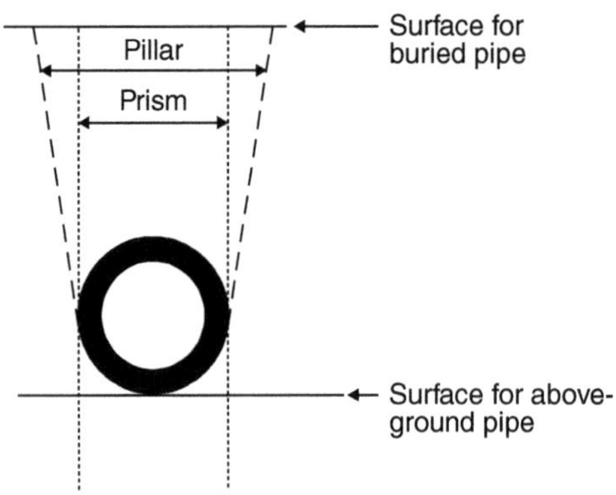

Assumptions

1. The soil is hard flat undisturbed earth (worst case).
2. W_e, the prism load on the pipe is zero for pipe with minimal surface contact.
3. A_p, the pipe circumference bearing against the soil is considered zero, the worst case, or least resistance.
4. If the resistance of the section normal to the river is less than the thrust, the pipe discharge must be anchored to prevent pipe whip. Therefore, the resistance of the section parallel to the river is not considered. It is the worst case or least resistance.

Solution

$$F_s = A_p C + W\tan\delta$$

$\delta = f_\phi \phi$, where ϕ is the internal friction angle of the soil.

$\phi = 30$ for sand or gravel w/silt and

$f_\phi = 0.75$ From Table 3, Suggested Values for Soil Parameters and Reduction Constant, K_n of the DIPRA standard[20]

$$W = 2W_e + W_p + W_w$$

where

$W_p = W \times L$ = weight of the pipe
W_w = weight of the water in the pipe per foot times L
$A_c C = 0$
$W = 2W_e + W_p + W_w$
$W = 0 + 18.97\ (50) + (8.34 \times 1.5 \times 50)$
$= 1574$ lbf
$F_s = 1574 \times \tan(30 \times 0.75)$
$= 1574 \times \tan 22.5 = 1574 \times 0.4145$
$\boldsymbol{F_s = 652}$ **lbf**

QED

This example illustrates that the weight of the pipe and water are sufficient to restrain the pipe. As the pipe size is enlarged, the standard achievable velocities will decrease as the flow increases. The corresponding weight will increase; therefore, the relationship between thrust and restraining forces will essentially remain the same.

The following is an example of the reaction force of the same pipe after the discharge has been throttled by a reducer:

Given

A 6 in × 2 in reducer is welded to the discharge end of the pipe in the previous problem. All other parameters remain the same. The head loss remains the same; however, P_1 and v_2 need to be calculated.

Find the Pipe Thrust

Velocity at the nozzle discharge,

$$v_2 = Q/A = 2.2/(3.36/144) = 94 \text{ ft/s}$$

Bernoulli's Equation

$$P_1 - P_2 = (v_2^2 - v_1^2)/2vg$$

$$P_1 = (94^2 - 11^2)62.4/64.4*144$$

$$= 59 \text{ psig}$$

Momentum Equation

$$T = A/vg\,(v^2_1 - v^2_2) - P_2A_2 + P_1A_1$$

$$= 0.022(62.4/32.2)(8714) - 0 + (59 \times 28.89)$$

$$= 372 + 1705$$

$$\mathbf{T = 2077\ lbf}$$

The modified pipe will whip and therefore must be restrained.

FLUSHING VELOCITY

Flushing of water process piping should be done at the maximum obtainable velocity. The flushing process is generally referred to as a high-velocity flush. However, the flushing usually never exceeds the normal operating velocity of the system because of system pump capacity. Table 17-3 provides normal velocities in piping systems.

A good point of reference is the AWWA Standard C651,[24] which states that the minimum velocity should not be less than 2.5 ft/s below which there will not be sufficient momentum to carry and flush debris from the system. Suction piping can be excluded but must have a suction screen until after the unit's 24-hr, full-load run. This minimum velocity should

TABLE 17-3 Normal Velocities in Piping Systems

Product	Velocity (ft/sec) Pipe Size (in)		
Water-Pump	*<2*	*2 – 6*	*≥8*
Condensate suction		1.5 – 2	2 – 2.5
Condensate discharge		4 – 7	5 – 7.5
BFP suction		2 – 3	2 – 3
BFP discharge		5 – 8	6 – 10
Service water	3 – 6	5 – 8	5 – 8.5
Circulating water		4 – 6	5 – 7
Steam			
under vacuum		33 – 50	50 – 67
Saturation 0–15 psig	33 – 42	42 – 75	50 – 67
Saturation 16–100 psig	33 – 50	50 – 83	75 – 125
Saturation>100	33 – 58	50 – 100	83 – 150
SH. 0–100 psig	42 – 67	50 – 117	100 – 167
SH. 101–500 psig	42 – 75	58 – 125	117 – 200
SH. 501–1000 psig	50 – 75	67 – 150	133 – 250
SH. >1000 psig	50 – 83	83 – 200	167 – 333

apply to all systems. Also, the commissioning engineer should not use temporary flush piping that has a smaller diameter than the process piping unless it is restrained, based on what was demonstrated in the preceding example.

HANGERS, SUPPORTS, AND RESTRAINTS

Transmission Lines

All pipelines used for the transport of fuel, steam, water, or other products are usually supported on the ground, above ground on sleepers, or buried. Nothing seems to evoke more controversy than the question of whether to bury a pipeline or install it on the ground. The choice of whether the pipe is above or below ground is only a matter of cost, security, and ambience. It is always cheaper to run uninsulated pipe on the ground without sleepers. If the line is to be installed in an inhabited area and/or will be an eyesore, it is best buried. If it crosses uninhabited desert, for example, it is cheaper to install it directly on the ground.

Pipelines resting directly on the ground have to be fully restrained at the ends and at the top and bottom of sharp inclines, to eliminate daily expansion–contraction movement, which will erode away the bottom of the pipe. Buckling restraints, which also act as longitudinal restraints, can be installed at short intervals under the pipe (10 to 25 m, depending on the pipe diameter). These restraints can be simple inverted U-girders embedded in a small trench filled with concrete. The restraining weight of the product in the line cannot be used in the buckling and expansion stress calculations because the maximum temperature and related stress will occur when the pipe is empty. Pipeline engineers have to select a tie-in temperature close to the average daily temperature. That temperature has to be as close as possible to the no-stress temperature at which any pipe repairs or tie-ins have to be made. The difficulty involving this temperature selection is compounded if drastic seasonal changes exist.

A problem that can confront commissioning engineers is improper design of a system that must be isolated periodically. Example, assume a fuel oil pipeline used for ignition and boiler warmup is isolated at night (Fig. 17-4). The line has a bellows expansion joint that fails the next day in the sun. The ambient temperature is 70°F but the black pipe temperature rose to 100°F in the sun. What is the root cause of the failure? Someone proposes to restrain the bellows to prevent further damage, stating that the line expansion caused the failure. Should the replacement bellows be restrained to prevent damage?

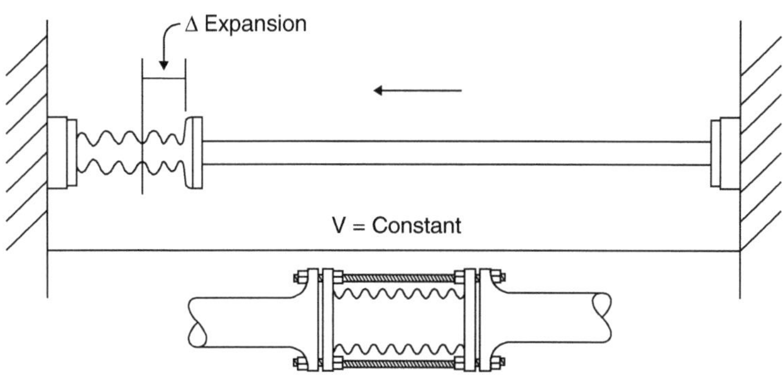

FIGURE 17-4 Restrained pipe bellows.

The line is 12-in pipe, 200 ft long. The bellows is 2 ft long. It is proposed to restrain the bellows with six, 1-in, A307 Grade A, allthread rods to limit its travel.

Given

Schedule 80 pipe cross-sectional area A = 0.7056 ft^2

Modulus of elasticity, E, for fuel oil = 38.9×10^6 lb/ft^2

Coefficient of expansion, β, for fuel oil = 0.00063

Tensile strength of the allthread = 55,000 psi

The bellows is rated at 1500 psi

Assume the bellows compressed 2 in, but the total system volume remained constant.

Find the maximum possible pressure in the system if the bellows had not failed:

Equations of State

$$\Delta V = V_i\,\beta\,(t - t_i)$$
$$= (200 \times 0.78) \times 0.00063\,(100 - 40)$$
$$= 5.897$$

Line Pressure Change = $\Delta P = E\,\Delta V / V_{\text{Initial}}$, $P_i = 0$

$$\Delta P = (38.9 \times 10^6 \times 5.8968)/(144 \times 200 \times 0.7056)$$
$$P_f = 10{,}211 \text{ psi}$$

The pressure in the line far surpasses the maximum allowable for the bellows; therefore, the bellows was not crushed but burst under extreme pressure. The root cause for the failure was extreme pressure caused by isolating the line. Therefore, a pressure relief valve must be installed to relieve the isolated line pressure back to a storage tank or other unrestricted line or vessel.

Find the maximum allowable line pressure and the stress on the proposed restraining rods.

$$\text{Stress on all-thread restraints} = S = (\Delta P/6) = 10{,}211/6 = 1702 \text{ psi}$$

The stress on the studs is within their tensile strength of 55,000 psi. Check the maximum allowable working pressure of the fuel oil line. The line is Schedule 80, ASTM A106 Grade B.

From ANSI B31.1[2]

$$P = \{2SE(t_m - A)\}/\{(D - 2y)/(t_m - A)\},$$
$$A = 0, \text{ for welded fittings}$$
$$y = 0.40$$
$$SE = 15{,}000 \text{ psi}$$

From B36.10M[10]

$$t_m = t \times 0.875\text{-}t \times (\text{mill tolerance})$$

Therefore

$$P = (2 \times 15{,}000 \times 0.688 \times 0.875)/(11.374 - 2 \times 0.40)(0.688 \times 0.875)$$
$$= 1472 \text{ psi}$$

QED

After the problem was fully analyzed, the conclusions were obvious. However, in no discipline, other than commissioning, are there more bystander distractions that can lead to false conclusions. These distractions are not always easily dismissed without detailed analysis.

Note

Pipelines can be fully restrained, but a detailed analysis must be conducted to ensure that they are not overpressurized due to temperature fluctuations. Generally, the maximum temperature rise for above-ground, cross-country pipelines occurs when the pipeline is empty. Therefore, at the maximum temperature rise, the internal pressure is minimum.

Underground Pipe

It is always cheaper to install insulated pipe above ground on sleepers, utilizing expansion loops. The pipe, however, can be installed underground with special considerations. A good source for information regarding underground insulated pipelines is *The District Heating Handbook*[25] published by the International District Heating Association. The following illustrations are from the handbook and serve as examples of five different methods for protecting insulated underground pipelines.

There are four types of conduit used to encase and protect underground insulated pipe Fig. 17-5. They are used on large-scale systems such as cogeneration steam piping that have to accommodate public right of ways. The four types are box, box with cover, solid pour, and factory prefabricated. An attempt to deviate from proven methods of running underground insulated pipe will amount to a research and development program.

These raceways must be checked for adequate, watertight protection and drainage. The pipe also must have adequate drains for warmup without water hammer. In subzero operation, adequate freeze protection must be provided and checked out. Expansion joints, as

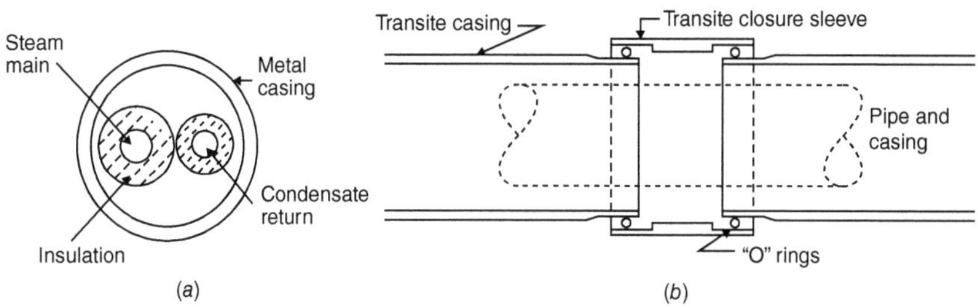

FIGURE 17-5 (*a*) Metal casting, prefabricated condult. (*b*) Cement pipe casting, prefabricated conduit. (*c*) Reinforced concrete tunnel. (*d*) Solid pour condult.

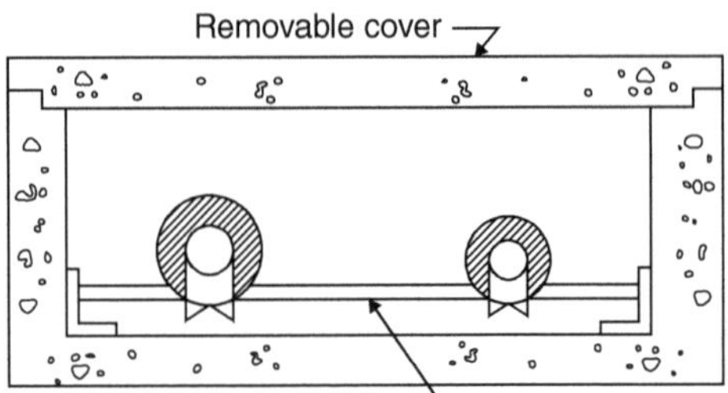

(*c*)

Removable cover on box type condult

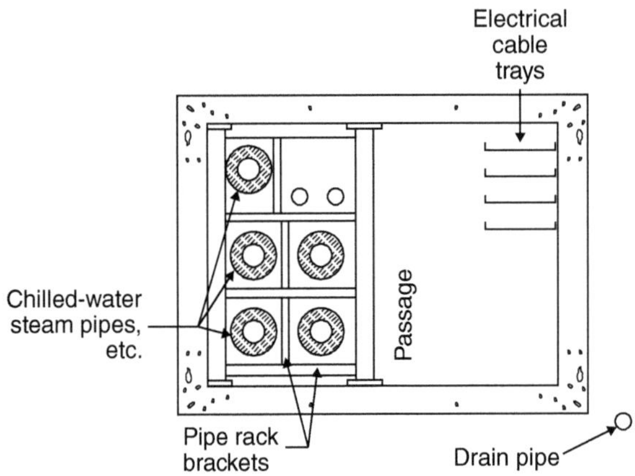

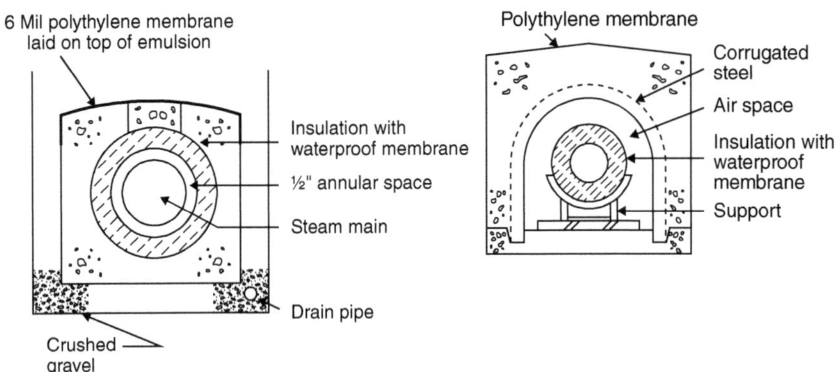

(*d*)

FIGURE 17-5 (*Continued*)

well the pipeline, must be checked for freedom of movement. Access to expansion joints, valves, drain traps, and other piping fixtures must be provided through manholes for startup, operations, and maintenance.

PLANT PIPING SYSTEMS

Pipe Supports

Installation and support of plant piping systems is governed by ANSI and AWWA standards as well as the MSS standard practices. The figures on the next page, from MSS 58[26] and 69,[27] represent typical hangers. Types 51, 52, and 53 are variable-spring hangers in which the load on the hanger varies with the vertical expansion of the line. Types 54, 55, and 56 are constant-support hangers in which the load on the hanger remains constant regardless of line movement. (If type 51 is used for "hot" pipe supports, the pipe clamp material grade must be compatible with the operating temperature of the line.) Type 39 and 40 protection saddles and shields are used to isolate the hot pipe surface from the supporting hangers.

Rigid Supports

Angle iron used for attaching and supporting hanger rod can be bored, but structural beams cannot. Attachment to structural members must be by welded or adjustable clamps manufactured for that purpose.

Rod nuts must be fully engaged with the rod. Jam nuts are required on all threaded connections except the left-hand thread end of turnbuckles and rod couplings. The nut should be a half nut or a full nut, with at least half of the threads engaged.

U-bolts and clamps supporting pipe with a service temperature above 200°F should have a 1/16-in clearance. Hanger rod and other rigid support assembles must be adjusted to support gravity loads only.

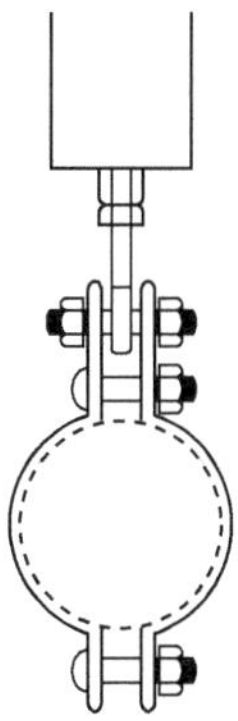

Variable-spring hanger type-51

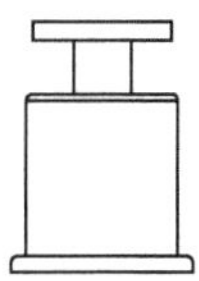

Variable-spring base support type-52

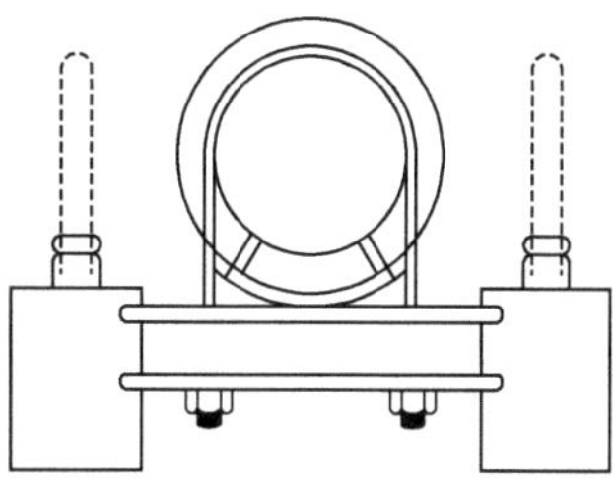

Variable-spring trapeze hanger type-53

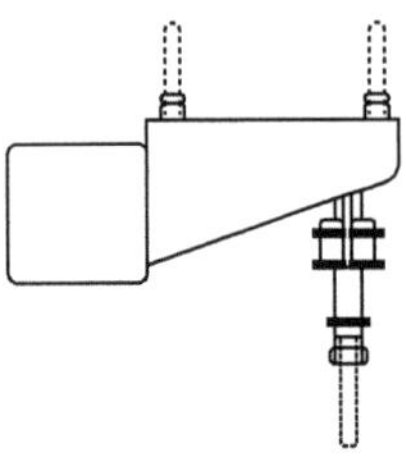

Constant-support hanger
horizontal type
type-54

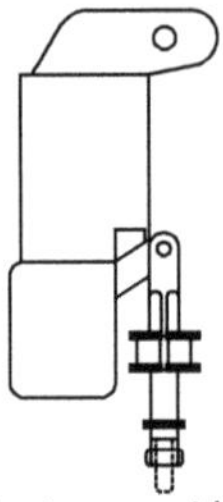

Constant-support hanger
vertical type
type-55

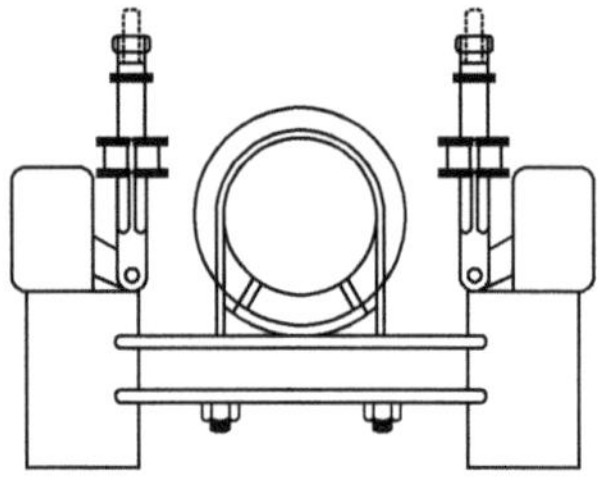

Constant-support hanger
trapeze type
type-56

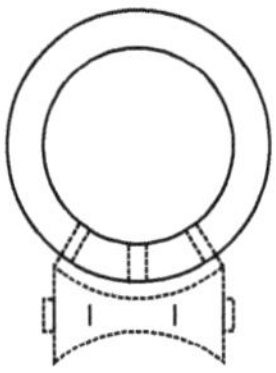

Steel pipe covering
protection saddle
type-39

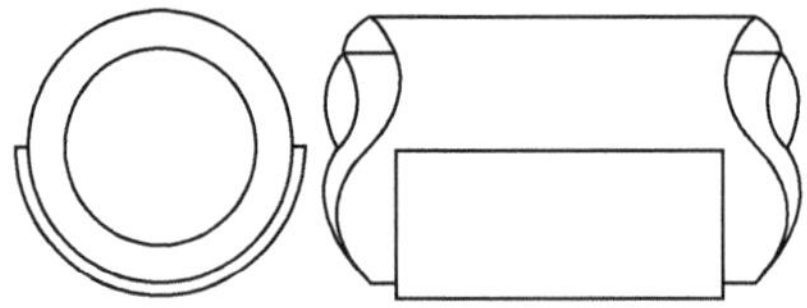

Protection shield
type-40

Spring Supports

Spring supports are ordered, for initial construction, pinned in the cold position. However, they can be ordered pinned in the hot position so that existing supports can be replaced with the system in operation. If supports are to be used in a salt air environment, they should be ordered hot-dip galvanized with an additional epoxy paint coating on the cans and a minimum 6-mil neoprene coating on the springs.

All spring supports must be set to carry the loads specified in the engineering drawings. With the gap turnbuckle, the support can be adjusted until the spring is supporting the design load at the design elevation. If additional adjustment is necessary to support the load, it can be done with the assembly load-adjustment bolt. If this is required, engineering must be notified. At the design load, a minor adjustment of the load bolt should release the shipping pin, which can be removed by hand. The pin should always be attached to, and left with, the hanger.

Some hangers are designed to carry design loads plus the hydrostatic test load. The commissioning engineer should always check the hanger design data sheet before re-installing pins for hydro testing.

As a hot unit or system is being placed in service, the piping expansion must be constantly checked for proper movement. Once the unit is at the design operating temperature the hanger hot settings have to be checked and recorded.

> V-6.1.1, Movements of critical piping systems from their design locations shall be used to access piping integrity. The Operating Company shall have a program requiring such movements be taken on a periodic basis along with a procedure precluding the unnecessary removal of insulation when measurements are taken. [*sic*] Piping system movement records shall be maintained. The Operating Company shall evaluate the effects of position changes on the safety of the piping systems and shall take appropriate corrective action.[2]

Before commissioning a system, the pipe supports must be checked and adjusted and the cold settings recorded. During warmup, the system must be watched carefully for proper expansion. Finally, the hot settings must be checked and recorded. Figure 17-6 is an ANSI B31.1 form for recording purposes.

Company ______________________ System ______________________ Prepared by ______________________

Plant Name ______________________ Unit No. ______________________ Sheet ____________ Of ____________

Date ______________________

Hanger Mk. No.	Hanger size[1]	Hanger type[2]	Dsgn. Mvt.	Constant support hanger data		Rigid & variable support data		Design elev. cold,[3] ft. in.	Hanger scale reading[4]							
									Date:		Date:		Date:		Date:	
				Design load, lb	Design total travel	Design load, lb										
						H	C	(4)	Cold	Read by	Hot	Read by	Cold	Read by	Hot	Read by

Notes:

(1) Hanger size to be taken from hanger fabrication drawing.

(2) For constant support and variable support types indicate by CS or VS. For rigid, anchor, guide, sliding, or other type support, indicate by letter R, A, G, or S, respectively.

(3) Elevation of center line of pipe after cold springing and final hanger settings with line cold.

(4) "o" indicated the highest scale position with "5" being the midpoint, and "10" being the lowest scale position.

FIGURE 17-6 Hamper record sheet from ANSI B31.1.[2]

High-pressure, high-temperature steam system commissioning is serious business that should not be taken lightly; above all, it should not be supervised by engineers without experience.

> All high-temperature components have finite design life and a practical operating life, because of the continual accumulation of creeps and fatigue damage. These two damage mechanisms may also combine synergistically (often called creep/fatigue interaction) to reduce operating life well below that predicted using creep or fatigue calculations alone.[28]

Steam and water hammer, quenching, and heating of headers, coupled with improperly set supports, can dangerously reduce the operating life of a piping system. There are too many stations, especially in developing countries, that are disasters waiting for the day of reckoning. This reminder is presented because this, as many other areas in commissioning, is often not completely understood and is taken too lightly.

PIPE FITTINGS, FLANGES, AND FASTENERS

Fittings

ASME/ANSI B16 Series, "Standards of Pipes and Fittings,"[29] provides dimensional data as well as weights and tolerances for pipe fittings. These are also the dimensional data provided in supplier catalogs.

The accompanying illustrations show the different fittings available.

Note

Long-radius elbows have a radius 1.5 times the diameter. For short-radius elbows, the radius is equal to the diameter (Fig. 17-7). Street elbows are elbows with one end male and the other female (Fig. 17-8). Regular butt-weld elbows have male connections at both ends. Socket-weld fittings have female ends (Fig. 17-9).

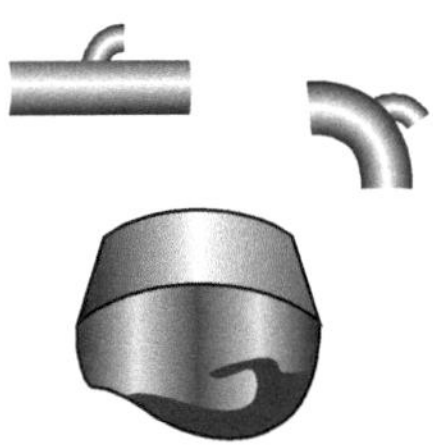

FIGURE 17-7 Typical sock-o-lets.

FIGURE 17-8 45° street elbow.

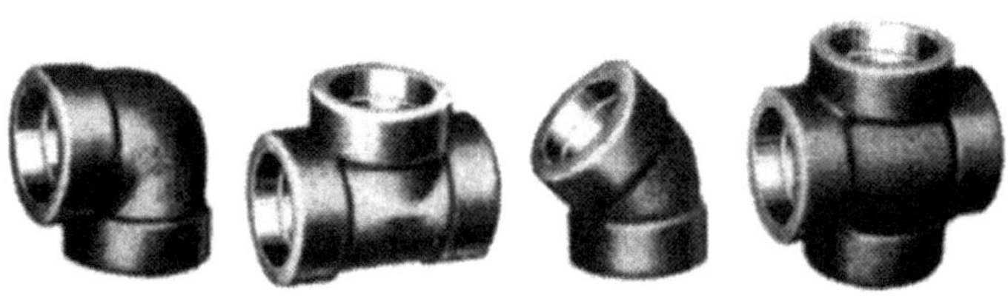

FIGURE 17-9 Typical socket-weld fittings.

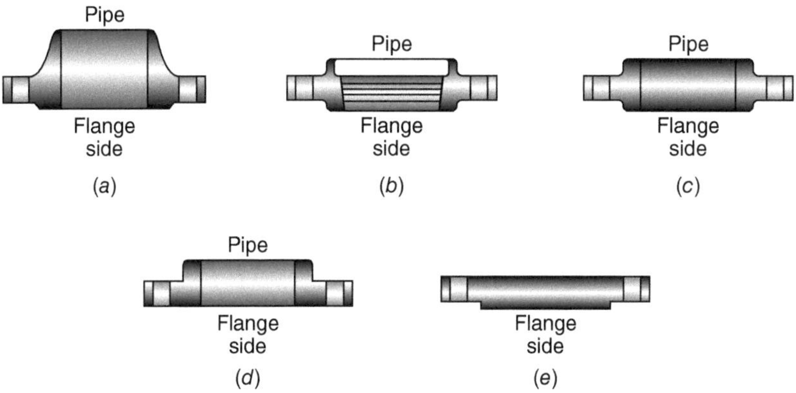

FIGURE 17-10 (*a*) Welding neck. (*b*) Threaded. (*c*) Slip-on. (*d*) Lap joint. (*e*) Blind.

Flanges

An illustration of typical flanges is provided in Fig. 17-10. The slip-on flange is welded on both sides, inside and out. A socket-weld flange is also available, which is welded only on the outside.

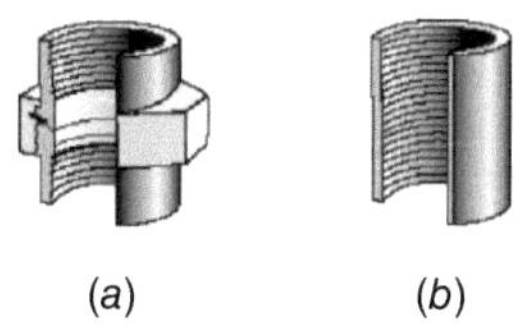

FIGURE 17-11 (*a*) Pipe union. (*b*) Coupling.

Couplings are used to join pipe where removal for maintenance is not a requirement. If a threaded strainer or other threaded piece of equipment must be removed for maintenance, one end of the equipment must have a union [Fig. 17-11(*a*)]. Otherwise, it can't be removed without cutting the line. Half couplings are frequently used in place of weld- o-lets (sock-o-lets) for pipe nipples [Fig. 17-11(*b*)].

Socket weld fittings are frequently used in small-bore piping systems. The reason is that fit-up is fast, and fillet welds are easier to make than butts welds. All socket welds have a mandatory 1/16-inch gap requirement between the pipe end and bottom of the socket. This can cause difficulties for commissioning engineers during system flushing. The gap, whether in a socket-weld flange or a fitting, provides a "hideout" for crud. This is especially bad in lube oil piping systems. In some cases where socket welded carbon steel was used, it is better for startup to scrap the system entirely and replace it with butt-welded stainless steel. The time and expense lost flushing can be far more costly than the replacement.

Backing rings also cause problems for commissioning because, like the socket weld, they can gather crud. Sometimes, the rings are burnt through and the ring itself contributes to the debris in the piping. The best pipe-welding techniques for critical systems, such as steam seal and lube oil, use consumable inserts. The insert provides the filler metal for the root pass as it is consumed. However, the procedure requires inert "backing" gas in the pipe if the system is stainless or nickel steel (Fig. 17-12). The root pass is welded using TIG (GTAW) or wire-feed inert gas welding (GMAW).[a] This technique provides the highest quality weld with no system contamination.

[a] Tungsten inert gas (gas tungsten arc welding); gas metal arc welding.

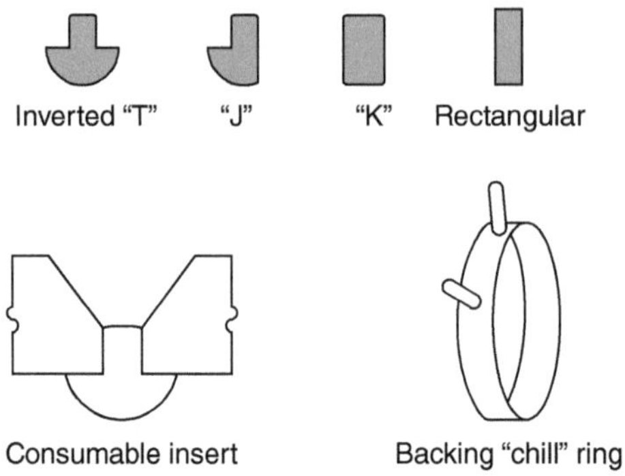

FIGURE 17-12 Consumable insert types versus backing ring.

> **Caution**
>
> If dams are used to contain backing gas, they must be removed after a weld is made. They can be pulled down to the next weld, but eventually, at the last tie-in weld, there must be a means of removing the dams or the entire system must be purged with gas without them.

Fasteners

Torque is the twisting action that is used to acquire the correct tension, or stretch, on bolts and bolt studs. The tension of common bolts is usually 75 percent of the minimum yield strength. Bolt studs (simply referred to as studs) are used for blind holes so that there is no chance of the fastener bottoming out in the hole and the applied torque compressing, rather than tensioning, the bolts and overstressing the equipment it is fastening to. Large, specialty studs are cut from "allthread" rod and, because of the labor involved, are more expensive than bolts. On the upside, there are no standard lengths because they are custom cut. At least one nut, on a stud used for flanges, often has to be tack welded to the stud in order for it to be removed. This should be a consideration when ordering studs versus bolts for pipe flanges. When ordering them, also note that the length of a bolt is shorter than a stud for the same flange dimension because the bolt length does not include the cap. The stud length must be a nut thickness longer than a bolt for the same application.

Torque is affected by the surface friction at the point of contact of the cap or nut with the mating surface. If an unlubricated torque value is used to torque a bolt that is lubricated, the bolt may break. The reason is, the best will be tensioned or stratched beyond its yield strength before reaching the specified torque. Bolt suppliers can provide the types of lubricants and their respective multiplication factors. Figure 17-13 is a graph from GRI/Durlon illustrating the effects of lubricants on friction factor.

Torque should be applied 20 percent at a time in a cross-rotation pattern until the full torque is reached. Common bolt torque for fasteners up to ¾ is usually calculated using the following equation:

$$T = K \times D \times P$$

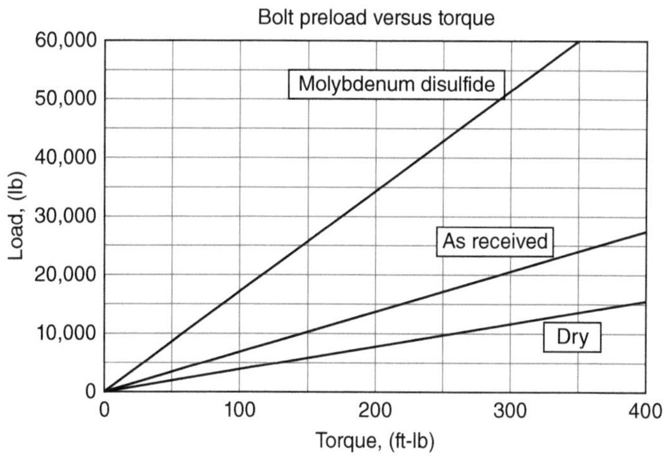

FIGURE 17-13 Torque difference for dry vrs. lubricated.

where:

f = coefficient of friction
D = bolt diameter
P = load = 75 percent of yield strength

The Table 17-4 (Hydratight RSL table) contains torque values for ASTM 193 Grade 7 bolts. Note the difference in torque values for the different lubricants.

The point to be made is that the bolt supplier must provide a torque chart for all bolts provided on the project, with and without lubricants, for gasketless joint assemblies. The most common lubricant is thread never-seize paste. When gaskets are specified, the gasket supplier must be consulted for its torque charts as well. Torque values for gaskets can be significantly less than for bolts. Also, if at least one flange is nonmetallic, the flange supplier must be contacted. ANSI/ASME PCC-1-2000 (R2005), "Guidelines for Pressure Boundary Bolted Flange Joint Assembly,"[30] *sets the guidelines for bolting of flanges.*

The power torque wrenches on the left are Hydratight RSL hydraulic tools. The illustrations are from their operation and maintenance instruction manual that can be found on the Internet. It contains useful information regarding torque. Remember, the manufacturer, or engineer of record, must provide the torque values. The Hydratight RSL tightening sequence is listed in Figs. 17-14 and 17-15, as well as one of the torque tables (Table 17-4).

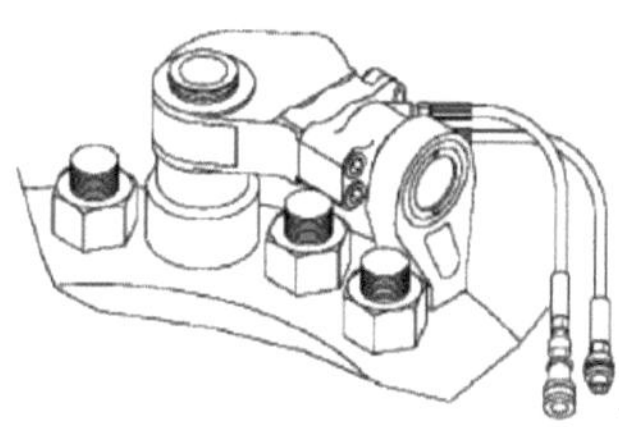

(*a*)

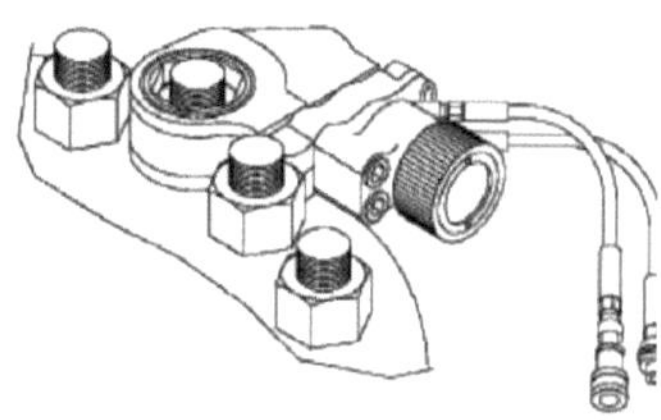

(*b*)

Impact tools and slugging wrenches should never be used on high-pressure system fit-up. There is no way to determine the exact torque nor is there a way to guarantee even torque distribution. This practice can lead to chronic steam leaks and wire drawn flanges. Valve caps and other components with gasketless surfaces are especially susceptible to leakage

TABLE 17-4 Chart of Torque Values in Imperial Units for ASTM A-193 Grade B7 with Stud Stress Equal to 50% of Minimum Yield Strength[31]

			Torque For Specified							
Bolt Size (IN.)	Minimum Yield Strength (PSI)	Bolt Tension (LBS.)	503 SWEENEY® f = .067	504 SWEENEY® f = .085	505 SWEENEY® f = .100	506 SWEENEY® f = .110	507 SWEENEY® f = .100	API SA2 f = .150	Machine Oil	Dry Steel on Steel f = .200
3/4	1,05,000	17,754	104	128	147	159	147	168	209	272
7/8	1,05,000	24,495	167	204	234	254	234	268	335	435
1	1,05,000	32,129	249	304	349	379	349	400	499	649
1-1/8	1,05,000	41,873	362	442	508	553	508	584	730	953
1-1/4	1,05,000	52,905	502	615	709	772	709	816	1023	1337
1-3/8	1,05,000	65,226	675	829	958	1043	958	1103	1386	1814
1-1/2	1,05,000	78,835	878	1082	1253	1366	1253	1446	1820	2387
1-5/8	1,05,000	93,733	1124	1388	1608	1755	1608	1858	2342	3076
1-3/4	1,05,000	1,09,920	1411	1746	2024	2210	2024	2340	2953	3883
1-7/8	1,05,000	1,27,395	1744	2160	2507	2739	2507	2901	3664	4822
2	1,05,000	1,46,158	2100	2611	3037	3320	3037	3519	4455	5874
2-1/8	1,05,000	1,66,210	2529	3148	3663	4007	3663	4247	5381	7099
2-1/4	1,05,000	1,87,551	3011	3751	4368	4779	4368	5067	6424	8480
2-3/8	1,05,000	2,10,180	3552	4429	5160	5648	5160	5989	7597	10,033
2-1/2	1,05,000	2,34,098	4154	5184	6042	6615	6042	7015	8903	11,764
2-3/4	95,000	2,58,580	5024	6278	7324	8020	7324	8508	10,807	14,291
3	95,000	3,10,020	6545	8189	9558	10,472	9558	11,111	14,124	18,689
3-1/4	95,000	3,66,124	8346	10,452	12,207	13,377	12,207	14,196	18,057	23,907
3-1/2	95,000	4,26,891	10,451	13,099	15,307	16,778	15,307	17,808	22,664	30,022
3-3/4	95,000	4,92,321	12,880	16,156	1887	20,707	1887	21,982	27,989	37,091
4	95,000	5,62,415	15,658	19,655	22,986	25,206	22,986	26,760	24,088	45,190
4-1/4	75,000	5,03,030	14,850	18,651	21,819	23,931	21,819	25,409	32,379	42,939
4-1/2	75,000	5,65,731	17,654	22,185	25,961	28,478	25,961	30,240	38,548	51,134

Note: 503 – MOLYBDENUM DISULIFIED BASED (FUNCTIONAL EQUIVALENT C670, MOLY-COTE G PASTE)
504 – MOLYBDENUM LEAD OXIDE & GRAPHITE BASED (FUNCTIONAL EQUIVALENT C100)
505 – COPPER & GRAPHITE BASED (FUNCTIONAL EQUIVALENT N1000)
506 – NICKEL & GRAPHITE BASED (FUNCTIONAL EQUIVALENT NICKEL EASE N5000, NEVER-SEEZ NICKEL SPECIAL)
507 – COPPER & GRAPHITE BASED (FUNCTIONAL EQUIVALENT C5A, NEVER SEEZ)

Tightening

1. Mark the correct tightening sequence on the studs in a clockwise direction with chalk as per diagram. For more explanation please read ASME-PPC-1-2000 or refer to the Hydratight joint integrity assurance procedures.

Criss-cross tightening sequence examples:

4 Bolt flange 1-3-2-4
8 Bolt flange 1-5-3-7-2-6-4-8
12 Bolt flange 1-5-9-3-7-11-2-6-10-4-8-12
16 Bolt flange 1-9-5-13-3-11-7-15-2-10-6-14-4-12-8-16
20 Bolt flange 1-13-5-17-9-3-15-7-19-11-2-14-6-18-10-4-16-8-20-12
28 Bolt flange 1-13-21-5-17-9-25-3-15-23-7-19-11-27-2-14-22-6-18-10-26-4-16-24-8-20-12-28
32 Bolt flange 1-17-9-25-5-21-13-25-3-19-11-31-7-29-15-27-2-18-10-30-6-22-14-26-4-20-12-32-8-24-16-28

Note: Correct numbering of bolts should result in all odd numbered bolts around one side of the flange and all even numbered bolts around the other side.

- Determine the torque value for the stud or bolt to be tightened
- First tightening stage should be limited to a maximum of 30% of the final torque setting
- Second tightening stage should be limited to a maximum of 60% of the final torque setting
- Third tightening stage should be carried out at the 100% torque setting
- On the fourth and final tightening stage, change from diagonal tightening to adjacent bolt-to-bolt tightening clockwise using the 100% torque setting and chase around flange until nuts finally stop rotating.
- Note: First, second and third stages should be tightened using the criss cross tightening sequence and the fourth stage should be tightened using the adjacent clockwise bolt to bolt sequence as shown in the diagrams above.

For more specific application based procedures please consult **Hydratight sweeny joint integrity assurance** procedures.

*Hydratight offers manual control pendant (standard) and digital control pendant (option).

Marked up 8 bolt flange

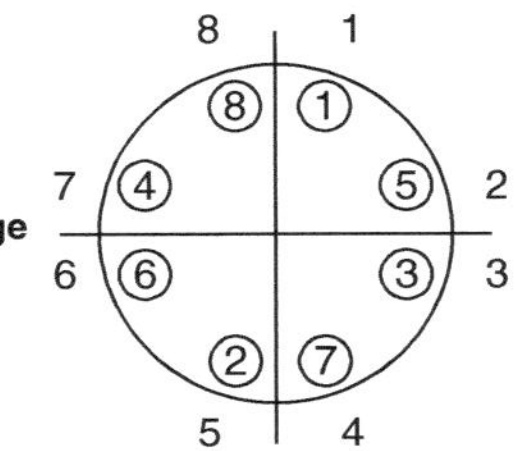

1st, 2nd, & 3rd stage
criss cross sequence

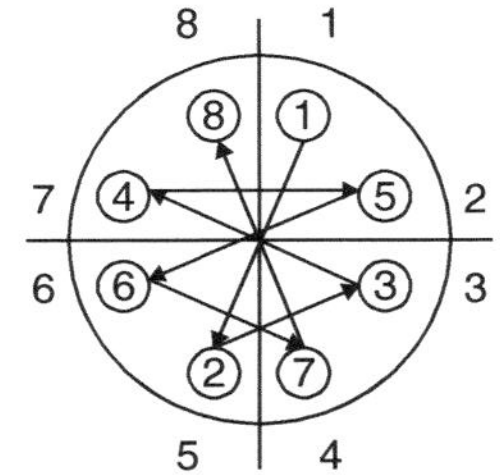

4th & final stage
adjacent bolt sequence

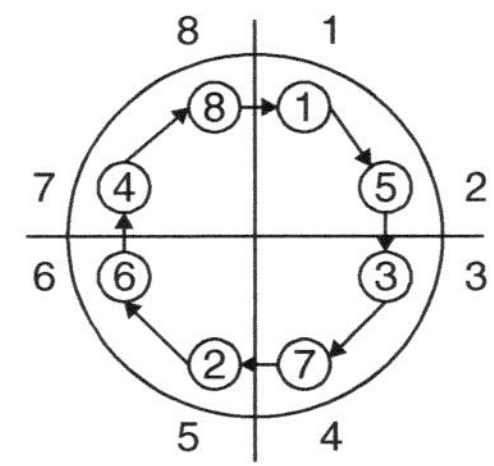

FIGURE 17-14 Fastening sequence for circular flanges.

24 Bolt	1-13-7-19 4-16-10-22 2-14-8-20 5-17-11-23 3-15-9-21 6-18-12-24
28 Bolt	1-15-8-22 4-18-11-25 6-20-13-27 2-16-9-23 5-19-12-26 7-21-14-28 3-17-10-24
32 Bolt	1-17-9-25 5-21-13-29 3-19-11-27 7-23-15-31 2-18-10-26 6-22-14-30 4-20-12-28 8-24-16-32
36 Bolt	1-2-3 19-20-21 10-11-12 28-29-30 4-5-6 22-23-24 13-14-15 31-32-33 7-8-9 25-26-27 16-17-18 34-35-36
40 Bolt	1-2-3-4 21-22-23-24 13-14-15-16 33-34-35-36 5-6-7-8 25-26-27-28 17-18-19-20 37-38-39-40 9-10-11-12 29-30-31-32
44 Bolt	1-2-3-4 25-26-27-28 13-14-15-16 37-38-39-40 5-6-7-8 29-30-31-32 17-18-19-20 41-42-43-44 9-10-11-12 33-34-35-36 21-22-23-24
48 Bolt	1-2-3-4 25-26-27-28 13-14-15-16 37-38-39-40 5-6-7-8 29-30-31-32 17-18-19-20 41-40-43-44 9-10-11-12 33-34-35-36 21-22-23-24 45-46-47-48
52 Bolt	1-2-3-4 29-30-31-32 13-14-15-16 41-42-43-44 5-6-7-8 33-34-35-36 17-18-19-20 45-46-47-48 21-22-23-24 49-50-51-52 25-26-27-28 9-10-11-12 37-38-39-40
56 Bolt	1-2-3-4 29-30-31-32 13-14-15-16 41-42-43-44 21-22-23-24 49-50-51-52 9-10-11-12 37-38-39-40 25-26-27-28 53-54-55-56 17-18-19-20 45-46-47-48 5-6-7-8 33-34-35-36
60 Bolt	1-2-3-4 29-30-31-32 45-46-47-48 13-14-15-16 5-6-7-8 37-38-39-40 21-22-23-24 53-54-55-56 9-10-11-12 33-34-35-36 49-50-51-52 17-18-19-20 41-42-43-44 57-58-59-60 25-26-27-28
64 Bolt	1-2-3-4 33-34-35-36 17-18-19-20 49-50-51-52 9-10-11-12 41-42-43-44 25-26-27-28 57-58-59-60 5-6-7-8 37-38-39-40 21-22-23-24 53-54-55-56 13-14-15-16 45-46-47-48 29-30-31-32 61-62-63-64
68 Bolt	1-2-3-4 37-38-39-40 21-22-23-24 53-54-55-56 9-10-11-12 45-46-47-48 29-30-31-32 61-62-63-64 17-18-19-20 57-58-59-60 33-34-35-36 5-6-7-8 41-42-43-44 13-14-15-16 49-50-51-52 25-26-27-28 65-66-67-68
72 Bolt	1-2-3-4 37-38-39-40 21-22-23-24 57-58-59-60 5-6-7-8 41-42-43-44 25-26-27-28 61-62-63-64 9-10-11-12 45-46-47-48 29-30-31-32 65-66-67-68 13-14-15-16 49-50-51-52 33-34-35-36 69-70-71-72 17-18-19-20 53-54-55-56

FIGURE 17-15 Bolt-tightening sequences per ASME PCC-1-200.

when the bolts are not tightened with specialty tools such as heaters, tensioners, or hydraulic torque devices.

> **Note**
>
> Ironworkers use pneumatic tools, but they "rattle up" using tension proofing washers to ensure correct and uniform torque is applied to the fasteners.

Heaters elongate the bolts to minimize torque. Tensioners stretch the bolts mechanically. Hydraulic torque wrenches torque fasteners using hydraulic jacks.

Nonmetallic flange joining must be done in accordance with the flange manufacturers' instructions. The minimum yield strength of carbon steel or stainless-steel fasteners is far greater than the strength of nonmetallic flanges.

HYDROSTATIC TESTS

Hydrostatic testing should be conducted on all fuel, chemical, high-energy, and underground systems because if these systems have leaks during initial service, there will be substantial commissioning delays. Systems such as low-pressure general service water and condensate can be initial service tested because leaks generally can be rapidly corrected in service without causing significant commissioning delays. Initial service testing is permitted by ANSI B31.1 Section 137.7[2] when approved by the owner.

Systems hydrostatically tested must be pressurized to 1.5 times the design pressure for at least 10 min. All systems exposed to the sun must be visibly inspected for leaks. Decay testing can be misguiding because exposure to the sun and climate radiant changes caused by blocking and clearing cloud cover can change the hydrostatic water temperature enough to cause expansion and contraction pressure changes not related to system integrity. The hydrostatic water temperature should be greater than 60°F to ensure that the metal is above its nil-ductility temperature unless otherwise specified by the engineer of record. Refer to App. O for details for hydro testing.

REFERENCES

1. "Instrumentation Symbols and Identification," 1SA-5.1-1984 (R1992).
2. "Power Piping," B31.1.
3. "Process Piping," B31.3.
4. "Pipeline Transportation Systems," B31.4.
5. "Refrigeration Piping," B31.5.
6. "Nuclear Piping," B31.7.
7. "Gas Transmission and Distribution Piping Systems," B31.8.
8. "Building Services Piping," B31.9.
9. "Slurry Piping System," B31.11.
10. "Welded and Seamless Wrought Steel Pipe," B36.10.
11. "Stainless Steel Pipe," B36.19.

12. "Standards of Pipes and Fittings," ASME/ANSI B16.
13. "Standard Specification for Seamless Ferritic Alloy-Steel Pipe for High-Temperature Service," ASTM A335/A335M-95a.
14. "Department of Transportation, Pipeline Safety Regulations," 49 CFR 190-199.
15. "Standard Practice for Pipe Hangers and Supports, Design and Manufacture," SP 58.
16. "Pipe Hangers and Supports-Selection and Application," SP 69.
17. American National Standards/American Water Works Association C950.
18. AWWA M9.
19. Ductile Iron Pipe and Fittings for Water and Other Liquid.
20. Thrust Restraint Design for Ductile Iron Pipe.
21. P 91 Industry Alert.
22. B&PV code sec. 18.
23. American National Standards/American Water Works Association C150/A21.50.
24. C651.
25. District Heating Handbook.
26. MSS 58.
27. MSS 69.
28. Cronin, M., and Lefton, S., "Guard Against High-Temperature Steam-Piping Failures," *Power*, June 1986.
29. "Standard of Pipes and Fittings," ASME/ANSI B16 series.
30. "Guidelines for Pressure Boundary Bolted Flange Joint Assembly," ANSI/ASME PCC-1-2000 (R2005).
31. Hydratight RSL table for ASTM 193, Grade7 bolts.

CHAPTER 18
PUMPS

CENTRIFUGAL PUMPS

A good field reference book for centrifugal pumps is the *Centrifugal Pump Clinic*, by the late Igor J. Karassik.[1] It is a good reference book for commissioning because it addresses the root causes for actual pump problems and failures that Karassik encountered in his lifetime.

Centrifugal pumps are categorized by discharge flow which can be radial, mixed, and axial. In other words they are categorized by how the discharge flow exits the impeller. For example, if the discharge is not completely accelerated 90 (radially) from the inlet flow, it is called a mixed flow pump. Sometimes a mixed flow pump is referred to as a turbine pump. But, it still belongs to the centrifugal pump category. A propeller pump can be purely an axial flow pump but it also belongs to the centrifugal pump category. Centrifugals are variable displacement pumps. That is they displace only the amount of fluid called for at the discharge. All residual energy not used to displace fluid is converted to heat. The other category discussed in this chapter is the positive displacement pump which as its name implies displaces the same amount of fluid no matter how much flow is required at the discharge. Therefore, fluid not needed at the discharge must be bypassed somewhere.

Pump laws, often called affinity laws, are based on dimensionless ratios that can be used to determine the effects that changing one parameter has on the other parameters within an equation. They are identical to fan laws. For example, if we change a centrifugal pump impeller diameter, we can find the change in flow and head. These ratios are:

$$W/ND^3,\ H/N^2D^2 \text{ and } hp/\varrho N^3D^5$$

where:

W = volumetric flow
N = speed
D = impeller diameter
H = head
ϱ = density

For example, we find that the head of a new pump is above the design pressure of a heat exchanger. Therefore, we would like to shave the impeller to lower the head instead of throttling the discharge valve. To find out how much reduction will be required in order to determine if it will be possible, we can solve for the new diameter as follows:

$$(H/N^2D^2)_{\text{new}} = (H/N^2D^2)_{\text{old}}$$
$$H_{\text{old}} = 175 \text{ psi}$$
$$H_{\text{new}} = 150 \text{ psi}$$
$$N = \text{constant}$$
$$D_{\text{old}} = 12$$

Find the New Impeller Diameter:

$$D_n = (D_o^2/H_o \times H_n)^{1/2}$$
$$= (144/175 \times 150)^{1/2}$$
$$D_n = 11.12 \text{ in}$$

How is flow affected? Assume 1000-gpm initial flow rate.

$$(W/ND^3)_{\text{new}} = (W/ND^3)_{\text{old}}$$
$$W = D^3_o/W/D^3_n$$
$$= 12^3/1000/11.12^3$$
$$= \mathbf{794\ gpm} \leftarrow$$

If a pump produces too much head, usually the pump discharge valve will be throttled to reduce the head to a desirable value. If the valve is throttled, a new system curve is created for the downstream portion. The throttling reduces the flow far more than if the pump impeller diameter is reduced. The pump now operates at a higher discharge pressure on a new system curve because the valve has artificially changed the system. Reducing the impeller diameter allows the discharge to be opened, restoring the old system curve (Fig. 18-1). Using the affinity laws, it is discovered that the lost capacity is less than that

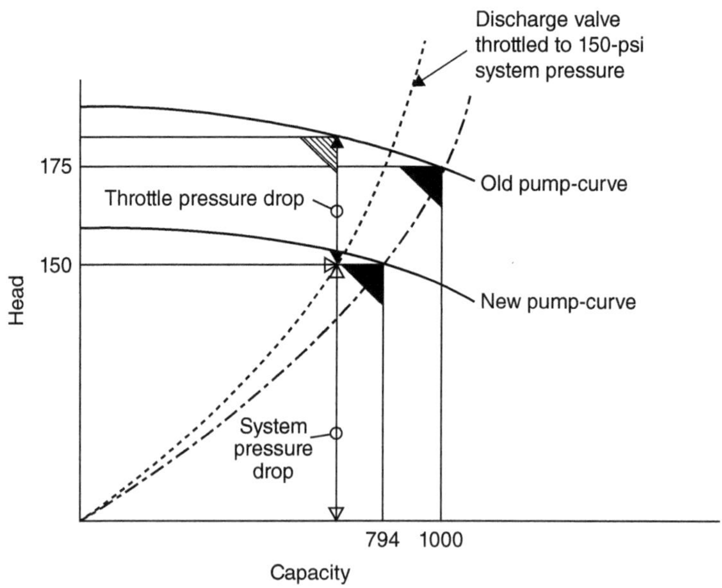

FIGURE 18-1 Curve change for impeller diameter change.

lost by throttling. However, diameter reduction has its limits, after which the affinity laws are no longer valid. The manufacturer will provide analysis and instructions for the reduction. This example serves only to illustrate what can be done if a reduction in either horsepower or head at rated capacity is required.

Specific speed is another tool that can be used in a similar fashion. It is defined as:

$$N_s = \text{rpm}(\text{gpm})^{1/2}/H^{3/4}$$

The higher the specific speed of the pump, the higher the capacity. Specific speed is used by designers but not in the field because field engineers will have a specific curve for each pump. Figure 18-2 is an illustration of different impeller designs with their respective specific speeds.[2]

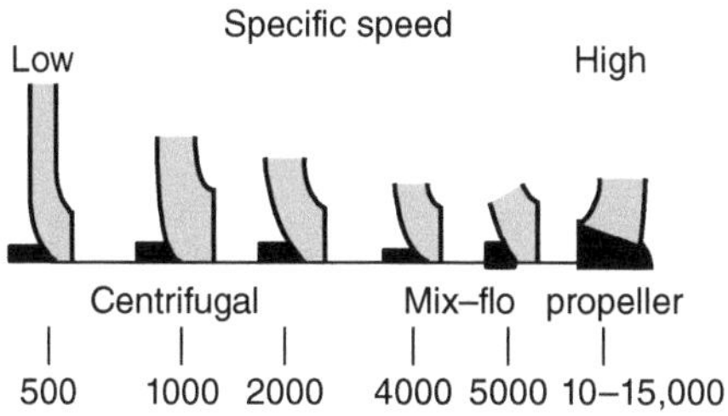

FIGURE 18-2 Centrifugal Impeller classification.

Note that the mix-flow impellers in Fig. 18-2 appear the same as the vertical turbine pump impellers in Fig. 18-4. The difference is that a vertical turbine pump is really a diffuser pump. This means a diffuser is used with multiple vanes into which the impeller discharges. The purpose of the diffuser is to create opposite and equal balancing forces within the pump. Therefore, the impeller of a vertical pump could be a mix-flow type. When coupled with diffusers, it is referred to as a vertical turbine (diffuser) pump. The vertical turbine pump impellers in Fig. 18-4 are mixed flow impellers. In the literal sense, a turbine pump is a high-head, low-capacity pump commonly found in breweries or other facilities that process liquids with a high, entrained vapor content. Figure 18-3 is an illustration of that type of turbine pump.

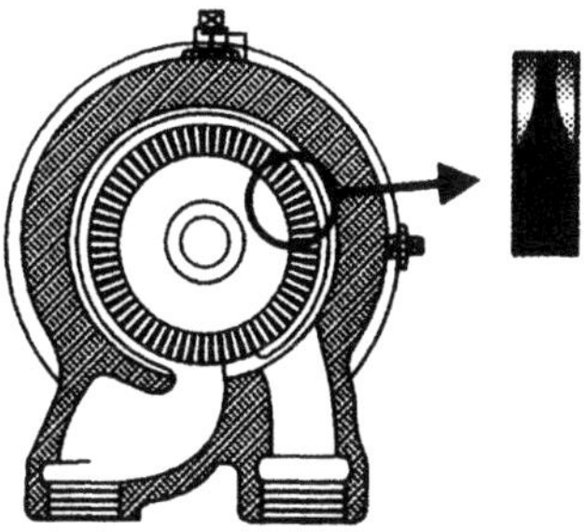

FIGURE 18-3 Turbine pump impellers.

Note the balance holes in the impeller on the right. Note also the small turbine pump in Fig. 18-3. It has no resemblance to the large vertical pump impeller in Fig. 18-4.

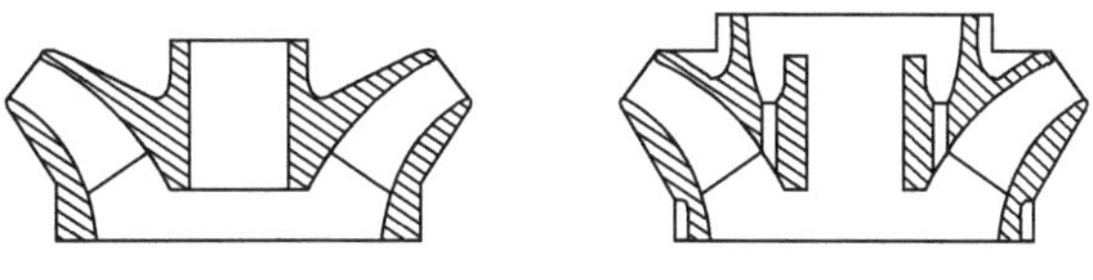

FIGURE 18-4 Turbine pump impellers.

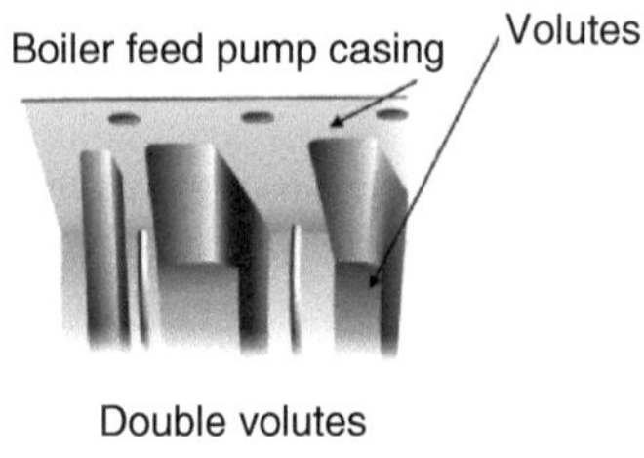

Large centrifugal pumps can have double volutes, diffusers, or both. Both volutes and diffusers are used to balance the radial forces within the pump. If the forces are not balanced, they will cause shaft deflection that can cause excessive wear on wear rings and other close tolerance fittings within the pump. These forces increase dramatically as flow is decreased below the design capacity. If a centrifugal pump is operated at dead head for an extended period of time, these forces (coupled with overheating) can cause the shaft to break. For constant speed pumps, energy that is not converted to flow "work" will be converted to "heat." The shaft of single-stage pumps will break almost exactly in the center of the impeller due to repeated bending cycles. Figure 18-5 is an illustration of the dynamic radial forces within a centrifugal pump.

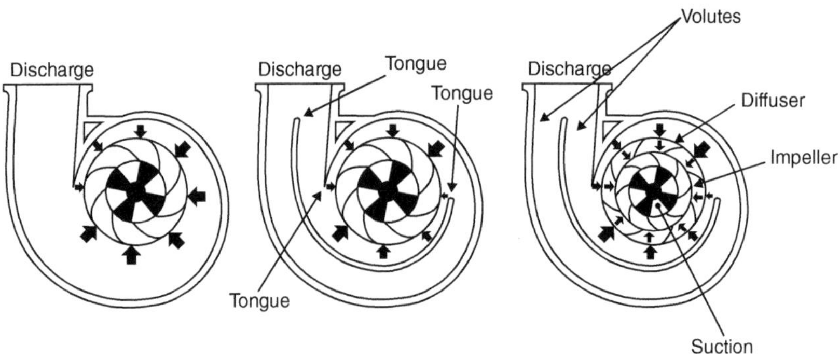

FIGURE 18-5 Radial force.

The axial flow, or propeller, type pump is sometimes used in high-capacity, low-head circulating water systems. They are very unstable at "dead head" and can vibrate violently at full speed, no flow. A common practice of design engineers is to configure the control system to start circulating water pumps, then open the discharge valve to 20 percent long enough to fill the system. In other words, the pump start command is a permissive for the discharge valve open command. This prevents water hammer, which can cause serious damage to a condenser tube sheet. However, an axial flow pump will vibrate violently until the discharge valve opens far enough to allow the pump to reach a stable flow region. Therefore, for a propeller pump, it is recommended that a 5 percent discharge valve travel position be made the permissive for a pump start command, after which the valve can travel to 20 percent to fill the system. This will be the trial operating point that can be adjusted open or close if needed, depending on the instability zone of the pump. See Chap. 13 for a typical axial machine curve. Axial pumps, compressors, and fans all exhibit the same low-flow, unstable-zone characteristics.

The drooping pump curve is typical of an axial flow impeller. Because the pump has two-head versus capacity points, the pump can only operate in parallel when there is enough flow to drive the head below the dead-head point on the curve. The flat curve is also troublesome for parallel-pump operation. If two pumps are operated in parallel on the flat part of the curve, one pump will always dominate the other. That means the dominant pump will back the other off to dead head. When that happens, the dead-headed pump shaft will break if the shutoff condition goes unnoticed for an extended period of time. If two pumps are to be operated in parallel—and are started automatically by discharge pressure—they must have steady- or steep-rising curves. If not, two-pump operations, with a flat curve, must be controlled to start and stop in a narrow range on the rising part of the curve. That can cause excessive start and stop cycling. The most common curve characteristic is the steady-rising curve, which gives good stable operation over the entire range of operation (see Fig. 18-6).

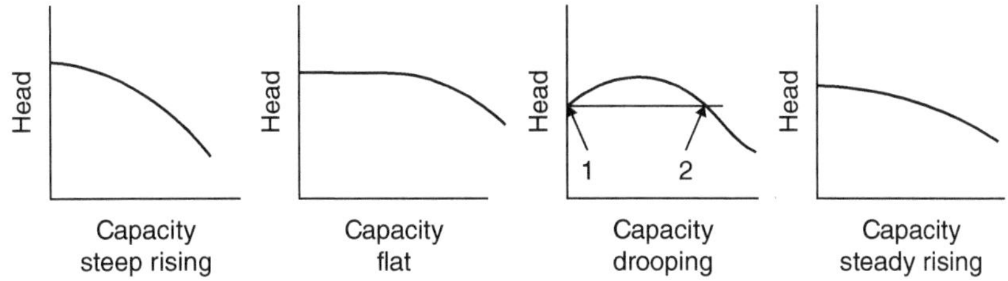

FIGURE 18-6 Pump curve characteristics.

Multistage pumps operate like pumps in series. That is, flow remains constant while the head is boosted in each consecutive stage. In parallel-pump operation, the head remains the same and the flow increases as each pump is placed in service. Refer to Fig. 18-7, which was made during startup in order to properly set the recirculation, surge protection, and service water supply controls. The pumps were multistage, vertical turbine pumps operating in parallel.

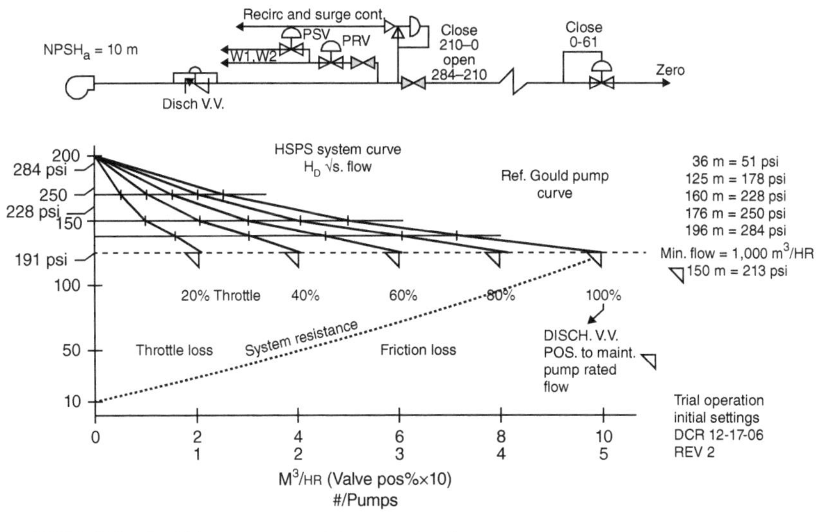

FIGURE 18-7 Parallel pump curves with system sketch.

Pump internal clearances are of the utmost importance to the startup engineer. The matching surfaces of the impeller and casing are capped with wear rings (called wearing rings by some experts) that are provided in matched sets. Refer to Fig. 18-9. They are usually flat or L-shaped and can be both a paired set for the impeller and casing or just for the casing.

The flat ring clearance can be checked with a feeler gauge, but the L ring clearance must be measured with dial indicators by lifting the shaft (Fig. 18-8). They are designed to take the wear between the casing and impeller and are, in that sense, re-machineable or disposable.

During commissioning, it is imperative that all foreign material is cleaned from the pump suction and suction piping. As a further precaution, suction screens should be installed to prevent any material that was not dislodged during initial flushing and cleaning from entering these passages. The screens are usually 100 mesh for initial operation up to 75 percent load. After substantial operation with the fine-mesh screens, before ramping to 100 percent, the screens should be changed to coarse mesh of 60 to reduce the screen pressure drop. This ensures that adequate net positive suction pressure is available. The engineer should specify the screens in accordance with the pump manufacturers' specification and the clearance specified in the drawings and the regured net positive suction head (NPSH).

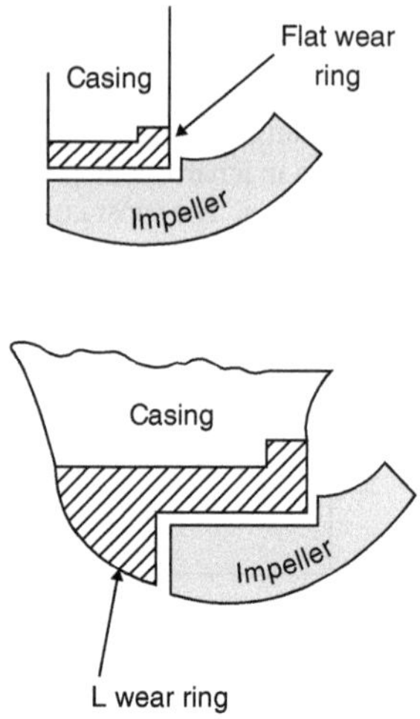

FIGURE 18-8 Wearing rings.

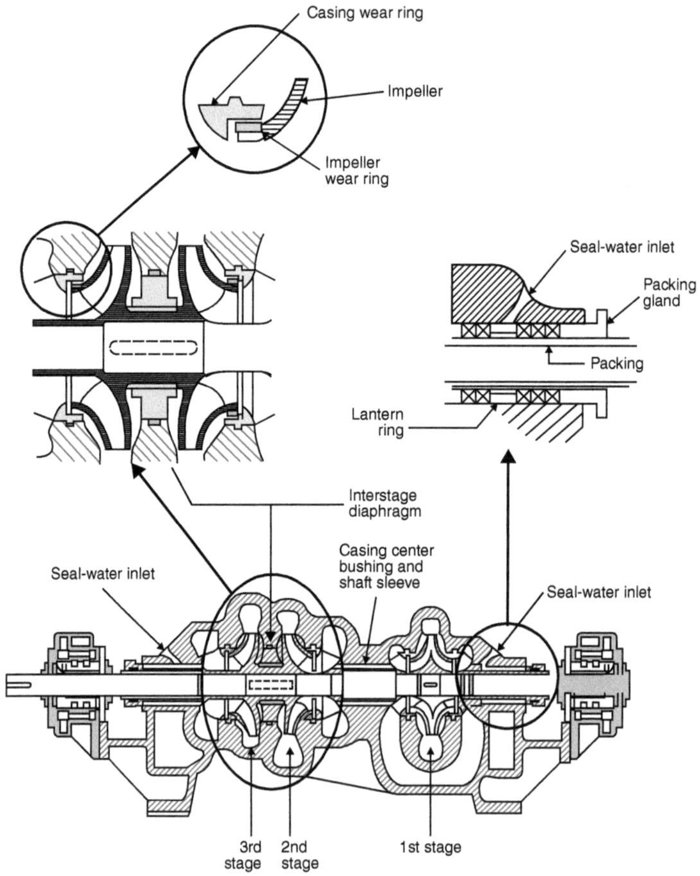

FIGURE 18-9 Horizontal split case pump internals.

Balancing devices, such as drums or disks, also have close clearances. See Figs. 18-11 and 18-12. In addition, all devices such as interstage diaphragms, bushings, or any interstage running joints will have close clearances. These tight clearances restrict flow; nonetheless, flow does occur, which produces a thrust in the opposite direction. Therefore, these reaction forces are used to balance the force produced by the impellers. This flow also cools and lubricates the joints. Placing an equal amount of stages, half opposed to the other half, does not balance all forces. The balance occurs across the interstage diaphragm, or bushings,

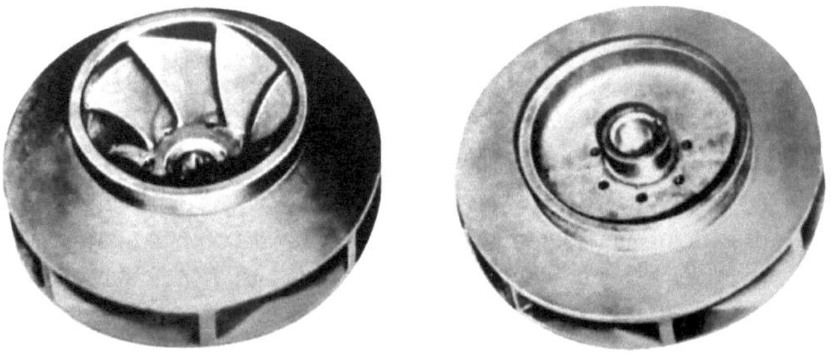

FIGURE 18-10 Impeller balance holes.

in the same manner as a balancing device. A balance device has a designed leakoff calculated to create a thrust reaction opposite to the direction of the flow in order to counter the impeller thrust. The leakoff is discharged to a chamber that is piped to the pump suction or to the first upstream pressure vessel. Details of reaction forces are described in more detail in Chap. 17. Figure 18-11 provides sketches of typical balancing devices.[1,4]

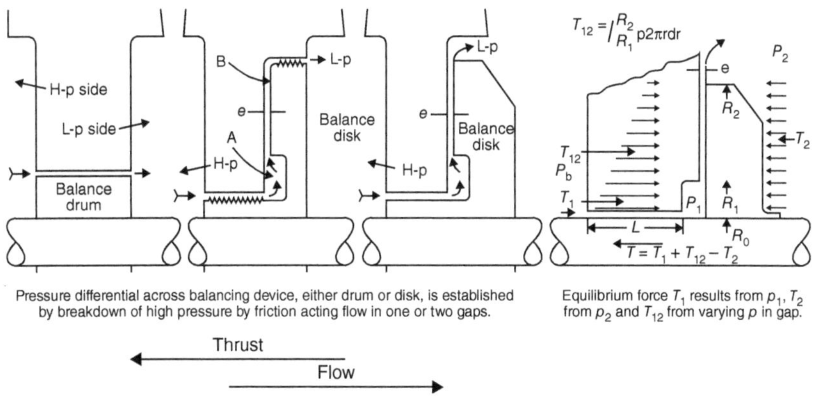

FIGURE 18-11 High pressure balancing devices.[3]

The single-suction impeller in Fig. 18-9 has holes to bleed pressure to the backside in order to balance the dynamic forces. If these holes become plugged, the thrust bearing can fail. They are clearly seen on the back side photo to the right. Double-suction centrifugals are balanced impellers and don't require bleed holes.

The most common multistage centrifugal pumps are built with either horizontal "axial" split cases or with radial split cases called barrel pumps. Both can have dual-volute and/or diffuser-stage discharges. The theory of operation is similar for both designs. Both can use

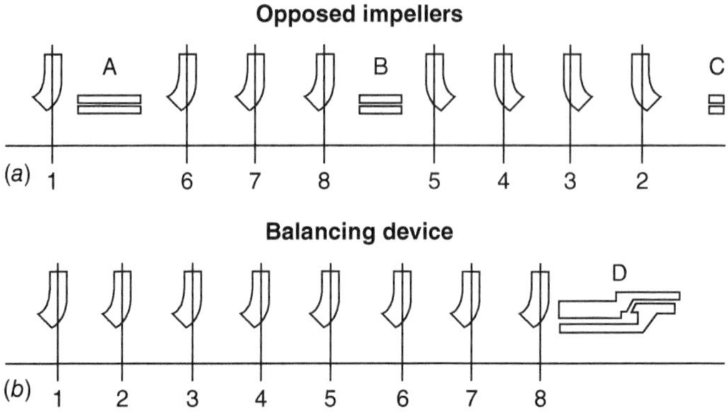

Joint A is subject to differential pressure of	4 stages
Joint B is subject to differential pressure of	3 stages
Joint C is subject to differential pressure of	1 stage
Sum of pressure differentials	8 stages
Joint D is subject to a differential pressure of	8 stages

FIGURE 18-12 Low pressure balancing devices.

the same dual-volute or diffuser discharge and impeller design. Interstage diaphragms and diffusers are stationary and can be designed as one piece or split. The stationary, one-piece elements are installed on the shaft with the impellers but are gibbed, or keyed, stationary in the casing. These elements have close tolerances that are very sensitive to foreign particles. Furthermore, they are lubricated by the process fluid that is being pumped. Therefore, pumps cannot be run dry even for a moment. Figures 18-13 and 18-14 show examples of both types.

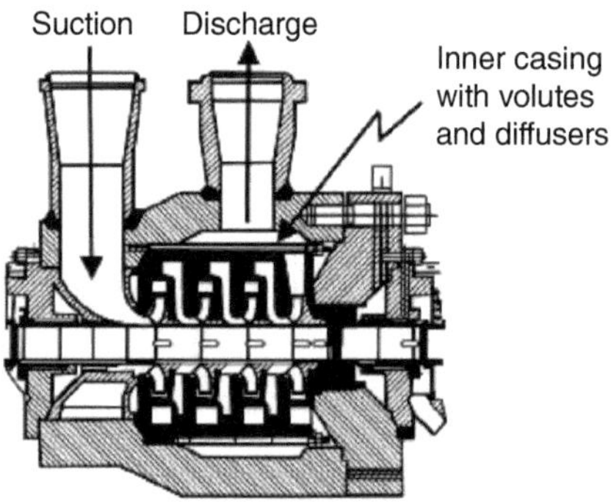

FIGURE 18-13 Radially split (barrel) pump.

When commissioning a rehabilitation project with large vintage pumps, such as boiler feed pumps, the availability of spare parts such as wear rings, bushings, impellers, and diffusers should be checked. If they are not available, it is advisable to send everything that could be damaged during commissioning to a foundry and machine works facility for replication.

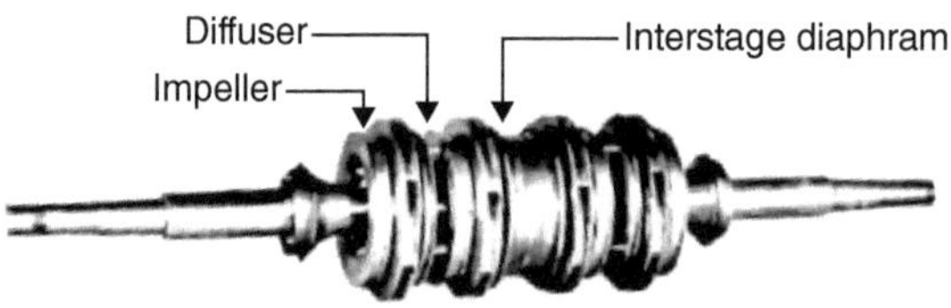

FIGURE 18-14 Rotating element of horizontal (axially) split pump.

The foundry can make new patterns and coreboxes for items such as impellers and then cast and machine them. Shafts can be machined from stock material. Make sure you have spare parts before operating the equipment. A spare pump usually is not required. For example, if the casing needs replacement because of damage due to oxygen pitting, then it is cheaper to buy a new pump.

Standby boiler feed pumps need warmup lines that keep the pumps up to operating temperature so they can be quickly placed in service. They are piped from the discharge back to the upstream pressure vessel. That is, flow from the upstream heater to the pump suction, through the pump, to the discharge, and back to the heater via the warmup line keeps the pump at process temperature. The balance leakoff is sometimes piped to the warmup line upstream of the warmup isolation valves. Ensure that the balance leakoff lines cannot be isolated. If the leakoff is isolated, the leakoff chamber can reach full discharge pressure, which is a dangerous situation.

POSITIVE-DISPLACEMENT PUMPS

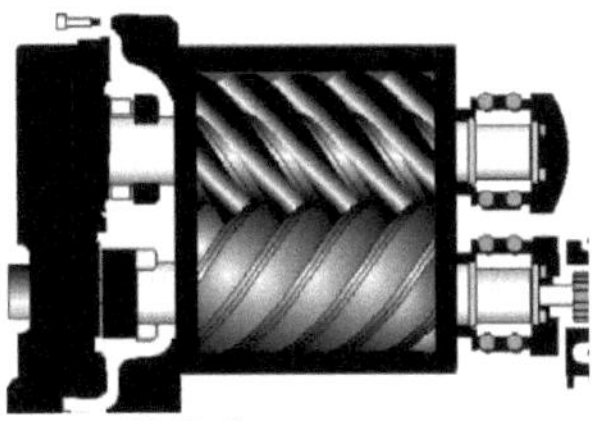
Helical or screw pump

Positive-displacement pumps, by their design, move a constant volume of fluid. Therefore, flow for a constant-speed pump is constant. The only way to vary the flow is to bypass fluid back to the source or suction of the pump or by varying the speed. Bypassing is inefficient because the pump uses 100 percent of its power, but the energy used to pump the bypass flow is wasted. In other words, if 50 percent of the work is used for bypass, only 50 percent is available for useful work. It can readily be seen that they cannot be deadheaded or they will rapidly build up pressure until they are destroyed.

The positive-displacement pump is used quite often for fuel oil handling. If the oil is very viscous, such as heavy residual fuels, the fuel must be heated to the design temperature before it can be pumped. This includes the suction piping and the contents of the storage tank. If it is not, the pump will deadhead and destroy itself. Relief valves will not function under these conditions either.

Positive-displacement pumps will self-prime, but there should not be an anti-syphon hole in the suction pipe. Anti-syphon holes are normally installed in a tank suction piping to prevent spilling the entire tank contents in case of a line break or similar accident.

PUMP SEALS

Pump seals (stuffing boxes) are either of the packing or the mechanical type. Most pumps are built so that either can be used without modifying the pump or seal. In other words, they are interchangeable. The type selected depends on owner preference, type service, and process fluid. Conventional packing is used to throttle water used to seal, cool, and lubricate the pump shaft or sleeve. Therefore, there is always a flow through the seal to drain. If leakage is not tolerable, a mechanical seal must be used. The sealing water is usually injected into a lantern ring, which distributes the water evenly around the shaft. Packing seal application is limited to relatively low pressures and nontoxic fluids such as water.

Leakage rates are important. The higher the PV value (pressure at the packing vs. the peripheral velocity) the higher must be the leakage rate. It depends more on sealed pressure than on shaft rpm. With pump suction from below atmospheric to 60 psig, the leakage rate must be at least 1 drop per second from each stuffing box. A liquid from an external source may be necessary. From 60 to 100 psig, 1 gal every 20 min is the requirement for each stuffing box. And for pressures on up to 250 psig, 1 gal every 8 min is necessary. Above 250 psig, the rate would be so high that it is easy to see why the mechanical face seal is needed.

A bent shaft, or one that is out-of-round by more than 0.003 in, will gyrate or orbit and cut packing life ... Care of packing can be decisive. A good lubricant can slash wear of stuffing-box parts, but water is not a good lubricant, especially if it is hot or laden with abrasives. Commercial compression packing contain some emergency lubricant, but it is not for continuous use—its purpose is to melt at excessive temperature, pass out of the stuffing box, and relieve gland pressure so some leakage flow can start again. Retightening the packing gland would repeat the process and eventually defeat the aim of the packing designer.

One exception to this is graphite-filament yarn and graphite-foil packing, in which the bulk of the material is itself a lubricant, and needs no extra lubricant except during break-in. Even this packing needs some coolant flow, however, unless conduction alone can take heat out of the stuffing box.

Stuffing box is home for compression packing. Static friction against non-rotating housing prevents packing from rotating. This is because the total compression load holding the packing against the surface is higher than that against the shaft sleeve surface, in direct ratio to surface areas. Stuffing-box surface finish is not as smooth as shaft sleeve surface, and so contributes more friction. Although static friction against stuffing-box surface is desirable, kinetic or sliding friction against shaft sleeve is not.[3]

Mechanical seals are generally used for high-pressure and toxic service (Fig. 18-15). The mechanical seal consists of two sealing parts. Each has a highly polished face—one stationary and the other rotating—that are forced together by a spring and fluid pressure. The stationary face is held in the casing and is usually made from highly polished cast iron. The rotating piece or seal ring is usually made from carbon and is also highly polished. The material used, of course, depends on the service, but for water cast iron and carbon are usually used; for other

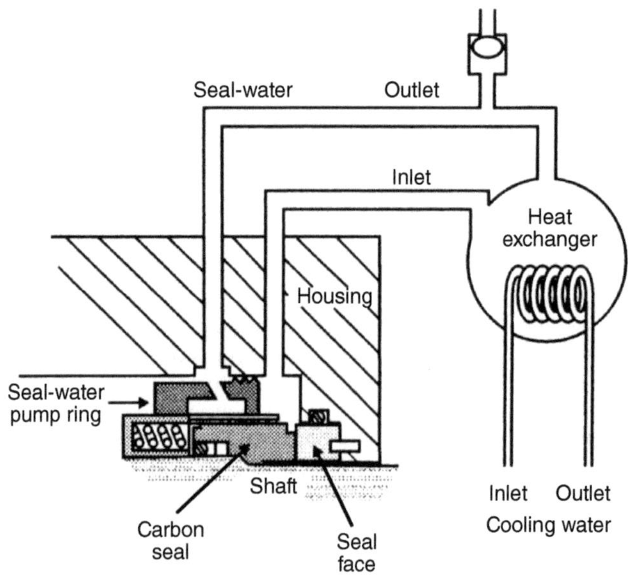

FIGURE 18-15 Mechanical seals.

service, a wide variety are available, but carbon against ceramic and satellite are common. These seals require perfect alignment because the sealing surfaces must match perfectly. The surfaces can also be easily damaged or scoured by debris. Therefore, the sealing system must be thoroughly cleaned before the pump can be operated. A small, almost insignificant, scratch on the carbon seal surface will cause a very significant leak. Other sources for leaks are O-rings, gaskets, or other secondary seals for the gland.

A crucial part of the mechanical seal is the seal-water cooler. It must have an air vent so that the cooling water side does not fill with air and prevent cooling. The cooler must keep the sealwater cool enough so it does not flash. The seal-water side should also have a vent so the cooler can be primed to prevent the pump starting with a dry seal.

EXPANSION JOINTS

Expansion joints are used to compensate for minor thermal expansion or vibration or are used as dielectric couplings. Usually, if a pump is placed outside, where the piping is above ground and subjected to thermal expansion caused by environmental changes, the connections are made with expansion joints. Frequently, the pumps (as well as other steel) are also isolated with expansion joints from underground piping for cathodic protection purposes. Therefore, the commissioning engineer must verify that there are no connections, such as electrical conduit or supports, bridging the joints and breaking the dielectric isolation.

PUMP SUCTION STRAINERS AND SCREENS

If strainers are used for the pump suction, they must have a tab with a tag visible from a distance. If the flanges have been insulated and lagged over, remove the insulation and lagging for inspection to see if a strainer was installed. Leave them visible with an identification tag. Strainer and screen types and sizes are illustrated in App. T. Ensure that eccentric suction reducers are installed in accordance with Fig. 18-16.

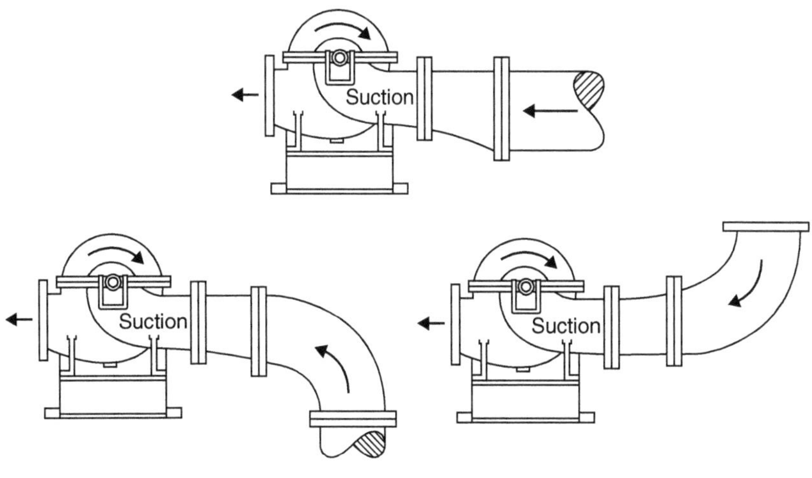

FIGURE 18-16 Correct installation of eccentric reducers.

PUMP MOUNTING

Mounting is a point often overlooked by startup personnel. It is often considered solely as a construction function, but it is really a vital aspect of startup. To a startup engineer, mounting should mean proper leveling and arrangement of equipment on their bases with respect to system components such as piping.

Another point to remember is that grouting is not cosmetic. It is used as the sole support of the equipment. Anchor bolts are used to hold it down, not up. For rotating equipment, the anchor bolts must be designed to hold the equipment down for the worst-case uplift scenario of a "full-load locked rotor." Pressure vessels must be held in place in the event of steam and water hammer, surge shock waves, and/or seismic and wind loads. Shims, wedges, and leveling bolts are devices used only to align and level; structural support is provided entirely by grouting. Also, a point to remember is that shims and wedges used to level equipment generally are not removed from centrifugal machines because of their low-vibration characteristics. For reciprocating machines, however, where vibration is relatively high, shims and wedges should be removed after the major section of the equipment frame is grouted. The shim and wedge spaces are then also filled with grout.

Pumps must be aligned so that piping does not impart stress on pump nozzles. Therefore, flange faceup must be within tolerances that do not make it necessary for gasket thickness to compensate for misalignment. If a pump cannot be aligned without pulling the pipe in, or otherwise make faceup (parallel matching of flange faces), the pipe must be cut and refit. The final weld in any piping system must be at the final vessel or pump flange or nozzle. If the first weld is made at the pump, for example, and a final weld is made in the middle of a run where the pipe can be easily pulled into place for welding, it can cause huge stresses at the other end connected to the pump. In other words, pulling the pipe, with a long run acting as a lever arm, can impart tremendous stress on the pump. In many cases, it can twist the pump and cause serious driver–pump misalignment and stress. If pipe hanger positions appear abnormal, it is a sign that the line may have been pulled for weld alignment.

If the pump is flanged when turned over to commissioning, the pump, drive-coupling, and pipe flanges must be broken for realignment. The coupling-alignment dial indicators or lasers are not removed until the pipe-to-pump flanges are made up again. Any change in dial-indicator readings during the flange-up process means the pipe is imposing stress on the pump.

PUMP ALIGNMENT

Alignment is the "fine tuning" of the mounting process. To a commissioning engineer, alignment means proper arrangement of a driver with its machine. A pump, for example, is mounted level and the pipe aligned to it. The driver is then aligned to the pump. Because the pump is the focal point of the alignment, it can be mounted, leveled, aligned, grouted, and doweled. Final alignment is accomplished by moving only the driver. For equipment with three elements (e.g., pump, gearbox, and driver), the gearbox should be aligned to the pump and then the driver to the gearbox.

Belt-driven equipment with sheaves should be leveled and aligned by the sheaves. Both pieces of equipment are leveled using the sheaves as reference and then aligned. Remember that belts supplied in matched sets for multiple-belt sheaves are not interchangeable.

To more fully understand alignment, it is important that the commissioning engineer also understand how motors are constructed to allow for alignment. Large horizontal motors (200 hp and larger) that have plane bearings do not have thrust bearings. These motors are built with sleeve shoulders and a ½-in end-float clearance. This float is limited to 3/16 in by couplings (see Fig. 18-17).

Large, horizontal motors are first centered mechanically by moving the motor toward the stationary pump. Once the motor is centered properly, it can be aligned.

Pump thrust clearance, established by the manufacturer to maintain proper hydraulic thickness of the lubricant, will be adjusted by shimming the thrust bearing. Improper thrust clearance will cause overheating, vibration, or both.

Another consideration is the motor magnetic center. Motors are bumped for rotation, and sometimes the magnetic center is scribed on the shaft prior to alignment. This is done because the magnetic center is rarely the same as the mechanical center. Some motors are aligned by magnetic center in order to minimize loading of the machine thrust bearing. Others may be aligned off magnetic center to preload the thrust bearing in order to minimize pump buffeting during starts by slightly loading the bearings with the magnetic centering force.

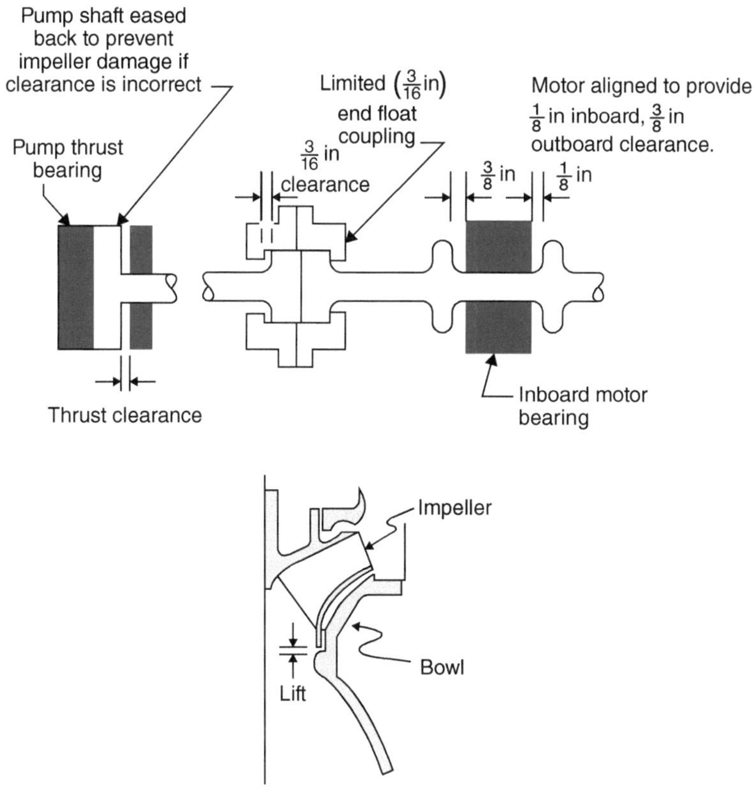

FIGURE 18-17 Horizontal and vertical pump clearance settings

Some motors are aligned low because motor internal heating causes them to rise while reaching operating temperature. The amount the motor is lowered to compensate for heating can depend on the pumping fluid temperature because high-temperature fluids will also cause the pump to rise. For example, if a motor is expected to rise 0.002 in and the pump 0.007 in, the motor will be set 0.005 in high. If, however, the pump operating temperature is the same as its idle temperature, the motor will be set 0.002 in low. If unsure of the motor expansion characteristic, final alignment should be checked at operating temperature regardless of pumping fluid temperature.

Pumps and motors also expand lengthwise; therefore, both are dowelled on the inboard side to allow expansion outward. Motors are not dowelled until after final alignment. Expansion and rise of equipment is proportional to temperature rise and are dependent on the coefficient of expansion of the material of which it is constructed. Therefore, rise can be easily calculated and predicted in advance by the manufacturer.

Vertical pumps are centered along the shaft the same as horizontal pumps. End float, however, is established in the pump rather than the motor and is called "pump lift." The total lift is the distance the impeller can be lifted in the pump bowls (see Fig. 18-17). In other words, lift is the distance the impeller is raised by the shaft, at the coupling, to obtain adequate impeller to bowl clearance. If the pump impeller is allowed to rest on the bowl (zero lift), or if sufficient lift is not provided to compensate for shaft elongation during operation, the pump and motor can be severely damaged. If there is excessive clearance, recirculation can occur, thus reducing pump head and capacity. This will be indicated by a low shutoff head on initial operation. Also, excessive clearance will not allow compensation for the upward surge of the impeller caused by *initial* unbalanced internal dynamic forces during starts.

PUMP ANALYSIS

Pump Curves

Before commissioning a pump, the pump curve should be analyzed to determine the expected operating parameters. Therefore, all the instrumentation related to the pump and process system must have been calibrated beforehand. This will eliminate the uncertainty of instrumentation error versus pump problems when operating data do not meet the design criteria (Fig. 18-18).

Pump performance curve terms are as follows:

TH = total head = discharge head plus or minus the suction head without consideration for velocity head

TDH = total dynamic head = discharge head plus or minus the suction head including the differential velocity heads

$NPSH_r$ = net positive suction head required = suction head, determined by the manufacturer, required to maintain proper pump operation

$NPSH_a$ = net positive suction head available = suction pressure that is actually available; should be checked by the startup engineer

NPSH is always given in absolute pressure. Therefore to determine what is needed to meet the "required" NPSH, reading plant instrumentation, the atmospheric and velocity heads must be subtracted and the vapor pressure added to the manufacturer's $NPSH_r$. The differential static head, suction screen differential, and instrument height must then be determined to find the NPSH that is available. The suction screen must have a differential

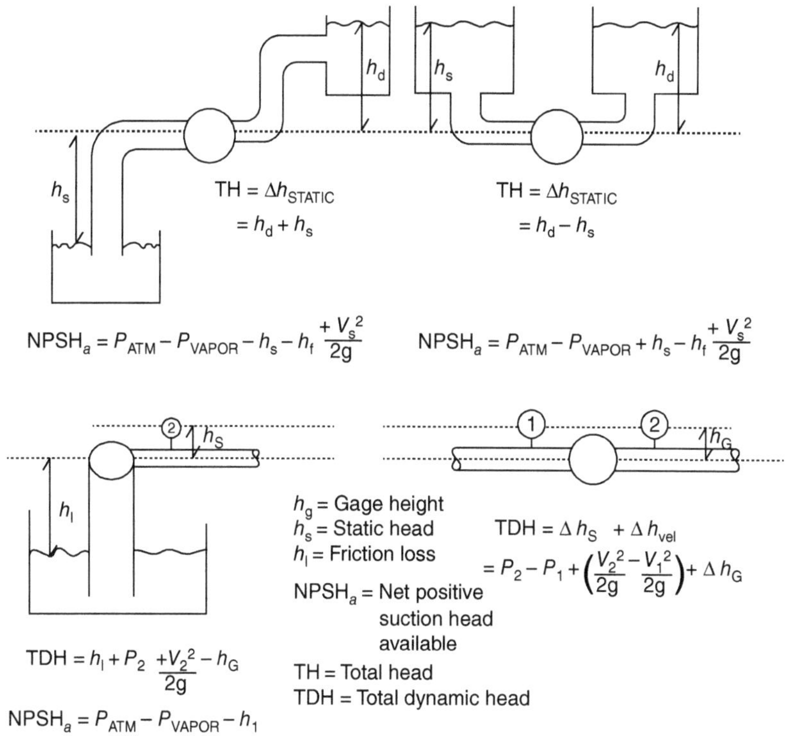

FIGURE 18-18 Pump head illustraion.

pressure transmitter to alarm and trip before the suction pressure drops below that which is necessary to meet the manufacturer's NPSH requirement.

To properly analyze the pump performance for total head, suction and discharge pressure gauges must be installed close enough to inlet and outlet flanges to eliminate friction-head loss errors. If the gauges are installed at the same height, the height will cancel in the calculation, therefore, only the differential pressure is required for performance testing. The same holds true for atmospheric and vapor pressure. For initial trial operational purposes, differential gauge height and velocity head can be ignored. To determine the expected discharge pressure, the suction head available must be added to the TH or TDH from the pump curve.

The following method for determining the maximum allowable suction screen differential pressure is a good application of available versus required NPSH. In this example, the static head was determined from the hotwell level to the inlet of a condensate pump. The maximum allowable differential is the difference between the $NPSH_a$ and $NPSH_r$. $NPSH_a$ is the static head plus condenser pressure minus vapor pressure which was considered negligible. The suction screen differential pressure gauge may have to read inches of H_2O if a psi gauge does not have an adequate range. The suction strainer pressure differential alarm must be set at the difference between $NPSH_a$ and $NPSH_r$ plus a safety margin.

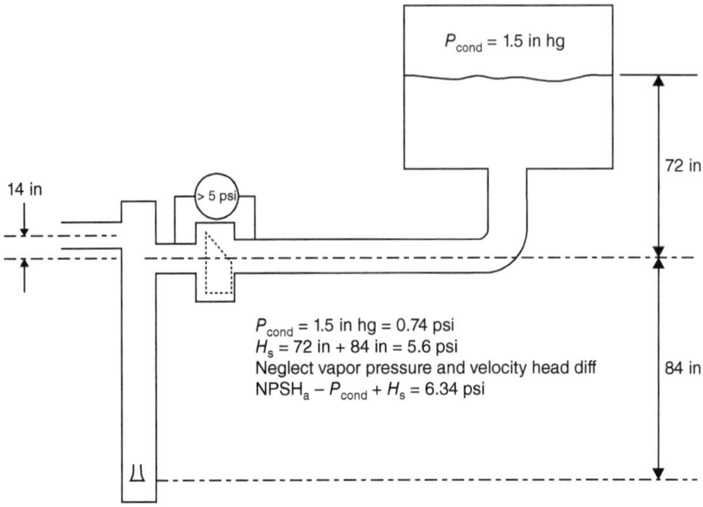

The following is a plant engineer's analysis of a steam generator circulating-water-pump performance:

Assume viscosity does not significantly affect the process:

Viscosity @ 60°F, $\mu = 1.10$ centipoises
Viscosity @ 543°F, $\mu = 0.08$ centipoises

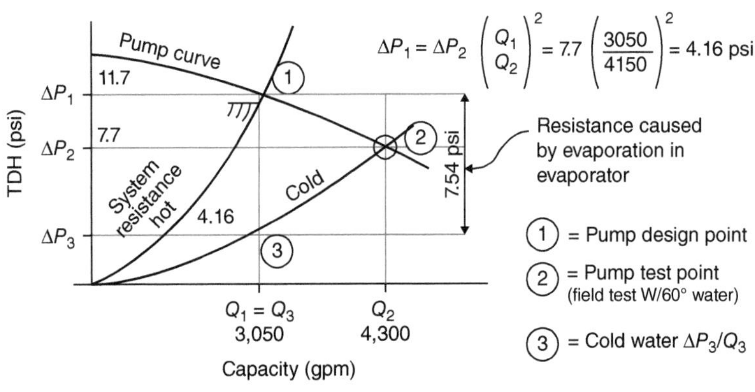

Initial testing of the BWCP must be performed using cold water; therefore, a new system curve must be established because the pump curve only represents performance at the design operating temperature. By pumping cold water, rather than water at operating temperature, the system resistance decreases because friction losses are less.

When the boiler is fired, evaporation takes place; therefore, the specific volume of the water–steam mixture inside the evaporation section will be much greater than initially with cold water. The steam–water mixture velocity will represent additional resistance that was nonexistent during initial pump testing.

It is expected that when evaporation takes place, the additional resistance caused by the flow of a water–steam mixture through the evaporator will create the design conditions indicated on the manufacturer's pump curve as illustrated here.

Converting gpm to lb/hr (643°F Water)

$$(\text{lb/hr}) = (\text{gpm}) \times (\text{lb/gal}) \times (\text{min/hr})$$
$$1000 = \text{gpm} \times 1/0.1615 \times 60$$
$$\text{gpm} = 1000 \times 0.1615 \times 1/60 = 2.695 \text{ gpm per } 1000 \text{ lb/hr}$$

Therefore: 2695 gpm = 1.0×10^6 lb/hr

Brake Horsepower at the Design Point

$$\text{bhp} = (\text{gpm} \times \text{th} \times \upsilon)/(3960 \times \eta) = (3050 \times 36 \times 0.747)/(3960 \times 0.80) = 25.9 \text{ hp}$$

where

$$\upsilon = \text{specific gravity and } \eta = \text{pump efficiency}$$

Cold-Water System Resistance

$$\Delta P = \Delta P_2\,(2000/4150)^2 = 7.7 \times 0.232 = 1.81$$

Hot-Water System Resistance

$$\Delta P = \Delta P_1\,(2000/3050)^2 = 7.7 \times 0.4300 = 5.04$$
$$\Delta P = \Delta P_1\,(1000/3050)^2 = 7.7 \times 0.1077 = 1.28$$

The calculated horsepower corresponds with the manufacturer's pump curve. The new, cold-water curve is plotted on the manufacturer's curve.

Check the Motor Power for Cold Water (Standby Pump)

First check motor data for maximum current. Full-load amps = 55; with a 1.15 service factor, maximum amps = 63.25.

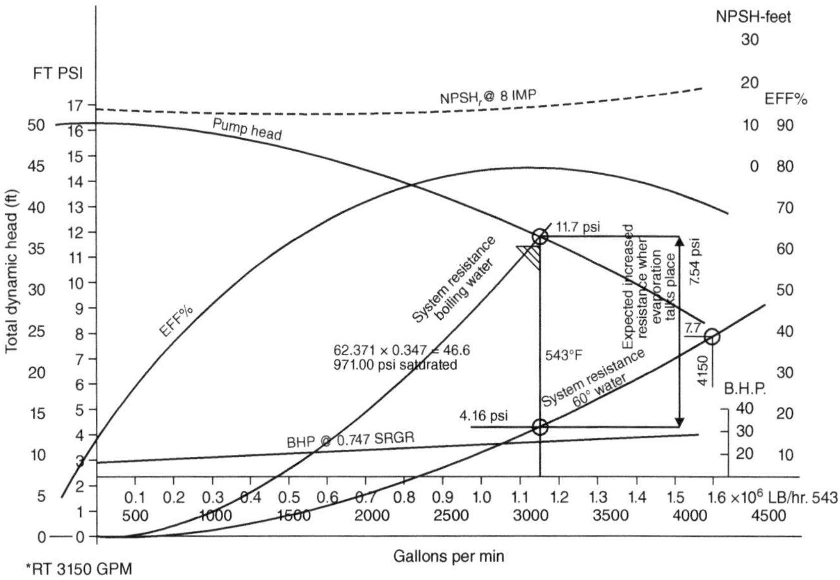

Measured Test Data

Amps	A	B	C	Volts	
1	57	58	58	AB	482
2	56	57	57	AC	484
				BC	485

$$\text{watts} = 3^{1/2} \times \text{I} \times \text{E} \times pf \times eff = 3^{1/2} \times 57 \times 484 \times 0.71 \times 0.9 = 30.51\ \text{kW}$$
$$\text{hp} = 30.51 \times 1.341 = 45.5\ \text{(electrical)}$$
$$\text{bhp} = 45.5\ eff = 45.5 \times 0.90 = 40.9$$

The test is possible without overloading the motor.

Pump Minimum Flow

Years ago, rules of thumb were established to determine minimum flow. A common figure was 30 gpm per 100 hp to limit the temperature rise within the pump to 15°F. Another rule—more folklore than real—was 25 percent of rated flow. Today's pump manufacturers use more precise analysis for determining the minimum flow which can yield temperature rises far in excess of 15°F and therefore lower minimum flows.

Pump problems & cavitation

> The point of highest temperature rise is in the balancing device leak off chamber and it is this area which must be protected against flashing when determining the recommended minimum flow. . . . Most installations made today provide for the return of the balancing leak-off back to the suction side of the casing. Empirical considerations dictate that the vapor pressure of the feedwater within the balancing device leak-off chamber not be permitted to exceed a value equivalent to the pump suction pressure less 10 psi.[5]

The vapor pressure in the balancing leakoff chamber should be $\leq P_{\text{suction}} - 10$ psi.

Because the total temperature rise is a function of head and efficiency, the minimum permissible efficiency can be computed as follows:

$$\text{Minimum } \eta = \text{TH}/(\max \Delta\text{T} \times 778)$$

From the pump curve, the flow rate at the minimum efficiency can be found, which will represent the pump minimum flow.

Of course, temperature rise is not the only factor in determining minimum flow. If the commissioning engineer finds that a pump running at minimum flow causes pump pulsation, noise, vibration or piping system sway or vibration, he or she must increase the flow until the critical pump flow zone or system harmonic zone is passed.

Also, some pump minimum flows are established by the downstream heat exchangers, not the pump. This is the case for condensate pumps, where the gland steam condenser dictates the minimum flow. Flat curves also mean higher minimum flows in order to operate on the rising part of the curve. The best rule of thumb is always to follow the manufacturer's instructions, unless the actual operating conditions dictate higher flows. A corollary to this rule of thumb is never decrease the minimum flow, always increase it if conditions dictate.

Remember, minimum flow is just that. If system piping swings or vibrates the pump impeller, pulsations may be in resonance with the piping system. The pump-blade pass frequency may be generating low-frequency harmonic vibration in the piping. In this case, the minimum flow must be increased to get out of the resonance range of the piping. On some systems, the vibration can be subtle and in other cases very violent. Read more in Chap. 21.

Pump Problems and Cavitation

Low-frequency pulsation and impeller stall plagued pump manufacturers during the 1970s and early 1980s, as power station units increased in size twofold from those built in the 1960s. Some of the problems faced by commissioning engineers are discussed here.

Hydraulic pulsation in the range of 1 to 8 cycles per second was reported at flows below 60 percent of rated, but within the minimum flow range, in boiler feed pumps. This pulsation preceded back-of-the-vane cavitation damage. This is different from classic cavitation, which is caused by operating with a lower-than-required NPSH. The pulsation and later cavitation were caused by recirculation in the impeller.[6] Classic cavitation damage caused by operating at lower than design NPSH is always on the front side of the inlet vane. Classic cavitation, from reference 2, is illustrated here.

During startup of the George Neil Unit 3, the ash sluice pumps were heavily damaged on the backside of the discharge vanes. It was discovered that the sluice timing sequence overlapped between the bottom-ash and one fly-ash hopper for approximately 10-min. This caused the pump to run out, off the pump curve, during the 10-min time frame. Correcting the sequence solved the problem. It was believed that boundary-layer separation and stall (discharge recirculation) occurred due to the high flow.

Vibration trouble of the boiler feed booster pumps occurred at the Navajo Power Station within a short time after startup. The station discovered that axial vibration was so heavy when the boiler feed booster pumps ran near minimum flow (26 percent) that the minimum flow requirement had to be increased much closer to best efficiency point (BEP) flow. This problem was fixed by closing the gap between the impeller and volute.[7]

In another case, the volute tongues were cut back to increase the gap between the volute and impeller. This solution remedied the pulsation and cavitation damage to the diffuser which occurred by operating at lower than 60 percent capacity.[8] Refer to Fig. 18-19.

> It is now well established that surging and extensive cavitation damage of the inlet vanes not associated with inadequate NPSH can be traced directly to operation of the pump in the suction recirculation zone. It has also been established that cavitation damage to the discharge tips of the impeller vanes can be traced directly to the operation of the pump in the discharge recirculation zone...It is of interest to note that the flow mechanism is identical for suction and for discharge recirculation. A flow reversal occurs initially at either the suction or the discharge as the flow is reduced, or, as has been shown for inlet to outlet diameter ratios in excess of 0.5, the discharge may impinge on the suction and produce higher values of suction recirculation. To demonstrate the occurrence of recirculation and to confirm the validity of the equations [*sic*] there is one direct and a number of indirect procedures that can be used in evaluating the characteristic of any centrifugal pump.[9]

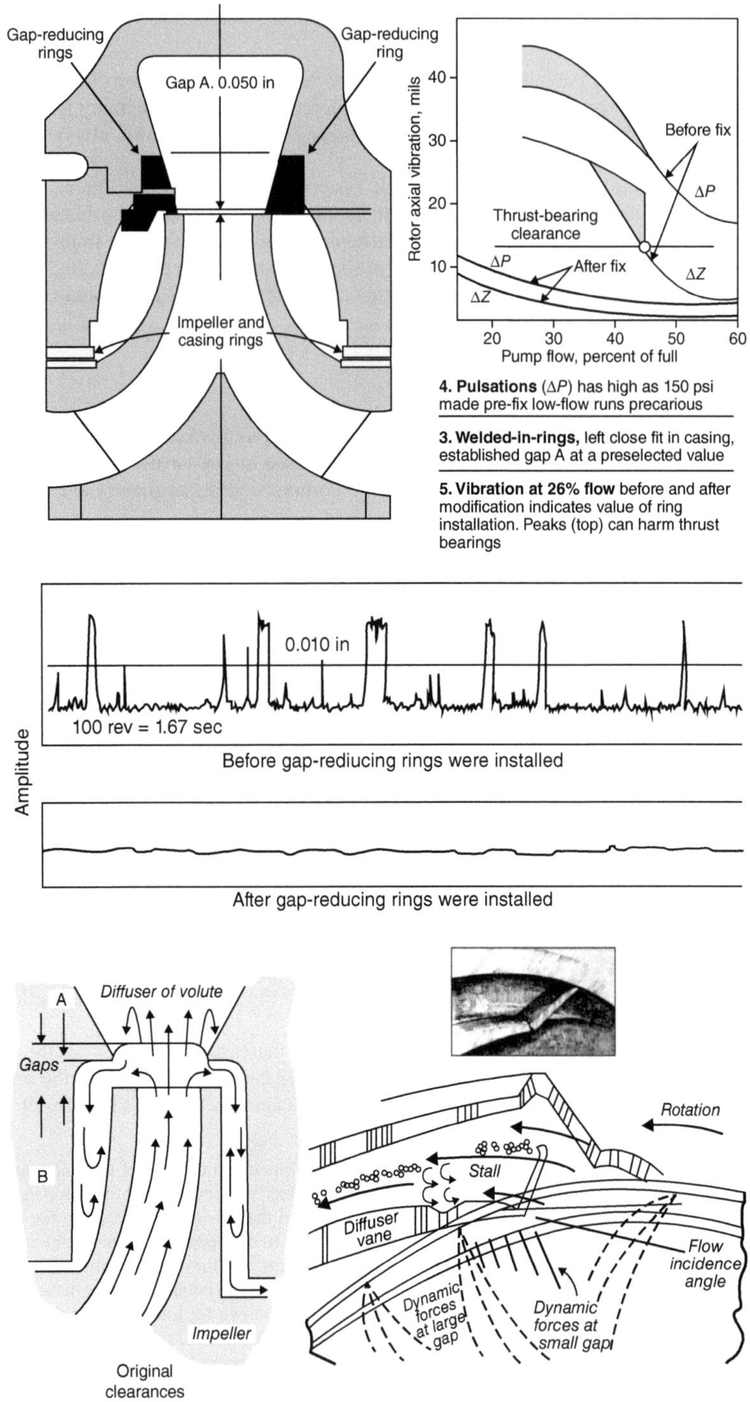

FIGURE 18-19 Analysis of cavitation damage.

TRIAL OPERATION PREPARATION TIPS

Ensure that the lube oil system has been flushed. Open and inspect the bearings internally to ensure that debris has not been inserted by disgruntled factory employees or construction workers. Ensure that all lube oil passages are bored through to the correct location by blowing lightly with instrument air or squeeze bulb and stopper (air blow continuity check). For bearing and lubrication refer to Chap. 4.

Check the seal-water system, and ensure mechanical seal-water coolers are primed and vented. Ensure that the seal-water supply lines have been flushed and cleaned. Check to ensure seal-water passages go to the right places by injecting clean water into the seal.

Break the pump and driver coupling, and install an alignment dial or laser indicator. Then break the suction and discharge pipes free and ensure there is no movement of the dial and no movement of the piping. Make sure the suction strainer has a commissioning tag stating "DO NOT REMOVE OR COVER THIS TAG."

Ensure that the instrumentation has been calibrated. Calculate the expected discharge pressure by adding the available NPSH from the TH at the deadhead, minimum, and design flow on the pump curve. Ensure that there is a differential pressure reading and alarm available for the strainer.

Example

Pump Seizure: You have successfully run a "barrel-type" boiler feed pump, and it rolls down normally after a normal shutdown. Several days later, you cannot turn it by hand because it apparently has seized. A common startup problem is overpassivation of boilers after chemical cleaning. This causes large amounts of black magnetite to sluff off the boiler internal surfaces. This black magnetic mud can plug suction strainers, cause water conductivity problems, and seize pumps.

The magnetite is so fine it can take many hours to settle. This can happen inside a pump. The fine particles will gradually settle in the lowest part of the pump. The accumulation from the piping above the pump and within the pump itself can cause the pump to seize. This is especially true for barrel-type pumps, where the suction and discharge piping extend vertically above the pump. The solution to the problem is to uncouple the pump and remove the thrust bearing pads. With the pads removed, bump the pump shaft coupling toward the thrust bearing. This will free the impellers from the mud. Immediately couple and restart the pump.

REFERENCES

1. Igor J Karassik, *Centrifugal Pump Clinic*, McGraw-Hill, New York, NY 1989.
2. Worthington, H. R., Rotary and Centrifugal Pump Theory and Design, Worthington Pump Corp., East Orange, New Jersey, 1971.
3. Makey, E., and Saez, J., "High-Pressure Boiler Feed Pump Troubles Clear Up after Balancing-Disk Changes," Power, June 1974.
4. Karassik, I. J., and Valentin, John E. C. Jr., "Steam Power Plant Clinic Part 80," Combustion, July 1976.
5. Karassik, I. J., and Valentin, John E. C. Jr., "Boiler Feed System Design Parameters—Effect on Pump Life," (presented at American Power Conference, Chicago, Illinois, April 18–20, 1977), Combustion, January 1978.
6. Makay, E., "Better Understanding of Sources of Feed Pump Damage Boosts Performance, Reliability," Power, June 1979.
7. Makay, E., "Enlarged Radial Gap in Feed Pumps: A Remedy with Minimal Side Effects," Power, March 1981.
8. Fraser, W. H., "Recirculation in Centrifugal Pumps," (presented at the American Society of Mechanical Engineers Annual Meeting, November 15–20, 1981).

CHAPTER 19
TURBINES

This chapter covers the ABCs of steam turbines. The design philosophies presented here do not reflect that of any particular design or manufacturer. The principal goal is to clarify manufacturing and engineering literature that quite often assumes the field engineer understands how turbines react during startup, shutdown, and other extreme transit conditions. In other words, often the ABCs, or "the obvious," are left out as far as commissioning and plant engineering are concerned.

For design details and fundamentals of steam turbines and auxiliaries, see excellent material in the manufacturers' manuals.

TERMINOLOGY

Turbine Types

Tandem-Compound, Double-Flow Reheat Turbine

Tandem refers to the fact that the high-pressure (HP), intermediate-pressure (IP), and low-pressure (LP) turbines and generator are connected in tandem to form a single, rotating shaft. In contrast, a *cross-compound* turbine will have a separate, independent low-pressure turbine-generator set. In the cross-compound turbine, steam flows from the exhaust of the intermediate-pressure turbine by way of a *crossover* pipe to the LP turbine. Generally, the LP turbine for a cross-compound unit will rotate at half the speed of the HP turbine.

Compound refers to, in simple terms, a multistage design versus a single-stage design. Compounding is derived from the dynamics of fluid flow through the turbine stages. A fully *pressure-compounding turbine* has a full pressure drop from the inlet to the exhaust in a series of small pressure drops across each stage with no decrease in velocity. A fully *velocity-compounding turbine* would have a decrease in velocity from the inlet to exhaust with no drop in pressure. Realistically, high-pressure turbines cannot be completely one or the other; they must be a combination of the two or pressure-velocity compounding.

Double flow generally refers to turbines where the steam flows through two identical turbines opposed to each other, with a common inlet and casing, and separate exhausts at each end. The *single flow*, in contrast, will have a single exhaust.

Reheat refers to the IP and LP turbines, indicating that the exhaust from the HP is returned to the boiler for reheating before it enters the IP turbine, which exhausts into the LP and finally the condenser.

Reaction-Impulse Turbine

Reaction turbine refers to turbines whose blades are shaped like airfoils in which the motive force is produced by the velocity of the steam over the airfoil by creating a low-pressure zone on the upper side of the blade. Thereby, the higher pressure on the underside of the blade exerts the motive force. This is the same force that creates the lift on aircraft wings. Reaction turbines are pressure-compounding turbines . Figure 19-1(*a*) illustrates the dynamics of reaction blades.

Impulse turbine refers to turbines whose blades are shaped like buckets in which the motive force is direct steam impingement in the "bucket." Impulse turbines are velocity-compounding turbines. Figure 19-1(*b*) illustrates the dynamics of reaction blades.

Bucket is a term that has become a universal name for the rotating blades of steam turbines whether they are of the impulse or impulse/reaction types. In theory, no high- or intermediate-pressure turbine stages contain rotating blades that are completely of the reaction or impulse type but do contain a combination of the two. In other words, they are velocity-pressure compounding turbines. *Diaphragms* are the stationary blades or nozzles. Low-pressure turbine blades are usually reaction blades and are called just that, "blades"—although many experts also refer to them as "bucket."

Figure 19-2 is a depiction of how steam expands through a simple hypothetical turbine. Note that the enthalpy is constantly dropping. Note that the simplest test performed to monitor turbine performance is called the *enthalpy drop test*. It is used during commissioning as well as during preventive maintenance. In the hypothetical turbine, the pressure and temperature drop from the turbine inlet conditions to the exhaust temperature and pressure intersection on the Mollier diagram. In this example, the exhaust enters the LP without reheating. Note that points 2 and 3 in Fig. 19-1 are separated to better illustrate the equations. In actuality, they would be the same point for this turbine.

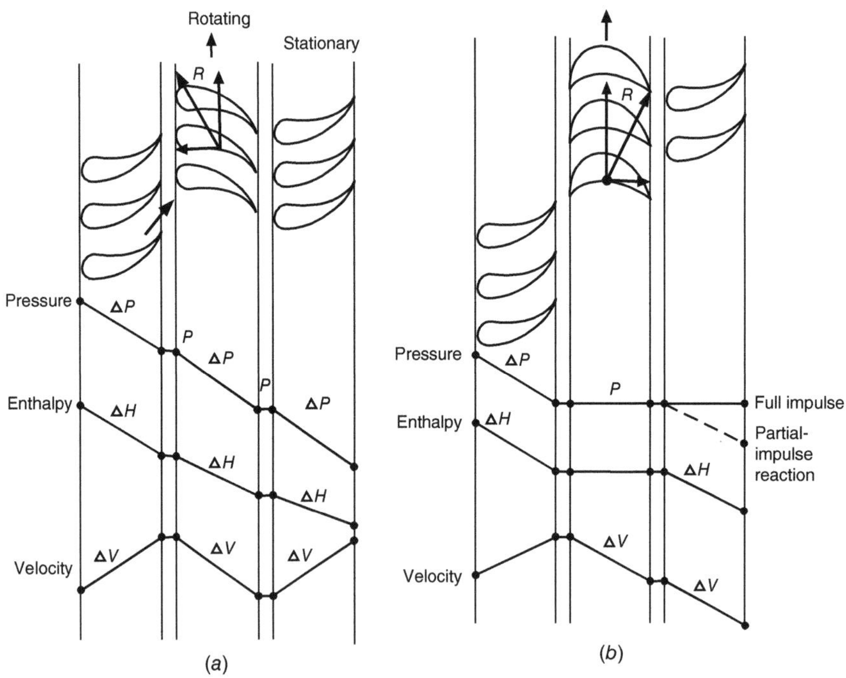

FIGURE 19-1 (*a*) Reaction turbine force diagram. (*b*) impulse turbine force diagram.

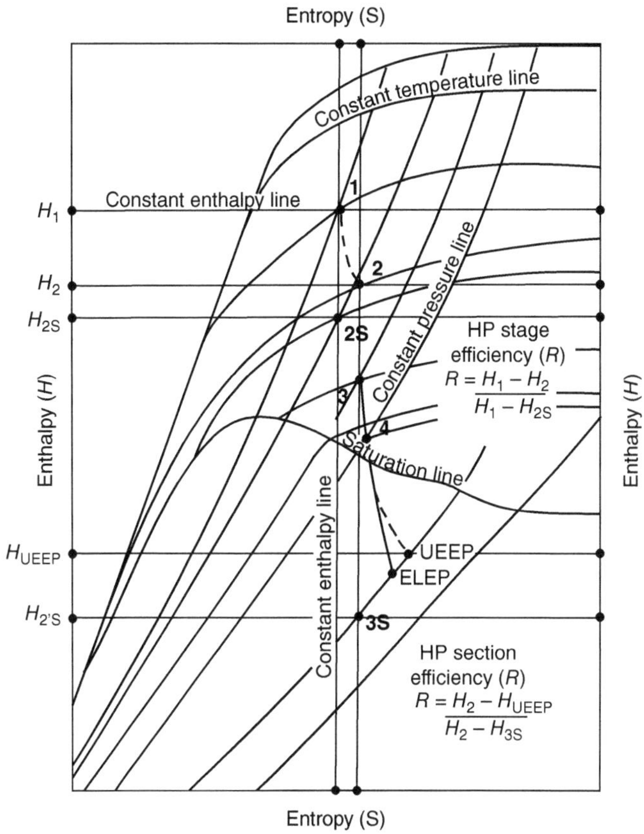

FIGURE 19-2 Graph depicting how steam expands through a simple hypothetical turbine.

The steam exhausting into the condenser ends at an *expansion line end point* (ELEP). Because there is no temperature line intersecting the pressure line, the end point is established by continuing the shape of the curve from above the saturation line down into the vapor zone until it hits the pressure line corresponding to the turbine backpressure at the exhaust. This point is not the true end point. To find the true end point, the exhaust losses, which include a correction factor for moisture, must be found. The turbine manufacturer, in what is called a *thermal kit,* provides these corrections as well as the curves for the expansion line. Using the exhaust loss, the *used energy end point* (UEEP) is found. The thermal kit also provides gland leakage and mechanical losses for true performance testing. The enthalpy drop test does not go to such extensive elaboration. A typical form used for enthalpy drop testing is illustrated in Fig. 19-3.

TURBINE COMPONENTS

Bucket and Blades

All impulse or partial-impulse buckets have shroud caps and seals to prevent leakage because of pressure differential between stages. Buckets and blades are installed in sets and are tied together.

DCR ENGINEERS, INC.

PROJECT ____________________
EQUIP/SYSTEM NAME ____________________
DATE ____________________

TURBINE EFFICIENCY
Enthalpy Drop Method

PLANT: ______________________________
UNIT: ______________________________
TEST DATE: ______________________________
GROSS GENERATION: ____________________

READINGS

T_1 = Throttle steam temperature	____________	°F
F_1 = Throttle steam pressure	____________	PSIA
T_1 = H.P. Turbine exhaust temperature	____________	°F
F_2 = H.P. Turbine exhaust pressure	____________	PSIA
T_3 = Reheat steam temperature	____________	°F
P_3 = Reheat steam pressure	____________	PSIA
T_4 = I.P. Turbine exhaust temperature	____________	°F
P_4 = I.P. Turbine exhaust pressure	____________	PSIA

CALCULATIONS

H.P. TURBINE EFFICIENCY

H_1 = Enthalpy @ and	____________	BTU/lb
S_1 = Enthalpy @ and	____________	BTU/lb °F
H_{e1} = Enthalpy @ and	____________	BTU/lb
H_2 = Enthalpy @ and	____________	BTU/lb
H.P. Turbine efficiency $= \frac{H_1 - H_2}{H_1 - H_{e1}} =$	____________	%

FIGURE 19-3 Sample form used for an enthalpy drop test.

The rotating HP and IP buckets are inserted one by one through notches in the rotor. They are held in place by dovetail hooks that match cutouts in the wheel. The dovetails resemble Christmas trees cut out by children; thereby the root is sometimes referred to as the Christmas tree. They can be male or female configured. Some long, LP bucket dovetails are slipped over the wheel steeple axially and then anchored individually in place. These blades are not grouped and shrouded because they are reaction blades and thereby have no pressure drop between stages. See a typical reaction blade grouping in Fig. 19-4.

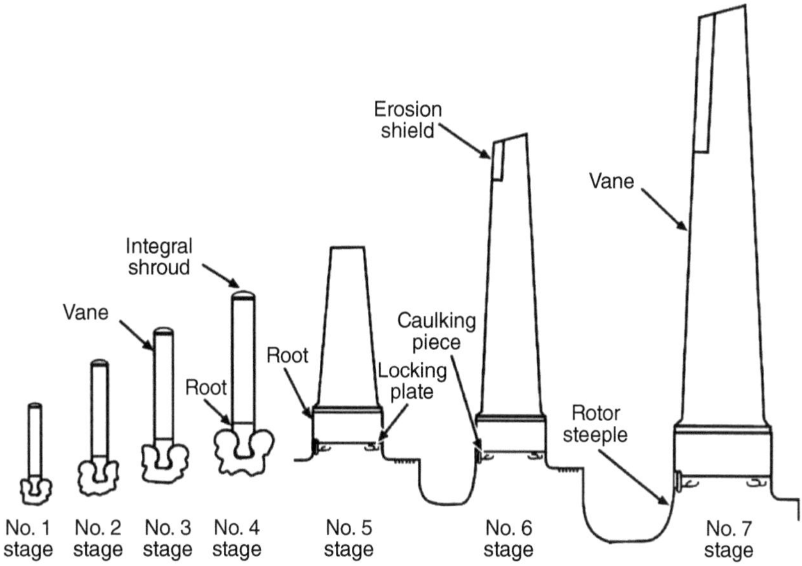

(*a*) Small turbine bucket groups.

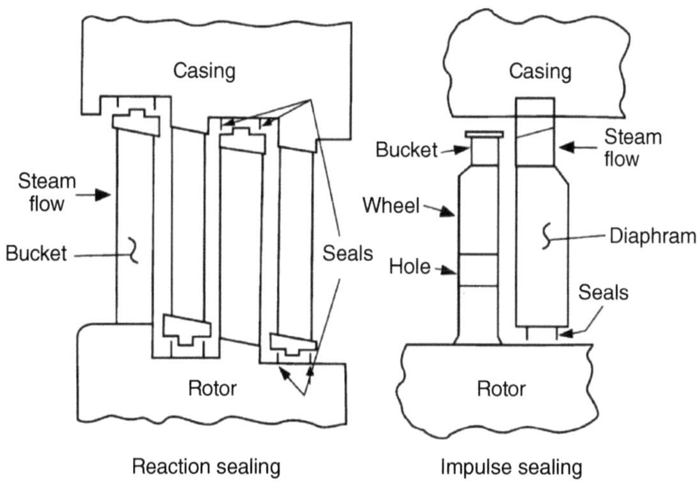

(*b*) Steam trubine blade seals.

FIGURE 19-4 Both reaction blade diaphragms and buckets are sealed; but only impulse diaphragms (stationary) are sealed.

Balance Drums

As steam is exhausted from the impulse stages, an opposite reaction drives the rotor assembly back into the thrust bearing. To counteract this thrust, balance drums are installed that allow steam to exhaust through an annulus in the opposite direction from a high-pressure section to a low-pressure section. The exhaust from the drum creates a reaction force in the opposite direction of the propulsion steam flow, which reduces thrust bearing loading. Examples can be found in Chap. 18.

Gland Steam

Casing glands are incorporated to prevent steam from escaping from around the rotor shaft. The glands contain labyrinth seals that prevent the steam from exiting to atmosphere around the shaft and casing in normal operation (See Fig. 19-5). Sealing steam from a gland-steam pressure regulator enters the gland and flows into the turbine and the gland-steam condenser, thus preventing air from entering during startup. As the turbine is ramped up, the flow reverses and the process steam from the turbine enters the seals and exhausts into the gland-steam condenser rather than to atmosphere. The labyrinth seals are designed to gradually reduce the steam pressure entering or exiting the turbine so that the steam is recovered in the sealing-steam condenser.

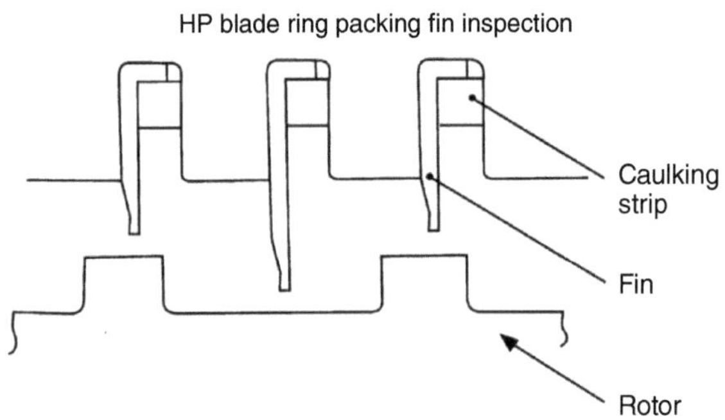

FIGURE 19-5 Labyrinth seals.

Lubricating Oil System

Bearings and lubrication, as well as lube oil system flushing, are covered in separate chapters; therefore, only the basic design for steam turbines is covered in this section. The most obvious basic difference between a steam-turbine lube oil system and others is the provisions in the design for steam conditions. The first difference is the vapor extractor on the lube oil reservoir. This is installed to remove water as well as hot oil vapor that enters the system via the turbine shaft. Water vapor comes from shaft steam-seal leakage. Thus, this provision is twofold. First, it presents pressurization of the reservoir with oil vapor and emulsification of the oil with water. The reservoir pressure should be only slightly negative, but with sufficient vacuum to exhaust the vapor. The pressure is established by the manufacturer, but it is generally no more than –10 in H_2O to minimize pulling vapor and dirt into the system. The second difference is that the high-pressure turbine lube oil supply lines are always encased in the return lines. This is a fire protection consideration.

If a supply line were to develop a leak, such as a mist spray, the oil could be ignited by the turbine and piping surfaces.

Another consideration for large steam turbines is they must always be provided with emergency lube oil pumps powered from the station batteries in case of a power failure, such as a black plant trip. They must be capable of supplying bearing lube oil until the turbine rolls to rest. For some turbines, this takes hours.

Casings

Small or intermediate- and low-pressure turbine stationary shells are made of a single casing. Large, high-pressure turbines are made with inner and outer casings. The steam supply piping is attached to the outer casing. Nozzles on the inside of the outer casing slip into ports in the inner casing for steam admission and exhaust. The turbine stationary blades and seals are mounted in the inner casing. The purpose of the double casing is to reduce the inner-to-outer pressure and temperature differentials. For a double-casing machine, the differential is divided between the two casings, thus enabling thinner design, which reduces manufacturing difficulties and operational stresses.

Gibs and Keys

Turbines must be *allowed* to freely expand and contract. Gibs and keys are used to guide this expansion and contraction. *Keys* are similar in function to boiler buckstay guides. Inner-casing gibs anchor the casing, and the keys guide its expansion relative to the outer casing.

The outer casing must also have gibs for anchoring, as well as a key guiding system. Most casings are anchored at the generator end and are free to expand in the direction of the front standard. To limit total growth of the rotor toward the generator and turning gear, the HP-IP turbines can be anchored between them.

When the IP is connected, or integral, to the HP casing, and the front standard is attached to the outer casing, the standard will move, taking the rotor shaft—anchored by the thrust bearing—with it. The rotor expands away from the generator because the casing expands in that direction, taking the trust bearing with it. This helps considerably to reduce casing–rotor expansion differential. As the front standard moves, it is usually guided by keys in the standard and turbine baseplate. Casing anchor bolts that are used as keys must have loose washers to guarantee freedom of movement. In fact, they are used to determine casing freedom of movement. See the example in Fig. 19-6.

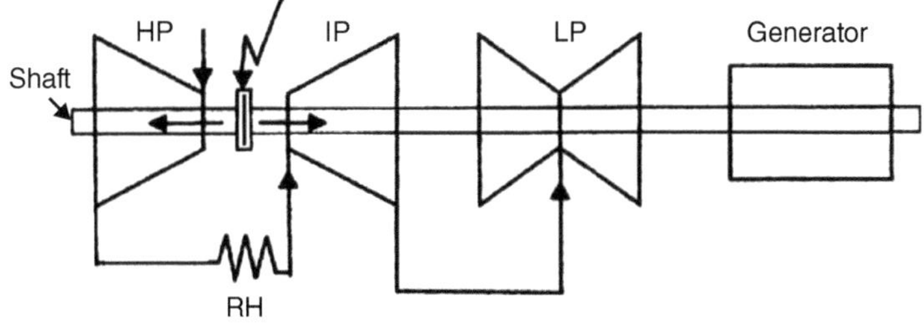

FIGURE 19-6 Trust bearing between HP and IP turbines.

Front Standard

The front standard contains the turbine rotor first, or "front," journal bearing; it also contains the shaft thrust bearing, instrumentation, and mechanical and hydraulic turbine trip devices. Because the rotor shaft is "anchored" at its hot end by the thrust bearing, it must expand with the standard away from the generator. If the front standard is designed to move with the turbine casing as it expands, it will be attached to the outer casing. This balances casing–rotor expansion. Extreme caution must be exercised to ensure that the rotor and casing expand together by controlling their differential temperature and, thereby, their differential expansion. The rotor will have a tendency to expand faster during startup warming and transient loading than the casing because of its relatively lower mass.

There are three major turbine components that must expand together differentially when heated: the outer casing, the inner casing, and the rotor. The most massive of the elements to be heated is the section of steam chest between the stop valves and governor valves on the outer casing. Therefore, its inner-to-outer differential temperatures usually set the pace for warmup of the HP casing. Periodically opening and closing the stop valves controls the steam chest differential temperature. Controlling steam admission by opening and closing the governor valves controls rotor expansion.

Note

If turbine roll is delayed for a long period of time with the condenser vacuum pulled, the sealing steam can heat the rotor excessively, causing a high differential expansion.

Note

Turbine casing–rotor expansion control varies by manufacturer and with the particular turbine designs. Therefore, the field engineer must determine where all anchor and guide points are, as well as expected movements. This information is given in the manufacturer's literature for every machine. Lessons learned from boiler startup and commissioning can be carried over as fundamentals for steam turbine commissioning. That is, observe and control metal differential temperatures and, thereby, their expansion and contraction.

Drains

Drains are a vital part of the turbine design both for startup–warmup and during normal operation. During startup–warmup, it is obvious that the steam warming the cold turbine components will condense and must be removed. Startup drains are used for that purpose. Also required are continuous drains, or orifice drains, that must be open at all times. Usually, engineers drill holes in the gates of drain valves to control the drain flow and prevent it from being completely blocked during normal operation. The valves are fully open during startup and are closed to control the drain flow during normal operation. If left fully open during normal operation, the high-velocity steam can cause erosion and failure of the drain pipelines in a very short period of time. The continuous drains are used to drain condensate that accumulates in the casing condensate collection channels in the lower-pressure sections of the turbine.

DISASSEMBLY/ASSEMBLY DETAILS

If a turbine requires inspection during commissioning, it must be done correctly as if the inspection was a major overhaul. Because massive steam turbines must be treated with "kid gloves," clearances are meticulously measured and maintained. Stage rotating element-to-casing and shaft-to-casing clearances are maintained by delicate sealing systems and, thereby, are extremely sensitive to damage. Damage to a delicate seal can, in turn, cause severe damage to massive turbine components. In order to understand the meticulous nature in which turbines must be treated, some disassembly details are provided in this section. Refer to Fig. 19-7 as a guide.

Disassembly

Occasionally, a turbine must be internally inspected because of rubs or unusual noises accompanied by high vibration during initial roll. This presents an opportunity to collect signature data for all internal clearances that may differ from factory readings, which may have changed during shipment or field assembly. Also, the inspection checks for any anomalies, including foreign material, which may have been introduced at the time of assembly. Often, a turbine is opened for inspection during commissioning for the sole purpose of locating a seal that has rub marks or other obvious indicators of a vibration root cause.

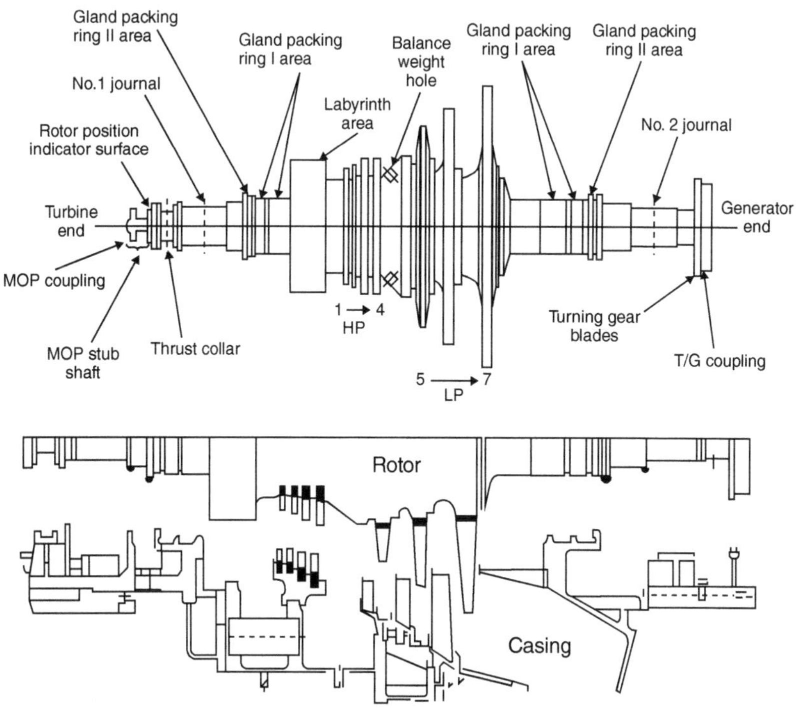

FIGURE 19-7 Simplified internal illustration of turbine to be disassembled.

Unfortunately, no other troubleshooting is performed other than visual observation. The following brief disassembly-assembly procedure illustrates a more comprehensive method of investigation. A small geothermal turbine is selected, as an example, for simplicity. This turbine uses a shaft driven pump as the main oil pump (MOP).

Preparation

All equipment and procedures must be available as would be for a major turbine overhaul. The crane used must be of sufficient capacity to lift the heaviest piece, but it should not be used to break free pieces that are hung up and, therefore, must have the capability for monitoring load. Bags, crates, and an organized storage space must be available for collection of all the cleaned pieces removed from the turbine.

Upper-Bearing Half Oil-Deflector Ring Removal

Before removing the oil-deflector rings, measure and record the fin to rotor clearance. Then remove rings.

Front Bearing Pedestal Covers

Install rotor centering fixtures and center rotor before removing pedestal covers; then remove covers and measure and record the side radial clearances.

Turning Gear Bearing Pedestal Covers

Install the rotor turning fixture at the turbine generator coupling on the rear pedestal. Measure and record the clearance between the fixture and the bore parts of the bearing and turbine gland seals.

Journal Bearing Disassembly

Measure and record all oil-def-ector rings to rotor clearances.

Disconnect thermocouples and connecting tubing. Lift upper bearing shafts to clear rotor. Place lead wire on top of rotor. Lower the upper half and rebolt it. Remove again and record lead wire thickness for clearance.

Thrust Bearing

After the thrust clearance checks and before removing the thrust bearing, uncouple the generator. After disassembly of the thrust bearing and installation of a temporary thrust restraint, check the turbine-to-generator rotor coupling alignment. Push rotor to stop and record thrust clearance.

Turbine Upper Casing Removal

The upper casing must be removed, monitoring crane load. This helps ensure that if it is hung up on internal components, no damage will occur. If it is hung up, use jacking bolts and/or port-a-power jacks to lift it free. Remove the casing, set on cribbing, lift, rotate, then reset on cribbing upside down. Immediately inspect and photograph everything inside the upper and lower casings; then remove the upper stationary blade groups from the lower casing.

Measure and record clearances for the packing glands, labyrinth rings, stationary blades to rotor fins, and the rotor blades to stationary fins. Measure rotor bow using multiple runout measurements.

Turbine Rotor Removal

Prepare to remove the rotor by removing the temporary braces and centering fixtures. Prepare wooden-wedge guide bars to guide the rotor out of the lower casing without swinging it and causing damage. Immediately inspect the lower-casing internals and steam paths for foreign material and obvious damage. Photograph everything regardless of condition, for later verification.

Remember that the secondary objective is to measure and record all clearances (the first objective being to find the root cause of the problem). That means the upper casing and rotor will be removed, the bearings "blued," and lead wire installed over seals. Then the rotor and upper casing will be reinstalled in order to record bearing-to-journal pattern using Prussian blue ink and lead-wire thickness to indicate seal clearances. The rotor and upper casing are removed a second time to recover the lead wire and photograph the bearing pattern before final reassembly.

TURBINE FIRST ROLL

Steam blows and lube oil flushing must have been completed prior to turbine roll. (See Chap. 6 for steam blows and lube oil flushing.) Before steam can be admitted to the turbine, all insulation must have been installed on all steam admission lines and turbine components. However, if it is safe to have them exposed, it is not a bad idea to *not* insulate drain line elbows until after turbine roll because it makes locating steam and vacuum leaks easier. The insulation can hide the leaks or make them appear to be originating from a different location. Differential temperature for small-bore piping is not a concern. Keep in mind that if the entire drain line is bare, condensation may reduce flow and back water up into the turbine. Large-bore piping must be insulated to minimize condensation during warm up.

Caution

If a turbine stops dead on rolldown, continued rotation should never be forced. Inadequate clearances or foreign material may have caused the rotor to freeze. Whatever the reason, forcing the rotor will cause serious internal damage. If this happens, the turbine must be disassembled to determine the cause of the locked rotor.

All nonreturn valves (NRVs) and water-induction-prevention valves must be fully tested and commissioned. All feedwater heaters should be ready for service. If the startup goes smoothly during initial turbine-generator testing, they can be continuously loaded after synchronization tests.

The boiler isolation valve should be open and initial flow controlled by turbine bypass, startup vents, and drains. If a second boiler, on a two-boiler, single-turbine unit, is to be brought into service, make sure the boiler isolation valve is open before firing. Let the boiler NRV balance the two boilers with the main steamline pressure.

Placing the heaters in service and monitoring them during ramp-up takes a dedicated startup team not involved with boiler and turbine trial operation. Therefore, the heaters are more often than not ignored until after the generator is at full load. This is not good practice.

The seal-steam system must be placed in service while pulling vacuum on the condenser. Sealing the glands with steam will prevent dirt or debris from being pulled into the seals under a vacuum. The turbine exhaust-hood sprays must be in service because during turbine no-load testing, the turbine is not absorbing energy from the steam, and therefore, the steam can overheat the low-temperature blades at the exhaust. At full-speed no-load, they can be overstressed at the elevated temperature.

PREPARATION

Condenser, Containment, and Fire Protection

The condenser must have been hydrostatically tested to above the neck expansion joint. This is the initial vacuum leak test. All turbine drain valves to the condenser must be open if the drains initiate above the expansion joint. If a condenser pit is part of the design, the pit drains and fire deluge system must be tested. The pits have a double function: (1) to act as containment for fire protection and (2) to lower the turbine pedestal. The pedestals are massive in order to withstand the uplift force of a full-power, instantaneous, turbine-generator locked rotor. Therefore, any reduction in height translates into substantial cost reduction. If the pedestal is on grade, it will have oil containment walls for fire protection, and the same procedures apply. The fire protection system must be tested as well as the containment. The containments allow two-dimensional controlled firefighting, thereby preventing a fire from becoming a three-dimensional, out-of-control conflagration. Therefore, the deluge spray system and drains must be tested to ensure the deluge is within the confines of the containment.

Steam-Seal System

The steam-seal system and steam supply must have been blown clean and placed into service. Vacuum on the condenser should never be pulled without sealing steam. The steam will ensure that no foreign material will be pulled into the turbine gland seals. After the seal system is ready for operation, vacuum can be pulled on the condenser and leakage testing can begin.

> **Note**
>
> The seal steam is the first source of heat for the turbine and will help the prewarming process. However, the rotor will heat faster than the casing. If no-load testing is prolonged, differential expansion may require that vacuum be broken and the seal-steam system shut down until the differential is restored to acceptable values. This generally applies to very large turbines.

Ensure that the condensate system minimum flow is set up using the gland-steam condenser minimum flow specification. This is the system minimum flow, which is not to be confused with the condensate pump minimum flow (which will be less). The gland steam condenser drain trap and loop seal to the main condenser should be checked for proper type and installation. Float-type traps will be held closed because the float buoyancy will not overcome differential pressure under vacuum. This causes the gland-steam condenser to flood, preventing condensation, which results in a loss of vacuum.

All condenser air ejectors or vacuum pumps will be equipped with meters to check the exhaust gas flow, which is equivalent to the condenser air leakage plus other noncondensable gases. The meter will be located in a bypass for the main discharge line. Close the main-line isolation valve so that all the flow is through the meter. The reading should be midscale or less. If the reading is "pegged out" or in the upper part of the scale, there is excessive air leakage into the main condenser. If the meter is flooded with water, the condenser is flooded and not condensing; check the trap and drain system. If the exhausters are the steam-ejector type and the reading fluctuates between pegged out and normal, accompanied by a loss in condenser vacuum, check the steam-temperature setting. Ejectors will not function with saturated steam. The steam-temperature setpoint must be well into superheat. If the setpoint is too close to saturation, the ejectors will exhibit an on/off malfunction phenomena.

Exhaust Hood Sprays

Exhaust hood sprays must be in service because, during initial roll and checkout, the unloaded HP/IP turbines will absorb little energy from the steam; therefore, hotter than normal steam will be entering the LP turbine. This steam can overheat the last-stage blades beyond design limits. This can damage the blades because they will be experiencing the same centrifugally imposed stresses as they would during full-load operation, but at elevated temperatures.

Feedwater Heaters

The feedwater heaters must be ready for service. The water induction and nonreturn valves must be checked and placed in service prior to turbine roll. Remember, NRVs are check valves designed to prevent heater-stored energy from reversing flow, reentering the turbine, and overspeeding the turbine after a generator trip. They are not water-induction-prevention valves. Rising water will not provide a differential pressure across the valve disc and, therefore, will not block water induction. The water-induction valves are gate valves designed for that purpose. The valve interlocks are checked by raising and lowering the water in the level columns on the heaters. It is not necessary to actually raise and lower the heater levels. Check the NRV counterweights. They must be set for close assist without restricting extraction flow.

Sometimes there are not enough operators or startup personnel to adequately cover turbine roll and heater commissioning simultaneously. Therefore, the heater commissioning is left until after the turbine is commissioned. This is not an acceptable practice because, with proper planning and preparation, this operation can be adequately staffed. Running a generator to full load without the heaters in service can seriously overload IP and LP stages. Steam that was designed to exit the turbine is now passed to downstream stages. This cascades from each consecutive stage, compounding the extra load to the IP and LP turbines and the condenser.

Lube Oil System

Reservoir heaters and coolers must be ready for service. If there is a large difference in daily ambient temperatures, these will play an important role in startup. For example, if the temperature drops considerably, it may be necessary to warm the lube oil prior to roll. If the oil is too cold, the turbine will vibrate excessively at half-speed, requiring a shutdown. For more information about lube oil systems refer to Chap. 4.

Steam Quality

Steam quality refers to the moisture content of the steam. The quality must be within the parameters set by the turbine manufacturer. The standard normally used requires at least 100°F superheat for turbine roll.

Note

Many LP turbines have suffered extensive blade damage because of reheat spray operation at low steam flow. The damage usually does not surface for many hours of operation, after which cracked blades finally fail during normal operation. At low steam flow, the water does not attemperate the steam and can accumulate and flash. The flashing sends slugs of dense steam through the IP and LP turbines.

Steam Purity

Steam purity refers to the concentration of impurities in the steam. The standards are established by the American Boiler Manufacturer's Association (ABMA). The latest standard and the turbine manufacturer's instructions should be consulted. Monitoring steam purity is essential. If the boiler uses phosphates for chemical control, sodium monitoring is also essential. If sodium sampling and monitoring are not part of the original design, temporary sodium monitors should be rented or purchased for commissioning. Here is a good reason for sodium detectors: If high drum level or foaming occurs, sodium from the phosphates will be entrained in the water and detected by the monitors. *Remember, poor boiler circulation can cause high-level swell in the center of a drum while the end level instruments read normal.* Table 19-1 lists several types of impurities and their treatment chemicals.

TABLE 19-1 Impurities and Their Associated Chemical Treatments

Deposit-Forming Impurities	Corrosion-Causing Impurities
Calcium salts	Ammonia
Copper salts	Carbon dioxide
Iron salts	Chlorides
Magnesium salts	Oxygen
Sodium salts	Sodium hydroxide
Bicarbonates	Sodium carbonates
Carbonates	Sulfides
Nitrates	
Organic matter	
Silica	
Sulfates	

pH Treatment Chemical	Oxygen Treatment Chemical	Hardness Treatment Chemical
Ammonia	Hyrazine	Sodium phosphate
Morpholine	Sodium sulfite	Chelants
Sodium hydroxide		
Sodium phosphate		
Trisodium phosphate		

Note that some chemicals used as inhibitors can also form impurities that can cause serious damage to boilers, turbines, and condensers. The strictest use of inhibitors must be exercised at all times. "Sodium hydroxide has probably damaged more turbines than any other single contaminant."[1] Lindsay goes on to illustrate that high sodium hydroxide concentrations can occur in a turbine as the steam expands following the expansion line. These concentrations can reach 20 to 90 percent in solution.

By combining these two sets of lines, on a Mollier chart, it can be seen how a caustic contamination incident that would put a quantity (say 100 ppb) of sodium hydroxide in the steam could produce a 90% concentrate of sodium hydroxide as a liquid somewhere in the IP turbine, as indicated by the intersection of the 100 ppb solubility limit and the IP steam expansion line. Damage could also occur in the LP turbine, but caustic contamination effects are not limited to the interiors of turbines. For example, crossover piping has suffered from caustic damage.[1]

A point to remember when inspecting a turbine is this: Sodium hydroxide deposits will form a white powder (sodium carbonate) when exposed to air as the turbine is disassembled.

Water Treatment Systems

Water treatment systems must be fully functional and closely monitored. Chemicals used to treat impurities can also become an impurity that can cause serious damage if overdosed.

Turbine Backpressure

Turbine backpressure must be controlled within the turbine manufacturer's guidelines. If the turbine is operated below the design backpressure, damage to the LP blades and condenser tubes is possible. The low pressure causes higher-than-normal velocities that can induce vortex shedding in LP blades and condenser tubes. The shedding causes the blades and tubes to vibrate, fatigue, and fail. This failure can also occur at higher-than-design backpressures. The LP blades can flutter, fatigue, and fail also at high backpressures, as reported after extensive laboratory and field testing.[2]

Note

In Fig. 19-5, the blade above the one with separation is the stalled blade because flow is blocked on the under, or thrust side. This is caused by boundary layer separation and vortex shedding. The separation causes backflow and stall. Refer to Chap. 21 for further information.

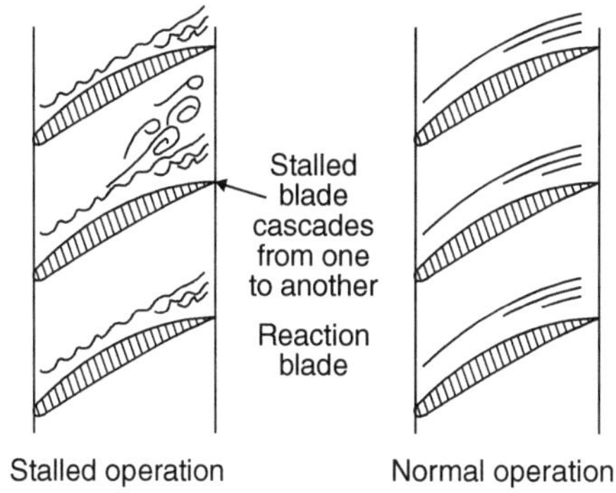

FIGURE 19-5 Blade stall.

REFERENCES

1. Lindsay, W. T., "Behavior of Impurities in Steam Turbines," *Power Engineering*, May 1979.
2. Silvestri, G. J., "Progress in the Development of Reliable and Economical Steam Turbines for Dry Cooling Application," *Combustion*, March 1981.

CHAPTER 20
VALVES

A good field reference book for valves is the *Control Valve Handbook* by Emerson Process Management.[1] It is an excellent comprehensive compilation of information and data by Fischer Controls and is available via the Internet: www.documentation.emersonprocess.com/groups/public/documents/book/cvh99.pdf.

ISOLATION VALVES

Gate valves are used exclusively as isolation valves. All hand valves such as the gate valve can be purchased with either rising or nonrising stems as illustrated in Fig. 20-1. The gate is always a round, wedge-shaped disc that fits tightly into the valve seats. Hand valves are always opened clockwise and closed counter-clockwise. (The memory tip is righty tighty–lefty loosey, the same as right-hand threaded fasteners.) When the valve is fully open or closed, the hand wheel should be broken free from the stem so that expansion or contraction of the valve does not jam the stem so tightly the valve cannot be broken free without a cheater.

Gate-valve flow characteristics make them inadequate for flow or pressure control, and therefore, they should never be used to throttle flow or regulate pressure. Throttling will cause cavitation seat and disc damage due to the high velocities encountered at 20 percent or less valve position. Gate valves will not start to throttle until they are 80 percent closed. Refer to the flow versus valve position curve in Fig. 20-2.

High-pressure valve seals are illustrated in Fig. 20-3.[2] They should be studied prior to boiler firing and steam pressure buildup. Many of these valves can cause a forced outage during commissioning due to steam leaks. Also, plan ahead and be prepared to bring in a leak repair contractor. That is, determine if there is a contractor within a short traveling distance from the plant and contact him or her ahead of time. Proper torquing and retorquing at temperature are essential to prevent leakage and damage.

Damage sometimes can get so severe the unit will require a shutdown. However, many times the boiler pressure can be lowered to lessen wire-drawing damage to the valve. Then, the leak can be sealed online, and an outage can be scheduled for the valve replacement.

Remember, when opening and closing boiler and pressure vessel, double-isolation valves, always open the inboard valve first and close it last. That is, open and close the inboard valve with no flow. There are two reasons for this:

1. The inboard valve is governed by the ASME boiler and pressure vessel codes and the outboard by the ANSI B31 series piping codes.
2. Repair of the inboard valve requires a unit shut down but the outboard does not.

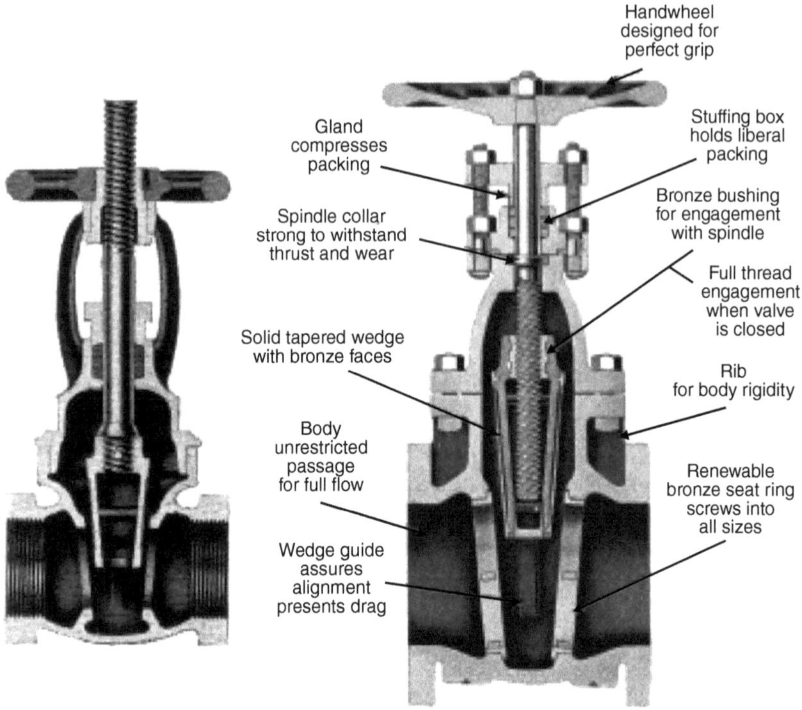

FIGURE 20-1 Jenkins gate valves with rising and nonrising stems.

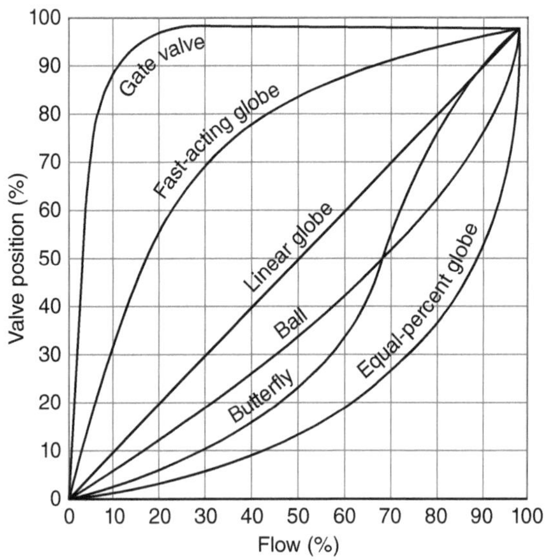

FIGURE 20-2 Valve characteristics.

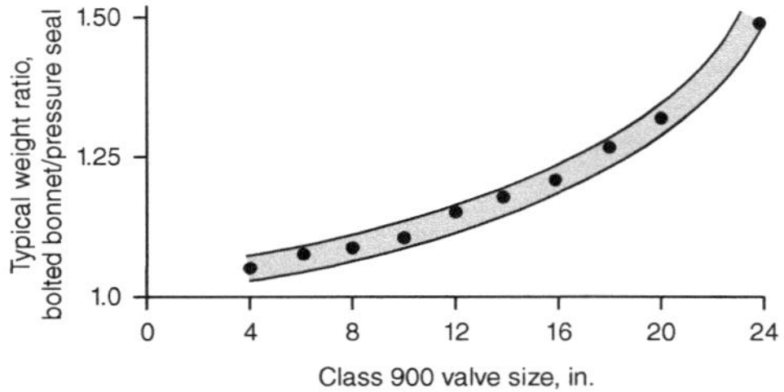

Weight advantage of pressure-seal valve becomes more helpful as line size (and pressure class) of the valve increases.

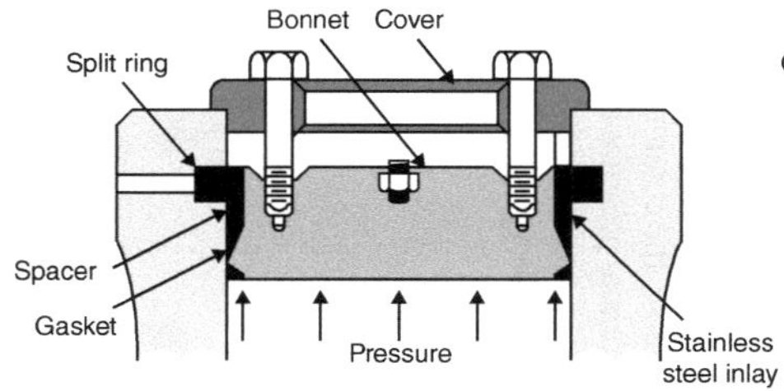

Stainless-steel body inlay in a check valve, typetying design of the pressure seal, prevents corrosion of body and gasket

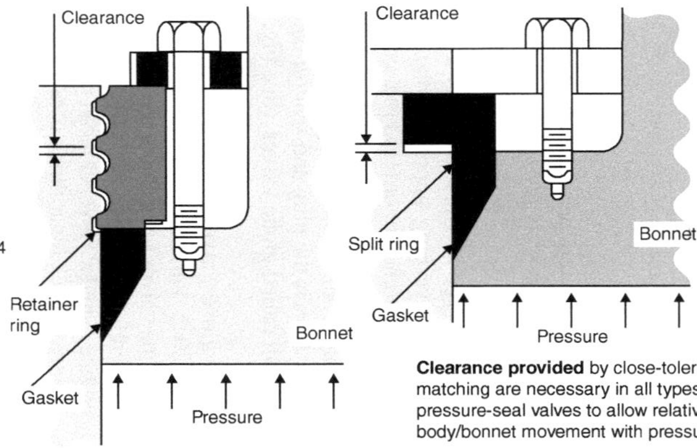

Clearance provided by close-tolerance matching are necessary in all types of pressure-seal valves to allow relative body/bonnet movement with pressure

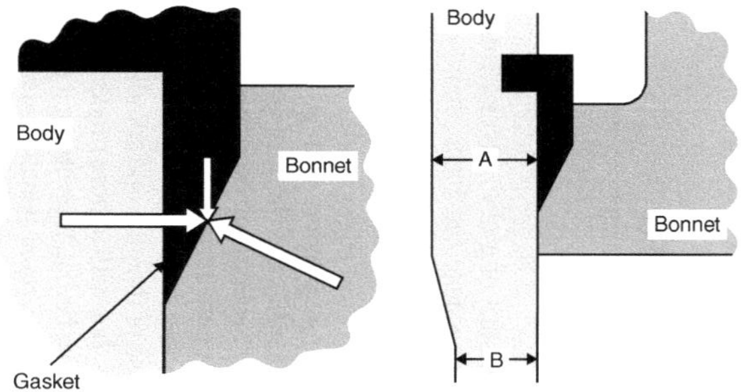

Wedge configuration of tapered seal transmits radial component two to three times higher than pressure load (left), so body top must be thicker to withstand this

FIGURE 20-3 High pressure valve seals.

BRASS VERSUS BRONZE

Brass is an alloy of copper that contains zinc and is softer than bronze, whose principle alloy is tin. The primary use for brass is ornamental. Valves are made from bronze because of its strength; however, most people call all copper alloy valves "brass." Recently, brass can be found with as much as 40 percent tin (plus zinc) because tin is cheaper than copper. Bronze can have as much as 20 percent tin.

CONTROL VALVES

Globe Valves

The characteristics of globe valves are determined by the type of trim used. They are called *globe valves* because the basic, hand-operated globe valve plug appears globe-shaped. Actually, there are many different types of trim that are designed in order to reduce control the gain (proportional band) requirement, noise, and cavitation. Some are simple globe and seat, while others use cages made of complex stacks of serrated disks, instead of seats, with plugs resembling cylinders with geometric openings. *Trim* is the plug and seat (or cage) in the valve. Regardless of the trim geometry, they are all called globe valves. (See Fig. 20-4 for examples of several globe valves.)

The flow characteristics of linear, equal-percent globe valves, as well as fast-acting globe valves, are determined by their trim. See Fig. 20-2. *Valve lift* is a term used for the percentage rise of the plug and spindle. Fast-acting globe valves have characteristics similar to gate valves, but with a more moderate curve and, therefore, are used for "on–off" control.

The plug can be flat, resembling a combustion engine cylinder valve, or it can resemble the engine piston. The equal-percent valve is actually a valve with a logarithmic characteristic. If graphed on semi-log paper, the slope would look like a linear globe valve. Algorithms that take the anti-log of the slope do just that. A flow coefficient, C_v, is given by the manufacturer to control valves for the full-open position or at 1 psi pressure drop across the valve. Knowing the C_v and the differential pressure across the valve, the commissioning engineer can determine the flow rate. Knowing the flow rate and differential pressure for a particular valve position, the engineer can determine the C_v for any valve position. This can be very helpful in processes with limited flow instrumentation. Refer to the C_v versus valve position curves for water in Fig. 20-5.

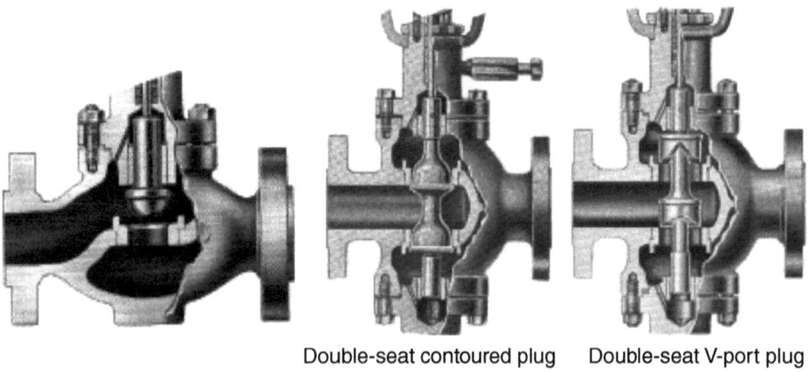

FIGURE 20-4 Masoneillan single-seat and double-seat globe valves.

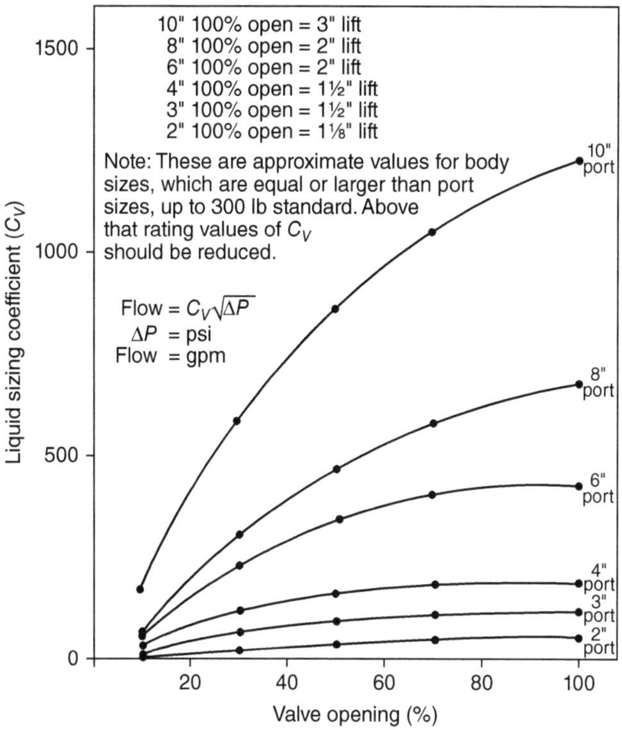

FIGURE 20-5 C_v versus valve position curves for water.

Butterfly Valves

Butterfly valves can be used as control valves, but conventional wafer disc butterflies have limited range. At valve positions less than 20 percent, the flow velocity can damage the seals and seat if maintained for extended periods. Between approximately 30 and 70 percent, there is only about a 10 percent change in flow. Because of the narrow control range, butterfly valves are best suited to control fixed, steady-load processes. They are also used as isolation valves in low- to medium-pressure processes because of their narrow profile compared with gate valves of similar size and pressure rating. Butterfly valves usually weigh five times less than gate valves and cost half as much. They also can be fully opened and closed with a quarter turn of the stem. Plug, ball, and butterfly valves are called rotary-type valves. Because of the short stroke, they should be opened and closed slowly to prevent process shock or surge. See water hammer in Chap. 17.

Plug Valves

Plug valves have the same characteristics as a ball valves. As a matter of fact, the ball valve is a special variation of the plug valve. Designers can develop body passages that make three-, four-, and five-way valves possible. The port in the plug can have a full port, where the inside port diameter is the same as the inside pipe diameter. This has the advantage of allowing probes to be inserted. For example, stop cocks through which a pitot lance is inserted for flow measurement are special applications of the plug valve. Plug and ball valves are used extensively in the gas and petrochemical industry because of their tight sealing characteristics.

Pinch Valves

Pinch valves are used in slurry processes or other systems that handle abrasive or corrosive fluids. They are made of rubber or synthetic tubes or sleeves in a metal body. The sleeves completely protect the valve body from corrosion and erosion. They have excellent flow characteristics, similar to equal-percent globe valves, and therefore can be used for throttling. They are available in sizes from as small as $\frac{1}{8}$ in to as large as 72 in in diameter.

Check Valves

Check valves are obviously used to prevent reverse flow. They can be spring-loaded ball checks for use in hydraulic lines or swing checks used in parallel-pump operation. Check valves used for steam-turbine extraction and in main steamlines for parallel-boiler operation are called nonreturn valves. Most check valves must function without external devices, indicators, or other appurtenances that isolation and control valves have. The exceptions are valves similar to the turbine extraction and boiler NRVs. The turbine NRVs have weighted levers for open-assist and cylinders for power-assist closing during turbine trips. The boiler NRVs are usually dual-purpose stop-checks; therefore, they have actuators.

Because there is no exterior indication regarding the internal condition of these valves, the commissioning engineer must use some ingenuity to determine if they are operating properly. Some tips are as follows:

- Turbine NRVs can be checked in service by pulling the weighted lever arm down toward the close position. If there is no resistance, the disc has fallen off and the extraction must be taken out of service. For spring close–assist checks, if the lever arm is in the closed position and the system is in service with normal flow, again the disc has fallen off the hinge.
- Check valves without external lever arms can be checked only when they are in the closed position. For example, if they are used in pump discharges for parallel-pump operation, there should be no pressure on the pump side of the check valve. If there is, the disc has fallen off, is damaged, or is stuck in the open position.
- In hydraulic systems, flow can be determined by feeling if there is a temperature difference on the inlet and outlet lines. If there is a temperature difference, the check valve is preventing flow. If there should be flow, the valve is stuck in the closed position. If there shouldn't be flow and the inlet and outlet temperatures are the same, the check is stuck in the open position.

Pulsating or sudden acceleration or deceleration of fluid can cause hammering of check valves. Repeated hammering can cause internal damage. Therefore, the valve orientation in the line can be critical. If a heavy check valve is installed in the vertical position in a pump discharge, it will slam shut if the pump trips. Repeated slamming and resulting water hammer can cause serious problems. Also, discs are usually loose enough to seat properly but must be retained so they don't spin off their hinge bolt. Turbulence in the valve can cause that to happen.

Some lines may have check valves that have spring-loaded split discs hinged in the middle. The double disc, in this case, is much lighter than single-disc checks. This design allows the valve to be installed in any position and reduces slamming that can occur with large, heavy, single-disc swing checks. The springs must be sized with enough force to close the valve with no flow but be light enough to pass fluid in the direction of flow. This is obviously more important if the valves are installed in a vertical position. That makes spring condition a critical determination. The disadvantage is that their internal condition cannot be determined without removing the valve. However, if they are used as turbine-extraction nonreturn valves, with the extraction in service, and there is a temperature difference between the upstream and downstream sides, the valve is stuck in the closed position. An advantage is they are relatively light and easy to remove for inspection.

Backflow Preventers

Backflow preventers are not check valves. They are valves designed for zero backflow leakage of contaminated water, where the makeup supply comes from a potable water source—for example, makeup water for heating boilers and irrigation systems with potable water makeup or supply. (See Fig. 20-6.)

Solenoid Valves

Mainly used for controls, solenoid valves can be found in compressor control systems and process valve actuation for literally every system in a power or process plant. The basic valve is a three-port, normally open, normally closed or universal valve. The universal can be made either normally open or closed by plugging a selected port. An Asco valve flow diagram is illustrated in Fig. 20-7.

Never assume that instrument lines connected to the valve were blown clean during the instrument air system commissioning. Dirt in solenoids is one of the main causes for their malfunction; the second is improper application. The valves must be compatible with the system operating pressure or they will not function properly. That is, if the valve differential pressure is not set per the manufacturer's specification, the valves may not operate. If they are clean and don't function properly, always check the solenoid specification against the system pressure. Valve positioners operate with a specific pressure range—for example, 0 to 20 psi, matching a 0- to 20-mA signal. If the solenoid valve is rated at an operating

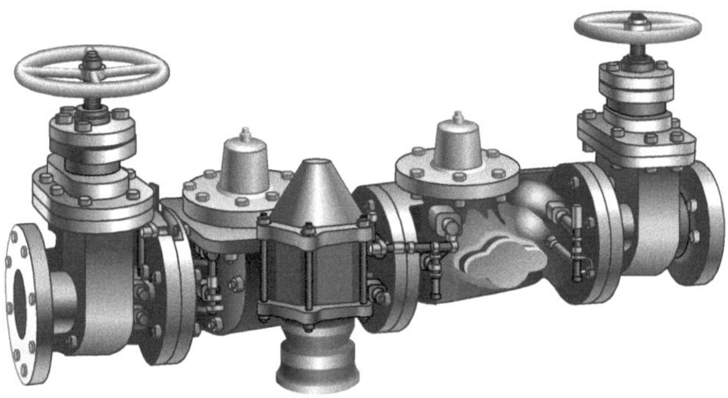

FIGURE 20-6 Backflow preventer valve.

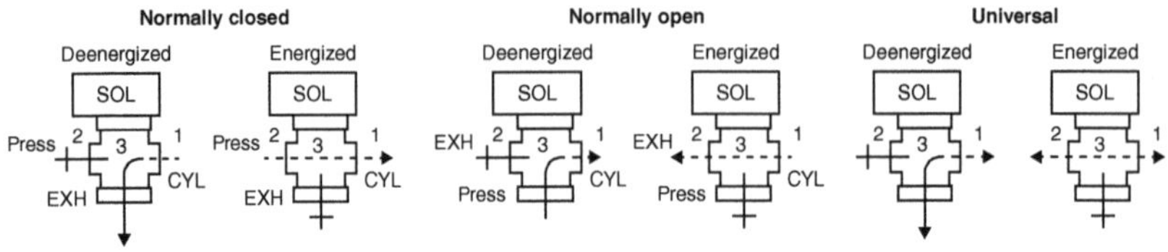

FIGURE 20-7 Asco flow valve diagram.

pressure of 100 psi and is using a supply pressure of 20 psi, it will not function. Refer to the troubleshooting guide in Table 20-1.

Relief Valves

Relief valves are designed to prevent overpressure by gradually opening proportional to overpressure. They should not be confused with safety valves. The primary function of relief valves is to prevent safety valves from relieving.

Safety Valves

Safety valves are designed to open instantly when the set pressure is reached. The following example provides a note of caution regarding stored energy in piping, boilers, and pressure vessels:

Stored Energy in Power and Process Plant Systems

Safety valves are common in process system design and are a key element for safe operation. Safety valve maintenance must be maintained in accordance with Section VI, "Recommended Rules for the Care and Operation of Heating Boilers" or Section VII, "Recommended Rules for the Care and Operation of Power Boilers." ASME Boiler and Pressure Vessel Code Sections I, III, and VIII cover power boilers, nuclear power components, and unfired

TABLE 20-1 Solenoid Troubleshooting Guide

Problem	Check
Valve unable to operate	1. Feel solenoid—if cold, check power supply; if warm, lift the coil to see if the magnetic field is active. *Do not remove coil if it is energized or it will burn out.* 2. Check coil to see if it is burned out. 3. Check solenoid voltage rating against supply voltage. 4. If coil is hot and magnetic field is active, check solenoid pressure rating against supply pressure. Solenoids must be rated for the design conditions or they will not operate properly. 5. If the solenoid rating is compatible with the design conditions, check for dirt or debris in the valve diaphragm. 6. Check the internals for a bent or damaged armature tube or diaphragm.
Valve partially open or closes	Check steps 4 through 6.

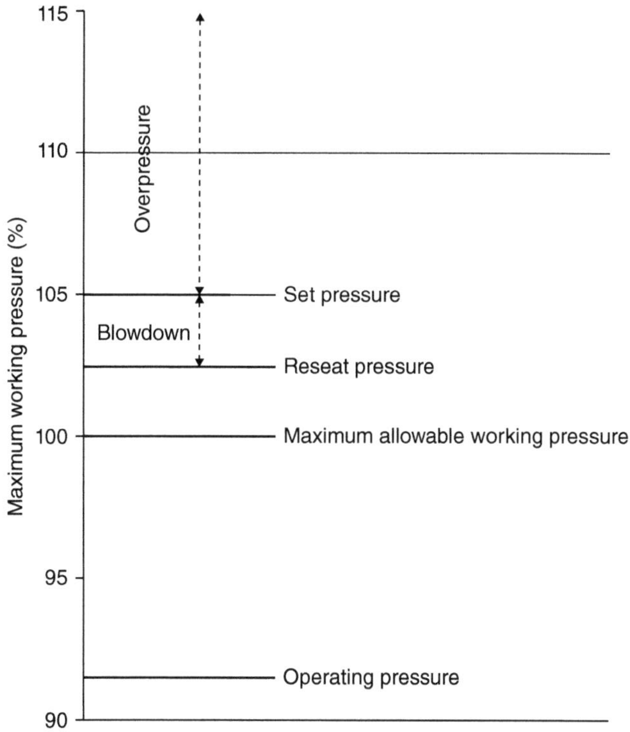

FIGURE 20-8 Pressure settings % of maximum allowable working pressure.

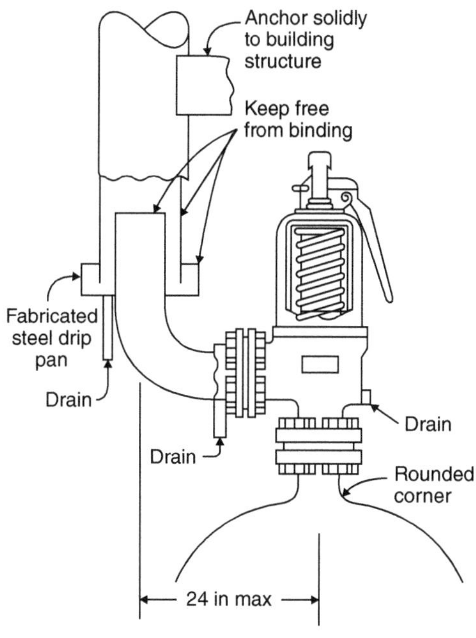

FIGURE 20-9 Typical safety valve installation.

Customer ______________ Test sheet 1586 hydro set Date ______________

Station ______________ Unit no. ______________ Service order ______________

Hydro set | Steam test

Type	Size	Serial no.	Nameplate setting	Steam pressure in vessel	Pressure differential take from diagram	Actual set pressure	Actual steam pop	No. of POPS	Blow down	Comp. steam adjust.	Adjust lower ring		Adjust upper ring		Overlap		Valve tight-ness	Remarks is valve operate
											POS.	ADJ.	POS.	ADJ.	POS.	ADJ.		

Use N – for NOTCHS, rings & overlap F – for flats. compression screw arrows for direction

Remarks ______________________________

____
Service representative

Use ledger ring to reduce simmer only
Raise↑ one notch at a time only.
This will increase blow down

Blow down adj: Upper rign
Raise↑ upper ring to shorten blow down
Lower↓ upper ring to increase blow down

Blow down adj: Overlap
Raise↑ overlap collar to increase blow down
Lower↓ overlap collar to shorten blow down

FIGURE 20-10 Hydro set test form.

pressure vessels. American Petroleum Institute (API) RP 520 part 1 covers design and part 2 covers installation for refineries.

The installation of the valves should be checked to confirm proper mounting. Adjusting ring pins should be locked and sealed. When separate piping is used for long blowpipes, it must be rigidly anchored to structural columns, and the valve must be free to move. Drains must be piped to a safe location. Drip pans must be free of debris.

Manual operation of the valve should never be performed at less than 75 percent of set pressure. *When operating a valve manually, always use a cord at least 20 ft long in order to actuate it from a safe distance. High-pressure steam and hot water can flash from the drip pan and seriously injure personnel in the vicinity.* Also, never manually operate a safety valve before steam blows or vessel and pipe cleaning are completed.

Safety valves must be gagged during hydrostatic testing. It is extremely important that they are not clamped down with excessive force. Overgagging is one of the most common causes of bent spindles, which will cause valve binding during operation. Ensure that all gags are removed after hydro and before trial operation.

Many safety and relief valves can be bench tested at facilities set up with the proper testing equipment. It is always better to bench test them instead of field test them. Check with local facilities for their valve-testing limits before field test them. Also, high-pressure valves can be "hydro set," which minimizes the blows required for full pressure "popping." Valve manufacturers have pressure diagrams for use with hydraulic setting devices. They use pressure in hydraulic cylinder versus differential between desired set pressure and steam pressure for given valve orifice numbers. These curves are used to set the valve using lower steam pressure. Using a hydraulic jack to preload the valve spring, the stem does not fully lift; the valve will whisper steam rather than popping. Once the valves are adjusted, they must be tested using full-steam pressure. An example of hydro set test form is illustrated in Fig. 20-10.

REFERENCES

1. "Control Valve Handbook," *Emerson Process Management.*
2. Hickenkamp, F. W., "Optimizing the Pressure-Seal Valve," *Power*, July 1978.

CHAPTER 21
VIBRATION

ROTATING EQUIPMENT VIBRATION

State-of-the-art analyzers make balancing rotating equipment easier than the vector-analysis methods used years ago. The primary reason for this is that the analyzers now perform the vector analysis for you. However, it would seem that nothing has been learned in all these years, or perhaps some of the art has been lost. Here are a few examples.

An old startup manager watched as three young instrument technicians from the owner's maintenance group analyzed the vibration of an ID fan. The vibration readings from the outboard motor bearing to the outboard fan bearing were 8, 6, 4, and 2 mils. The concrete pedestal was on a steep slope. The outboard motor bearing was 15 ft above grade, and the outboard fan bearing 5 ft above grade.

The following day, a young man came to the site with a new, state-of-the-art vibration analyzer. Asking what conclusion the balance expert had reached, the old startup manager was told the state-of-the-art analyzer readout was that the motor had an imbalance caused by a cracked rotor. The old manager explained that the vibration was most probably caused by inadequate distribution of mass in the pedestal, causing resonance, and not a motor imbalance. To prove his point when his comment was rejected by the group technicians, he had ironworkers tie 14-in beams, as temporary knee braces, from the spread footing to the top of the motor pedestal. With the anchors tight at the bottom, he called for the technician to take vibration readings as the top anchors were being tightened. When tightened, the vibration readings dropped in half at each bearing.

The next day, the startup manager watched as the motor was removed for delivery to a motor shop for the rotor repair. It was the same shop that had sent the balancing expert.

There were two errors committed here: (1) When the owner's technicians became involved with the vibration problem, they essentially absolved the Engineering, Procurement, and Construction (EPC) contractor of any errors or omissions. (2) They then forced the unit to reduce capacity during a worthless exercise.

In another incident, eight 4,000 hp fans were vibrating at 6 to 8 mils. All exhibited the same vibration. The problem was complicated by the fact that they were ID fans downstream of a wet scrubber. When the fans were shut down, wet deposits caused an imbalance when the deposits slid down the fan to the bottom side of the wheel. Also, rivets were showing signs of fretting corrosion, leading the investigation down false trails. This problem was analyzed and the fans balanced for a week as the unit load was being increased during unit trial operation. One observation that should have been immediately noted was that all the fans exhibited the same vibration when trial operated during cold-air testing. The vibration was amplified as the load was increased with hot gas. It was highly unlikely all the fans could have had the same unbalance. Therefore, it should have been immediately concluded that unbalance was not causing the vibration. Pedestal resonance was later determined to be the cause of the vibration. The resolution was to increase the mass of the fan pedestals, which was a very simple solution.

A steam turbine was being trial operated at a plant rehabilitation project after an overhaul. The turbine exhibited high vibration at one bearing. A vibration expert was called in to balance the turbine. He worked on the turbine balance for a week before another man from his office was called in. If he had analyzed the problem before attempting to balance the turbine, he would have found that the bearing had failed twice due to fatigue before he was called in. The vibration was caused by improper leveling of the rotor. The bearing level was too high and was preloaded to the point of overload, resulting in high vibration and ultimately bearing-fatigue failure. The bearings had been relined, but the turbine journal to bearing fit had not been checked during reassembly. If the phase angle of a trial weight keeps shifting, the cause of the vibration may be bearing misalignment.

In yet another incident, resonance vibration in fans was caused by disc wobble.

Fan balancing is usually a fairly straightforward job, but it can be made virtually impossible by a problem known as disc-wobble resonance. This resonance is caused by flexing of the central support disc (see sketch) which makes it almost impossible to locate balance weights by the usual single-plane balancing method.

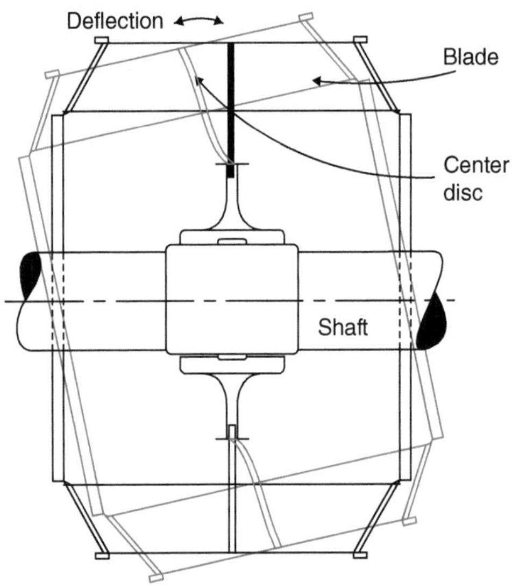

Distortion of center disc may cause fan wheel to vibrate as shown here

Bearing vibration amplitude plotted against speed during startup and shutdown tests (see curve) showed that the fan was operating near resonance. Normally, vibration caused by unbalance alone increases as the speed squared, but, in this case, it increased much faster than that, especially over 900 rpm. The dramatic 180-deg shift did not appear in the bearing-housing vibrations, indicating that the resonance was probably not a shaft critical speed.

To locate the cause of the unknown resonance, a large variable speed shaker was bolted to the concrete foundations so as to apply horizontal vibrations. Shaker speed was varied from well below, to 50% above, fan speed, and vibration data were taken from bearing housings, shaft, and foundations.

The disc resonance was identified by measurements from inside the wheel. When the shaker speed reached 930 rpm (15.5 Hz), the whole wheel vibrated at 20 mils or more, while the shaft remained relatively still. Measurements of the vibration at selected locations around the wheel showed that the center disc was wobbling, as shown in the sketch.[1]

Probability of Success

When called out to balance fans, I find when I arrive that unbalance is the problem in only 50 percent of the cases. The other causes of excessive vibration on fans are resonance, eccentric pulleys, looseness, motors, and belts or couplings. The common reaction to a fan vibration

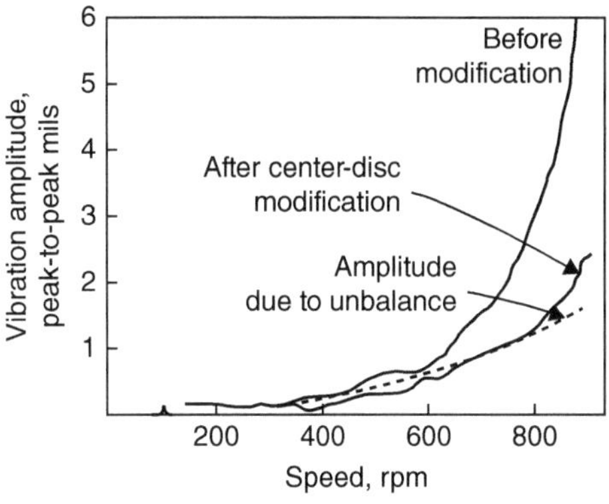

Vibration amplitude increases much more rapidly than bearing vibration

problem is to balance it. Be prepared to do a little analysis when arriving on the scene to at least verify unbalance before starting into the procedure. If you discover that unbalance is not the problem after 4 h of balancing with little improvement, you will be embarrassed and the owner will doubt your credibility.[2]

Remember shaft unbalance vibration is always at a 1 × rpm (frequency). Also, shaft vibration can be generated by blade passing. The frequency of blade pass vibration is *N* (Number of blades) × rpm.

AERODERIVATIVE VIBRATION

One comment about gossip and folklore in the power industry regarding aircraft-derivative engines is noteworthy. It is a totally false statement that aircraft engines do not have vibration trips because aeronautical engineers would rather the engine fail and destroy itself than have the engine trip and lose thrust. If you have an aircraft-derivative engine and it has vibration problems, you should first be looking at your instrumentation or resonance-generating sources before sending it to a repair shop for balancing. Aircraft engines generally do not vibrate because they are lightweight and easily balanced in the factory. They also have precision, machined rotating elements as well as the highest-quality antifriction bearings.

> *Lightweight gas turbines, such as aircraft engines and their derivatives have a [sic] stator/rotor mass ratio in the realm of 1 to 3, and, in general, anti-friction bearings are used. With very small bearing clearances and light weight casings, these machines are characterized by much higher stator responses than Class I and II type machinery. It is unlikely that shaft relative motion will be very meaningful for this class of machinery. In addition, stator vibration limits established for the other classes of equipment do not, and cannot, be applied.*[3]

Note

Only 50 percent of all vibration problems are unbalance related.

BALANCING BASICS

Balancing basics referenced here concern centrifugal fans. Turbines and other equipment that require more complex analysis are not mentioned because they are normally balanced by experts. Most fans can be balanced using single-plane balancing techniques and, more specifically, the "one-run" technique. This will require a trial run with a trial weight. The angles are determined by strobe light.

One-Run Balancing, Trial Weight

The first step in one-run balancing is to determine the weight of the trial balance shot.

> A common criteria in popular use is to select a trial weight that produces a *force* of 10 percent of the rotor weight. Using this criteria, the additional dynamic load due to the trial weight will not exceed 10 percent of the static load ... The equation below uses the 10 percent criteria to select a safe trial weight.

$$W_t = 56{,}375\ W_R/N^2{}_r$$

where

W_t = trial weight, oz
W_R = static weight of rotor, lb_f
N^2 = speed of rotor, rpm
r = radius at which the trial weight will be attached, in

> The trial weight is placed in the radius location that the balance correction weight will be attached. The location around the circumference of the wheel is a guess at best. If the trial weight is installed to close to the unbalance the vibration will increase; if it is 180° out, the phase angle will flip. What is most desired in the trial run is to get a phase change of 90°. This will produce widely separated vectors and will result in the most accurate calculations. It is not uncommon for experienced balancers to stop the rotor and move the trial weight so as to achieve a phase change closer to a 90° separation.[2]

Modern analyzers make the calculations more accurately than you can. This example is an illustration of what the microcomputer is doing for you.

The example problem is as follows:

Vector original = 5 mils at 180°

Vector trial run = 2 mils at 270°

If we now attached a trial weight of 10 g at 90° and the vector flipped to 270°, this means we need to calculate a new weight, but we'll leave it in the same position. From the vector diagram in Fig. 21-1, and an online trig calculator, $T = 5.4$. The trial weight will be changed as follows:

$$\text{(Correction weight) } CW = \text{(Trial Weight) } TW \times O/T$$
$$CW = 10 \times 5/5.4 = 9.3 \text{ g, a reduction}$$

What if a trial weight of 10 g were attached at 60°? The correction weight must then be moved. The phase shift was clockwise, so the weight must be moved counter-clockwise 22° (angle OT) from the trial run position $O + T$. Remember, if the weight is to be welded in place, the filler metal (welding rod) weight must be part of the correction weight.

These, of course, are fictitious examples to illustrate the vector analysis. Modern analyzers will make these calculations for you.

If you are not an experienced balancer, but are required to balance a fan, it is highly recommended that you practice with your analyzer beforehand. That is, if you are in an inhospitable place where a balancing expert won't go, acquire a fractional horse motor of the same speed as the fan; couple a pulley or cart wheel to the motor and use it for a lab vibration experiment. Use a shaft with bearing supports on both sides of the wheel and properly align it with the motor. If the pulley or wheel is balanced, deliberately unbalance it by adding weld metal of known weight, for example $\frac{1}{4}$ welding rod, to one spot on the circumference. Use the 10 percent formula for determining

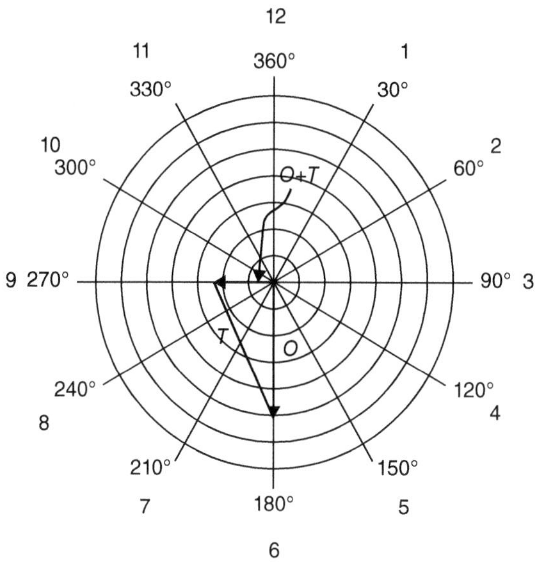

FIGURE 21-1 Vector diagram one-run balancing.

your safe, trial-run weight. Using the analyzer instruction manual, practice balancing the wheel to get a feel for the instrument sensitivity and accuracy. It is not a bad idea to purchase a balancing textbook to learn the theories behind balancing. You know where the imbalance is, so you will learn very quickly what the vector analysis is all about as you follow examples in textbook.

TUBE VIBRATION

Boiler and condenser tube can be damaged by vibration caused by what is known as a Karman vortex street.

Wake behind a Cylinder: The wake behind a cylinder is of particular interest and has been studied experimentally and theoretically at great length. Initially, a pair of vortices form in the separated region, as shown in the figure. These vortices act like "aerodynamic rollers" over which the main stream flows. One of the vortices will always break away first and be washed downstream in the wake, whereupon another vortex will begin growing in its place. Later, the second mature vortex will break away, and so a process is started whereby vortices break away alternately from the cylinder and move downstream, as shown diagrammatically below.[4]

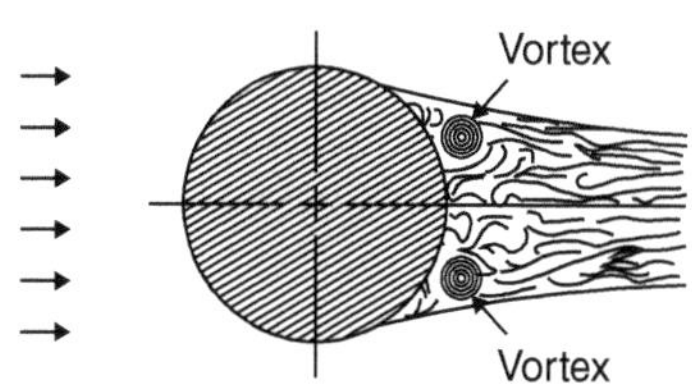

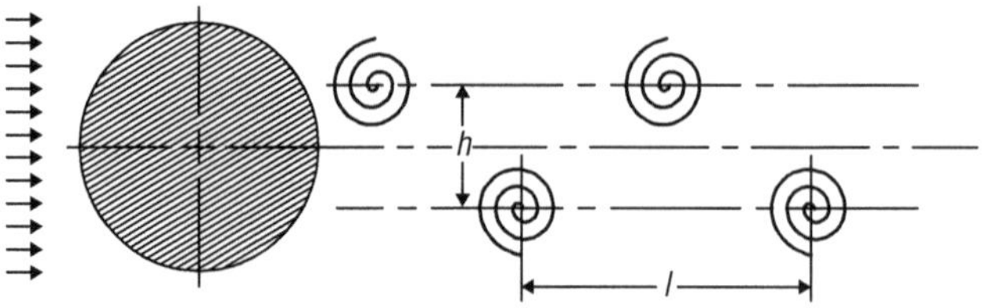

FIGURE 21-2 Height versus length for vortex shedding.

The vortex shedding will cause the tube bundles to vibrate. In boilers and HRSGs, this will happen in the laminar-flow gas paths. The vibration will cause a low-frequency humming noise that can be heard, at intervals, miles from the plant. Note that the noise a mile from the plant may be of the same intensity as at the back of the boiler.

Condensers are under vacuum, and thus emit no noise and so the presence of tube vibration must be checked after the first 72 h full-load run and turbine fine-mesh screen run.

Two things must be present for the phenomena to occur: (1) tube symmetry and (2) laminar flow. A random tube pattern will cause turbulence, which will break up the "street." If tube symmetry exists, because it is not feasible to change the tube bundles, baffles are installed to create turbulence. This is done in boilers, HRSGs, and condensers. The noise is often ignored at first because it is not loud enough to be objectionable. However, this situation cannot be ignored because the vibration will eventually cause tube failures.

When inspecting condenser tubes for vibration, look for shiny spots on the tubes midway between support sheets directly under the turbine exhaust. The shiny spots are the points where the tubes rubbed together as they vibrated. The shiny spots will eventually be the points of failure for the tubes. Baffle plates must be installed to create turbulence and break up the vortex street.

The stability of the pattern is very delicate and can be broken very easily. For example:

> Von Kàrmàn was able to show that, for stability of the pattern, the configuration must have a geometry such that
>
> $$h/l = 0.281$$
>
> where h and l are dimensions shown in Fig. 21-2.[4]

FURNACE VIBRATION

Furnace vibration can be very violent to the point that structural damage can occur if it continues. This can be a dangerous situation. At one station, a gas-fired furnace started to vibrate violently when the last burner was placed in service. When the startup engineer changed the position of one air register, the vibration stopped immediately. This indicated the burner system had been in resonance with the natural frequency of the furnace. The following are discussions of such phenomena.

Excitation by the Burner[5]

Excitation by the burners is usually due to the instability of the combustion gas swirl as emitted from the burner. Chen reports on a classic example (Reference 6): Flue gas vibrations were observed in a forced draft boiler having three oil burners at 90% load with all three burners in operation.

The pressure amplitudes were measured under 1 and 2-burner operation at about ±80 mm H_2O, however under 3-burner operation they increased to ±920 mm H_2O (figure below). The frequency of this vibration was measured at 100 Hz. This was equivalent to half a standing wave. The excitation was blamed on the swirl flow emitted from the burner. Swirl flow is very sensitive to any disturbance and numerous tangential wave movements are induced as shown below.

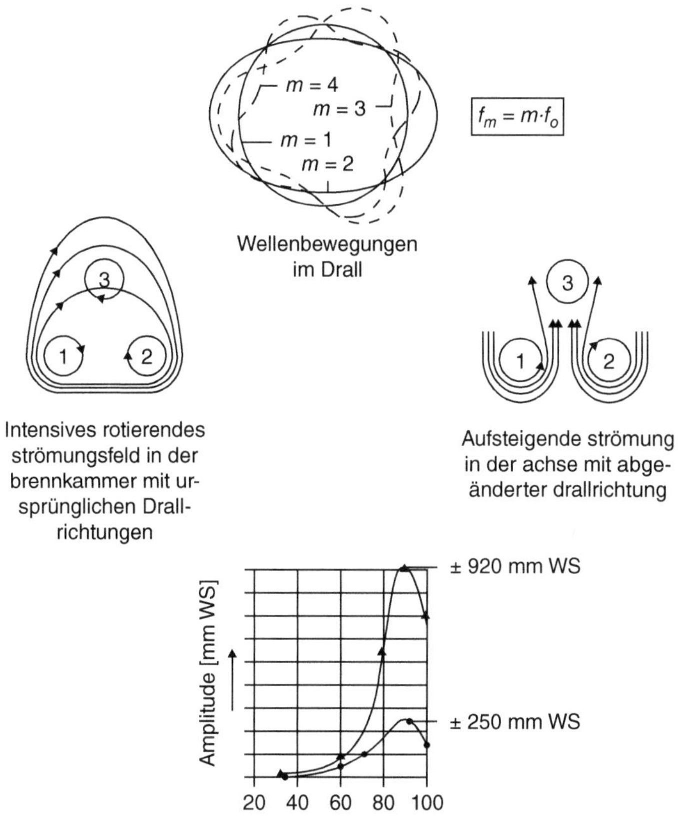

The wave frequency results from the m-fold swirl frequency:

$$f_m = m \cdot f_o$$

where f_m = wave frequency and f_o = swirl frequency. If one studies the swirl flow emitted from the lower burners it becomes obvious that the upper burner is subjected to a very unequal flow pattern. This causes tangential wave movement.

A wave whose frequency is equal to the natural frequency of the flue gas column (here the 1/2-fold standing wave) will create resonance. Feedback coupling will increase this wave and a strong self-excited gas vibration will occur.

The remedy here was a change in the direction of the swirl of burner 1, improving the flow over burner 3. The flow over burner 3 was now almost parallel, so that the disturbance of the swirl emitted from burner 3 was negligible. After this change an amplitude of ±250 mm H_2O resulted or an improvement of factor of 3.7.

A similar condition was observed in a gas fired boiler with six burners (boiler capacity 180 t/hr). During startup a standing wave with a frequency of 24 Hz occurred at 88 t/hr load, figure below.

The pressure amplitude was $\Delta p = \pm 40$ mm H_2O. The wave was a 1/2-fold standing wave in the longitudinal direction of the boiler, where its frequency could be calculated from the dimensions of the furnace...

For a vibration which originates in a burner, one of the following changes would affect a prevention of resonance:

1. *Impeller dimension (O.D., impeller blades, flow area, etc.)*
2. *Position of impeller in burner throat,*
3. *Arrangement of nozzle opens,*
4. *Position of swirl vanes,*
5. *Proportioning of primary and secondary air.*

The success from the utilization of the above indicated remedies cannot be calculated but must be determined by experiments. Often a great number of tests are required before the desired result is achieved.[5]

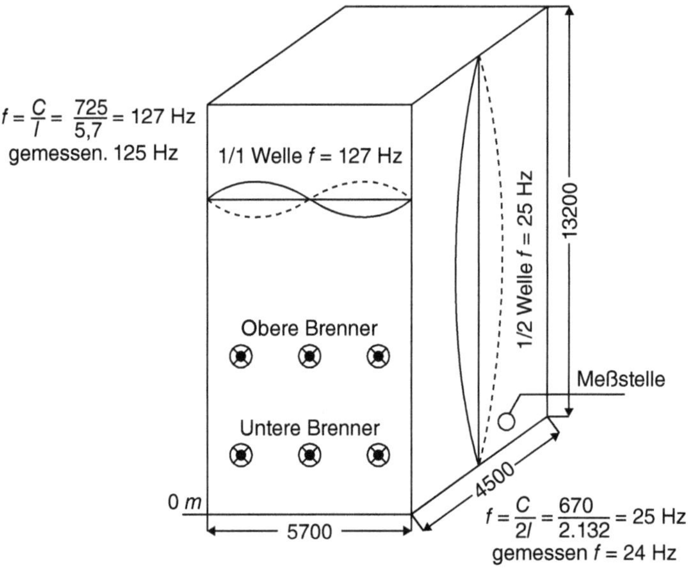

DUCT VIBRATION

Excessive duct vibration can be analyzed and elaborate engineered solutions can be implemented to reduce the turbulence in the duct that is exciting the vibration. However, from a practical point of view, the quickest and cheapest way to remedy the problem is to stiffen the ductwork. That is, add more stiffeners. This does not require an engineering solution. Install stiffeners until the vibration is eliminated. A rule of thumb is to double the number of original stiffeners with stiffeners of the same size.

FAN NOISE

Engineers can choose low fan speeds in order to reduce noise, or they can choose high fan speeds to reduce costs. This is usually a problem with EPC contractors. They will choose speeds to reduce costs and the fans will scream. The best and most cost-effective remedy, to at least mitigate the problem, is insulation. Do not place insulation directly on the fan or ductwork using the stiffeners as framework.

Have a sheet metal shop build a lightweight secondary sheet metal frame using the fan and duct stiffeners as an attachment point. Make sure the frame minimizes contact with the ductwork. Build the frame for 6 in of hardboard insulation and cover with sheet metal lagging. The trick is to provide an air space between the insulation and duct while minimizing metal frame contact that will act as a transmitter of the noise. Most people use the stiffeners as the framework for insulation. That method does not work.

PIPING SWING

Piping swing can occur if the pressure pulsations emitted by a pump are in resonance with the reflective pulsations traveling back to the pump.

> *Hydraulic resonance in a piping system is produced when reflected pulses from a piping discontinuity travel back to and arrive at the pump in time to join in phase with the next pump pulse. The resultant larger pulse travels down the pipe and is reflected back to the pump again to be increased by the addition of the next pulse, and so on. The final amplitude of this reinforcing process is limited by the system dynamic friction losses. Pulsation levels can be increased 50 times or more due to such resonance phenomena. A normally operating pump may produce a pulsation level of 5 psi and be considered a satisfactory installation. However, pipe resonance could increase this value to 250 psi peak-to-peak or more at the impeller vane frequency.*[7]

If swing occurs in the piping system, the easiest and quickest way to correct it is to change the flow. In a condensate system at one plant, the system started swaying at the system minimum flow. The resolution was to simply increase the minimum flow setpoint.

However, if this occurs at high flows there will be problems. The first step to resolve the problem is to determine what is the reflection point; then determine if it can be changed.

COMPRESSOR BLADE-PASS VIBRATION

Blade-pass vibration frequency is $N \times$ rpm; where N = number of blades. However, the energy of this high-frequency vibration can generate a whole family of lower-frequency vibrations. For example, the impulse coupling of impeller pulsations in pumps can cause low-frequency piping vibrations. This type vibration can cause pipe swaying or can be extremely violent, causing serious structural damage. This can also happen in centrifugal compressors. In one particular case, a third-stage bypass to the first-stage inlet in a booster air compressor caused high-frequency vibrations that caused high pipe vibrations and internal failures to guide vanes and a flow restriction orifice.

In the aforementioned case of blade-pass vibration problems started when a temporary startup strainer was removed because it was found damaged. The fine-mesh screen had broken loose and was wadded up in the bottom of the strainer. Also, the perforated backing plate was polished thin where the fine screen had been vibrating against it. With the screen installed, machine vibration had been normal. Because the startup strainer was temporary, it was removed and the compressor was returned to service. After removing the screen, the machine vibration went from 1 mil to 4 mils, and severe piping vibration occurred. Because the first-stage inlet elbow did not have flow straightening vanes, turbulence was thought to be causing the machine and piping system vibration. Therefore, the elbow was replaced with one with vanes and the strainer was removed and replaced with a flow orifice, upstream of the surge control valve, to recreate the screen backpressure.

The frequency of the vibration was high—much too high to have been generated by low-frequency turbulence. That is, the lack of the vanes would have caused low-frequency vibration; therefore the two vibrations, blade pass and turbulence, were not related. Nevertheless, when the compressor was returned to service, the machine vibration returned to normal; however, there was a high-frequency, low-amplitude vibration of the piping system. The first fix also included the installation of rigid pipe supports. Within 3 days, the orifice failed and a large piece was ingested by the first stage, severely damaging the impeller. The orifice was bowed toward the direction of flow and showed signs of embrittlement. The vanes in the elbow also had severe cracking damage and would have failed eventually.

It was learned that the piping system for this compressor had been modified by the manufacturer. The bypass piping on predecessor units had the connection much further upstream on the first-state suction line than these units. The temporary fix was to install a thick, perforated plate at the discharge of the surge control regulator. The final fix was to return the piping to its original design and replace the control valve with one with noise-reducing trim. The lesson learned here was that resonance is as fickle as fate and is almost impossible to design into a system; but it invariably surfaces with new designs.

Remember, one high-frequency machine pulsation can excite a vibration of lower frequency, such as an acoustical vibration, which can in turn can generate a low-frequency system vibration to form a family of vibrations. If any of them finds a resonance system frequency, there will be a problem.

TWICE-LINE-FREQUENCY VIBRATION

A power supply produces an electromagnetic attracting force between the stator and the rotor which is at a maximum when the magnetizing current flowing in the stator is at a maximum either positive or negative at that instant in time. As a result there will be 2 peak forces during each cycle of the voltage or current wave reducing to zero at the point in time when the current fundamental flux wave pass through zero. This will result in a frequency of vibration equal to 2 times the frequency of the power source. This particular vibration is extremely sensitive to the motor's foot flatness, frame and base stiffness and how consistent the air gap is between the stator and rotor, around the stator bore.

As load is applied to the motor, both stator and rotor currents increase together and balance one another, therefore, there are no significant changes in flux. This means that the basic magnetic forces are independent of load current and are nearly the same at no load or full load. Therefore, the main component of twice-line-frequency vibration which is created by an unbalanced magnetic pull due to air gap dissymmetry and does not change with load.[8]

FURTHER STUDY

Vibration should be carefully monitored during turbine initial roll and overspeed testing. Do not roll a turbine and attempt to push it through high-vibration point, even at critical speeds. Rubs, lack of lube oil, and blade failure can all be causes for the high vibration. Don't make an assumption that it is a critical-speed phenomena that will go away once passed through.

Similar failures have been attributed to motoring of generators. In fact, the motoring of a synchronized generator is of no consequence to a generator, but the motorizing and windmilling of the turbine can be disastrous. Blades can fail due to turbulent vibration and overheating in less than a minute of no-load, no-steam-flow operation. The result can be a locked rotor that can cause a turbine-generator failure similar to the one depicted in Fig. 21-4.

Turbine back pressure must be controlled within the turbine manufacturer's guidelines. If the turbine is operated below the design backpressure, damage to the LP blades and condenser tubes is possible. The low pressure causes higher-than-normal velocities that can induce vortex shedding in LP blades and condenser tubes. The shedding causes the blades and tubes to vibrate, fatigue, and fail. See Fig. 21-3 example of votex shedding on turbine blades. This failure can also occur at higher-than-design backpressures. The LP blades can flutter, fatigue, and fail also at high backpressures as reported after extensive laboratory and field testing.[10]

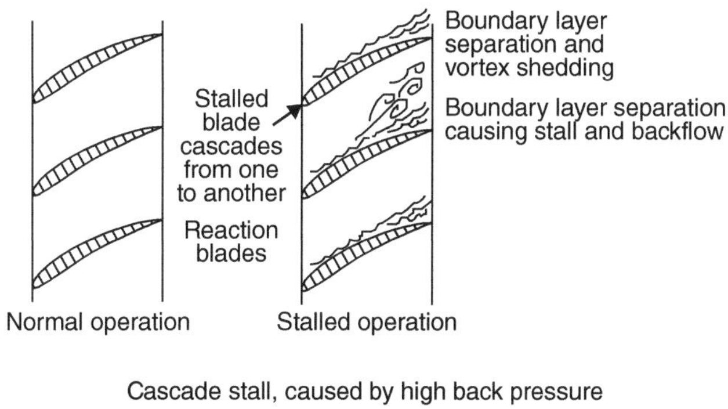

FIGURE 21-3 Blade stall example.

FIGURE 21-4 Turbine locked rotator damage.[9]

Note

In the illustration in Fig. 21-2, the blade above the one with separation is the stalled blade because flow is blocked on the under, or thrust, side. This is caused by boundary layer separation and vortex shedding. The separation causes backflow and stall.

Blade stalling can alternate or cascade from one blade to another, which will set up a cascade vibration that can cause blades to fail. Higher than design or no flow (motored turbine) can set up such vibration.

VIBRATION SEVERITY CHART

The vibration severity chart in Fig. 21-5 is from ASME Publication 67-PEM-14. It is no longer in print but appears in many handbooks and articles. It should be used with caution and only as a general reference when the manufacturer's data are not available. This applies generally to small rotating equipment. Suppliers of large rotating machinery always provide vibration limits.

VIBRATION TERMS[a]

Aliasing—causes high-frequency signals to appear at low frequencies.

Asymmetrical support—rotor support system that does not provide uniform restrain in all radial directions.

Asynchronous vibration—vibration components that are not related to rotating speed, such as aerodynamic related vibration.

Band-pass filter—a filter with a single transmission band extending from lower- to upper-cutoff frequencies. The width of the band is determined by the separation of frequencies at which amplitude is attenuated by 3 dB (0.707).

Bandwidth—the spacing between frequencies at which a band-pass filter attenuates the signal by 3 dB. In a dynamic signal analyzer, measurement bandwidth is equal to [(Frequency span)/(Number of filters) × (Window factor)]. Window factors are 1 for uniform, 1.5 for Hanning, and 3.63 for flat top.

Blade-passing frequency—a potential vibration frequency on any bladed machine (turbine, axial compressor, fan, etc.). It is represented by the number of blades times shaft rotating frequency (rpm).

Bode—rectangular coordinate plat of 1× component amplitude and phase (relative to a keyphasor) versus running speed.

BPFO, BPFI—common abbreviations for ball-pass frequency of defects on outer and inner bearing races, respectively.

Campbell diagram—also known as an interference diagram. A mathematically constructed diagram used to check for coincidence of vibration sources (i.e., 1 × imblance,

[a]Visit www.technicalassociates.net/glossary.html for a complete glossary of terms.

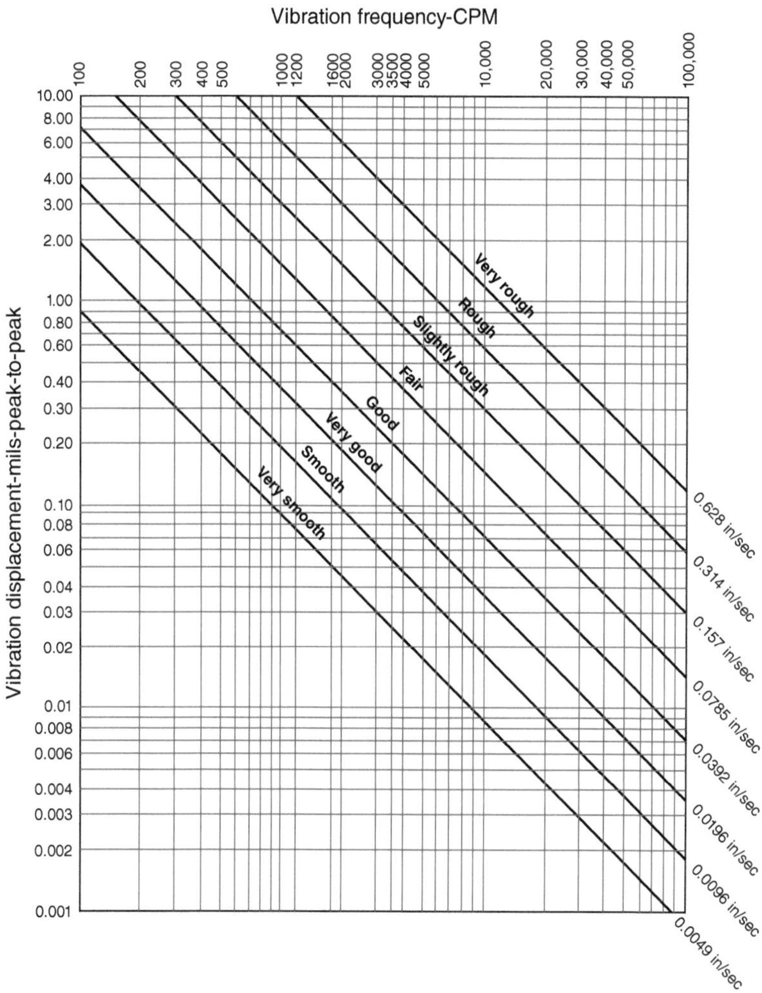

FIGURE 21-5 Vibration severity chart.

2 × misalignment) with rotor natural resonances. The form of the diagram is a rectangular plot of resonant frequency (y-axis) versus excitation frequency (x-axis).

Critical speed—in general, any rotating speed that is associated with high-vibration amplitude, often, the rotor speeds that correspond to natural frequencies of the system.

Damping—the quality of a mechanical system that restrains the amplitude of motion with each successive cycle. Damping of shaft motion is provided by oil in bearings, seals, and so on. The damping process converts mechanical energy to other forms, usually heat.

Damping critical—the smallest amount of damping required to return the system to its equilibrium position without oscilation.

Decibel (dB)—a logarithmic representation of amplitude ratio, defined as 20 times the base 10 logarithm of the ratio of the measured amplitude to a reference, dBv readings, for example, are referenced to 1-V rms. Decibel amplitude scales are required to display the full dynamic range of a dynamic signal analyzer.

Displacement transducer—a transducer whose output is proportional to the distance between it and the measured object, usually the shaft.

Dynamic signal analyzer (DSA)—vibration analyzer that uses digital signal processing and the fast fourier transform to display vibration frequency components.

Eccentricity, mechanical—the variation of the outer diameter of a shaft surface when referenced to the true geometric centerline of the shaft (shaft bow).

Eddy current—electrical current that is generated and dissipated in a conductive material in the presence of an electromagnetic field.

Filter—electronic circuitry designed to pass or reject a specific frequency band.

Flat-top filter—dynamic signal analyzer window function that provides the best amplitude accuracy for measuring discrete-frequency components.

Forced vibration—the oscillation of a system under the action of a forcing function. Typically, forced vibration occurs at the frequency of the exciting force.

Gear-mesh frequency—a potential vibration frequency on any machine that contains gears; equal to the number of teeth multiplied by the rotational frequency of the gear.

Hanning window—dynamic signal analyzer window that provides better frequency resolution than the flat-top window, but with reduced amplitude accuracy.

Harmonic—frequency component at a frequency that is an integer multiple of the fundamental frequency.

Hysteresis—also called *deadband*, or that portion of a system's response where a change in input does not produce a change in output.

Micron (μm)—one millionth (0.000001) of a meter (1 micro = 1×10^{-6} = 0.04 mil).

Mil—one thousandth (0.001) of an inch (1 mil = 25.4 μm).

Oil whip—occurs at the oil-flow velocity within a plane bearing, usually at 40 to 49 percent shaft speed. Oil whip occurs when the whirl frequency coincides with (and becomes locked to) a shaft-resonant frequency. It usually occurs when the oil is colder than the design startup temperature.

Period—the time required for a complete oscillation or for a single cycle of events, the reciprocal of *frequency*.

Preload—the dimensionless quantity that is typically expressed as a number from 0 to 1, where a preload of 0 indicates no bearing load upon a shaft, and 1 indicates the maximum preload (i.e., line contact between shaft and bearing).

Preload external—any of several mechanisms that can externally load a bearing. This includes soft preloads such as process fluids or gravitational forces as well as hard preloads from gear contact forces, misalignment, rubs, and so on.

Spectral map—a three-dimensional plot of the vibration amplitude spectrum versus another variable usually time or rpm.

Spectrum analyzer—an instrument that displays the frequency spectrum of an input signal.

REFERENCES

1. Reason, J., "Fans," *Power*, Special report, September 1983.
2. Wowk, V., *Machine Vibration Balancing*, McGraw-Hill, New York, 1995.
3. Ludwig, G. A., and Erdmann, O. D., "Gas Turbine Vibration Limits—A Fundamental View," ASME Publication 73-GT-48, American Society of Mechanical Engineers, 1973.
4. Shames, I. H., *Mechanics of Fluids*, McGraw-Hill, New York, 1962.
5. Berg, M., "Furnace Vibration: Causes and Remedies," *Combustion*, December 1973.
6. Chen, Y. N., "Causes and Prevention of Flue Gas Vibrations in Boilers Due to Instability of Combustion Gas Swirl and Kàrmàn Vortex Streets," *Mitteilungen der VGB51*, Issue 2, April 1971.
7. Shuey, S. J., and Templin, K. W., "Vibration and Noise in Piping Systems," *Power Engineering*, June 1973.
8. "An Analytical Approach to Solving Motor Vibration Problems," IEEE PCIC-99-20, Institute of Electrical and Electronics Engineers, 1999.
9. Adams, M. L., "Protect Against Large Rotor Unbalance," *Power*, July 1981.
10. Silvestri, G. J., "Progress in the Development of Reliable and Economical Steam Turbines for Dry Cooling Application," *Combustion*, March 1981.
11. Baxter, R. E., and Bernard, L. O., ASME Publication 67-PEMA 14, 1967.

CHAPTER 22
WATER TREATMENT

This chapter is water treatment made simple. It attempts to make simplified explanations for startup professionals who will utilize the services of a water treatment engineer or technician. A good source for this type of discussion can be found in Ref.1. Also, *Power* magazine has special reports on water treatment that are excellent for the layperson as well as professionals.

Basically, there are four important steps in water treatment: clarification, filtration, demineralization, and chemical treatment. All four are almost always required for treatment of surface raw-water sources such as lakes and rivers. Groundwater sources, such as wellwater, are the only sources that "in general" will not require a clarification step. Each step treats the water to the level required by the process in which it will be used. For example, for potable water production, only clarification, filtration, and chemical treatment are required. For industrial use, the degree and type of ion exchange—or demineralization, in general—is dictated by the purity requirements of the process. The process pressure and temperature requirements of boilers and the cyclic concentrations of impurities, which affect cooling towers, determine the process requirements.

In all the steps involved in water treatment, particle charge is a key factor. This is true in the case of clarification where the attraction or repulsion of particles is of principal importance or in demineralization, where ion exchange is the key with negatively charged anions and positively charged cations. Cathodes and anodes are also proponents in corrosion.

CLARIFICATION

Clarification involves coagulation, flocculation (coagulants and coagulant aids are known as flocculants), and settlement. Alum is a common coagulant, and polymers are common flocculants—both of which are used to aid settlement.

Coagulation and Flocculation

> Water treatment parishioners once used the word *coagulation* to refer to an entire process that included addition of chemical coagulants, dispersion, flocculation, and settling. Today, the chemist's definition of coagulation is used: it is the driving together of colloidal particles by chemical force. The process is speedy and occurs within seconds of the application of the coagulating reagent to the water. Because of this, intense mixing is necessary at the point of chemical application in order to insure uniform chemical distribution and exposure of the fine particles in the water to the coagulating agent before the coagulation reaction is completed. This is the work of rapid mixing.

The term *flocculation* refers to the assembling of coagulated particles into floc particles. Flocculation may be partly a chemical bridging mechanism, enhanced by the use of substances like poly-electrolytes, but it is much slower, and more dependent on time and amount of agitation than is coagulation.[2]

Cloudy or murky water, categorized by turbidity, is caused by fine, suspended particles that do not have sufficient weight to settle out and, therefore, remain suspended. Observe a swimming pool with cloudy water. The finely divided particles have become suspended. These particles are so fine that they will not settle out while the pool is being agitated by normal use. Adding sodium hypochlorite (for chlorination) to high alkaline water generally causes milky swimming pool water. Sodium hypochlorite raises the pH; thus dissolved salts, such as calcium carbonate, begin to precipitate, causing the milkiness. The particles are so fine they cannot be filtered. Particles of bacteria and viruses also contribute to water turbidity. These fine particles are practically impossible to remove by filtration. Both temperature and pH affect particle precipitation. For swimming pools, bacteria and virus growth are prevented by chlorination and particle precipitation by proper pH control.

Experience has demonstrated that few bacteria and soil particles and almost no viruses are removed by passing unflocculated water through rapid sand filters.[3] Table 22-1 illustrates typical particle sizes.

TABLE 22-1 Typical Particle Sizes

Material	Particle Diameter (Microns)
Sand	500
Soil	1–100
Bacteria	0.3–3.0
Viruses	0.005–0.01
Floc particle	100–2000

This phenomenon, flocculation, can be explained by the fact that each particle has an electrical surface charge.

The charge attracts a compact layer of oppositely charged ions (counter-ions) in very close proximity, and a more diffuse layer further away. The combination of the two layers forms an electrostatic potential—called the zeta potential—around the particle, which repels other particles similarly charged.[2,4]

Chemicals called coagulants are added to the water to break down this barrier. They introduce ions of opposite charge—positive in this instance—which act to compress this double electrical layer, thus reducing the barrier to agglomeration. Actually a second force exists—the van der Waals force—which is attractive in nature; it is the algebraic sum of these two forces—the net repulsive force—that must be overcome to coalesce individual particles. This can be considered as neutralization of particles to allow attachment to each other.[2]

Coagulation and flocculation are greatly influenced by physical and chemical forces such as electrical charges on particles, exchange capacity, particle size and concentration, pH, water temperature, and electrolyte concentration.[3]

Startup engineers must be aware that seasonal changes that alter water temperature, particle size, concentration, and pH can alter a clarification process so greatly that the design limitations of the equipment can be exceeded.

Note

Coagulation will not occur unless the pH is maintained within the narrow range for that particular coagulant, no matter how much reagent is added to the water. For example, the aluminum sulfate effective pH range is 6.5–8.0 with an optimum range of 6.8–7.0. The effective pH range of iron coagulants, such as ferric chloride, is 4.0–11.0 with an optimum of 5.0–8.0. The reason that coagulation does not occur outside these ranges is that the coagulants are soluble outside these ranges. Also, be aware that the reaction of the coagulant lowers the pH so it may be necessary to raise the pH with another chemical such as soda ash.

Note

"Alum" is actually "aluminum potassium sulfate" but is often incorrectly used as the term for its constituent, aluminum sulfate, which is known as "filter alum" and many other trade names. Both are used in the water treatment industry as coagulants.

Settlement

As previously mentioned in the swimming pool example, undissolved solids in water create turbidity and color. The solids include inorganic material such as sand and organic solids such as bacteria. Settlement should remove all visible color and turbidity from the raw water. Settlement can be accomplished in steps such as plain sedimentation or sedimentation accelerated by coagulation and flocculation. The remaining solids that are not visible to the naked eye can be removed by filtration if they are coagulated and flocculated.

All particles will eventually settle in undisturbed water, but possibly not within a practical time frame for water treatment. For example, viruses may take decades to settle and bacteria years, while sand may take only seconds. The velocity of settlement can be calculated using theoretical assumptions. However, because the consistency of particles is never uniform, using such calculations is at best an estimate. That is, there is no such thing as the theoretical discrete particle.

> *Newton's law leads to the conclusion that when the vertical forces acting on a particle in suspension in water remain balanced, there is no vertical acceleration and the vertical settling velocity of the particle remains constant. Rarely, however, do the vertical forces remain balanced. The balance of forces is disrupted by such phenomena as: (1) flocculation of particles, (2) density, wind, and thermal currents, (3) short circuiting, and (4) the irregular shape of particles.*
>
> *A discreet particle is one that does not change its size, shape, or weight as it settles. The vertical forces acting on such a particle in water can be resolved into two forces: a downward impelling force WW, equal to the weight of the particle in water, and an upward resisting or drag force FD, which is the resultant of the other vertical forces....*
>
> *In the case of flocculating particles and precipitating particles, both the density and volume of particles change as they become attached to each other through the complex mechanisms of flocculation and chemical precipitation. Consequently W_W, and F_D, both change, and the balance of vertical forces is upset.*[3]

As a consequence, the design of water treatment plants depends greatly on the designer experience and sometimes personal preference. For example, square versus circular rapid mixers would be a personal preference, and the detailed design would be based on personal experience. The type and quantity of chemicals used for coagulation and flocculation must be field tested using raw water samples representative of the actual raw water, including pH and temperature.

The settling velocity of a particle varies inversely with kinematic viscosity, which is related to water temperature as shown in Table 22-2. The ratio of the kinematic viscosity at 50°F to the kinematic viscosity at any other temperature will indicate the effect of the corresponding change of the water temperature on the settling velocity of a particle. For example, changing the water temperature from 50°F (Table 22-2), to 86°F would increase the settling velocity of a particle by 1.310/0.804, or 1.63 times. Similarly, reducing the temperature from 50°F to 32°F would reduce the settling velocity to 1.310/1.792, or 0.73 times its former value. Thus, it is easy to see that water temperature has an important effect on sedimentation tank design. With cold waters, tank overflow rates should be lower than with warmer waters. Water temperature has been found to be a very important design criteria and, in fact, may be ever more important than specific gravity of the particle.[3]

TABLE 22-2 Relationship of Kinematic Viscosity to Water Temperature[3]

Temperatures		Kinematic Viscosity, v $(10)^{-5} ft^2/sec$
°C	°F	
0	32	1.792
5	41	1.519
10	50	1.310
15	59	1.146
20	68	1.011
30	86	0.804

A Mason jar is a jar commonly used to collect samples and should not be confused with "jar testing" (see Fig. 22-1). Jar testing is a dynamic settlement test, not a static one, used to determine the optimum chemical additive used for coagulation and flocculation. It is an extremely important test that is carried out in a laboratory to reduce costly trial-and-error operation in the field.

In the jar tests (Fig. 22-1), six, 1-l samples are tested simultaneously. The samples are stirred to simulate field conditions in the rapid mixer and the gentle stirring in the clarifier. The rapid mix is generally 1 to 2 min with gentle stirring 10 to 20 min afterward, depending on the rapid-mixer design and clarifier-mixer speed setting. Start with a factory setting for one complete revolution in the clarifier for the gentle-mix phase. If that doesn't work well in the lab, it probably won't work well in the field. Therefore, the clarifier speed must be changed. Test the different chemicals to be used first. Then, tests should be run to determine the optimum quantity of chemicals to be added. If a jar clears halfway down within 2 min, on the gentle stir cycle, it is a good indication of the correct proportion. Don't forget how important temperature and pH are for clarification.

FIGURE 22-1 Typical jar tester.

Note

A cationic polymer is normally sufficient by itself for clarifying water with turbidity exceeding 60 JTU (Jackson turbidity units, roughly equivalent to parts per million of suspended solids), and can satisfy part of the demand for inorganic chemicals for turbidities 10–60 JTU.[3,5]

Sometimes it is easier to clean high-turbidity water than low-turbidity water. Water with few fines may not coagulate well. If all other conditions are satisfactory, clay fines may have to be added as a coagulant aid to the water in order for the attraction caused by coagulants and flocculants to occur.

FILTRATION

Rapid filters can be gravitation or pressure filters. Rapid filters are multi-media filters that disperse the filtered particles throughout the depth of the filter media. Slow filters are filters using only sand as a filter media and are generally not used where there is fine clay to be filtered. Rapid filters can have as many as three types of filter media or more. Generally, they can be identified by color. Anthracite coal is black. Garnet (a natural mineral gem) is red, and sand is tan colored. Thus, the types can be readily identified when loading a filter. The reason for the different media is to distribute the filtered particles throughout the media depth. This creates longer run cycles, and then the limitations of the slow filters can be easily seen.

Garnet is generally used as the fine filter with sand next and coal as the coarse filter. Back flushing the filters will effectively stratify the media according to density. The dense garnet will settle the quickest, with sand next, and the light coal last—creating a perfectly stratified multimedium. Flow is from the coarse to fine direction. The break in this trend would be when coarse gravel is used as an underbed. Uniform-media sand filters (filters using coarse and fine sand mixed together) with the sand interspersed uniformly in the vessel is another example. The main point being that the color will determine how much of each is to be preloaded. The order is not important because immediate backflushing afterward is always required to clean the media. This will stratify the media. A scum can float to the surface of gravity filters if backflushing is not performed to clean the media prior to water production.

Gravity, as well as some pressure filters, will use a low-rate backwash with air injection as an agitator to break loose the filtered material; then a high-rate backwash is used to carry the material to waste. For gravity filters, it is extremely important that the water levels for backwash and normal operation be set properly to end cycles by properly opening or closing valves.

The trial operation of the filters must be performed in the automatic mode using the PLC or DCS system. It is very important that all pneumatic valves function in proper sequence and the electro-pneumatic system for pressure filters functions as designed. All valves for gravity filters must be stroked and the limits properly set such that there is no leakage. Gravity-filter low-rate, high-rate, and normal-level switches must be set properly.

Note

The predemineralization filter system must be in commercial operation before the demineralization equipment can be trial operated. The lessons learned commissioning the filter system will help prevent delays or mishaps when trial operating the demineralization system. The experience learned during commissioning and the study of the control system logic is an important process. Proof of the PLC/DCS logic and the interconnection to the process system will provide valuable lessons learned for the next phase of the water treatment plant startup. Do not settle for manual filter operation.

DEMINERALIZATION

An important prerequisite for demineralizer trial operation is that the filtration system has been trial operated and is in fully automatic, commercial operation. This is due to the fact that mishaps in the demineralization system can be costly. Therefore, full attention must be given to the system trial operation—without disruptions upstream. Remember, reverse-osmosis (RO) membranes and cation/anion resins can cost more than the vessels that contain them.

Demineralizers can remove as much or more minerals from water than distillation, but much more economically. They comprise trains of flash desalination units followed by RO units. Or they can be designed with zeolite softeners and RO units or acid-regenerated softeners and caustic-regenerated anion exchangers. All of these, and other combinations, can be used depending on the chemistry of the raw water. Basically, to remove water hardness in resin-based demineralization, the hardness-causing minerals must be exchanged for acids that have to be taken out in another step.

Hydrogen detaches from the resin and is replaced by calcium, magnesium, and iron in the feed stream, and as a result, acids are formed in the effluent stream. The detached hydrogen combines with chlorides, sulfates, and so on in the feed stream to form strong acids such as hydrochloric acid. The basis of the reaction is ionization. An *electron* has an assigned charge of −1. Therefore, an atom that receives an electron will have a negative charge. An atom that gives up a charge will have a positive charge. These charged particles are called *ions*. The positive ions are called *cations* and the negative ions *anions*. The process is *ion exchange*.

Cation Exchanger

If a softener is used, it is always the first stage. It removes hardness associated with alkalinity derived from such dissolved minerals as calcium and magnesium. This stage uses a zeolite resin that can be regenerated either by salt or sulfuric acid. The acid-regenerated resin can be a weak- or a strong-acid cation. The acid-based resins have the advantage of removing hardness while not adding salt to the water. The weak acid has the advantage of not adding significantly to the acidity of the effluent or regenerative waste. Free mineral acidity (FMA) is measured to determine the regenerative cycle.

Note

An internal inspection should be made of the vessels prior to loading resins. Rubber linings should be checked for quality of application. Check support angles for voids or spalding. Lateral flanges should be checked for proper alignment. Flanges out of parallel can cause leakage and loss of resin. If they are not parallel, they can be aligned quickly by applying heat to one side of the pipe. In nuclear plants, temperature measurements taken on the inside of the pipe must be taken and documented to ensure the metal temperature is within strict procedural requirements. Only moderate heating of stainless steel will pull the flange into alignment. Check lateral openings for proper sizing against the design drawings. If higher-than-expected differential pressure is encountered, this information will be important. Having made a thorough inspection—documented with photos—before resin is loaded, will eliminate doubt if problems occur during operation.

In essence, the cation exchanger exchanges hydrogen ions for the hardness-producing minerals, forming strong acids in the exchange process. Therefore, the vessels must be made of stainless steel or be rubber lined.

Anion Exchanger

The cation exchanger produces low pH water with a high carbon dioxide level. Therefore, the anion exchanger is used to remove the acids and carbon dioxide. If a strong-acid resin is used, it will remove silica as well as carbon dioxide. The strong-based acid resin came into being because of high-pressure boiler needs, which require the removal of silica as well as carbon dioxide and acid. A weak-base acid resin has a higher capacity to remove strong mineral acids such as sulfite, chloride, nitrate, and phosphate. Therefore, sometimes they are used together with the strong-acid resin (silica scavenger) in the same vessel. In this case, the weaker resin is placed on top to remove the strong acids first, and the strong resin is placed on the bottom to remove the weaker acids, carbon dioxide, and silica.

Decarbonator

If an anion exchanger is not used, as such in some softener RO units, a decarbonator is used to scavenge the carbon dioxide. It is basically an aerator that blows air up through falling droplets of water. Examples of cations and anions are illustrated in Table 22-3.

A demineralizer train that does not incorporate a decarbonator, after the cation unit, would be a good candidate for an anion bed consisting of a combination of weak and strong base resins. The weak base resin would be loaded on top of the strong base resin during the initial installation. During normal operation, the resin bed would be regenerated prior to backwashing. This is necessary to prevent the mixing of the two resins during backwashing. In their exhausted state, the density of the resins is very nearly the same. In the regenerated state, their densities are different enough for the resins to separate during backwashing. Consequently, the weak base resin is always on top. This is the preferred position, because it insures that the greatest capacity for the strong acid anions receives the effluent from the strong acid cation first. Another advantage of this arrangement is that, like the strong base anion resins, the weak base resins also absorb organics; however, the organics absorbed by a weak resin are removed during standard regeneration. This helps protect the strong base resin from irreversible organic fouling....

Mixed-bed Polishers

In very high-pressure boiler operations, even trace contaminants are extremely detrimental. Demineralized water requirements are more stringent than can be produced by the two-bed unit. To cope with these specifications, many operations follow the two-bed demineralizer with a mixed-bed polisher. The mixed-bed polisher consists of a special vessel into which are loaded both strong base anion and strong acid cation resins. The cation resin, being denser, is always on the bottom of the vessel after backwashing. The regenerant piping is designed so that caustic enters the top of the vessel, flows down through and regenerates the strong base anion resin, and exits the side of the vessel near the resin interface and flows downward through the strong acid cation resin, regenerates it, and flows out of the bottom to the vessel to the sewer. Prior to the addition of the caustic, the vessel is filled with water and kept under pressure (blocking flow) to insure that no caustic channels down to the cation resin before it leaves the waste line at the resin interface.

Condensate Polishers

Another version of the standard demineralizer system is the condensate polisher. Many high-pressure boiler systems incorporate a vessel loaded with strong acid cation resin or a combination of cation and anion resin to polish the condensate stream coming from the turbine surface condenser. Because many of these systems return a high percentage of their steam condensate as boiler feed water makeup, the polishers serve to insure that any contaminants that might enter the condenser (as from the condenser leaks) are removed prior to reuse.[1]

TABLE 22-3 Examples of Cations and Anions

Cations	Anions
Sodium	Chloride
Magnesium	Bicarbonate
Calcium	Sulfate
Potassium	Nitrate

> **Note**
>
> The following discussion of service cycles is taken in its entirety from GC3 Specialties technical database because of its clarity and direct applicability to power plant operations.

Strong Base Anion. Service flow rate through strong base anion resins typically runs at 2–14 gpm /ft^2 of resin bed cross-sectional area. The conductivity and silica level in the treated water are the parameters normally used to monitor the performance of these units. Many industrial demineralizer systems are supplied with automatic silica analyzers and on-line conductivity analyzers with alarms and adjustable setpoints.

Strong base anion resins are generally taken out of service for regeneration when the silica level in the treated water exceeds 0.1 ppm and/or the conductivity exceeds 10 micromhos. Most strong base anion units are designed to provide specification water for a period of about 24 hours.

Resin degradation due to organic fouling and resin loss due to attrition are the two main reasons why anion train service cycles tend to decrease with time. Most resin manufacturers quote a useful life of about 3 years for a load of strong base resin. It is not unusual to see a strong base resin lose up to 30% of its salt-splitting capacity in the first of year operation. Other factors (inlet water characteristics, channeling, improper regeneration, etc.) can affect the service run length of any strong base resin. All of these factors must be taken into consideration when the production capacity of a strong base anion unit falls below its specified capacity.

As discussed under the section on strong acid cation unit operation, the actual ion exchange capacity of a strong base anion resin depends on a number of factors. Some of these include regeneration level (lb/ft^3 of resin), caustic quality, regeneration temperature, the type and quantity of exchangeable anions, etc. Again, a complete description of the process by which strong base anion resin capacity is determined is beyond the scope of this book.

Backwash

Strong Base Anion. There is one phenomenon that occurs at the end-of-run on a strong base anion unit that helps the operator determine the proper time to get the unit off-line for regeneration. This phenomenon is associated with the conductivity of the effluent water and is used to catch a unit before it breaks on silica. Just prior to silica breakthrough, the conductivity of the effluent water from the anion units takes a noticeable drop: a short time later, it increases sharply, indicating breakthrough. This decrease in conductivity is clearly evident on a plot of the readings taken from the conductivity probe in the bottom of the anion vessel. During the normal service cycle, sodium that leaks from the cation unit combines with OH- from the anion unit to produce NaOH, which is conductive. As silica first starts to break through, the silica combines with the sodium leakage from the cation to produce sodium silicate, which is less conductive than the NaOH it replaces. Therefore, a dip in the effluent conductivity occurs.

Once the anion unit is taken out of service, it is made ready for the regeneration procedures. First, the block valves are manipulated to isolate the unit from the treated water system. Water from the companion cation unit is directed up from the bottom of the unit at a rate of 2–3 gpm/ft^2 of vessel surface area. Backwashing is continued for 5–10 minutes with the effluent going to the sewer. A bed expansion of 50% is desired.

Regeneration

Strong Base Anion. The strong base anion unit is regenerated with a 4% solution of rayon grade caustic and cation free water. Caustic (50%) is pumped from the storage area and mixed with dilution water of the required quality. The 4% caustic solution passes through a heater (steam or electric) and on to the regenerant distribution system in the anion vessel.

The effluent from the caustic heater is controlled at 120°F (48.89°C). The elution of the exchanged silica from the strong base resin is greatly enhanced by the hot caustic. If the available time and heater capacity exist, it is advisable to preheat the anion bed to the desired caustic temperature before caustic injection is started. This insures that the resin bed operates at 120°F (48.89°C) during the full regeneration cycle. If hot caustic is not used for the regeneration of the strong base resin, silica removal can be incomplete. This leads to reduced capacity on the next service cycle. Silica remaining in the resin can also polymerize and permanently foul the resin.

Commonly, regeneration levels of 4–6 lbs/ft^3 of resin are used for anion units. The warm caustic is introduced to the unit at a rate of 0.25–1.0 gpm/ft^3 of resin volume for 30–80 minutes.

Slow Rinse

Strong Base Anion. Once the full charge of caustic has been injected, the caustic pump is turned off and a slow rinse with dilution water only is started at the same flow rate as that used for caustic injection. Slow rinsing is conducted for a minimum of 15 minutes. Slow rinsing moves the regenerant solution through the resin in "plug" flow fashion and helps remove some of the exchanged anions.

Fast Rinse

Strong Base Anion. Once the slow rinse cycle is completed, the rinse rate is increased to 1.0–2.0 gpm/ft^3 of resin for 1–2 hours to scour all of the regenerant and exchanged anions from the resin. The end of the fast rinse step is generally determined to be the point at which the effluent conductivity reaches 10 micromhos.

If the anion unit is placed on standby for a significant period of time, it should receive a short fast rinse prior to being put back in service. This insures that any silica or other anions that have eluted from the resin back into the stored water are swept to the sewer and not the treated water storage system.

Return to Service or Standby

Strong Base Anion. Based on need, the anion unit is either returned to service or placed on standby until needed.

Other Anion Resin Types

The regeneration procedures enumerated above apply generally to other anion resin types. The one notable exception is the so-called stratobed anion unit, which has a combination of weak and strong base resins. With this type of unit, regeneration precedes backwashing to insure that the resins don't mix during backwashing.[1]

Troubleshooting Tips for the Demineralizer

Strong Acid Cation

A. *Problem: High Sodium Leakage (low FMA).*
   1. *Check service flow rate... poor flow distribution may result from high or low flow rates.*
   2. *Has the unit been returned to service from standby? A short rinse before service can help.*
   3. *Is the regeneration level satisfactory for the raw water characteristics? Refer to resin data; increase acid by 30% at next regeneration.*

4. *Is unit regenerated or backwashed at least once every 24 hours? Extended service runs can result in channeling of flow.*
5. *Was the unit run to complete exhaustion during the last service cycle? It should receive a double regeneration before the next service cycle to achieve maximum capacity.*
6. *Is a valve leaking? Leaking block valves can introduce a small amount of raw water into the treated water.*
7. *Are the acid concentrations satisfactory? High acid concentrations during initial phases of regeneration can result in $CaSO_4$ precipitation in the resin bed. Low acid concentrations result in incomplete removal of sodium ions from the bottom of the resin bed and sodium leaking during the following service.*
8. *Are the regeneration flow rates and times satisfactory? Check the manufacturer's recommendations. Excessive backwash flow rates may result in resin loss while low backwash flow rates may lead to channeling during subsequent steps. Low flow rates during acid injection may result in $CaSO_4$ precipitation. High acid flow rates lead to poor regeneration due to insufficient contact time.*

B. *Problem: High Free Mineral Acidity*
1. *Is the rinse of adequate flow and duration? If the previous rinse was inadequate, the high FMA (higher than during previous service cycle) is due to the incomplete removal of spent acid from the resin after the regeneration. Additional rinsing will bring it down.*
2. *Does the effluent contain hardness? Adjust sample pH to 4.3–10.0 with 1N NaOH prior to running hardness. If calcium is present, the precipitation of $CaSO_4$ should be suspected. Check acid concentrations during regeneration and flow rates used.*
3. *Are the acid regenerant valves leaking acid into the unit during the run? If the FMA from the unit in service increases only when a parallel unit is being regenerated, there is a good chance that a valve is not holding the pressure of the acid line and acid is entering the operating unit.*

C. *Problem: Low Capacity*
1. *Has the raw water cation loading increased? Check the raw water total cation loading against the manufacturer's design loading. To determine the total raw water cation loading, add the sodium content to the total hardness. A good estimate can be made in the field by adding the M alkalinity of the raw water to the FMA at the cation effluent.*
2. *Is the service flow rate satisfactory? Poor flow distribution may result from either high or low flow rates. Channeling will cause only a small area of the resin bed to receive a large portion of the raw water.*
3. *Is the unit regenerated or backwashed at least every 24 hours? Extended service runs can lead to channeling of the flow.*
4. *Was the unit overrun during the previous service cycle? A unit run to exhaustion must be regenerated twice before full capacity will be restored.*
5. *Are the acid concentrations satisfactory? Insufficient acid feed to the unit during regeneration will lead to lower than expected capacity. Check the actual acid fed against the manufacturer's recommendations.*
6. *Are the regeneration flow rates and times satisfactory? Excessive flow rates during regeneration will reduce the acid/resin contact time required for good regeneration and reduced capacity will result.*

D. *Inspect the Unit*

When all of the troubleshooting tips are checked out and the problem is not resolved, a physical examination of the vessel interior will often reveal a mechanical problem. Check the raw water inlet distributor for missing spray heads, the regenerant distributor for damaged or missing parts, the surface of the resin for unevenness or signs of resin degradation, etc.

Strong Base Anion

A. *Problem: High Effluent Conductivity*
   1. *What is the cation sodium leakage? High cation sodium leakage is indicated by a low FMA at the cation outlet. Each ppm of sodium leakage from the cation adds about 10 micromhos of conductivity to the effluent of the strong base anion unit. The effluent pH from the strong base anion will also be high (over 8.0). If high sodium leakage is the problem, refer to the troubleshooting tips for the strong acid cation unit for a resolution.*
   2. *Was the anion unit rinsed sufficiently at the end of the last regeneration? Good anion rinse usually requires 100–150 gallons of rinse water per cubic foot of resin in the unit. Check rinse flow rates and times to confirm.*
   3. *Is a block valve leaking? Check water samples upstream and downstream of any suspect valve.*
   4. *Was* $Mg(OH)_2$ *or* $CaCO_3$ *precipitated on the resin during the last regeneration? Cation treated water should be used for backwashing the anion resin and in the preparation of caustic regenerant solutions. The use of raw water for either will result in precipitation of* $Mg(OH)_2$ *and* $CaCO_3$ *in the resin bed. Elution of these compounds during the service cycle can produce high effluent conductivities.*

B. *Problem: High Silica Leakage*
   1. *Is the service flow rate satisfactory? High or low service flow rates may produce channeling.*
   2. *Is the regeneration level satisfactory for the raw water characteristics? Refer to the resin manufacturer's strong base anion resin leakage curves. Increase the caustic used for regeneration by 30% to see if the silica leakage decreases during the following service cycle.*
   3. *Is the unit regenerated or backwashed at least every 24 hours? Extended service runs can result in channeling of the flow.*
   4. *Was the unit overrun during the last service cycle? If the unit ran to exhaustion during the last run (the unit broke on conductivity as opposed to silica), a double regeneration will be required to restore full operating capacity.*
   5. *Is the caustic concentration and temperature satisfactory? In order to insure good regeneration (silica removal), the caustic must be fed at a concentration of 4% and a temperature of 120°F (48.89°C).*
   6. *Are the regeneration flow rates and times satisfactory? Variations from the recommended procedures will lead to unsatisfactory operation.*

C. *Problem: Low Capacity*
   1. *Has the raw water changed with respect to the design water analysis? Have a thorough analysis made of the raw water and check the total exchangeable anion content against that used for the unit design.*
   2. *Determine whether the cation or anion is exhausting first. If the cation unit exhausts first, the conductivity and pH of the anion effluent water will increase. If the anion unit is exhausting first, the conductivity of the anion effluent water will rise and the pH will drop.*[1]

REVERSE-OSMOSIS DEMINERALIZATION

Osmosis is the chemical process through which two solutions of different concentrations obtain equilibrium when separated by a semipermeable membrane. If, for example, a membrane in an aquarium separates two solutions, the solvent from the solution of higher concentration will move to the lower, concentrated side, raising the water level until the osmotic pressure equilibrium is achieved—that is, the solution concentration on each side is equal. If the cycle is reversed by forcing the solution through the membrane, in the opposite direction, the process is known as reverse osmosis. In this case, the reverse happens; the pure solute passes through the membrane, leaving the concentrated solution behind. This is one way to extract pure water from a saltwater solution. The product water is demineralized, and the wastewater is more concentrated than the original solution.

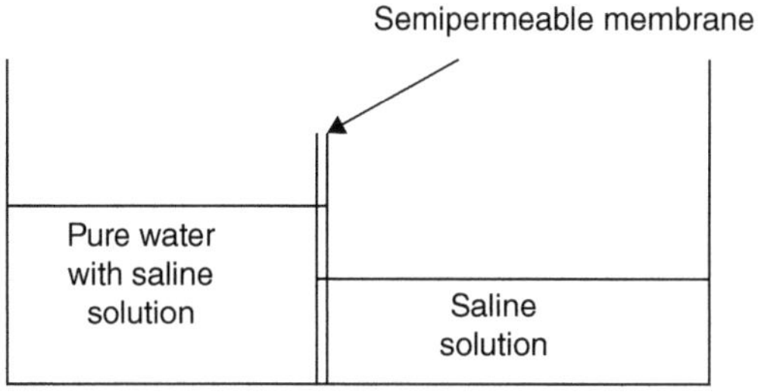

Originally, the pure-water level was equal to the saline-solution level. Having reached osmotic equilibrium, the pure-water side rose as the water from the higher concentrated side migrated through the membrane to the less concentrated side. That is, both sides now have equal percent concentration. The natural osmotic pressure of seawater is ≈350 psi (24 bar), which has to be overcome in the reverse-osmosis process.

It can be seen, or should be obvious, that pressure control in addition to chemistry control is a very important part of the RO process.

Pretreatment

Pretreatment is an essential part of RO operation. The process should eliminate total suspended solids and hardness in the raw water. The process includes a prefilter, softener, and after-filter prior to entering the RO unit. The prefilter will be a fast filter and the softener a zeolite, salt, or weak-acid cation exchanger. The afterfilter will be a micron filter, which is the RO prefilter.

Acid is injected to adjust the pH of the water entering the RO to prevent calcium carbonate precipitation. Also, if chlorine was used as a disinfectant in the clarification–filtration phase, an activated carbon filter must be used, or a chemical chlorine scavenger must be injected, for chlorine removal. The chlorine must be removed to avoid attack of the membranes.

The membrane cartridges are manufactured in standard sizes, and the pressure housings (cylinders) are manufactured as dictated by the standard membrane size. The membranes are delivered separately and must remain in their liquid-filled protective containers until they are to be installed. The membranes must be kept moist at all times whether in service or not. Biological attack will occur in the membranes if they are left dry for only a short period of time.

When installing membranes, remember that the flow-directional arrow on the membranes must be adhered also. In addition, note on spiral-wound membranes that, at the ends of the pressure housing and the extreme ends of the membrane sets, one end of the permeate outlet tube is for discharge and the opposite end (opposite to direction of flow) is merely a thrust-retainer tube. The center permeate tube, between membranes, is a spacer tube that creates a waterbox for the inlet to the next membrane. The flow of saline solution is through the membrane tube, not around it. The membrane seals (O-rings) must be installed in accordance with flow direction. The seals prevent short-circuiting upstream membranes.

To install the membranes with the flow direction or the seals reversed can damage the seal rings. The seals are V-shaped and must be pointing in the direction of flow. Especially note if it appears to take excessive force to insert the membranes into the housings.

RO Problems

The following are examples of early problems encountered with RO units in the late 1970s that might shed some light on operating problems today.

Take the experience of Central Illinois Public Service Company at Coffeen power station, as described by A. E. Toomey:

> Problems at the plant stemmed from rising levels of dissolved solids in Coffeen Lake, which provides both water and coolant for the plant. TDS increase in the water was due to sulfates in the fly ash (sluice water is recirculated via the lake), and a dropping water level in the lake. A reverse-osmosis unit was ordered as the only practical solution. The unit comprised three modules, using cellulose triacetate hollow-fiber membranes, with eight first-stage and three second-stage permeators per module; it was designed to produce 400 gpm of product water. Also included in the system were a heat exchanger, three cartridge filters, four h-p pumps, two pretreatment clarifiers with 700-gpm combined capacity, and associated feeders, cleaners, and system controls. Test performance on municipal water for one month was satisfactory: Although pressures varied somewhat from design values, the unit operated at design capacity, producing water at 20-micromho conductivity from 460-μmho feed. But when operated briefly on clarified lake water and then returned to municipal water, effluent conductivity increased to 30 μmho. In the same period, operating pressure dropped from 350 psi to 250 psi, due to an increase in water temperature; conductivity then continued to increase to 40–45 μmho, indicating an 85% rejection rate. Aside from temperature effects, other contributing factors were found to be iron, silt, and alum, all of which were present during the break-in period, and had apparently fouled the membranes.
>
> Laboratory inspection revealed that the membranes were intact and could be readily restored to a rejection rate of 92%. Cleaning in the field, however, was more involved. It required initial flushing to remove concentrated brine solution from the permeator when taken out of service, to prevent calcium scaling; this was followed by a low-pH permeator flush. Iron- and silt-cleaning procedures were ineffective. The result of the corrective actions was to restore system pressures, but not the TDS-removal capability. Returned to operation on clarified lake water, after four weeks the RO system produced 700-μmho effluent from feed that now had a conductivity of 1600 μmho; this performance remained unchanged over the following two months.
>
> Inspection at the end of that time showed damage by hydrolysis to the inner wrap of two permeators. Some silt fouling was also evident. Although the cause of the problem was thought to be the large quantities of iron present in the permeators, this was never proven. The first-stage permeators were replaced in two RO modules, and the staging was changed in an attempt to increase the inlet pressure. The new design featured six permeators in the first stage and three in the second stage of one module, five first-stage and two second-stage permeators in the second. This almost met the design flow and pressure requirements, while the pressure drop was 200 psi better and design rejection rate was achieved. At present, the permeators are being monitored for evidence of new trends; feedwater brine and product conductivity and complete chemical analysis are checked routinely.

Problems of the same type were echoed by Michael Scherm in describing results with RO at a Union Carbide plant:

> It was a pilot project (as the membrane-supplier's commentator correctly pointed out), recycling wastewater containing suspended solids, residual organics, and dissolved salts. The final treatment step before testing as boiler feedwater was reverse osmosis followed by ion exchange. The unit

selected was designed for automatic batch-mode operation and 90% demineralization. And adaptable to a variety of membrane types. Mixed-media and cartridge filters protected the membranes from solids in the influent or precipitated during pH adjustment; temperature, pH, and conductivity instruments monitored water variables.

Two membrane configurations were tested: (l) tubular type with cellulose acelate (CA) membranes, and (2) spiral wound units with either CA or polyamide elements. Operating ranges for the tubular units were 40–110 F, pH 3.0–7.0, pressure below 600 psi; design salt rejection was 96–97% at 90% water recovery. Conditions for the spiral wound CA units were 100 F, 500 psi, and pH of 4.0–6.0 with 95% rejection at 90% water recovery; same for the polyamide modules, except for 135 F temperature and pH 3.0–9.0.

Performance Wanting

Rejection rates for the tubular/CA units were below design, ranging from 67% to 89% over a three-month period. Permeate conductivity averaged 1000 µmho, whereas a range of 200–400 µmho was essential for satisfactory resin-bed life in the downstream ion-exchange columns. Examination showed the membranes to have been heat-damaged, either from high circulating-water temperature or exposure to the sun. Closing of membrane pores had reduced flux levels, raised operating pressures, and encouraged brine leakage.

Spiral/CA units were substituted, and a heat exchanger added to keep the influent from causing damage. Suspended solids below 10 ppm were considered low enough to avoid particulate fouling of the modules. Initial conductivity rejection was 95%, with brine pressure at 420 psi and permeate at 390 psi. After two weeks, however, rejection dropped to 85% and permeate flow declined. Examination of the modules revealed rust-like particles and blackened membrane edges on the inlet side. Cleaning with solutions used for heavy metals or organics did not result in flow improvement, indicating permanent deterioration. The cause was attributed to particulates and membrane compaction.

High circulating water temperature also contributed to the problem despite addition of the heat exchangers with recirculation rapidly increasing it above 110 F. Mostly for this reason, the membrane was changed to the polyamide type, having capability up to 135 F. A 10-micron cartridge filter was added on the suction side of the feed pump as an added precaution against particulate fouling. The new membranes gave more uniform performance and better rejection rates than either of the others. Although it declined with time, conductivity rejection averaged 92% over a 7-wk period, with 75% to 80% water recovery.

Batch-mode operation of the system with the CA membranes made pH control difficult, undoubtedly contributed to the problems encountered with these membranes in both configurations. On the other hand, with the broader pH tolerance of the polyamide, this control was not necessary.

Reduction to 5.0 was found to improve conductivity rejection, but this resulted in precipitation of a gel-like material that plugged the cartridge filter. To avoid filter fouling, the unit was operated without pH control.

Brine leakage into the product water occurred continuously with the tubular/CA system. Source of the problem was the O-ring seals at the ends of the fiberglass tubes, as well as the grommets connecting modules in series. Leakage was also noted with the spiral modules, although to a far lesser extent. It usually occurred at the O-rings used at the module interconnections, particularly when the operator allowed system pressure to build too rapidly during startup, unseating the rings.

The Union Carbide experience showed that membrane fouling did not hamper operations with tubular cellulose-acetate RO modules. With large flow channels (1/2-in), and therefore high fluid velocities compared to spiral-wound units, the configuration is well-suited to influent water containing suspended solids. As indicated, fouling by iron and carbon like fines was a real problem with the spiral/CA units, not even remedied by chemical cleaning. It necessitated addition of the 10-micron cartridge filter for removal of the small particulates.

The filter apparently aided operation of the polyamide for a while, but an eventual decline in flux rates and permeate quality indicated that fouling had set in here as well. The situation was improved by rinsing the membranes in deionized water, followed by soaking for several days, then a final flush with deionized water. Unfortunately, this required considerable downtime.

The lesson learned here was that development of an effective cleaning procedure for a specific application—well before the start of operations—is a difficult challenge that must be met to insure success with reverse osmosis.

Other lessons afforded by these experiences were related to the effect of wastewater corrosivity on piping, valves, pumps, etc., resulting in precipitation of metal salts and scouring of membranes if not filtered out. The need for temperature control is also apparent. While polyamides have wider tolerance, new membrane materials with less-severe temperature and pH requirements would optimize rejections and water recoveries at lower cost. With it all, Scherm points out it will still be difficult to achieve recoveries above 70% when treating variable wastewaters with dissolved solids above 400 mg/l for recycle to steam generators.[6]

Electrodialysis Reversal (EDR)

Principles of Electrodialysis, Effect of DC Potential on Ions

Electrodialysis (ED) is an electrochemical separations process in which ions are transferred through membranes from a less concentrated to a more concentrated solution as a result of the flow of direct electric current. To understand this process more clearly, it is important to know the effect of a DC potential on an ionic solution. Assume a rectangular tank with an electrode at either end and filled with a sodium chloride (NaCl) solution.

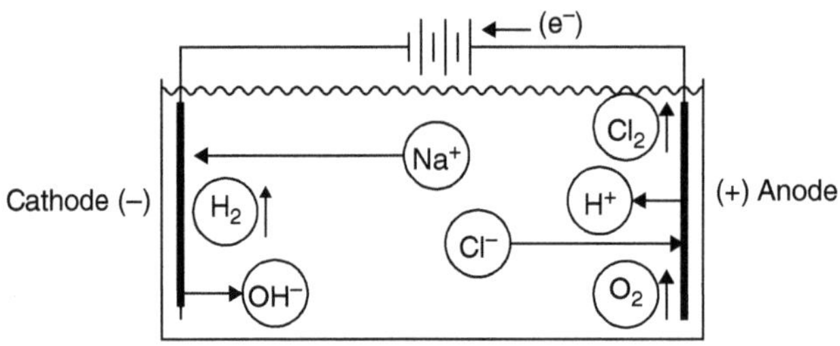

When a DC potential is applied across the electrodes, the following occurs:

1. Cations (Na*) are attracted to the negative electrode called the cathode.
2. Anions (Cl^-) are attracted to the positive electrode called the anode.
3. The water dissociation reaction involving the gain of electrons (reduction) occurs at the cathode.

$$2H_2O + 2e \rightarrow 2(OH^-) + H_2 \uparrow$$

4. The water disassociation reaction involving the loss of electrons (oxidation) occurs at the anode.

$$2H_2O + 4H^+ \rightarrow O_2 \uparrow + 4e^-$$

5. A reaction involving the formation of chlorine gas may also occur at the anode.

$$2Cl^- \rightarrow Cl_2 \uparrow 2e^-$$

Effect of Anion and Cation Transfer Membranes

To control the movement of ions in the tank containing the ionic solution and electrodes, several membranes can be added to form water-tight compartments as shown in the Fig. 22-2.

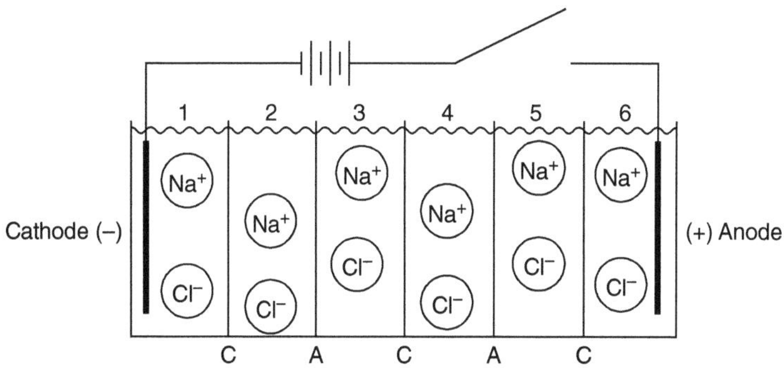

FIGURE 22-2 Electrodialysis open circuit example.

In the figure two types of membranes are used.

1. Anion Transfer Membrane: (which allows only the passage of negatively charged anions). This membrane is electrically conductive and is essentially water impermeable under pressure. (Labeled A in the above figure.)
2. Cation Transfer Membrane: (which allows only the passage of positively charged cations). This membrane is also electrically conductive and essentially water impermeable under pressure. (Labeled C in the above figure.)

In Fig. 22-2, no DC potential is being applied, resulting in no ion transfer. In Fig. 22-3 it is illustrated what occurs when a DC potential is applied. Observe compartments 1 and 6: These compartments are unique in that they contain the metal electrodes. Chlorine gas, oxygen gas, and H^+ ions are produced at the anode or positive electrode. Hydrogen gas and OH^- ions are produced at the cathode, or negative electrode (shown in the diagram below).

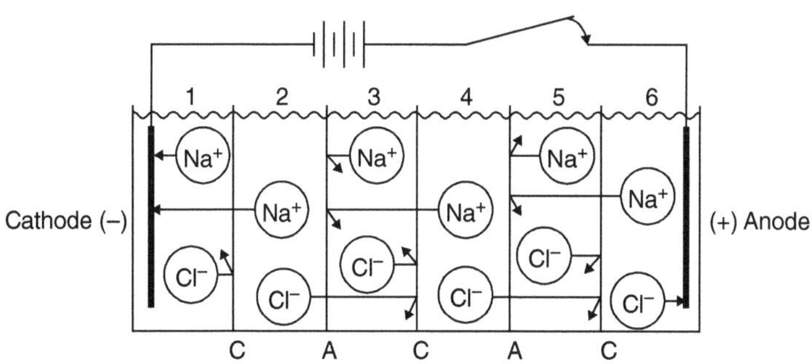

FIGURE 22-3 Electrodialysis closed circuit example.

Compartment 2:

1. Cl^- ions pass through anion membrane (A) into compartment 3.
2. Na^+ ions pass through cation membrane (C) into compartment 1

Compartment 3:

1. The Na^- ions cannot pass through the anion membrane and remain in compartment 3.
2. The Cl^- ions cannot pass through the cation membrane and remain in compartment 3.

Compartment 4:

1. Cl^- ions pass through the anion membrane into compartment 5.
2. Na^+ ions pass through the cation membrane into compartment 3.

Compartment 5:

1. The Na^+ ions cannot pass through the anion membrane and remain in compartment 5.
2. The Cl^- cannot pass through the cation membrane and remain in compartment 5.

The overall effect shows that compartments 2 and 4 have been depleted of ions and the ions have been concentrated in compartments 3 and 5. As shown in last figure, alternating compartments of demineralized solutions and concentrated solutions are formed in this membrane cell when a DC potential is applied across the electrodes. When properly manifolded, this unit will yield two major and separate streams; demineralized and concentrate waters, and two minor streams from the electrode compartments. In normal applications several hundred of these demineralizing and concentrating compartments are assembled into a "membrane stack" to obtain the desired water flow rate. These membrane stacks are the heart of the ED system.[7]

BOILER WATER CHEMISTRY CONTROL

It probably should be noted here that the evolution of modern water treatment in high-pressure boilers took a dramatic change in the 1960s and early 1970s when the term "coordinated control" was coined. In the 1960s, a step change was taking place in boiler pressure design. Boiler pressures were rising from 1800 psi to 2800 psi in subcritical units, and 3800-psi supercritical units were being developed. Boiler water chemical treatment for the lower-pressure boilers was primarily pH control. Phosphate injection was used as an emergency injection system that was utilized in the event of a condenser tube leak, which was detected by a sudden rise in condensate conductivity.

However, it was recognized that this was an after-the-fact procedure and really did not take into consideration undetected contamination. This could not be tolerated in the higher-pressure units. Also recognized was the need for chemical oxygen scavenging in addition to the standard deaeration systems. Therefore, the transition was made to continuous phosphate and oxygen scavenger injection and thereby the term *coordinated control of water chemistry* came into being.

Chemical feed system commissioning tips:

- Air-blow all chemical feedlines to verify that they are going to the proper injection point.
- Never use an injection point that is under vacuum. The vacuum can cause the entire contents of the feed tank to be pulled, uncontrolled, into the system.
- Chemical injection pumps require that the pump casing be primed manually to purge internal air. Air will compress, effectively blocking pumping action similar to that of diesel engine injection systems.
- Ensure that all the proper chemical laboratory equipment and chemicals are available.
- Ensure the lab exhaust system purges air to a safe area outside the building, not just from the lab into the demineralizer building itself.
- Air-blow all sample lines and ensure they go to the proper sampling point.
- Ensure that the continuous monitoring system is placed in service using the manufacturers' instructions.

OPEN COOLING-WATER SYSTEMS

Scaling

Closed-loop cooling water systems using evaporative-type cooling towers can create a rapid concentration of dissolved solids and an associated increase in pH as the unit load is increased during trial operation. That, coupled with the increase in water temperature, can rapidly create carbonate scaling in the hot section of the turbine condenser and the hot-water distribution section of the cooling tower. The concentrations must be limited by either makeup and blowdown or by softening. To control the increase in pH, sulfuric acid must be added.

Remember that carbonates will come out of solution at high concentrations. Once calcium carbonate precipitates, it will adhere to hot surfaces, forming a greenish crystalline scale. If ignored, the scale can build up and break off, plugging condenser tubes. At this point, scale removal is very difficult, and usually, the condenser cleanliness factor cannot be brought back to the like-new condition using acid cleaning only. Mechanical scraper pigs may have to be pulled through the tubes to completely remove the scale.

Scaling tendency of calcium carbonate can be estimated by using the Langelier Index. The index is based on the work of Professor Langelier, published in 1936. The index is basically the pH of the solution minus the pH at saturation. Determining the saturation pH makes the calculation somewhat laborious. However, nomograms can be found in water chemistry handbooks for the purpose of estimating the scaling tendency of calcium carbonate. An index of zero and minus will indicate no scaling tendency but may be corrosive as the index decreases. Refer to scaling index in Table 22-4, which was provided in *Power* magazine.[8]

Fouling

The following is an excerpt, a *Power* magazine Special Report,[7] which can be obtained from the *Power* Reprint Department for a nominal fee.

TABLE 22-4 Scaling Index[7]

LSP[a]	RSP[b]	Condition
3.0	3.0	Extremely severe scaling
2.0	4.0	Very severe
1.0	5.0	Severe
0.5	5.5	Moderate
0.2	5.8	Slight
0.0	6.0	Stable water[c]
–0.2	6.5	No scaling, very slight tendency to dissolve scale
–0.5	7.0	No scaling, slight tendency to dissolve scale
–1.0	8.0	No scaling, moderate tendency to dissolve scale
–2.0	9.0	No scaling, strong tendency to dissolve scale
–3.0	10.0	No scaling, very strong tendency to dissolve scale

[a]Langelier Saturation index.
[b]Ryznar Stability index.
[c]No scaling or tendency to dissolve scale.

Control programs aim for 99% reduction of organism populations. This is done by using one or more biocides. Some combinations have shown synergistic effects: The use of quaternary ammonium salts with organo-tin toxicants has yielded better overall biological control than either one alone. Since a small number of organisms will be resistant to specific biocide, complete elimination of every strain cannot be expected.

Selection of microbiocides involves several factors. They must be effective in eliminating microbial activity, of course, but they must also make for an economical program. It may be possible to combine a highly effective but expensive biocide with a less expensive type for broad control at reasonable cost. The use of oxidizing biocides is quite common, but their efficiency hinges on careful control of their addition to cooling water. If this cannot be ensured, they may have to be discarded to avoid problems. Wood delignification is an example: Since this results from over application of an oxidant, use of a nonoxidizing biocide would be preferred.

Environmental effects are another factor. Disposal problems due to persistent toxicity have limited the use of certain biocides in many regions. Other situations require selection of a biocide that can be easily detoxified before cooling-system bleedoff can contaminate receiving stream.[7]

Bear in mind that while a biofouling control program may provide a high-percentage reduction in fouling, it should be relatively economical to be acceptable. Thus, it must provide an optimum period between successive dosings. (See chart below.) If a program gives adequate control but allows rapid organism buildup to the tolerance level, more biocide must be added to maintain control. In such a situation, the system may have to be dosed as often as once a day, which can be very costly. Ideally, normal slug doses should be applied an average of one to three times a week, unless fouling is particularly severe.[a] Above all else, matching the biocide to the problem organism is the key: Use of a biocide to which the organism is not especially sensitive has obvious drawbacks. Bacterial counts are constantly at dangerously high levels, and biofouling is likely.[2]

These systems should be ready for service as soon as the circulating water system is ready for trial operation.

Chlorine

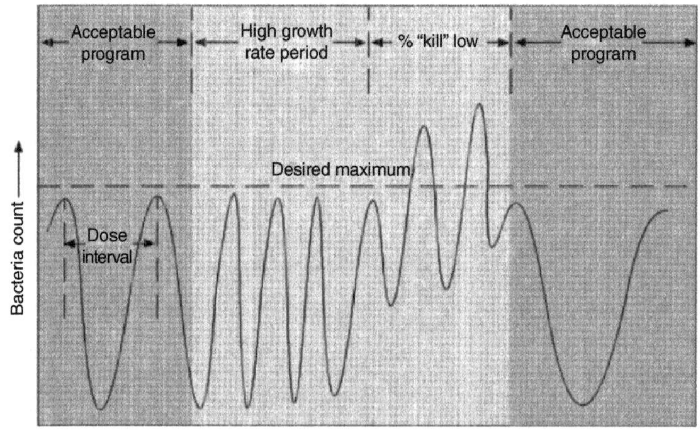

Acceptable biological control program (dark shaded area) gives high count reduction with suitable intervals. Otherwise, control is either costly or inadequate.

[a]G. A. Cappeline, J. G. Carroll, and S. D. Strauss, Enhance Your Cooling System's Performance through Proper Use of Microbiocides, *Power*, October 1977, p 56.

Chlorine is the most extensively used biocide and disinfectant for both the power and water treatment industries. Chlorinators can consist of cylinder or ton containers with a vacuum injector located on the container or at a remote panel. Ton containers are mounted on weighing devices to determine when the container is empty. Ton containers are most common for potable water treatment because of the higher demand than for cooling tower requirements, which will generally use upright cylinders. The American Water Works Association (AWWA) sets the standards for chlorine dosages for sterilization and disinfection. Local authorities will have stringent requirements governing chlorine residuals in discharge streams.

There are three basic types of chlorine: liquid form, containing 100% chlorine; calcium hypochlorite known as HTH (high-test hypochlorite), containing 65% chlorine; and sodium hypochlorite, containing 5, 10, and 15 percent chlorine. The liquid form usually comes in ton or upright cylinder containers. HTH and sodium hypochlorite come in both solid granular or tablet form. Sodium hypochlorite is the most commonly used disinfectant for residential swimming pools for safety reasons. HTH is a very strong chlorite that will give off chlorine gas when opened. Therefore, it is not recommended for residential use.

Caution

Liquid chlorine shall be used only (1) in combination with appropriate gas-flow chlorinators and ejectors to provide a controlled high-concentration solution feed to the water to be chlorinated; (2) under the direct supervision of a person who is familiar with chlorine's physiological, chemical, and physical properties, and who is trained and equipped to handle any emergency that may arise; and (3) when appropriate safety practices are observed to protect working personnel and the public.[8]

Chlorine enhancers are available that are blends of penetrants, surfactants, and dispersants. They are used in circulating water systems to reduce the amount of residual chlorine required to kill microorganisms. Remember, slime is a organism that does not produce its own food like algae and does not require sun light and therefore can grow anywhere in the system.

Zero Discharge

The ideal zero-discharge plant is one in which Total Suspended Solids (TSS) and Total Dissolved Solids (TDS) are reduced to solid industrial waste to the point that the wastewater has a quality equal to or better than the plant raw-water supply. Water that is raw-water quality such as boiler blowdown and sample water should not be discharged or evaporated. All power plant water is treated by clarification, filtration, and ion exchange in some form or other. All streams will concentrate dissolved solids whether it is in the boiler or cooling tower. These concentrated solids have to be removed.

The processes mentioned in this chapter outline the methods. Remember, some of the equipment used depend on water parameters such as temperature and flow stability. For example, clarifiers are generally steady-state operating equipment and cannot tolerate rapid changes in temperature and flow. Ion exchangers should not be expected to filter suspended solids.

Designers will select the type of equipment and process according to personal preference, experience, water availability, and location of the site. But the basic principles of water chemistry and its control will not change. Zero discharge is no more difficult than plant-water quality control. However, the commissioning engineer should be keenly aware of the water balances in the plant as well as quality. Record all flows and check them against design parameters. See Chap. 9 for more information on zero discharge.

Oily-Water Separators

Follow the manufacturer's instructions for commissioning oily-water separators, but make certain that they are precharged with water before any oily-water sources are connected. If they are not, oil can contaminate the clean-water discharge chamber. Closely monitor the oily-water inlet for a reasonable period whenever lube oil system is placed in service to determine that oil is not inadvertently being discharged from a lube oil system via incorrect piping.

Outside spill containments will usually have their own separators that discharge rain or fire system deluge water to a drain system while blocking the discharge of oil. These containments are usually located a long distance from a central-plant oily-water separator and can also add excessive size and cost for simply handling rainwater. Small units for remote locations can discharge rainwater, and if there was a oil spill, it can be extracted with submersible pumps placed in sumps. Figure 22-4 is an example of a remote self contained oily water separator.

Hazardous Waste

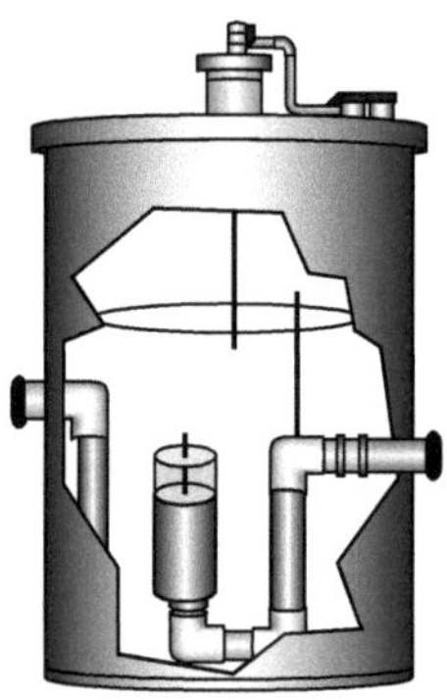

FIGURE 22-4 Oily water separator for outside containments (provided by AFL Industries Inc.) blocks flow of oil but allows water to pass.

The key to hazardous waste disposal is reducing the hazardous wastewater to industrial solid waste plus water with a water quality acceptable for reuse. For example, before disposing of a boiler chemical cleaning solution, it should be neutralized and the metal removed. After neutralizing the wastewater, iron can be removed by aeration, sedimentation, and filtration. Copper should not be present during a new plant startup. The iron is then an industrial waste, and the water is a suitable raw water source.

Chemical Inventory

Ensure that your chemical inventory is available long before trial operation begins. This is as important as your spare parts and tools inventory.

REFERENCES

1. Gulf Coast Chemical Commercial Inc., 1995; revised June 29, 1998. Search this publication and others by Gulf Coast, at www.gc3.com, for more guides to Water Treatment (search by Water Treatment topic).
2. Strauss, S. D., "Removing Suspended Solids from Water for Utility and Industrial Steam Generation," *Power* (Special Report), June 1981.
3. "Water Treatment Plant Design," American Water Works Association Inc., 1969.
4. Kenner, F. J., *The Nalco Handbook*, McGraw-Hill, New York, 1979.
5. *Handbook of Industrial Water Conditioning*, 8th ed., Betz Laboratories Inc., 1980.
6. Strauss, S. D., and Puckorius, P. R., "Cooling-Water Treatment for Control of Scaling, Fouling, Corrosion," *Power* (Special Report), June 1984.
7. Meller, F. H., *EDR*, March 1984.
8. ANSI/AWWA C652-02, "Standard for Disinfection of Water-Storage Facilities," August 1, 2002. This precaution is stated in all AWWA standards relating to chlorination.

APPENDIX A
ACRONYMS AND GLOSSARY

AMCA Air Movement and Control Association International Inc.

AC actual cost

ACWP actual cost of work performed

AD activity description

ADM arrow diagramming method

AF actual finish date

AGR acid gas removal

AOA activity-on-arrow

AON activity-on-node

AS actual start date

ASHRAE American Society of Heating, Refrigeration, and Air Conditioning Engineers

ASU air separation unit

BAC budget at completion

BCWP budgeted cost of work performed

BCWS budgeted cost of work scheduled

BFP boiler feedwater pump

bunker C early steam ship term for residual fuel of the lowest quality "bunkered"; No. 6 fuel oil

BWF black-water flash

CAP control account plan (previously called cost account plan)

CCB change control board

combined cycle combined Brayton (gas) and Rankine (steam) cycles

CMU concrete masonry unit, correct term for concrete block, cinder block, etc.

COS molecular formula for carbonyl sulfide (significant impurity in syngas)

CPFF cost-plus-fixed-fee

CPI cost performance index

CPIF cost-plus-incentive-fee

CPM critical path method

CRO control room operator

CRT cathoray tube

CST condensate storage tank

CTG combustion turbine generator

CV cost variance

CWD control wiring diagram

DCS distributed control system

DD data date

DDL device description language

DEMA Diesel Engine Manufacturers Association

DLL data link layer

dry contacts contacts that are *not* powered by the same voltage source as the coil

DOE Department of Energy

DSA dynamic signal analyzer

DU duration

EAC estimate at completion

EF early finish date

EHC electrohydraulic control

EPRI Electric Power Research Institute

ERP enterprise resource planning

ES early start date

ETC estimate to complete

EV earned value

EVM earned value management

FAS fieldbus access sublayer

FMS fieldbus message spec

FERC Federal Energy Regulator Commission

FF free float or finish-to-finish

FFB flexible function block

FFP firm fixed-price

Fieldbus bus for connecting man–machine interfaces and programmable logic controllers to field devices using a fiber-optic highway

FPIF fixed-price-incentive-fee

FS finish-to-start

green rot high-temperature corrosion in gas turbines

GERT graphical evaluation and review technique

H1 fieldbus that connects field devices (31.25 kbit/s)

HDMI high definition multimedia interface (connector type)

HSE high-speed ethernet that provides integration of PLCs, H1, servers, and workstations (100 kbit/s)

HRSG heat recovery steam generator ("hersig")

HVAC heating, ventilation, and air conditioning

IGCC integrated gasification combined cycle

interlocks devices or actions that permit or prohibit an operation; e.g., a motor start

IEC International Electrotechnical Commission

IFB invitation for bid

interposing relay a relay that is used to power another relay

ISA International Society of Automation

ISO International Organization for Standardization

ladder logic PLC logic drawn vertically rather than horizontally

LAS link active scheduler

LF late finish date

LOE level of effort

LS late start date

MAC main air compressor

MCC motor control center

MOC mechanically operated contact

MODBUS communication protocol originally published by Modicon used to connect programmable logic controllers to field devices—openly published and royalty free

MIS management information system

MVC multi-variable container (object), serves to encapsulate multifunction block parameters

motor starters circuit interrupters for a low-voltage, frequently energized apparatus

NAS National Aerospace Standard (for cleanliness of lubricants)

NPI new product introduction (GE)

NRO nuclear reactor operator

OBS organization(al) breakdown structure

OCB oil circuit breaker

open cycle simple or Brayton cycle

OSI open systems interconnect

PC percent complete

PDM precedence diagramming method

PERT program evaluation and review technique

PF planned finish date

pH short for potentiometric hydrogen ion concentration. 0–14, 7 = neutral (acidic < 7 < basic or alkaline).

PID proportional-integral-derivative

PLC programmable logic controller

PM project management or project manager

Profibus process field bus; Profibus is a fieldbus technology; see *fieldbus* and *modbus*; not openly published nor royalty free

PS planned start date

putrescibles green waste such as fruit and vegetable waste with high inherent moisture content

PURPA Public Utility Regulator Policies Act (1978)

PV planned value

QA quality assurance

QC quality control

QED *quod erat demonstrandum* (that which was to be demonstrated), needs no further proof

RAM responsibility assignment matrix

RDU remaining duration

RFP request for proposal

RFQ request for quotation

RO reverse osmosis (unit)

RSC radiant syngas cooler

RTU remote terminal unit

SCR selective catalytic reduction

SF scheduled finish date or start-to-finish

simple cycle Brayton (gas) cycle as with simple-cycle gas turbine

SF6 sulfur hexafloride circuit breaker, gas insulated CB

SMACNA Sheet Metal and Air Conditioning Contractors National Association Inc.

SOW statement of work

SPI schedule performance index

SS scheduled start date or start-to-start

SRU sulphur recovery unit

standards set tolerances, whereas procedures define the method

STG steam turbine generator

SSR steam seal regular

SV schedule variance

T leads conductors from a circuit interrupter to a motor

TC target completion date

TDDO time-delay drop-out relay

TDPU time-delay pick-up relay

TF total float or target finish date

TGU tail gas (treatment) unit

TOC truck-operated contact; a contact that interlocks a circuit breaker's position in a cubicle

TO thermal oxidizer

TQM total quality management

tram point(er) indicator used to indicate expansion movement

TS target start date

VE value engineering

WBS work breakdown structure

APPENDIX B

ASME BOILER AND PRESSURE VESSEL CODE LIST

Section I **Power Boilers**
Section II **Materials**
Part A Ferrous Material Specifications
Part B Nonferrous Material Specifications
Part C Specifications for Welding Rods, Electrodes and Filter Metals
Part D Properties (Customary)
Part D Properties (Metric)
Section III **Rules for Construction of Nuclear Facility Components**
Subsection NCA
General Requirements for Division 1 and Division 2
Division 1
Subsection NB
Class 1 Components
Subsection NC
Class 2 Components
Subsection ND
Class 3 Components
Subsection NE
Class MC Components
Subsection NF
Supports
Subsection NG
Core Supports Structures
Subsection NH
Class 1 Components in Elevated Temperature Service
Appendices
Division 2
Code for Concrete Reactor Vessels and Containments
Division 3
Containment System and Transport Packaging for Spent Nuclear Fuel and High Level Radioactive Waste
Section IV **Heating Boilers**
Section V **Nondestructive Examination**
Section VI **Recommended Rules for the Care and Operation of Heating Boilers**
Section VII **Recommended Guidelines for the Care of Power Boilers**

Section VIII Pressure Vessels
Division 1
Division 2 Alternative Rules
Division 3 Alternative Rules for Construction of High Pressure Vessels
Section IX Welding and Brazing Qualifications
Section X Fiber-Reinforced Plastic Pressure Vessels
Section XI Rules for In-Service Inspection of Nuclear Power Plant Components
Section XII Rules for Construction and Continued Service of Transport Tanks

ASME Authorization Stamp List

A	A assembly of power boilers
HV	HV heating boiler safety valves
E	E electric boilers
M	M miniature boilers
H	H cast iron sectional heating boilers
HLW	HLW lined potable water heaters
N	N certificate of authorization and certificate of accreditation
NA	NA certificate of authorization and certificate of accreditation
NPT	NPT certificate of authorization and certificate of accreditation
NS	NS certificate of authorization and certificate of accreditation
U	U pressure vessels (DIV.1)
UM	UM miniature pressure vessels
U2	U2 pressure vessels (DIV.2)
UD	UD rupture disk devices
U3	U3 pressure vessels (DIV.3)
UV	UV pressure vessels safety valves
V	V boilers safety valves
UV3	UV3 pressure vessels safety valves
NV	NV nuclear safety and pressure relief valves
RP	RP fiber-reinforced plastic vessels
T	Transport tanks
N3	N-type certificates of authorization certificates of accreditation: N3
TD	Transport tanks rupture disk devices
PP	PP pressure piping
S	S power boilers
TV	Transport tanks safety valves

APPENDIX C

ASME PERFORMANCE TEST CODE LIST

PTC 1	General Instructions
PTC 2	Definitions and Values
PTC 3.1	Diesel and Burner Fuels
PTC 3.2	Solid Fuels
PTC 3.3	Gaseous Fuels
PTC 4.1	Steam-Generating Units (with 1968 and 1969 Addenda) Diagram for Testing of Steam Generator, Fig. 1 (Pad of 100) Heat Balance of a Steam Generator, Fig. 2 (Pad of 100)
PTC 4.1a	ASME Test Form for Abbreviated Efficiency Test—Summary Sheet (Pad of 100)
PTC 4.1b	ASME Test for Abbreviated Efficiency Test—Calculation Sheet (Pad of 100)
PTC 4.2	Coal Pulverizers
PTC 4.3	Air Heaters
PTC 4.4	Gas Turbine Heat Recovery Steam Generators
PTC 5	Reciprocating Steam Engines
PTC 6A	Appendix A to Test Code for Steam Turbines (with 1958 Addenda)
PTC 6	Guidance for Evaluation of Measurement Uncertainty Report in Performance Test of Steam Turbines
PTC 6S	Producers for Routine Performance Tests Report of Steam Turbines
PTC 6.1	Interim Test Code for and Alternative Procedure for Testing Steam Turbines PTC 6 on Steam Turbines—Interpretations 1977–1983
PTC 7	Reciprocating Steam-Driven Displacement Pumps
PTC 7.1	Displacement Pumps
PTC 8.2	Centrifugal Pumps
PTC 9	Displacement Compressors, Vacuum Pumps, and Blowers (Keith 1972 Errata)
PTC 10	Compressors and Exhausters
PTC 11	Fans
PTC 12.1	Closed Feedwater Heaters
PTC 12.2	Steam—Condensing Apparatus
PTC 12.3	Deaerators
PTC 14	Evaporating Apparatus
PTC 16	Gas Producers and Continuous Gas Generators
PTC 17	Reciprocating Internal-Combustion Engines
PTC 18.1	Hydraulic Prime Movers
PTC 18.1	Pumping Mode of Pump/Turbines

PTC 19.1	Measurement Uncertainty
PTC 19.2	Pressure Measurement
PTC 19.3	Temperature Measurement
PTC 19.5	Application, Part II of Fluid Meters: Interim Supplement on lnstruments and Apparatus
PTC 19.5.1	Weighing Scale
PTC 19.6	Electrical Measurements in Power Circuits
PTC 19,7	Measurement of Shaft Power
PTC 19.8	Measurement of Indicated Horsepower
PTC 19.11	Flue and Exhaust Gas Analyses
PTC 19.11	Water and Steam in the Power Cycle (Purity and Quality, Lead Detection and Measurement)
PTC 19.12	Measurement of Time
PTC 19.13	Measurement of Rotary Speed
PTC 19.14	Linear Measurements
PTC 19.16	Density Determinations of Solids and Liquids
PTC 19.17	Determination of the Viscosity of Liquids
PTC 19.22	Digital Systems Techniques
PTC 19.23	Guidance Manual for Model Testing
TPC 20.1	Speed and Load Governing Systems for Steam Turbine-Generator Units
PTC 20.2	Overspeed Trip Systems for Steam Turbine-Generator Units
PTC 20.3	Pressure Control Systems Used on Steam Turbine-Generator Units
PTC 21	Particulate Matter Collection Equipment
PTC 22	Gas Turbine Power Plants
PTC 23	Atmospheric Water Cooling Equipment
PTC 23.1	Spray Cooling Systems
PTC 24	Ejectors
PTC 25.3	Safety and Relief Valves
PTC 26	Speed Governing Systems for Internal Combustion Engine-Generator Units
PTC 28	Determining the Properties of Fine Particulate Matter
PTC 29	Speed Governing Systems for Hydraulic Turbine-Generator Units
PTC 30	Air Cooled Heat Exchangers
PTC 31	Ion Exchange Equipment
PTC 32.1	Nuclear Steam Supply Systems
PTC 32.2	Methods of Measuring the Performance of Nuclear Reactor Fuel in Light Water Reactors
PTC 33	Large Incinerators
PTC 33a	Appendix to PTC 33-1978—ASME Form for Abbreviated Incinerator Efficiency Test (PTC 33[a]–1980)
PTC 36	Measurement of Industrial Sound
PTC 38	Determining the Concentration of Particulate Matter in a Gas Stream
PTC 39.1	Condensate Removal Devices for Steam Systems
PTC 40	Flue Gas Desulfurization Units
PTC 42	Wind Turbines
PTC 46	Performance Test Code on Overall Plant Performance

APPENDIX D

MANUFACTURERS STANDARDIZATION SOCIETY (MSS) STANDARDS LIST

SP-6-1996	Standard Finishes for Contact Faces of Pipe Flanges and Connecting-End Flanges of Valves and Fittings
SP-9-1997	Spot Facing for Bronze, Iron, and Steel Flanges
SP-25-1998	Standard Marking System for Valves, Fittings, Flanges, and Unions
SP-42-1990	(R 95) Class 150 Corrosion Resistant Gate, Globe, Angle, and Check Valves with Flanged and Butt Weld Ends
SP-43-1991	(R 96) Wrought Stainless Steel Butt-Welding Fittings
SP-44-1996	Steel Pipeline Flanges
SP-45-1992	Bypass and Drain Connections
SP-51-1991	(R 95) Class 150LW Corrosion Resistant Cast Flanges and Flanged Fittings
SP-53-1995	Quality Standard for Steel Castings and Forgings for Valves, Flanges, and Fittings and Other Piping Components—Magnetic Particle Examination Method
SP-54-1995	Quality Standard for Steel Castings for Valves, Flanges, and Fittings Other Piping Components—Radiographic Examination Method
SP-55-1996	Quality Standard for Steel Castings for Valves, Flanges, and Fittings Other Piping Components—Visual Method for Evaluation of Surface Irregularities
SP-58-1993	Pipe Hangers and Supports—Materials, Design, and Manufacture
SP-60-1991	Connecting Flange Joint Between Tapping Sleeves and Tapping Valves
SP-61-1992	Pressure Testing of Steel Valves
SP-65-1994	High Pressure Chemical Industry Flanges and Threaded Stubs for Use with Lens Gaskets
SP-67-1995	Butterfly Valves
SP-68-1997	High Pressure Butterfly Valves with Offset Design
SP-69-1996	Pipe Hangers and Supports—Selection and Application
SP-70-1990	Cast Iron Gate Valves, Flanged and Threaded Ends
SP-71-1997	Gray Iron Swing Check Valves, Flanged and Threaded Ends
SP-72-1992	Ball Valves with Flanged or Butt-Welding Ends for General Service
SP-73-1991	(R 96) Brazing Joints for Wrought and Cast Cooper Alloy Solder Joint Pressure Fittings
SP-75-1993	Specification for High Test Wrought Butt-Welding Fittings

SP-77-1995	Guidelines for Pipe Support Contractual Relationships
SP-78-1987	(R 92) Cast Iron Plug Valves, Flanged and Threaded Ends
SP-79-1992	Socket-Welding Reducer Inserts
SP-80-1997	Bronze Gate, Globe, Angle, and Check Valves
SP-81-1995	Stainless Steel, Bonnetless, Flanged, Knife Gate Valves
SP-82-1992	Valve Pressure Testing Methods
SP-83-1995	Class 3000 Steel Pipe Unions, Socket-Welding and Threaded
SP-85-1994	Cast Iron Globe and Angle Valves, Flanged and Threaded Ends
SP-86-1997	Guidelines for Metric Data in Standards for Valves, Flanges, Fittings, and Actuators
SP-87-1991	(R 96) Factory-Made Butt-Welding Fittings for Class 1 Nuclear Piping Applications
SP-88-1993	Diaphragm Type Valves
SP-89-1998	Pipe Hangers and Supports—Fabrication and Installation Practices
SP-90-1986	(R 91) Guidelines on Terminology for Pipe Hangers and Supports
SP-91-1992	(R 96) Guidelines for Manual Operation of Valves
SP-92-1987	(R 92) MSS Valve User Guide
SP-93-1987	(R 92) Quality Standard for Steel Castings and Forgings for Valves, Flanges, and Fittings and Other Piping Components—Liquid Penetrant Examination Method
SP-94-1992	Quality Standard for Ferritic and Martensitic Steel Casting for Valves, Flanges, and Fittings and Other Piping Components—Ultrasonic Examination Method
SP-95-1986	(R 91) Swage (d) Nipples and Bull Plugs
SP-96-1996	Guidelines on Terminology for Valves and Fittings
SP-97-1995	Integrally Reinforced Forged Branch Outled Fittings—Socket Welding, Threaded and Buttwelding Ends
SP-98-1996	Protective Coating for the Interior of Valves, Hydrants, and Fittings
SP-99-1994	Instrument Valves
SP-100-1997	Qualification Requirements for Elastomer Diaphragms for Nuclear Service Diaphragm Type Valves
SP-101-1989	Part-Turn Valve Actuator Attachment—Flange and Driving Component Dimensions and Performance Characteristics
SP-102-1989	Multi-Turn Valve Actuator Attachment—Flange and Driving Component Dimensions and Performance Characteristics
SP-103-1995	Wrought Copper and Copper Alloy Insert Fittings for Polybutylene Systems
SP-104-1995	Wrought Copper Solder Joint Pressure Fittings
SP-105-1996	Instruments Valves for Code Applications
SP-106-1990	(R 96) Cast Copper Alloy Flanges and Flanged Fittings, Class 125, 150, and 300
SP-107-1991	Transition Union Fittings for Joining Metal and Plastic Products
SP-108-1996	Resilient-Seated Cast Iron-Eccentric Plug Valves
SP-109-1997	Welded Fabricated Copper Solder Joint Pressure Fittings
SP-110-1996	Ball Valves Threaded, Socket-Welding, Solder Joint, Grooved and Flared Ends
SP-111-1996	Gray-Iron and Ductile-Iron Tapping Sleeves
SP-112-1993	Quality Standard for Evaluation of Cast Surface Finishes—Visual and Tactile Method. [This SP must be sold with a 10-surface, three-dimensional Cast Surface Comparator, which is a necessary part of the Standard. Additional Comparators may be sold separately at $19.00 each. Same quantity discounts apply on total order.]

SP-113-1994	Connecting Joint between Tapping Machines and Tapping Valves
SP-114-1985	Corrosion Resistant Pipe Fittings Threaded and Socket Welding, Class 150 and 1000
SP-115-1995	Excess Flow Valves for Natural Gas Service
SP-116-1996	Service Line Valves and Fittings for Drinking Water Systems
SP-117-1996	Bellows Seals for Globe and Gate Valves
SP-118-1996	Compact Steel Globe and Check Valves—Flanged, Flangeless, Threaded, and Welding Ends (Chemical and Petroleum Refinery Service)
SP-119-1996	Belled End Socket-Welding Fittings, Stainless Steel and Copper Nickel
SP-120-1997	Flexible Graphite Packing System for Rising Stem Steel Valves (Design Requirements)
SP-121-1997	Qualification Testing Methods for Stem Packing for Rising Stem Steel Valves
SP-122-1997	Plastic Industrial Ball Valves

(P-YEAR) Indicates year standard reaffirmed without substantive changes. A large number of former MSS practices have been approved by the ANSI or ANSI Standards, published by others. In order to maintain a single source of authoritative information, the MSS withdraws its Standard Practices in such cases.

Manufacturers Standardization Society of the Valve and Fittings Industry Inc., 127 Park Street, N.E., VA 22180-4602 / (703) 281-6613 / Fax (703) 281-6671.

APPENDIX E
ASTM SPECIFICATIONS LIST

A 36	Structural Steel
A 37	Ferritic Malleable Iron Castings
A 48	Gray Iron Castings
A 53	Pipe, Steel, Black and Hot-Dipped, Zinc Coated, Welded and Seamless
A 105	Forgings, Carbon Steel, for Piping Components
A 106	Seamless Carbon Steel Pipe for High-Temperature Service
A 126	Gray Cast Iron Castings for Valves, Flanges, and Pipe Fittings
A 134	Pipe, Steel, Electric-Fusion (Arc)-Welded (Sizes NPS 16 and Over)
A 135	Electric-Resistance-Welded Steel Pipe
A 139	Electric Fusion (Arc)-Welded Steel Pipe (NPS 4 and Over)
A 167	Stainless and Heat-Resisting Chromium-Nickel Steel Plate, Sheet and Strip
A 179	Seamless Cold-Drawn Low-Carbon Steel Heat-Exchanger and Condenser Tubes
A 181	Forgings, Carbon Steel for General Purpose Piping
A 182	Gorged or Rolled Alloy-Steel Pipe Flanges, Forged Fittings, and Valves and Parts for High-Temperature Service
A 197	Cupola Malleable Iron
A 202	Pressure Vessel Plates, Alloy Steel, Chromium-Manganese-Silicon
A 203	Pressure Vessel Plates, Alloy Steel, Nickel
A 204	Pressure Vessel Plates, Alloy Steel, Molybdenum
A 216	Steel Castings, Carbon, Suitable for Fusion Welding for High-Temperature Service
A 217	Steel Castings, Martensitic Stainless and Alloy, for Pressure-Containing Parts Suitable for High-Temperature Service
A 234	Piping Fittings of Wrought Carbon Steel and Alloy Steel for Moderate and Elevated Temperatures
A 240	Heat-Resisting Chromium and Chromium Nickel Stainless Steel Plate, Sheet, and Strip for Pressure Vessels
A 268	Seamless and Welded Ferritic Stainless Steel Tubing for General Service
A 269	Seamless and Welded Austenitic Stainless Steel Tubing for General Service
A 278	Gray Iron Castings for Pressure-Containing Parts for Temperatures Up to 650°
A 283	Low and Intermediate Tensile Strength Carbon Steel Plates, Shapes, and Bars
A 285	Pressure Vessel Plates, Carbon Steel, Low- and Intermediate-Tensile Strength
A 299	Pressure Vessel Plates, Carbon Steel, Manganese-Silicon
A 302	Pressure Vessel Plates, Alloy Steel, Manganese-Molybdenum, and Manganese-Molybdenum-Nickel
A 312	Seamless and Welded Austenitic Stainless Steel Pipe
A 333	Seamless and Welded Steel Pipe for Low-Temperature Service

A 334	Seamless and Welded Carbon and Alloy-Steel Tubes for Low-Temperature Service
A 335	Seamless Ferritic Alloy Steel Pipe for High-Temperature Service
A 350	Forgings, Carbon and Low-Alloy Steel Requiring Notch Toughness Testing for Piping Components
A 351	Steel Castings, Austenitic, Austenitic-Ferritic (Duplex) for Pressure-Containing Parts
A 352	Steel Castings, Ferritic and Martensitic, for Pressure-Containing Parts Suitable for Low-Temperature Service
A 353	Pressure Vessel Plates, Alloy Steel, 9 Percent Nickel, Double Normalized and Tempered
A 358	Electric-Fusion-Welded Austenitic Chromium-Nickel Alloy Steel Pipe for High-Temperature Service
A 369	Carbon Steel and Ferritic Alloy Steel Forged and Bored Pipe for High-Temperature Service
A 376	Seamless Austenitic Steel Pipe for High-Temperature Central-Station Service
A 381	Metal-Arc-Welded Steel Pipe for Use with High-Pressure Transmission Systems
A 387	Pressure Vessel Plates, Alloy Steel, Chromium-Molybdenum
A 395	Ferritic Ductile Iron Pressure-Retaining Castings for Use at Elevated Temperatures
A 403	Welded Austenitic Stainless Steel Piping Fittings
A 409	Welded Large Diameter Austenitic Steel Pipe for Corrosive or High-Temperature Service
A 420	Piping Fittings of Wrought Carbon Steel and Alloy Steel for Low-Temperature Service
A 426	Centrifugally Cast Ferritic Alloy Steel Pipe for High-Temperature Service
A 451	Centrifugally Cast Ferritic Alloy Steel Pipe for High-Temperature Service
A 479	Stainless and Heat-Resisting Steel Bars and Shapes for Use in Boilers and Other Pressure Vessels
A 487	Steel Castings Suitable for Pressure Service
A 494	Castings, Nickel and Nickel Alloy
A 515	Pressure Vessel Plates, Carbon Steel, for Intermediate and Higher-Temperature Service
A 516	Pressure Vessel Plates, Carbon Steel, for Moderate and Lower-Temperature Service
A 524	Seamless Carbon Steel Pipe for Atmospheric and Lower Temperatures
A 537	Pressure Vessel Plates, Heat-Treated, Carbon-Manganese-Silicon Steel
A 553	Pressure Vessel Plates, Alloy Steel, Quenched and Tempered 8 and 9 Percent Nickel
A 570	Hot-Rolled Carbon Steel Sheet and Strip, Structural Quality
A 571	Austenitic Ductile Iron Castings for Pressure-Containing Parts Suitable for Low-Temperature Service
A 587	Electric-Welded Low-Carbon Steel Pipe for the Chemical Industry
A 645	Pressure Vessel Plates, 5 Percent Nickel Alloy Steel, Specially Heat Treated
A 671	Electric-Fusion-Welded Steel Pipe for Atmospheric and Lower Temperatures
A 672	Electric-Fusion-Welded Steel Pipe for High-Pressure Service at Moderate Temperatures
A 691	Carbon and Alloy Steel Pipe, Electric Fusion-Welded for High-Pressure Service at High Temperatures
A 789	Seamless and Welded Ferritic/Austenitic Stainless Steel Tubing for General Service

A 790	Seamless and Welded Ferritic/Austenitic and Martensitic Stainless Steel Fittings
A 815	Wrought Ferritic, Ferritic/Austenitic Stainless Steel Pipe
B 21	Naval Brass Rod, Bar, and Shapes
B 26	Aluminum-Alloy Sand Castings
B 42	Seamless Copper Pipe, Standard Sizes
B 43	Seamless Red Brass Pipe, Standard Sizes
B 61	Steam or Valve Bronze Castings
B 62	Composition Bronze or Ounce Metal Castings
B 68	Seamless Copper Tube, Bright Annealed
B 75	Seamless Copper Tube
B 88	Seamless Copper Water Tube
B 96	Copper-Silicon Alloy Plate, Sheet, Strip, and Rolled Bar for General Purposes and Pressure Vessels
B 98	Copper-Silicon Alloy Rod, Bar, and Shapes
B 127	Nickel-Copper Alloy (UNS N04400) Plate, Sheet, and Strip
B 133	Copper Rod, Bar, and Shapes
B 148	Aluminum-Bronze Castings
B 150	Aluminum-Bronze Rod, Bar, and Shapes
B 152	Copper Sheet, Strip, Plate, and Rolled Bar
B 160	Nickel Rod and Bar
B 161	Nickel Seamless Pipe and Tube
B 162	Nickel Plate, Sheet, and Strip
B 164	Nickel-Copper Alloy Rod, Bar, and Wire
B 165	Nickel-Copper Alloy (UNS N04400) Seamless Pipe and Tube
B 166	Nickel-Chromium-Iron Alloy (UNS N06600) Rod, Bar, and Wire
B 167	Nickel-Chromium-Iron Alloy (UNS N06600-N06690) Seamless Pipe and Tube
B 168	Nickel-Chromium-Iron Alloy (UNS N06600-N06690) Plate, Sheet, and Strip
B 169	Aluminum Bronze Plate, Sheet, Strip, and Rolled Bar
B 171	Copper-Alloy Condenser Tube Plates
B 187	Copper Bar, Bus Bar, Rod, and Shapes
B 209	Aluminum and Aluminum-Alloy Sheet and Plate
B 210	Aluminum-Alloy Drawn Seamless Tubes
B 211	Aluminum-Alloy Bars, Rods, Wire, Shapes, and Tubes
B 221	Aluminum-Alloy Extruded Bars, Rods, Wire, Shapes, and Tubes
B 241	Aluminum-Alloy Seamless Pipe and Seamless Extruded Tube
B 247	Aluminum-Alloy Die, Hand and Rolled Ring Forgings
B 280	Seamless Copper Tube for Air Conditioning and Refrigeration Fluid Service
B 283	Copper and Copper-Alloy Die Forgings (Hot-Pressed)
B 265	Titanium and Titanium Alloy Strip, Sheet, and Plate
B 333	Nickel-Molybdenum Alloy Plate, Sheet, and Strip
B 335	Nickel-Molybdenum Alloy Rod
B 337	Seamless and Welded Titanium and Titanium Alloy Pipe
B 345	Aluminum-Alloy Seamless Extruded Tube and Seamless Pipe for Gas and Oil Transmission and Distribution Piping Systems
B 361	Factory-Made Wrought Aluminum and Aluminum Welding Fittings
B 366	Factory-Made Wrought Nickel and Nickel-Alloy Welding Fittings
B 381	Titanium and Titanium Alloy Forgings
B 407	Nickel-Iron-Chromium Alloy Seamless Pipe and Tube
B 409	Nickel-Iron-Chromium Alloy Plate, Sheet, and Strip
B 423	Nickel-Iron-Chromium-Molybdenum-Copper Alloy (UNS N08825 and N08221) Seamless Pipe and Tube

B 424	Nickel-Iron-Chromium-Molybdenum-Copper Alloy (UNS N08825 and N08221) Plate, Sheet, and Strip
B 425	Nickel-Iron-Chromium-Molybdenum-Copper Alloy (UNS N08825 and N08221) Rod and Bar
B 435	UNS N06022, UNS N06230, and UNS R30556 Plate, Sheet, and Strip
B 443	Nickel-Chromium-Molybdenum-Columbium Alloy (UNS N06625) Plate, Sheet, and Strip
B 444	Nickel-Chromium-Molybdenum-Columbium Alloy (UNS N06625) Seamless Pipe and Tube
B 446	Nickel-Chromium-Molybdenum-Columbium Alloy (UNS N06625) Rod and Bar
B 462	Forged or Rolled UNS N08020, UNS N08024, UNS N08026, and UNS N08367 Alloy Pipe Fittings, and Valves and Parts for Corrosive High-Temperature Service
B 463	Forged or Rolled UNS N08020, UNS N08026, and UNS N08024 Alloy Plate, Sheet, and Strip
B 464	Welded Chromium-Nickel-Iron-Molbdenum-Coper-Columbium Stabilized Alloy (UNS N08020) Pipe
B 466	Seamless Copper-Nickel Pipe and Tube
B 467	Welded Copper-Nickel Pipe
B 491	Aluminum and Aluminum Alloy Extrude Round Tubes for General-Purpose Applications
B 493	Zirconium and Zirconium Alloy Forgings
B 514	Welded Nickel-Iron-Chromium Alloy Pipe
B 517	Welded Nickel-Chromium-Iron Alloy (UNS N06600, UNS N06603, UNS N06025, and UNS N06045) Pipe
B 523	Seamless and Welded Zirconium Alloy Tubes
B 547	Aluminum and Aluminum-Alloy Formed and Arc-Welded Round Tube
B 550	Zirconium and Zirconium Alloy Bar and Wire
B 551	Zirconium and Zirconium Alloy Strip, Sheet, and Plate
B 564	Nickel Alloy Forgings
B 574	Low-Carbon Nickel-Molybdenum-Chromium Alloy Rod
B 575	Low-Carbon Nickel-Molybdenum-Chromium Alloy Plate, Sheet, and Strip
B 581	Nickel-Chromium-Iron-Molybdenum-Copper Alloy Rod
B 582	Nickel-Chromium-Iron-Molybdenum-Copper Alloy Plate, Sheet, and Strip
B 584	Copper Alloy Sand Castings for General Applications
B 619	Welded Nickel and Nickel-Cobalt Alloy Pipe
B 620	Nickel-Iron-Chromium-Molybdenum Alloy (UNS N08320) Plate, Sheet, and Strip
B 621	Nickel-Iron-Chromium-Molybdenum-Copper Alloy (UNS N08320) Rod
B 622	Seamless Nickel and Nickel-Cobalt Alloy Pipe and Tube
B 625	Nickel Alloy Plate and Sheet
B 649	Ni-Fe-Cr-Mo-Cu Low Carbon Alloy (UNS N08904) and Ni-Fe-Cr-Mo-Cu-N Low Carbon Alloy (UNS N08925, UNS N08031, and UNS N08926) Bar and Wire
B 658	Zirconium and Zirconium Alloy Seamless and Welded Pipe
B 675	UNS N08366 and UNS N08367 Welded Pipe
B 688	Chromium-Nickel-Molybdenum-Iron (UNS N08366 and UNS N08367) Seamless Pipe and Tube
B 705	Nickel-Alloy (UNS N06625 and N08825) Welded Pipe

B 725	Welded Nickel (UNS N0220/UNS N02201) and Nickel-Copper Alloy (UNS N04400) Pipe
B 729	Seamless UNS N08020, UNS N08026, UNS N08024 Nickel-Alloy Pipe and Tube
B 804	UNS N08367 Welded Pipe
E 112	Methods for Determining Average Grain Size API
5L	Line Pipe

General Note: It is not practical to refer to a specific edition of each standard throughout the Code text. Instead, the approved edition references, along with the names and addresses of the sponsoring organizations, are shown in Appendix E (Appendix F, this publication).

APPENDIX F
SOCIETY CONTACTS LIST

ASME B31.3-2004 APPENDIX E

Specifications and standards of the following organizations appear in Appendix E of ASME B 31.3-2004:

API American Petroleum Institute
Publications and Distribution Section
1220 L Street, NW
Washington, DC 20005-4070
202 682-8375
www.api.org

ASCE The American Society of Civil Engineers
1801 Alexander Bell Drive
Reston, Virginia 20191-4400
703 295-6300 or 800 548-2723
www.asce.org

ASME ASME International
Three Park Avenue
New York, New York 10016-5990
212 591-8500 or 800 843-2763
www.asme.org

ASME Order Department
22 Law Drive
Box 2900
Fairfield, New Jersey 07007-2300
973 882-1170 or 800 843-2763

ASNT American Society for Nondestructive Testing, Inc.
P.O. Box 28518
1711 Arlingate Lane
Columbus, Ohio 43228-0518
614 274-6003 or 800 222-2768
www.asnt.org

ASQ	American Society for Quality 611 East Wisconsin Ave. Milwaukee, Wisconsin 53202 800 248-1946 www.asq.org
ASTM	American Society for Testing and Materials 100 Barr Harbor Drive West Conshohocken, Pennsylvania 19428-2959 610 832-9500 www.astm.org
AWWA	American Water Works Association 6666 W. Quincy Avenue Denver, Colorado 80235 303 794-7711 or 800 926-7337 www.awwa.org
AWS	American Welding Society 550 NW LeJeune Road Miami, Florida 33126 305 443-9353 or 800 443-9353 www.aws.org
CDA	Copper Development Association, Inc. 260 Madison Avenue, 16th Floor New York, New York 10016 212 251-7200 or 800 232-3282 www.copper.org
CGA	Compressed Gas Association, Inc. 1725 Jefferson Davis Highway, Suite 1004 Arlington, Virginia 22202-4102 703 412-0900 www.cganet.com
CSA	CSA International 178 Rexdale Boulevard Etobicoke (Toronto), Ontario M9W 1R3, Canada 416 747-2620 or 800 463-6727 www.csa-international.org
EJMA	Expansion Joint Manufacturers Association 25 North Broadway Tarrytown, New York 10591 914 332-0040 www.ejma.org

MSS	Manufactures Standardization Society of the Valve and Fittings Industry, Inc. 127 Park Street, NE Vienna, Virginia 22180-4602 703 281-6613 www.mss-hq.com
NACE	NACE International 1440 S. Creek Drive Houston, Texas 77084 281 228-6200 www.nace.org
NFPA	National Fire Protection Association 1 Batterymarch Park Quincy, Massachusetts 02269 617 770-3000 or 800 344-3555 www.nfpa.org
PFI	Pipe Fabrication Institute 655-32nd Avenue, Suite 201 Lachine, Quebec H8T3G6 Canada 514 634-3434 www.pfi-institute.org
PPI	Plastics Pipe Institute 1801 K Street NW, Suite 600K Washington, DC 20006-1301 202 974-5318 or 800 541-0736 www.plasticpipe.org
SAE	Society of Automotive Engineers 400 Commonwealth Drive Warrendale, Pennsylvania 15096-0001 724 776-4970 or 800 832-6723 www.sae.org

General Note to List of Organizations: Some of the organizations listed above publish standards that have been approved as American National Standards. Copies of these standards may also be obtained from:

ANSI American National Standards Institute, Inc.
11 West 42nd Street
New York, New York 10036
212 642-4900
www.ansi.org

APPENDIX G
ELECTRICAL DEVICE NUMBERS AND FUNCTIONS

This device list was extracted from the Silent Sentinels Publication (Westinghouse), Applied Protective Relaying, 1982.

Silent Sentinel publications from 1960–1989 have been formatted on CD and are available free. Readers can e-mail tyler.j.kerner@us.abb.com to request a free copy. Or, browse for "abb silent sentinels" on the Internet.

The devices in switching equipment are referred to by numbers, with appropriate suffix letters when necessary, according to the functions they perform.

These numbers are based on a system adopted as standard for automatic switchgear by IEEE, and incorporated in American Standard C37.2-1970. This system is used in connection diagrams, in instruction books, and in specifications.

1 **Master element** is the initiating device, such as a control switch, voltage relay, float switch, etc., that serves directly, or through such permissive devices as protective and time-delay relays, to place an equipment in or out of operation.

2 **Time-delay starting, or closing, relay** is a device that functions to give a desired amount of time delay before or after any point of operation in a switching sequence or protective relay system, except as specifically provided by device functions 48, 62, and 79 described later.

3 **Checking or interlocking relay** is a device that operates in response to the position of a number of other devices (or to a number of predetermined conditions), in an equipment, to allow an operating sequence to proceed, to stop, or to provide a check of the position of these devices or of these conditions for any purpose.

4 **Master contactor** is a device, generally controlled by device function 1 or equivalent, and the required permissive and protective devices, that serves to make and break the necessary control circuits to place an equipment into operation under the desired conditions and to take it out of operation under other or abnormal conditions.

5 **Stopping device** is a control device used primarily to shut down an equipment and hold it out of operation. [This device may be manually or electrically actuated, but excludes the function of electrical lockout (see device function 86) on abnormal conditions.]

6 **Starting circuit breaker** is a device whose principal function is to connect a machine to its source of starting voltage.

7 **Anode circuit breaker** is one used in the anode circuits of a power rectifier for the primary purpose of interrupting the rectifier circuit if an arc-back should occur.

8 Control power disconnecting device is a disconnective device—such as a knife switch, circuit breaker, or pullout fuse block—used for the purpose of connecting and disconnecting the source of control power to and from the control bus or equipment. *Note*: Control power is considered to include auxiliary power, which supplies such apparatuses as small motors and heaters.

9 Reversing device is used for the purpose of reversing a machine field or for performing any other reversing functions.

10 Unit sequence switch is used to change the sequence in which units may be placed in and out of service in multiple-unit equipments.

11 Reserved for future application.

12 Overspeed device is usually a direct-connected speed switch that functions on machine overspeed.

13 Synchronous-speed device, such as centrifugal-speed switch, a slip-frequency relay, a voltage relay, an undercurrent relay, or any type of device, operates at approximately synchronous speed of a machine.

14 Underspeed device functions when the speed of a machine falls below a predetermined value.

15 Speed or frequency, matching device functions to match and hold the speed or the frequency of a machine or of a system equal to, or approximately equal to, that of another machine, source or system.

16 Reserved for future application.

17 Shunting or discharge switch serves to open or to close a shunting circuit around any piece of apparatus (except a resistor), such as a machine field, a machine armature, a capacitor, or a reactor. *Note*: This excludes devices that perform such shunting operations as may be necessary in the process of starting a machine by devices 6 or 42, or their equivalent, and also excludes device 73 function, which serves for the switching of resistors.

18 Accelerating or decelerating device is used to close or to cause the closing of circuits that are used to increase or decrease the speed of a machine.

19 Starting-to-running transition contactor is a device that operates to initiate or cause the automatic transfer of a machine from the starting to the running power connection.

20 Electrically operated valve is an electrically operated, controlled, or monitored valve in a fluid line. *Note*: The function of the valve may be indicated by the use of suffixes.

21 Distance relay is a device that functions when the circuit admittance, impedance, or reactance increases or decreases beyond predetermined limits.

22 Equalizer circuit breaker is a breaker that serves to control or to make and break the equalizer or the current-balancing connections for a machine field, or for regulating equipment, in a multiple-unit installation.

23 Temperature control device functions to raise or lower the temperature of a machine or other apparatus, or of any medium, when its temperature falls below, or rises above, a predetermined value. *Note*: An example is a thermostat that switches on a space heater in a switchgear assembly when the temperature falls to a desired value as distinguished from a device that is used to provide automatic temperature regulation between close limits and would be designated as 90T.

24 Reserved for future application.

25 Synchronizing or synchronism-check device operates when two AC circuits are within the desired limits of frequency, phase angle, or voltage to permit or to cause the paralleling of these two circuits.

26 **Apparatus thermal device** functions when the temperature of the shunt field or the armortisseur winding of a machine, or of a load-limiting or load-shifting resistor, or of a liquid or other medium exceeds a predetermined value, or if the temperature of the protected apparatus, such as a power rectifier, or of any medium decreases below a predetermined value.

27 **Undervoltage relay** is a device that functions on a given value of undervoltage.

28 **Flame detector** is a device that monitors the presence of the pilot or main flame in such apparatus as a gas turbine or a steam boiler.

29 **Isolating contactor** is used expressly for disconnecting one circuit from another for the purposes of emergency operation, maintenance, or test.

30 **Annunciator relay** is a nonautomatically reset device that gives a number of separate visual indications upon the functioning of protective devices and that may also be arranged to perform a lockout function.

31 **Separate excitation device** connects a circuit, such as the shunt field of a synchronous converter, to a source of separate excitation during the starting sequence; one that energizes the excitation and ignition circuits of a power rectifier.

32 **Directional power relay** is one that functions on a desired value of power flow in a given direction or upon reverse power resulting from arc-back in the anode or cathode circuits of a power rectifier.

33 **Position switch** makes or breaks contact when the main device or piece of apparatus, which has no device function number, reaches a given position.

34 **Master sequence device** is a device such as a motor-operated multi-contact switch, or the equivalent, or a programming device, such as a computer, that establishes or determines the operating sequence of the major devices in an equipment during other sequential switching operations.

35 **Brush-operating, or slip-ring-short-circuiting, device** is used for raising, lowering, or shifting the brushes of a machine; for short-circuiting its slip rings; or for engaging or disengaging the contacts of a mechanical rectifier.

36 **Polarity or polarizing voltage device** operates or permits the operation of another device on a predetermined polarity only or verifies the presence of a polarizing voltage in an equipment.

37 **Undercurrent or underpower relay** functions when the current or power flow decreases below a predetermined value.

38 **Bearing protective device** functions on excessive bearing temperature, or other abnormal mechanical conditions, such as undue wear, that may eventually result in excessive bearing temperature.

39 **Mechanical condition monitor** is a device that functions upon the occurrence of an abnormal mechanical condition (except that associated with bearings as covered under device function 38), such as excessive vibration, eccentricity, expansion, shock, tilting, or seal failure.

40 **Field relay** functions on a given or abnormally low value or failure of machine field current, or on an excessive value of the reactive component of armature current in an AC machine indicating abnormally low field excitation.

41 **Field circuit breaker** is a device that functions to apply, or remove, the field excitation of a machine.

42 **Running circuit breaker** is a device whose principal function is to connect a machine to its source of running or operating voltage. This function may also be used for a

device, such as a contactor, that is used in series with a circuit breaker or other fault-protecting means, primarily for frequent opening and closing of the circuit.

43 **Manual transfer or selector device** transfers the control circuits so as to modify the plan of operation of the switching equipment or of some of the devices.

44 **Unit sequence starting relay** is a device that functions to start the next available unit in a multiple-unit equipment on the failure or on the nonavailability of the normally preceding unit.

45 **Atmospheric condition monitor** is a device that functions upon the occurrence of an abnormal atmospheric condition, such as damaging fumes, explosive mixtures, smoke, or fire.

46 **Reverse-phase, or phase-balance, current relay** is a relay that functions when the polyphase currents are of reverse-phase sequence or when the polyphase currents are unbalanced or contain negative phase-sequence components above a given amount.

47 **Phase-sequence voltage relay** functions upon a predetermined value of polyphase voltage in the desired phase sequence.

48 **Incomplete sequence relay** is a relay that generally returns the equipment to the normal, or off, position and locks it out if the normal starting, operating, or stopping sequence is not properly completed within a predetermined time. If the device is used for alarm purposes only, it should preferably by designated as 48A (alarm).

49 **Machine, or transformer, thermal relay** is a relay that functions when the temperature of a machine armature, or other load-carrying winding or element of a machine, or the temperature of a power rectifier or power transformer (including a power rectifier transformer) exceeds a predetermined value.

50 **Instantaneous overcurrent, or rate-of-rise relay** is a relay that functions instantaneously on an excessive value of current or an excessive rate of current rise, thus indicating a fault in the apparatus or circuit being protected.

51 **AC time overcurrent relay** is a relay with either a definite or inverse time characteristic that functions when the current in an AC circuit exceeds a predetermined value.

52 **AC circuit breaker** is a device that is used to close and interrupt an AC power circuit under normal conditions or to interrupt this circuit under fault or emergency conditions.

53 **Exciter or DC generator relay** is a relay that forces the DC machine field excitation to build up during starting or that functions when the machine voltage has built up to a given value.

54 Reserved for future application.

55 **Power factor relay** is a relay that operates when the power factor in an AC circuit rises above or below a predetermined value.

56 **Field application relay** is a relay that automatically controls the application of the field excitation to an AC motor at some predetermined point in the slip cycle.

57 **Short-circuiting or grounding device** is a primary circuit switching device that functions to short-circuit or ground a circuit in response to automatic or manual means.

58 **Rectification failure relay** is a device that functions if one or more anodes of a power rectifier fail to fire or detect an arc-back, or on failure of a diode to conduct or block properly.

59 **Overvoltage relay** is a relay that functions on a given value of overvoltage.

60 **Voltage or current balance relay** is a relay that operates on a given difference in voltage or on the current input or output of two circuits.

61 Reserved for future application.

62 **Time-delay stopping or opening relay** is a time-delay relay that serves in conjunction with the device that initiates the shutdown, stopping, or opening operation in an automatic sequence.

63 **Pressure switch** is a switch that operates on given values or on a given rate of change of pressure.

64 **Ground-protective relay** is a relay that functions on failure of the insulation of a machine, transformer, or other apparatus to ground, or on flashover of a DC machine to ground. *Note*: This function is assigned only to a relay that detects the flow of current from the frame of a machine or enclosing case or structure of a piece of apparatus to ground, or detects a ground on a normally ungrounded winding or circuit. It is not applied to a device connected in the secondary circuit or secondary neutral of a current transformer or in the secondary neutral of current transformer, connected in the power circuit of a normally grounded system.

65 **Governor** is the assembly of fluid, electrical, or mechanical control equipment used for regulating the flow of water, steam, or other medium to the prime mover for such purposes as starting, holding speed or load, or stopping.

66 **Notching or jogging device** functions to allow only a specified number of operations of a given equipment, or specified number of successive operations within a given time of each other. It also functions to energize a circuit, periodically or for fractions of specified time intervals, that is used to permit intermittent acceleration or jogging of a machine at low speeds for mechanical positioning.

67 **AC directional overcurrent relay** is relay that functions on a desired value of AC overcurrent flowing in a predetermined direction.

68 **Blocking relay** is a relay that initiates a pilot signal for blocking of tripping on external faults in a transmission line or other apparatus under predetermined conditions, or cooperates with other devices to block tripping or block reclosing on an out-of-step condition or on power swings.

69 **Permissive control device** is generally a two-position, manually operated switch that, in one position permits the closing of a circuit breaker or the placing of an equipment into operation and, in the other position, prevents the circuit breaker or the equipment from being operated.

70 **Rheostat** is a variable-resistance device used in an electric circuit, which is electrically operated or has other electrical accessories, such as auxiliary, position, or limit switches.

71 **Level switch** is a switch that operates on given values, or on a given rate of change, of level.

72 **DC circuit breaker** is used to close and interrupt a DC power circuit under normal conditions or to interrupt this circuit under fault or emergency conditions.

73 **Load-resistor contactor** is used to shunt or insert a step of load limiting, shifting, or indicating resistance in a power circuit; to switch a space heater in circuit; or to switch light, or regenerative load resistor, of a power rectifier or other machine in and out of circuit.

74 **Alarm relay** is a device other than an annunciator, as covered under device function 30, that is used to operate, or operate in connection with, a visual or audible alarm.

75 **Position-changing mechanism** is a mechanism that is used for moving a main device from one position to another in an equipment; for example, shifting a removable circuit breaker unit to and from the connected, disconnected, and test positions.

76 DC overcurrent relay is a relay that functions when the current in a DC circuit exceeds a given value.

77 Pulse transmitter is used to generate and transmit pulses over a telemetering or pilot-wire circuit to the remote indicating or receiving device.

78 Phase-angle measuring or out-of-step protective relay is a relay that functions at a predetermined phase angle between two voltages, between two currents, or between voltage and current.

79 AC reclosing relay is a relay that controls the automatic reclosing and locking out of an AC circuit interrupter.

80 Flow switch is a switch that operates on given values, or on a given rate of change, of flow.

81 Frequency relay is a relay that functions on a predetermined value of frequency—either under or over or on normal system frequency—or rate of change of frequency.

82 DC reclosing is a relay that controls the automatic closing and reclosing of a DC circuit interrupter, generally in response to load circuit conditions.

83 Automatic selective control or transfer relay is a relay that operates to select automatically between certain sources or conditions in an equipment or that performs a transfer operation automatically.

84 Operating mechanism is the complete electrical mechanism or servo-mechanism—including the operating motor, solenoids, position switches, etc.—for a tap changer, induction regulator, or any similar piece of apparatus that has no device function number.

85 Carrier or pilot-wire receiver relay is a relay that is operated or restrained by a signal used in connection with carrier-current or DC pilot-wire fault directional relaying.

86 Locking-out relay is an electrically operated, hand or electrically reset, relay that functions to shut down and hold an equipment out of service on the occurrence of abnormal conditions.

87 Differential protective relay is a protective relay that functions on a percentage or phase angle or other quantitative difference of two currents or of some other electrical quantities.

88 Auxiliary motor or motor generator is one used for operating auxiliary equipment such as pumps, blowers, exciters, rotating magnetic amplifiers, etc.

89 Line switch is used as a disconnecting load-interrupter or isolating switch in an AC or DC power circuit when this device is electrically operated or has electrical accessories, such as an auxiliary switch, magnetic lock, etc.

90 Regulating device functions to regulate a quantity, or quantities, such as voltage, current, power, speed, frequency, temperature, and load, at a certain value or between certain (generally close) limits for machines, tie lines, or other apparatus.

91 Voltage directional relay is a relay that operates when the voltage across an open circuit breaker or contactor exceeds a given value in a given direction.

92 Voltage and power directional relay is a relay that permits or causes the connection of two circuits when the voltage difference between them exceeds a given value in a predetermined direction and causes these two circuits to be disconnected from each other when the power flowing between them exceeds a given value in the opposite direction.

93 Field-changing contactor functions to increase or decrease in one step the value of field excitation on a machine.

94 Tripping or trip-free relay functions to trip a circuit breaker, contactor, or equipment; to permit immediate tripping by other devices; or to prevent immediate reclosure of a circuit interrupter in case it should open automatically, even though its closing circuit is maintained closed.

95–97 Used only for specific applications on individual installations where one of the assigned numbered functions from 1 to 94 is suitable.

DEVICES PERFORMING MORE THAN ONE FUNCTION

If one device performs two relatively important functions in an equipment so that it is desirable to identify both of these functions, this may be done by using a double function number and name such as:

50/51 Instantaneous and Time Overcurrent Relay

SUFFIX NUMBERS

If two or more devices with the same function number and suffix letter (if used) are present in the same equipment, they may be distinguished by numbered suffixes as, for example, 52X-1 52X-2, and 52X-3, when necessary.

SUFFIX LETTERS

Suffix letters are used with device function numbers for various purposes. In order to prevent possible conflict, each suffix letter should have only one meaning in an individual equipment. All other words should use the abbreviations as contained in ANSI Y1.1, latest revision, or should use some other distinctive abbreviation or be written out in full each time they are used. The meaning of each single suffix letter, or combination of letters, should be clearly designated in the legend on the drawings or publications applying to the equipment.

Lowercase suffix letters are used in practically all instances on electrical diagrams for auxiliary, position, and limit switches. Capital letters are generally used for all other suffix letters.

The letters should generally form part of the device function designation and are usually written directly after the device function number; for example, 52CS, 71W, 49D. When it is necessary to use two types of suffix letters in connection with one function number, it is often desirable for clarity to separate them by a slanted line or dash, as for example, 20D/CS or 20D-CS.

The suffix letters that denote parts of the main device, and those that cannot or need not form part of the device function designation, are generally written below the device function number on drawings; for example, 52/CC or 43/A.

AUXILIARY DEVICES

X	
Y	Auxiliary relay[a]
Z	
R	Raising relay
L	Lowering relay
O	Opening relay or contactor
CS	Control switch
CL	Auxiliary relay, closed (energized when main device is in closed position)
OP	Auxiliary relay, open (energized when main device is in open position)
U	"Up" position-switch relay
D	"Down" position-switch relay
PB	Pushbutton

ACTUATING QUANTITIES

These letters indicate the condition or electrical quantity to which the device responds, or the medium in which it is located, such as:

A	Air, amperes, or alternating
C	Current
D	Direct or discharge
E	Electrolyte
F	Frequency, flow, or fault
H	Explosive
J	Differential
L	Level or liquid
P	Power or pressure
PF	Power factor
Q	Oil
S	Speed, suction, or smoke
T	Temperature
V	Voltage, volts, or vacuum
VAR	Reactive power
VB	Vibration
W	Water or watts

MAIN DEVICES

These letters denote the location of the main device in the circuit, the type of circuit, the type of circuit in which the device is used, or the type of circuit or apparatus with which it is associated, when this is necessary, such as:

A	Alarm or auxiliary power
AN	Anode

[a] In the control of a circuit breaker with a so-called X-Y relay control scheme, the X relay is the device whose main contacts are used to energize the closing coil or the device that, in some other manner (such as by the release of stored energy), causes the breaker to close. The contacts of the Y relay provide the anti-pump feature for the circuit breaker.

B	Battery, blower, or bus
BK	Brake
BL	Block (valve)
BP	Bypass
BT	Bus tie
C	Capacitor, condenser, compensator, carrier, current, case, or compressor
CA	Cathode
CH	Check (valve)
D	Discharge (valve)
E	Exciter
F	Feeder, field, filament, filter, or fan
G	Generator or ground[b]
H	Heater or housing
L	Line or logic
M	Motor or metering
N	Network or neutral[b]
P	Pump or phase comparison
R	Reactor, rectifier, or room
S	Synchronizing, secondary, strainer, sump, or suction (valve)
T	Transformer or thyratron
TH	Transformer (high-voltage side)
TL	Transformer (low-voltage side)
TM	Telemeter
U	Unit

MAIN DEVICE PARTS

These letters denote parts of the main device, divided in the two following categories:

1. All parts, except auxiliary contacts, position switches, limit switches, and torque limit switches.

BK	Brake
C	Coil, condenser, or capacitor
CC	Closing coil
HC	Holding coil
M	Operating motor
MF	Fly-ball motor
ML	Load-limit motor
MS	Speed adjusting, or synchronizing, motor
S	Solenoid
SI	Seal-in
TC	Trip coil
V	Valve

[b]Suffix "N" is generally used in preference to "G" for devices connected in the secondary neutral of current transformers or in the secondary of a current transformer whose primary winding is located in the neutral of a machine or power transformer, except in the case of transmission line relaying, where the suffix "G" is more commonly used for those relays that operate on ground faults.

2. All auxiliary contacts and position and limit switches for such devices and equipment as circuit breakers, contactors, valves, and rheostats and contacts of relays. These are designated as follows:

- a Contact that is open when the main device is in the standard reference position, commonly referred to as the nonoperated or deenergized position, and that closes when the device assumes the opposite position.
- b Contact that is closed when the main device is in the standard reference position, commonly referred to as the nonoperated or deenergized position, and that opens when the device assumes the opposite position.
- aa Contact that is open when the operating mechanism of the main device is in the nonoperated position and that closes when the operating mechanism assumes the opposite position.
- bb Contact that is closed when the operating mechanism of the main device is in the nonoperated position and that opens when the operating mechanism assumes the opposite position.

Standard reference positions of some typical devices are as follows:

Device	**Standard Reference Position**
Power circuit breaker	Main contacts open
Disconnecting switch	Main contacts open
Load-break switch	Main contacts open
Valve	Closed position
Gate	Closed position
Clutch	Disengaged position
Turning gear	Disengaged position
Power electrodes	Maximum gap position
Rheostat	Maximum resistance position
Adjusting means[c]	Low or down position
Relay[d]	Deenergized position
Contactor[d]	Deenergized position
Relay (latched-in type)	See 2-9.7.2 (C37.2-1970)
Contactor (latched-in type)	Main contacts open
Temperature relay[e]	Lowest temperature
Level detector[e]	Lowest level
Flow detector[e]	Lowest flow
Speed switch[e]	Lowest speed
Vibration detector[e]	Minimum vibration
Pressure switch[e]	Lowest pressure
Vacuum switch[e]	Lowest pressure, i.e., highest vacuum

Note: If several similar auxiliary switches are present on the same device, they should be designated numerically (1, 2, 3, etc.) when necessary.

[c] These may be speed, voltage, current, load, or similar adjusting devices comprising rheostats, springs, levers, or other components for the purpose.

[d] These electrically operated devices are of the non-latched-in type, whose contact position is dependent only upon the degree of energization of the operating, restraining, or holding coil or coils that may or may not be suitable for continuous energization. The deenergized position of the device that with all coils deenergized.

[e] The energizing influences for these devices are considered to be, respectively, rising temperature, rising level, increasing flow, rising speed, increasing vibration, and increasing pressure.

The simple designation "a" or "b" is used in all cases where there is no need to adjust the contacts to change position at any particular point in the travel of the main device or where the part of the travel where the contacts change position is of no significance in the control or operating scheme. Hence, the "a" and "b" designations usually are sufficient for circuit breaker auxiliary switches.

OTHER SWITCHES

These letters cover all other distinguishing features or characteristics or conditions, which serve to describe the use of the device or its contacts in the equipment such as:

A	Accelerating or automatic
B	Blocking or backup
C	Close or cold
D	Decelerating, detonate, down, or disengaged
E	Emergency or engaged
F	Failure or forward
H	Hot or high
HR	Hand reset
HS	High speed
L	Left, local, low, lower, or leading
M	Manual
OFF	Off
ON	On
P	Polarizing
R	Right, raise, reclosing, receiving, remote, reverse
S	Sending or swing
T	Test, trip, or trailing
TDC	Time-delay closing
TDO	Time-delay opening
U	Up

REPRESENTATION OF DEVICE CONTACTS ON ELECTRICAL DIAGRAMS

On electrical diagrams, the "b" contacts of all devices, including those of relays and those with suffix letters or percentage figures, should be shown as closed contacts, and all "a" contacts should be shown as open contacts.

For those devices that have no deenergized or nonoperated position, such as manually operated transfer or control switches (including those of the spring-return type) or auxiliary position indicating contacts on the housings or enclosures of a removable circuit breaker unit, the preferred method of representing these contacts is as an "a" switch. Each contact should, however, be identified on the elementary diagram as to when it closes.

In the case of latched-in or hand-reset locking-out relays, which operate from protective devices to perform the shutdown of an equipment and to hold it out of service, the contacts should preferably be shown in the normal non-locking-out position. In general, any devices, such as electrically operated latched-in relays, that have no deenergized or nonoperated position, which is most suitable for the ready understanding of the operation of the devices in the equipment, should contain sufficient description, as necessary, on the elementary diagram to indicate the contact operation.

APPENDIX H
NEMA ENCLOSURE LIST

DEFINITIONS[a]

Type 1 Enclosures constructed for indoor use to provide a degree of protection to personnel against access to hazardous parts and to provide a degree of protection of the equipment inside the enclosure against ingress of solid foreign objects (falling dirt).

Type 2 Enclosures constructed for indoor use to provide a degree of protection to personnel against access to hazardous parts; to provide a degree of protection of the equipment inside the enclosure against ingress of solid foreign objects (falling dirt); and to provide a degree of protection with respect to harmful effects on the equipment due to the ingress of water (dripping and light splashing).

Type 3 Enclosures constructed for either indoor or outdoor use to provide a degree of protection to personnel against access to hazardous parts; to provide a degree of protection of the equipment inside the enclosure against ingress of solid foreign objects (falling dirt and windblown dust); to provide a degree of protection with respect to harmful effects on the equipment due to the ingress of water (rain, sleet, snow); and that will be undamaged by the external formation of ice on the enclosure.

Type 3R Enclosures constructed for either indoor or outdoor use to provide a degree of protection to personnel against access to hazardous parts; to provide a degree of protection of the equipment inside the enclosure against ingress of solid foreign objects (falling dirt); to provide a degree of protection with respect to harmful effects on the equipment due to the ingress of water (rain, sleet, snow); and that will be undamaged by the external formation of ice on the enclosure.

Type 4X Enclosures constructed for either indoor or outdoor use to provide a degree of protection to personnel against access to hazardous parts; to provide a degree of protection of the equipment inside the enclosure against ingress of solid foreign objects (windblown dust); to provide a degree of protection with respect to harmful effects on the equipment due to the ingress of water (rain, sleet, snow, splashing water, and hose-directed water); that provides an additional level of protection against corrosion; and that will be undamaged by the external formation of ice on the enclosure.

Type 5 Enclosures constructed for indoor use to provide a degree of protection to personnel against access to hazardous parts; to provide a degree of protection of the equipment

[a]From NEMA 250-2003.

inside the enclosure against ingress of solid foreign objects (falling dirt and settling airborne dust, lint, fibers, and flyings); and to provide a degree of protection with respect to harmful effects on the equipment due to the ingress of water (dripping and light splashing).

Type 6 Enclosures constructed for either indoor or outdoor use to provide a degree of protection to personnel against access to hazardous parts; to provide a degree of protection of the equipment inside the enclosure against ingress of solid foreign objects (falling dirt); to provide a degree of protection with respect to harmful effects on the equipment due to the ingress of water (hose-directed water and the entry of water during occasional temporary submersion at a limited depth); and that will be undamaged by the external formation of ice on the enclosure.

Type 6P Enclosures constructed for either indoor or outdoor use to provide a degree of protection to personnel against access to hazardous parts; to provide a degree of protection of the equipment inside the enclosure against ingress of solid foreign objects (falling dirt); to provide a degree of protection with respect to harmful effects on the equipment due to the ingress of water (hose-directed water and the entry of water during prolonged submersion at a limited depth); that provides an additional level of protection against corrosion; and that will be undamaged by the external formation of ice on the enclosure.

Type 12 Enclosures constructed (without knockouts) for indoor use to provide a degree of protection to personnel against access to hazardous parts; to provide a degree of protection of the equipment inside the enclosure against ingress of solid foreign objects (falling dirt and circulating dust, lint, fibers, and flyings); and to provide a degree of protection with respect to harmful effects on the equipment due to the ingress of water (dripping and light splashing).

Type 12K Enclosures constructed (with knockouts) for indoor use to provide a degree of protection to personnel against access to hazardous parts; to provide a degree of protection of the equipment inside the enclosure against ingress of solid foreign objects (falling dirt and circulating dust, lint, fibers, and flyings); and to provide a degree of protection with respect to harmful effects on the equipment due to the ingress of water (dripping and light splashing).

Type 3S Enclosures constructed for either indoor or outdoor use to provide a degree of protection to personnel against access to hazardous parts; to provide a degree of protection of the equipment inside the enclosure against ingress of solid foreign objects (falling dirt and windblown dust); to provide a degree of protection with respect to harmful effects on the equipment due to the ingress of water (rain, sleet, snow); and for which the external mechanism(s) remain operable when ice laden.

Type 3X Enclosures constructed for either indoor or outdoor use to provide a degree of protection to personnel against access to hazardous parts; to provide a degree of protection of the equipment inside the enclosure against ingress of solid foreign objects (falling dirt and windblown dust); to provide a degree of protection with respect to harmful effects on the equipment due to the ingress of water (rain, sleet, snow); that provides an additional level of protection against corrosion; and that will be undamaged by the external formation of ice on the enclosure.

Type 3RX Enclosures constructed for either indoor or outdoor use to provide a degree of protection to personnel against access to hazardous parts; to provide a degree of protection of the equipment inside the enclosure against ingress of solid foreign objects (falling dirt); to provide a degree of protection with respect to harmful effects on the equipment due to the ingress of water (rain, sleet, snow); that will be undamaged by the external formation of ice on the enclosure that provides an additional level of protection against corrosion; and that will be undamaged by the external formation of ice on the enclosure.

Type 3SX Enclosures constructed for either indoor or outdoor use to provide a degree of protection to personnel against access to hazardous parts; to provide a degree of protection of the equipment inside the enclosure against ingress of solid foreign objects (falling dirt and windblown dust); to provide a degree of protection with respect to harmful effects on the equipment due to the ingress of water (rain, sleet, snow); that provides an additional level of protection against corrosion; and for which the external mechanism(s) remain operable when ice laden.

Type 4 Enclosures constructed for either indoor or outdoor use to provide a degree of protection to personnel against access to hazardous parts; to provide a degree of protection of the equipment inside the enclosure against ingress of solid foreign objects (falling dirt and windblown dust); to provide a degree of protection with respect to harmful effects on the equipment due to the ingress of water (rain, sleet, snow, splashing water, and hose-directed water); and that will be undamaged by the external formation of ice on the enclosure.

Type 13 Enclosures constructed for indoor use to provide a degree of protection to personnel against access to hazardous parts; to provide a degree of protection of the equipment inside the enclosure against ingress of solid foreign objects (falling dirt and circulating dust, lint, fibers, and flyings); to provide a degree of protection with respect to harmful effects on the equipment due to the ingress of water (dripping and light splashing); and to provide a degree of protection against the spraying, splashing, and seepage of oil and noncorrosive coolants.

APPENDIX I
KKS SYSTEM

The Siemens Kraftwerk-Kennzeichen-System (KKS), abridged version, "Identification System for Power Plants," is described in this appendix.

Source: Siemens abridged booklet for field use, 1995 edition.

FORMAT OF THE KKS

In consideration of the various requirements placed on the identification of plants, sections of plants, and items of equipment in power stations, KKS has three different types of code:

Process-related code Process-related identification of systems and items of equipment according to their functions in mechanical, civil, electrical, and control and instrumentation engineering.

Point of installation code Identification of points of installation of electrical and control and instrumentation devices in installation units (e.g., in cabinets, panels, consoles).

Location code Identification of locations in structures, on floors and in rooms and also of fire areas and topographical determinations (surface area grid).

These three types of code use the same identification scheme, which is subdivided into four breakdown levels and titled as follows:

Breakdown level	0	1	2	3
Process-related identification	Total plant	System code	Equipment unit code	Component code
Point of installation identification	Total plant	Installation unit code	Installation space code	
Location identification	Total plant	Structure code	Room code	

The individual breakdown levels are made up of classifying and numbering code elements which consist of data characters occupied by alpha numeric symbols:

Breakdown level	0	1	2	3
Designation	G	F_0 $F_1F_2F_3$ F_N	A_1A_2 A_N A_3	B_1B_2 B_N
Data character	A or N	N AAA NN	AA NNN A	A A N N

The following extract from the keys contains the most important classifying parts of the function, equipment unit, and component keys used in fossil-fueled power plants.

The numerical data characters of the identifier are normally not classifying and are employed on a plant and user-specific basis ("Project-specific establishments according to the KKS Guidelines").

FUNCTION KEY

Function Key, Main Groups

A	Grid and distribution systems
B	Power transmission and auxiliary power supply
C	Instrumentation and control equipment
E	Conventional fuel supply and residues disposal
G	Water supply and disposal
H	Conventional heat generation
L	Steam, water, gas cycles
M	Main machine sets
N	Process energy supply for external users (e.g., district heating)
P	Cooling water systems
Q	Cooling water systems
S	Ancillary systems
U	Structures
X	Heavy machinery (not main machine sets)

Function Key A

A	Grid and distribution systems
AB	> 420 kV systems
AC	380 (420) kV systems
AD	220 (245) kV systems
AE	110 (150) kV systems
AF	60 (72) kV systems
AG	45 (50) kV systems
AH	30 (35) kV systems
AJ	20 (25) kV systems
AK	10 (15) kV systems
AL	6 (5) kV systems
AM	1–3 kV systems
AN	< 1 kV systems
AP	Control consoles
AQ	Measuring and metering equipment
AR	Protection equipment
AS	Decentralized panels and cabinets
AT	Transformer equipment
AU	Open-loop control, checkback, and auxiliary equipment
AV	Marshalling racks
AW	Instrument panels
AX	Central equipment
AY	Communication equipment

Function Key B

B	Power transmission and auxiliary power supply
BA	Power transmission
BAA	Generator leads
BAB	Foundation cabinets
BAC	Generator circuit breaker, also commutating pole circuit breaker, including cooling system
BAT	Generator transformers, including cooling system
BAW	Earthing and lightning protection systems
BAX	Control air supply system
BAY	Control and protection equipment
BB	Medium-voltage distribution boards and transformers, normal system
BBA-BBS	MV distribution board
BBT	MV auxiliary power transformers
BBX	Fluid supply system for control and protection equipment
BBY	Control and protection equipment
BC	Medium-voltage distribution boards and transformers, general-purpose
BCA-BCS	MV start-up distribution board
BCT	Start-up, off site, general-purpose transformers
BCX	Fluid supply system for control and protection equipment
BCY	Control and protection equipment
BD	Medium-voltage emergency distribution boards and transformers, (diesel) emergency power system 1
BDA-BDS	MV emergency distribution board
BDT	MV transformer
BDX	Fluid supply system for control and protection equipment
BF	Low-voltage main distribution boards and transformers, normal system
BFA-BFS	LV main distribution board
BFT-BFW	LV auxiliary power transformer
BFY	Control and protection equipment
BH	Low-voltage main distribution boards and transformers, general-purpose
BHA-BHS	LV main distribution board
BHT-BHW	LV auxiliary power transformer
BHY	Control and protection equipment
BJ	Low-voltage sub-distribution boards and transformers, normal system
BJA-BJS	LV sub-distribution board
BJT	LV auxiliary power transformer
BJY	Control and protection equipment
BL	Low-voltage sub-distribution boards and transformers, general-purpose
BLA-BLS	LV sub-distribution board
BLT	LV auxiliary power transformer
BLY	Control and protection equipment
BM	Low-voltage distribution boards and transformers, (diesel) emergency power system 1
BMA-BMS	LV emergency distribution board
BMT-BMW	LV auxiliary power transformer
BMY	Control and protection equipment
BP	Power installations for variable-speed drives
BR	Low-voltage distribution boards, uninterruptible (converter) power supply
BRA-BRS	LV distribution board
BRT	Converter (rotary)

BRU	Converter (static), inverter
BRV	Emergency power generating equipment (if not under *X* = heavy machinery)
BRY	Control and protection equipment
BT	Battery systems
BTA-BTK	Batteries
BTL-BTV	Rectifiers, battery chargers
BTW-BTZ	Common equipment
BU	DC distribution boards, normal system
BUA-BUS	DC distribution board
BUY	Control and protection equipment
BV	DC distribution boards, emergency power system 1
BVA-BVS	DC distribution board
BVY	Control and protection equipment
BX	Fluid supply system for control and protection equipment
BY	Control and protection equipment

Function Key C

C	Instrumentation and control equipment
CA	Protective interlocks
CAA-CAQ	Cabinets for protective interlocks
CB	Functional group control, sub-loop control
CBA-CBN	Cabinets for functional group control
CBP	Cabinets for synchronization
CBQ	Cabinets for auxiliary power changeover
CC	Binary signal conditioning, drive control interface
CCA-CCQ	Cabinets for binary signal conditioning, drive control interface
CD	Drive control interface
CDA-CDQ	Cabinets for drive control interface
CE	Annunciation
CEA-CEH	Cabinets for annunciation systems
CEJ-CEQ	Fault recording
CF	Measurement, recording
CFA-CFF	Cabinets for measurement
CFQ	Cabinets for recording (meters, pen recorders)
CG	Closed-loop control (excluding power section)
CGA-CGH	Cabinets for closed-loop control
CH	Protection
CHA-CHD	Cabinets for generator and transformer protection
CJ	Unit coordination level
CJA	Unit control system (including cabinets)
CJD	Start-up control, set point control (unit) (including cabinets)
CJF	Boiler control system (including cabinets)
CJJ-CJN	Instrumentation and control cabinets for steam turbine set
CJP-CJT	Instrumentation and control cabinets for gas turbine set
CJU-CJY	Instrumentation and control cabinets for other main and heavy machinery
CK	Process computer system
CKA-CKH	Online supervisory and diagnostic computer
CKJ-CKM	Access control computer

CR	Programmable process control systems and associated cabinets
CRA	Cabinets for automation systems, e.g., unit control level, function unit 1
CRB	Cabinets for automation systems, function unit 2
CRC	Cabinets for automation systems, function unit 3
CRD	Cabinets for automation systems, function unit 4
CRE	Cabinets for automation systems, function unit 5
CRF	Cabinets for automation systems, function unit 6
CRG	Cabinets for automation systems, function unit 7
CRK	Cabinets for automation systems, e.g., several function units
CRN	Automation systems, e.g., operational
CRP	Automation systems, e.g., fail-safe
CRQ	Automation systems, e.g., AS620-T
CRR	Communication systems, e.g., terminal bus system
CRS	Communication systems, e.g., plant bus system
CRT	Communication systems, e.g., field bus system
CRU	Process control and information system, e.g., OM650
CRV	Configuring system, e.g., ES680
CRW	Diagnostics system, e.g., DS670
CRY	Common superordinate systems
CU	Closed-loop control (power section)
CUA	Cabinets for power adjuster 400V AC, e.g., function unit 1
CUB	Cabinets for power adjuster 400V AC, e.g., function unit 2
CUC	Cabinets for power adjuster 400V AC, e.g., function unit 3
CUD	Cabinets for power adjuster 400V AC, e.g., function unit 4
CUE	Cabinets for power adjuster 400V AC, e.g., function unit 5
CUF	Cabinets for power adjuster 400V AC, e.g., function unit 6
CUG	Cabinets for power adjuster 400V AC, e.g., function unit 7
CUL	Cabinets for interposing relays for 24 V DC solenoid valves
CUM	Cabinets for power supply of hardwired control room
CUN	Cabinets for 24 V DC fuses
CUP	Cabinets for 230 V DC fuses
CV	Marshalling racks
CW	Control rooms
CWA-CWE	Main control consoles
CWF-CWP	Main control panels
CX	Local control stations
CY	Communication equipment
CYA	Telephone system (PABX)
CYB	Control console telephone system
CYC	Alarm system
CYE	Fire alarm system
CYF	Clock system
CYG	Remote control system
CYH	Telemetry system
CYJ	Remote metering system
CYL	HF carrier telephone system
CYL	Staff paging system
CYP	Optical monitoring system
CYR	Pneumatic tube conveyor
CYS	Radiotelephone system
CYY	Telephone system for commissioning

Function Key E

E	Conventional fuel supply and residues disposal
EA	Unloading and storage of solid fuels
EAA	Ship unloading system
EAB	Rail wagon and truck unloading bay
EAC	Transport system
EAD	Stacking system
EAE	Bunker system, storage area (stockyard)
EAF	Bucket wheel system, reclaimer system
EAT	Weighing equipment
EAU	Sampling equipment
EAY	Control and protection equipment
EB	Mechanical treatment of solid fuels (also for gas generation and treatment) (crushing, mixing, drying, etc.)
EBA	Transport system
EBB	Mixing system
EBC	Crushing system, pulverizing system
EBD	Screening system
EBE	Separator system and discharge equipment thereof
EBF	Temporary storage system for milled raw coal
EBG	Predrying system
EBH	Main drying system
EBJ	Dried coal transport system including after cooling
EBK	Dried coal temporary storage system
EBL	Vapor compressor system
EBM	Exhaust system
EBR	Residues removal system
EBT	Weighing equipment
EBU	Sampling equipment
EBY	Control and protection equipment
EC	Distribution of solid fuels
ECT	Weighing equipment
ECU	Sampling equipment
ECY	Control and protection equipment
ED	Chemical treatment of solid fuels including residues removal (e.g., desulfurization plant)
EE	Conversion of solid fuels
EG	Supply of liquid fuels (light fuel oil)
EGA	Receiving equipment including pipeline
EGB	Tank farm
EGC	Pump system
EGD	Piping system
EGR	Residues removal system
EGT	Heating medium system
EGX	Fluid supply system for control and protection equipment
EGY	Control and protection equipment
EH	Chemical treatment of liquid fuels including residues removal (light and heavy fuel oil)
EHC	Additive dosing system
EHG	Chemical treatment of light fuel oil
EHN	Chemical treatment of heavy fuel oil

EHT	Heating medium system
EHX	Fluid supply system for control and protection equipment
EHY	Control and protection equipment
EK	Supply of gaseous fuels
EKA	Receiving equipment incl. pipeline
EKB	Moisture separation system
EKC	Heating system
EKD	Main reducing station, expansion turbine
EKE	Mechanical cleaning, scrubbing
EKF	Storage system
EKG	Piping system
EKH	Main pressure boosting system
EKR	Residues removal system
EKT	Heating medium system
EKX	Fluid supply system for control and protection equipment
EKY	Control and protection equipment
EL	Chemical treatment of gaseous fuels including residues removal
EM	Supply and treatment of fluxing agents
EN	Supply of other fuels (heavy fuel oil or crude oil)
ENA	Receiving equipment including pipeline
ENB	Tank farm
ENC	Pump system
END	Piping system
ENR	Residues removal system
ENT	Heating medium system
ENX	Fluid supply system for control and protection equipment
ENY	Control and protection equipment
EP	Treatment of other fuels (physical treatment of light and heavy fuel oil)
EPC	Pump system
EPG	Physical treatment of light fuel oil
EPN	Physical treatment of heavy fuel oil
EPT	Heating medium system
EPX	Fluid supply system for control and protection equipment
EPY	Control and protection equipment
EQ	Conversion of other fuels
ER	Ignition fuel supply
ERA	Pulverized coal supply system
ERB	Oil supply system
ERC	Gas supply system
ERY	Control and protection equipment
ET	Ash and slag removal system (from excl. removal equipment)
ETA	Wet ash conveying system
ETB	Storage or settling pond for wet ash
ETC	Wet ash dredger
ETD	Conveying system for granulate
ETE	Storage system for granulate
ETG	Conveying system for dry ash
ETH	Storage system for dry ash
ETK	Common conveying system for wet and dry ash
ETL	Common storage system for wet and dry ash
ETM	Settling plant for wet and dry ash

ETN	Forwarding, distribution, recovery, and disposal systems for flushing and ash water
ETP	Generation and distribution systems for carrier air
ETX	Fluid supply system for control and protection equipment
ETY	Control and protection equipment
EU	Treatment and transport system for combustion, fuel treatment, fuel conversion, flue gas cleaning, gas generation residues
EUA	Treatment system for fuel treatment residues
EUB	Treatment system for fuel conversion residues
EUC	Treatment system for fuel combustion residues
EUD	Treatment system for fuel flue gas cleaning residues
EUE	Treatment system for gas generation and treatment residues
EUF	Treatment system for fuel treatment residues
EUG	Treatment system for fuel treatment residues
EUH	Treatment system for fuel conversion residues
EUJ	Treatment system for fuel conversion residues
EUK	Treatment system for fuel combustion residues
EUL	Treatment system for fuel combustion residues
EUM	Treatment system for fuel flue gas cleaning residues
EUN	Treatment system for fuel flue gas cleaning residues
EUP	Treatment system for gas generation and treatment residues
EUQ	Treatment system for gas generation and treatment residues

Function Key G

G	Water supply and disposal
GA	Raw water supply
GAA	Extraction, mechanical cleaning
GAC	Piping and channel system
GAD	Storage system
GAF	Pump system
GAV	Lubricant supply system
GAX	Fluid supply system for control and protection equipment
GAY	Control and protection equipment
GB	Treatment system (carbonate hardness removal)
GBB	Filtering, mechanical cleaning system
GBC	Aeration, gas injection system
GBD	Precipitation system (e.g., for carbonate hardness removal)
GBE	Acid proportioning system (e.g., for carbonate hardness removal)
GBF	Ion exchange, reverse osmosis system (e.g., for carbonate hardness removal)
GBG	Evaporating system (e.g., for carbonate hardness removal)
GBH	Deaeration, drying system
GBK	Piping system, temporary storage system, pump system for main fluid
GBL	Storage system outside fluid treatment system (if not part of another system)
GBN	Chemicals supply system
GBP	Regeneration, flushing equipment
GBQ	Injection system for main fluid (for hardness stabilization)
GBR	Flushing water and residues removal system including neutralization
GBT	Heating, cooling, and flushing distribution system

GBX	Fluid supply system for control and protection equipment
GBY	Control and protection equipment
GC	Treatment system (demineralization)
GCB	Filtering, mechanical cleaning system
GCC	Aeration, gas injection system
GCD	Precipitation system (e.g., for carbonate hardness removal)
GCE	Acid proportioning system (e.g., for carbonate hardness removal)
GCF	Ion exchange, reverse osmosis system (e.g., for demineralization)
GCG	Evaporation system (e.g., for demineralization)
GCH	Deaeration, drying system
GCJ	Preheating, cooling system
GCK	Piping system, temporary storage system, pump system for main fluid
GCL	Storage system outside fluid treatment system (if not part of another system)
GCN	Chemicals supply system
GCP	Regeneration, flushing equipment
GCQ	Injection system for main fluid
GCR	Flushing water and residues removal system incl. neutralization
GCS	Sludge thickening system
GCT	Heating, cooling, and flushing fluid protection equipment
GCX	Fluid supply system for control and protection equipment
GCY	Control and protection equipment
GD	Treatment system (others)
GDB	Filtering, mechanical cleaning system
GDC	Aearation, gas injection system
GDD	Precipitation system (e.g., for carbonate hardness removal)
GDF	Ion exchange, reverse osmosis system (e.g., for demineralization)
GDG	Evaporation system (e.g., for demineralization)
GDH	Dearation, drying system
GDJ	Preheating, cooling system
GDK	Piping system, temporary storage system, pump system for main fluid
GDL	Storage system outside fluid treatment system (if not part of another system)
GDP	Regeneration, flushing equipment
GDQ	Injection system for main fluid
GDR	Flushing water and residues removal system including neutralization
GDS	Sludge thickening system
GDT	Heating, cooling, and flushing fluid distribution system
GDX	Fluid supply system for control and protection equipment
GDY	Control and protection equipment
GH	Distribution systems (not drinking water)
GHB	Distribution system after treatment (carbonate hardness removal)
GHC	Distribution system after treatment (demineralization) (demineralized water distribution system)
GHD	Distribution system after treatment (others) (mains water distribution system)
GK	Drinking water supply
GKA	Receiving point
GKB	Storage, forwarding, distribution system
GKC-GKZ	Assigned to structures
GM	Plant drainage system
GMA	Central disposal system
GMC-GMZ	Assigned to structures
GN	Plant drains treatment system

GNB	Filtering, mechanical cleaning system
GNC	Aeration, gas injection system
GND	Precipitation system (e.g., for carbonate hardness removal)
GNE	Acid proportioning system (e.g., for carbonate hardness removal)
GNF	Ion exchange system (e.g., for demineralization)
GNG	Evaporation system (e.g., for demineralization)
GNH	Dearation, drying system
GNJ	Preheating, cooling system
GNK	Piping system, temporary storage system, pump system for main fluid
GNL	Storage system outside fluid treatment (if not part of another system)
GNN	Chemicals supply system
GNP	Regeneration, flushing equipment
GNQ	Injection system for main fluid
GNR	Flushing water and residues removal system incl. neutralization
GNS	Sludge thickening system
GNT	Heating, cooling, and flushing fluid distribution system
GNX	Fluid supply system for control and protection equipment
GNY	Control and protection equipment
GQ	Domestic waste water collection and drainage systems
GQA	Central disposal system
GQB	Storage
GQC-GQZ	Assigned to structures
GR	Domestic waste water treatment system
GRB	Filtering, mechanical cleaning system
GRC	Aeration, gas injection system
GRD	Precipitation system (e.g., for carbonate hardness removal)
GRE	Acid proportioning system (e.g., for carbonate hardness removal)
GRF	Ion exchange system (e.g., for demineralization)
GRG	Evaporation system (e.g., for demineralization)
GRH	Dearation, drying system
GRJ	Preheating, cooling system
GRK	Piping system, temporary storage system, pump system for main fluid
GRL	Storage system outside fluid treatment system (if not part of another system)
GRN	Chemicals supply system
GRP	Regeneration, flushing equipment
GRQ	Injection system for main fluid
GRR	Flushing water and residues removal system incl. neutralization
GRS	Sludge thickening system
GRT	Heating, cooling, and flushing fluid distribution system
GRX	Fluid supply system for control and protection equipment
GRY	Control and protection equipment
GT	Water recovery from waste water
GU	Rainwater collection and drainage systems incl. treatment system
GUA	Central disposal system
GUB	Storage
GUC-GUZ	Assigned to structures

Function Key H

H	Conventional heat generation
HA	Pressure system, feedwater, and steam sections

HAA	LP part-flow heed heating system (flue-gas-heated)
HAB	HP part-flow feed heating system (flue-gas-heated)
HAC	Economizer system
HAD	Evaporator system
HAG	Circulation system (not used in natural-circulation boiler)
HAH	HP superheater system
HAJ	Reheat system
HAK	Secondary reheat system
HAM	Triflux system
HAN	Pressure system drainage and venting systems
HAX	Fluid supply system for control and protection equipment
HAY	Control and protection equipment
HB	Support structure, enclosure, steam generator interior
HBA	Frame including foundations
HBB	Enclosures
HBC	Brick linings including insulating brickwork
HBD	Platforms, stairways
HBE	Pressure vessel (supercharged boiler)
HBK	Steam generator interior
HC	Fireside heat transfer surface cleaning equipment
HCA	Air soot blowing system
HCB	Steam soot blowing system
HCD	Flushing equipment
HCE	Rapping gear
HCF	Shot cleaning system
HCW	Sealing fluid supply system
HCY	Control and protection equipment
HD	Ash and slag removal, particulate removal
HDA	Furnace ash removal, furnace slag removal, bed ash removal
HDB	Bed ash return system
HDC	Flue dust discharge and return system
HDD	Mechanical dust handling system
HDE	Electrostatic precipitator
HDF	Cyclone dust removal and return system
HDT	Fluid supply system for ash, slag, and dust moistening
HDU	Carrier air supply
HDW	Sealing fluid supply system
HDY	Control and protection equipment
HF	Bunker, feeder, and pulverizing system
HFA	Bunker for pulverizing system
HFB	Feeder system
HFC	Pulverizing system (including classifier)
HFD	Flue gas return system
HFE	Mill air system, carrier air system
HFF	Vapour/exhaust gas system
HFG	Pulverized coal temporary storage bunker after central pulverizing system (indirect firing)
HFW	Sealing fluid supply system
HFY	Control and protection equipment
HH	Main firing system (electric-powered as well)
HHA	Main burners
HHB	Retarded combustion grate

HHC	Grate combustion system
HHD	Other burner equipment (e.g., vapour burner, flue dust burner)
HHE	Pulverized coal bin, forwarding and distribution system
HHF	Oil temporary storage, pump and distribution system
HHG	Gas pressure reduction, distribution system
HHH	Temporary storage, forwarding and distribution system for other fuels, fluid 1
HHJ	Temporary storage, forwarding and distribution system for other fuels, fluid 2
HHK	Temporary storage, forwarding and distribution system for other fuels, fluid 3
HHL	Combustion air supply system
HHM	Atomizer medium supply system (steam)
HHN	Atomizer medium supply system (air)
HHP	Coolant supply system (steam)
HHQ	Coolant supply system (air)
HHR	Purging medium supply system (steam)
HHS	Purging medium supply system (air)
HHT	Purging medium supply system (steam)
HHU	Purging medium supply system (hot water)
HHV	Electric heating system
HHX	Fluid supply system for control and protection equipment
HHY	Control and protection equipment
HJ	Ignition firing equipment (if separate)
HJA	Ignition burners
HJE	Pulverized coal bin, forwarding and distribution system
HJF	Oil temporary storage, pump and distribution system
HJG	Gas pressure reduction, distribution system
HJL	Combustion air supply system
HJM	Atomizer medium supply system (steam)
HJN	Atomizer medium supply system (air)
HJP	Cooling supply system (steam)
HJQ	Cooling supply system (air)
HJR	Purging medium supply system (steam)
HJS	Purging medium supply system (air)
HJT	Heating medium supply system (steam)
HJU	Heating medium supply system (hot water)
HJX	Fluid supply system for control and protection equipment
HJY	Control and protection equipment
HL	Combustion air system (primary air, secondary air)
HLA	Ducting system
HLB	Fan system, forced-draught fan system
HLC	External air heating system (not flue-gas-heated)
HLD	Air heating system (flue-gas-heated)
HLU	Air pressure relief system
HLY	Control and protection equipment
HM	Gas heating system (for closed cycle)
HMA	Primary heater (primary convection section)
HMB	Radiation section
HMC	Secondary heater (second convection section)
HMD	Reheat system

HMY	Control and protection equipment
HN	Flue gas exhaust (without flue gas treatment)
HNA	Ducting system
HNC	Induced-draught fan system
HNE	Smoke stack system (chimney)
HNF	Flue gas recirculation system
HNU	Flue gas pressure relief system
HNY	Control and protection equipment
HR	Chemical flue gas treatment system incl. residues removal, adsorptive process (e.g., dry additives)
HS	Chemical flue gas treatment system incl. residues removal, catalytic process (e.g., DENOX)
HSA	Flue gas ducting system within *HS*
HSC	Flue gas-side heat exchanger, gas heater (not *HU*)
HSD	Reactor (reduction)
HSE	Converter (oxidation)
HSF	Flue gas-side cleaning equipment for reactor
HSG	Reduction agent dilution system
HSH	(Residues) separator
HSJ	Reduction agent supply system including storage
HSK	Reduction agent treatment and distribution system
HSL	Water supply and disposal system
HSM	Chemicals and additives supply system
HSN	Drainage systems
HSP	Flyash collecting system (including filtering) and removal system
HSQ	Sprinkler system including drainage
HSR	Oxidizing agent treatment and distribution system
HSS	(Residues) forwarding, storage, loading system
HST	Flushing fluid system including supply
HSU	Heating fluid system
HSW	Sealing fluid supply system
HSX	Fluid supply system for control and protection equipment
HSY	Control and protection equipment
HT	Chemical flue gas treatment system including residues removal, absorptive process (e.g., flue gas desulfurization plants)
HTA	Flue gas ducting system *HT*
HTB	Flue gas-side heat exchanger, gas heater (not *HU*)
HTC	Flue gas fan system
HTD	Flue gas scrubbing system
HTE	Flue gas cleaning and filtering system
HTF	Absorption cycle
HTG	Oxidation system including supply system
HTJ	Absorbent supply system including storage system
HTK	Absorbent preparation and distribution system
HTL	Piping system for discharge of solids
HTM	Thickening and solids dewatering system
HTN	Solids drying, compacting system
HTP	(Solids/product) forwarding, storage, loading system
HTQ	FGD plant water supply and disposal system
HTS	Chemicals and additives supply system
HTT	Drainage systems

HTW	Sealing fluid supply system
HTX	Fluid supply system for control and protection equipment
HTY	Control and protection equipment
HU	Flue gas reheating system
HY	Control and protection equipment

Function Key L

L	Steam, water, gas cycles
LA	Feedwater system
LAA	Storage, deaeration (including feedwater tank)
LAB	Feedwater piping system (excluding feedwater pump and feedwater heating system)
LAC	Feedwater pump system
LAD	HP feedwater heating system
LAE	HP desuperheating spray system
LAF	IP desuperheating spray system
LAH	Start-up and shutdown piping system
LAJ	Start-up and shutdown pump system
LAW	Feedwater sealing water system
LAX	Fluid supply system for control and protection equipment
LAY	Control and protection equipment
LB	Steam system
LBA	Main stream piping system
LBB	Hot reheat piping system
LBC	Cold reheat piping system
LBD	Extraction piping system
LBE	Back-pressure piping system
LBF	HP reducing station
LBG	Auxiliary steam piping system
LBH	Start-up steam system, shutdown steam system
LBJ	Moisture separator/reheater (MSR)
LBQ	Extraction steam piping system for HP feedwater heating
LBR	Piping system for branch or auxiliary turbine
LBS	Extraction steam piping system for LP feedwater heating (main condensate)
LBT	Emergency condensing system
LBU	Common dump line
LBW	Sealing steam system
LBX	Fluid supply system for control and protection equipment
LBY	Control and protection equipment
LC	Condensate system
LCA	Main condensate piping system (excluding main condensate pump system, LP feedwater heating system, condensate polishing plant)
LCB	Main condensate pump system
LCC	LP feedwater heating system
LCE	Condensate desuperheating spray system
LCF	Branch turbine condensate piping system
LCG	Branch turbine condensate pump system
LCH	HP heater drains system
LCJ	LP heater drains system

LCL	Steam generator drains system
LCM	Clean drains system (collecting and return system)
LCN	Auxiliary steam condensate system (collecting and return system)
LCP	Standby condensate system, including storage and pump system
LCQ	Steam generator blowdown system
LCR	Standby condensate distribution system
LCS	Reheater drains system (moisture separator/reheater)
LCT	Moisture separator drains system (moisture separator/reheater)
LCW	Sealing and cooling drains system
LCX	Fluid supply system for control and protection equipment
LCY	Control and protection equipment
LD	Condensate polishing plant
LDA	Fluid treatment extraction system (if not part of another system)
LDB	Filtering, mechanical cleaning
LDC	Aeration, gas injection system
LDD	Electromagnetic polishing system
LDE	Acid proportioning system (e.g., for carbonate hardness removal)
LDF	Ion exchange, reverse osmosis system (e.g., demineralization)
LDG	Evaporation system (e.g., demineralization)
LDH	Dearation, drying system
LDJ	Preheating, cooling system
LDK	Piping system, temporary storage system, pump system for main fluid
LDL	Storage system outside fluid treatment system (if not part of another system)
LDN	Chemicals supply system
LDP	Regeneration, flushing equipment
LDQ	Injection system for main fluid
LDR	Flushing water and residues removal system including neutralization
LDS	Sludge thickening system
LDT	Heating, cooling, and flushing fluid distribution system
LDX	Fluid supply system for control and protection equipment
LDY	Control and protection equipment
LF	Common installations for steam, water, gas cycles
LFC	Common drain and vent systems
LFG	Secondary-side steam generator tubesheet lancing system
LFJ	Steam generator layup system
LFN	Proportioning system for feedwater, condensate system, incl. proportioning in boiler and turbine area
LK	Gas system (closed cycle)
LKA	Storage system
LKB	Piping system
LKC	Compressor system (if separate from gas turbine)
LKD	Preheating system
LKE	Precooling system
LKF	Intercooling system
LKG	Pressurizing system
LKW	Sealing fluid supply system
LKX	Fluid supply system for control and protection equipment
LKY	Control and protection equipment
LL	Gas cleaning system (only for closed cycle)
LW	Sealing fluid supply system for steam, water, gas cycles
LY	Control and protection equipment

Function Key M

M	Main machine sets
MA	Steam turbine plant
MAA	HP turbine
MAB	IP turbine
MAC	LP turbine
MAD	Bearings
MAG	Condensing system
MAH	Motive water system (if separate from *MAJ*)
MAJ	Air removal system
MAK	Transmission gear between prime mover and driven machine, including turning gear
MAL	Drain and vent systems
MAM	Leak-off steam system
MAN	Turbine bypass station, including desuperheating spray system
MAP	LP turbine bypass
MAQ	Vent system (if separate from *MAL*)
MAV	Lubricant supply system
MAW	Sealing, heating, and cooling steam system
MAX	Non-electric control and protection equipment including fluid supply system
MAY	Electrical control and protection equipment
MB	Gas turbine plant
MBA	Turbine, compressor rotor with common casing
MBB	Turbine casing and rotor
MBD	Bearings
MBH	Cooling and sealing gas system
MBJ	Start-up unit
MBK	Transmission gear between prime mover and driven machine, including turning gear, barring gear
MBL	Intake air, cold gas system (open cycle)
MBM	Combustion chamber (gas heating, combustion)
MBN	Fuel supply system (liquid)
MBP	Fuel supply system (gaseous)
MBQ	Ignition fuel supply system (if separate)
MBR	Exhaust gas system (open cycle)
MBS	Storage system
MBT	Motive gas generator unit, incl. combustion chamber
MBU	Additive system
MBV	Lubricant supply system
MBW	Sealing oil supply system
MBX	Non-electric control and protection equipment, including fluid supply system
MBY	Electrical control and protection equipment
MBZ	Lubricant and control fluid treatment system
MK	Generator plant
MKA	Generator, complete, including stator, rotor, and all integral cooling equipment
MKB	Generator exciter set (use only if *MKC* is not sufficient for identification)
MKC	Generator exciter set, complete (10 auxiliary exitation, 20 main exitation unit)
MKD	Bearings
MKF	Stator/rotor liquid cooling system, including coolant supply system

MKG	Stator/rotor gas cooling system, including coolant supply system
MKH	Stator/rotor nitrogen cooling system, including coolant supply system
MKQ	Exhaust gas system (if separate from *MKG* and *MKH*)
MKU	Stator/rotor cooling oil system, including coolant supply system
MKV	Lubricant supply system (if separate system for generator)
MKW	Sealing oil system, including supply and treatment
MKY	Control and protection equipment
MP	Common installations for main machine sets
MPA	Foundation
MPB	Sheating
MPG	Frame, support structure
MPR	Forced cooling system
MPS	Drying and layup system
MV	Lubricant supply system
MX	Control fluid supply system
MY	Control and protection equipment

Function Key N

N	Process energy supply for external users (e.g., district heating)
NA	Process steam system (including condensate return)
NAA	Piping system (steam)
NAB	Piping system (condensate)
NAD	Process heat transmission system
NAE-NAJ	Piping system (steam)
NAK-NAM	Piping system (condensate)
NAX	Fluid supply system for control and protection equipment
NAY	Control and protection equipment
ND	Process hot water system
NDA	Piping system (forward)
NDB	Piping system (return)
NDC	Process hot water pump system
NDD	Process heat transfer
NDE	Hot water storage system
NDF-NDJ	Distribution systems
NDK	Pressurizing system
NDX	Fluid supply system for control and protection equipment
NDY	Control and protection equipment

Function Key P

P	Cooling water systems
PA	Circulating water system
PAA	Extraction, mechanical cleaning for direct cooling
PAB	Circulating water piping and culvert system
PAC	Circulating water pump system
PAD	Recirculation cooling system, outfall cooling system
PAE	Cooling tower pump system (if separate)
PAH	Condenser cleaning system, including all appurtenances
PAR	Make-up water piping system

PAS	Make-up water pump system
PAV	Lubricant supply system
PAX	Fluid supply system for control and protection equipment
PAY	Control and protection equipment
PB	Circulating water treatment system
PBA	Discharge from fluid treatment system (if not part of another system)
PBB	Filtering, mechanical cleaning system
PBC	Aeration, gas injection system
PBD	Precipitation system (e.g., for carbonate hardness removal)
PBE	Ion exchange system (e.g., for demineralization)
PBH	Evaporation system (e.g., for demineralization)
PBG	Deareation, drying system
PBJ	Preheating, cooling system
PBK	Piping system, temporary storage system, pump system for main fluid
PBL	Storage system outside fluid treatment system (if not part of another system)
PBN	Chemicals supply system
PBP	Regeneration, flushing equipment
PBQ	Injection system for main fluid
PBR	Flushing water and residues removal system, including neutralization
PBS	Sludge thickening system
PBT	Heating, cooling and flushing fluid distribution system
PBX	Fluid supply system for control and protection equipment
PBY	Control and protection equipment
PC	Service water system
PCA	Extraction, mechanical cleaning for direct cooling
PCB	Piping and culvert system
PCC	Pump system
PCD	Recirculation cooling system, outfall cooling system
PCH	Heat exchanger cleaning system
PCM	Service water system for generator, motor generator cooling
PCX	Fluid supply system for control and protection equipment
PCY	Control and protection equipment
PD	Service water treatment system
PDA	Discharge from fluid treatment system (if not part of another system)
PDB	Filtering, mechanical cleaning system
PDC	Aeration, gas injection system
PDD	Precipitation system (e.g., for carbonate hardness removal)
PDF	Ion exchange system (e.g., for demineralization)
PDG	Evaporation system (e.g., for demineralization)
PDH	Deaeration, drying system
PDJ	Preheating, cooling system
PDK	Piping system, temporary storage system, pump system for main fluid
PDL	Storage system outside fluid treatment system (if not part of another system)
PDN	Chemicals supply system
PDP	Regeneration, flushing equipment
PDQ	Injection system for main fluid
PDR	Flushing water and residues removal system, including neutralization
PDS	Sludge thickening system
PDT	Heating, cooling, and flushing fluid distribution system
PDX	Fluid supply system for control and protection equipment
PDY	Control and protection equipment
PG	Closed cooling water system

PGB	Closed cooling water system
PU	Common equipment for cooling water systems
PUE	Air removal system
PUN	Proportioning equipment
PUP	Fish barrier
PUR	Corrosion protection for mechanical equipment
PUS	Biocide generation system

Function Key Q

Q	Auxiliary systems
QC	Central chemicals supply
QCA	Hydrazine supply and distribution system
QCB	Corrosion inhibitor dosing system for open component cooling systems
QCC	Phosphate supply and distribution system
QCD	Ammonia supply and distribution system
QCH	Sodium hydroxide supply and distribution system
QCL	Oxygen dosing system (including H_2O_2 dosing)
QE	General compressed air and carrier air supply
QEA	Central compressed air and carrier air generation system
QEB	Central compressed air and carrier air distribution system
QF	General control air supply
QFA	Central compressed air and carrier air generation system
QEB	Central compressed air and carrier air generation system
QF	General control air supply
QFA	Central control air generation system
QFB	Central control air distribution system
QH	Auxiliary steam generating system
QHA	Pressure system
QHB	Support structure, enclosure, steam generator interior
QHC	Fireside heat transfer surface cleaning equipment
QHD	Ash and slag removal
QHE	Blowdown system, flash drain system
QHF	Bunker, feeder, and pulverizing system
QHG	Boiler water circulation system (also for electrode steam boiler)
QHH	Main firing system (electric powered as well)
QHJ	Ignition firing equipment (if separate)
QHL	Combustion air system (primary air, secondary air)
QHM	Gas heating system (for closed cycle)
QHN	Flue gas exhaust (without flue gas treatment)
QHP	Mechanical dust handling system
QHQ	Electrostatic precipitator
QHR	Chemical flue gas treatment system, including residues removal, adsorptive process
QHS	Chemical flue gas treatment system, including residues removal, catalytic process
QHT	Chemical flue gas treatment system, including residues removal, absorptive process
QHU	Flue gas reheating system
QHX	Fluid supply system for control and protection equipment
QHY	Control and protection equipment

QJ	Central gas supply, also inert gas
QJA	Gas supply and distribution
QJY	Control and protection equipment
QK	Chilled water systems
QKA	Central waste chiller
QKC-QKZ	Chilled water systems assigned to structures
QL	Feedwater, steam, condensate cycle of auxiliary steam generating and distribution system
QLA	Feedwater system
QLB	Steam system
QLC	Condensate system
QLD	Condensate polishing plant
QLF	Common equipment for auxiliary steam generation and distribution systems
QLY	Control and protection equipment
QM	Air humidifying system
QU	Sampling systems
QUA	for feedwater systems (*LA*)
QUB	for steam systems (*LB*)
QUC	for condensate systems (*LC*)
QUD	for auxiliary steam generation systems (*QH*, *QL*)
QUE	for heating and cooling medium systems (*SB*, *QK*)
QUG	for demineralized water systems (*GH*)
QUN	for district heating systems (*N*)
QUP	for cooling water systems (*P*)
QUS	for HVAC systems (*SA*)
QUW	for liquid waste (*GM* and similar)

Function Key S

S	Ancillary systems
SA	Heating, ventilation, air-conditioning (HVAC) systems
SAC-SAZ	Heating, ventilation, air-conditioning (HVAC) systems assigned to structures
SB	Space heating systems
SBA	Central heating systems
SBC-SBZ	Space heating systems assigned to structures
SC	Stationary compressed air supplies
SCA	Compressed air generation system
SCB	Compressed air distribution system
SD	Stationary cleaning systems
SE	Stationary welding systems
SF	Heating and fuel gas systems
SG	Stationary fire protection systems
SGA	Fire water system
SGC	Spray deluge systems
SGE	Sprinkler systems
SGF	Foam fire-fighting systems
SGG	Tank roof, tank shell cooling systems
SGJ	CO_2 fire-fighting systems
SGL	Powder fire-fighting systems
SGM	Inergen fire-fighting systems

SGX	Fluid supply system for control and protection equipment
SGY	Control and protection equipment
SM	Cranes, stationary hoists, and conveying appliances
SMA	Central equipment
SMC-SMZ	Cranes etc. assigned to structures
SN	Elevators
SNC-SNZ	Elevators assigned to structures
SP	Railway installations
SQ	Road installations
ST	Workshop, stores, laboratory equipment, and staff amenities
STA	Workshop equipment
STC	Maintenance areas
STE	Stores and filling station equipment
STG	Laboratory equipment
STP	Staff amenities

Function Key U

U	Structures
UA	Structures for grid and distribution systems
UAA	Switchyard structure
UAB	Grid systems switchgear building
UAC	Grid systems control building
UAE	Structure for pneumatic control system
UAG	Structure for transformers
UAH	Structure for supports and equipment
UAS	Shaft structure (shaft)
UAX	Special structure (plant-specific)
UAY	Bridge structure
UAZ	Ducting structure
UB	Structures for power transmission and auxiliary power supply
UBA	Foundation for Power Control Centre (switchgear building)
UBB	Foundation for Power Control Centre for boiler house
UBC	Structure for offsite system transformers
UBD	Structure for low-voltage auxiliary power transformers
UBE	Structure for medium-voltage auxiliary power transformers
UBF	Structure for generator transformers
UBG	Structure for start-up transformers
UBH	Structure for oil collecting pits
UBJ	Structure for transformer tracks
UBK	Transformer assembly building
UBL	Structure for busbars
UBM	Structure for transformer cooling systems
UBN	Structure for emergency power generating sets (without central water chiller)
UBP	Emergency power generating and central water chiller building
UBQ	Structure for emergency power generator fuel supply system
UBS	Shaft structure (shaft)
UBX	Special structure (plant-specific), e.g., Power Control Centre in desalination plant
UBY	Bridge structure
UBZ	Ducting structure

UC	Structures for instrumentation and control
UCA	Control room building
UCB	Control station building
UCL	Structure for measuring equipment
UCP	Cooling water monitoring structure
UCS	Shaft structure (shaft)
UCX	Special structure (plant-specific)
UCY	Bridge structure
UCZ	Ducting structure
UE	Structures for conventional fuel supply and residues disposal
UEA	Structure for unloading of solid fuels
UEB	Structure for storage of solid fuels
UED	Structure for transport of solid fuels
UEE	Structure for treatment of solid fuels
UEF	Structure for transfer (e.g., junction tower)
UEG	Structure for conversion of solid fuels
UEH	Structure for unloading and reception of liquid fuels
UEJ	Structure for storage of liquid fuels
UEL	Structure for forwarding of liquid fuels (e.g., pump house)
UEM	Structure for treatment and handling of liquid fuels
UEN	Structure for transfer, conversion, storage of gaseous fuels
UER	Structure for forwarding of gaseous fuels
UES	Shaft structure (shaft)
UET	Structure for ash storage
UEU	Structure for ash transport
UEV	Structure for ash settling pond
UEW	Structure for combustion residues handling
UEX	Special structure (plant specific) (e.g., manhole for gaspipe)
UEY	Bridge structure
UEZ	Ducting structure
UG	Structures for water supply and disposal
UGA	Structure for raw water supply
UGC	Structure for demineralized water supply
UGD	Structure for demineralization system
UGE	Structure for neutralization system
UGF	Structure for fire water supply
UGG	Structure for drinking water supply
UGH	Structure for rainwater
UGJ	Structure for cooling tower make-up water treatment
UGK	Flocculant mixing chamber
UGL	Flocculator structure, flocculator
UGN	Treated water basin
UGP	Sludge thickener
UGQ	Sludge dewatering building
UGR	Sludge storage structure
UGS	Shaft structure (shaft)
UGU	Structure for effluent disposal
UGV	Structure for sewerage plant
UGX	Piping and cable bridge (e.g., between *UG.* and *UM.*)
UGZ	Piping and cable duct (e.g., between *UG.* and *UM.*)
UH	Structures for conventional heat generation
UHA	Boiler house (steam generator enclosure, steam generator building)

UHB	Auxiliary boiler
UHF	Bunker bay
UHL	Structure for boiler compressed air supply
UHN	Bypass smoke stack (smoke stack including structure for flue gas circulation)
UHQ	Structure for flue gas filtering system
UHS	Shaft structure (shaft)
UHU	Structure for flue gas reheating system
UHV	Structure for combustion air circulation (e.g., for forced draft fan)
UHW	Boiler blowdown structure (plant-specific)
UHX	Special structure (plant specific)
UHY	Bridge structure
UHZ	Ducting structure
UL	Structures for steam-, water-, gas-cycles
ULA	Feedwater pump house
ULB	Emergency feed building
ULC	Structure for condensate system
ULD	Structure for condensate polishing plant
ULE	Structure for emergency supply systems
ULS	Shaft structure (shaft)
ULX	Special structure (plant-specific)
ULY	Bridge structure
ULZ	Ducting structure
UM	Structures for main machine sets
UMA	Steam turbine building
UMB	Gas turbine building
UMC	Gas and steam turbine building
UMD	Combined gas and steam turbine building with WHRB and auxiliary equipment
UMS	Shaft structure (shaft)
UMX	Special structure (plant-specific)
UMY	Piping and cable bridge (e.g., between *UMA* and *UHA*)
UMZ	Ducting structure
UN	Structures for process energy supply
UP	Structures for circulating water systems (e.g., circulating water intake)
UPA	Circulating water intake culvert
UPB	Service water intake culvert
UPC	Circulating water intake structure
UPD	Service water intake structure
UPH	Circulating water treatment structure
UPJ	Service water treatment structure
UPN	Circulating water inlet culvert
UPP	Service water inlet culvert
UPQ	Biocide treatment building
UPR	Debris trough
UPS	Shaft structure (shaft)
UPT	Screen wash water cleaning structure
UPX	Special structure (plant-specific)
UPY	Bridge structure
UPZ	Ducting structure
UQ	Structures for circulating water systems (e.g., circulating water pumps and outfall)

UQA	Circulating water pump building
UQB	Service water pump building
UQF	Circulating water storage pit
UQG	Circulating water overflow structure, surge tank
UQH	Screen wash water discharge culvert
UQJ	Circulating water seal pit, including circulating water aeration structure
UQK	Circulating water venting structure
UQL	Service water surge pond
UQM	Service water collecting pond
UQN	Circulating water outfall culvert
UQP	Service water outfall structure
UQQ	Circulating water outfall structure
UQR	Service water outfall structure
UQS	Shaft structure (shaft)
UQT	Service water discharge culvert
UQU	Circulating water spillway structure, including circulating water aeration structure
UQV	Structure for artificial circulating water aeration
UQW	Routing structure for circulating water discharge
UQX	Special structure (plant-specific)
UQY	Bridge structure
UQZ	Ducting structure
UR	Structures for circulating water systems (e.g., recirculation cooling)
URA	Cooling tower structure (circulating water)
URB	Cooling tower structure (service water)
URC	Structure for cooled condensor
URD	Cooling tower pump building (circulating water)
URE	Cooling tower pump building (service water)
URG	Cooling tower connecting structure
URH	Cooling tower outlet structure
URJ	Cooling tower return structure
URL	Cooling tower outfall culvert
URM	Circulating water distribution building
URN	Cooling tower bypass structure
URP	Cooling tower blowdown structure
URQ	Cooling tower blowdown culvert
URS	Shaft structure (shaft)
URX	Special structure (plant-specific)
URY	Bridge structure
URZ	Ducting structure
US	Structures for ancillary systems
USA	Structure for heating, ventilation, air-conditioning (HVAC) systems
USB	Structure for space heating systems
USC	Structure for stationary compressed air supply system
USG	Fire pump house
USR	Structure for waste treatment
USS	Shaft structure (shaft)
UST	Workshop and stores building
USU	Consumables storage building
USV	Laboratory building
USX	Special structure (plant-specific)
USY	Bridge structure

USZ	Ducting structure
UT	Structures for auxiliary systems
UTA	Supply systems building
UTB	Central water chiller building
UTF	Compressed air system building
UTG	Central gas supply systems building (no fuel supply)
UTH	Auxiliary steam generator building
UTJ	Smoke stack
UTR	Chemical storage tank pit
UTS	Shaft structure (shaft)
UTX	Special structure (plant-specific)
UTY	Bridge structure
UTZ	Ducting structure
UV	Structures for chemical flue gas treatment incl. residues removal (for *HR*, *HS*, *HT*)
UVA	Structure for flue gas-side heat exchanger
UVB	Structure for flue gas fan system
UVC	Structure for flue gas scrubber, reactor
UVD	Structure for adsorbent/absorbent circuit
UVE	Structure for reagent supply, including preparation/treatment, storage, forwarding
UVF	Structure for thickening and solids dewatering, solids drying, and compacting system
UVG	Structure for chemical flue gas treatment
UVH	Structure for (solids/product) forwarding, storage, loading
UVM	Structure for catalyst handling and storage
UVS	Shaft structure (shaft)
UVX	Special structure (plant-specific)
UVY	Bridge structure
UVZ	Ducting structure
UX	Structures for external systems (power plant-specific)
UXA	Structure for seawater desalination plant
UXS	Shaft structure (shaft)
UXX	Special structure (plant-specific)
UXY	Bridge structure
UXZ	Ducting structure
UY	General service structures
UYA	Office and staff amenities building
UYB	Staff amenities building
UYC	Administration building
UYD	Canteen
UYE	Gate house
UYF	Security gate house
UYG	Information center
UYH	Training facilities
UYJ	Medical center
UYN	Railway engine shed
UYP	Fire station
UYQ	Garage
UYR	Automobile workshop
UYS	Shaft structure (shaft)
UYT	Filling station

UYX	Special structure (plant-specific)
UYY	Bridge structure
UYZ	Ducting structure
UZ	Structures for transport, traffic, fencing, gardens, and other purposes
UZA	Works roadways, paths, including associated structures
UZC	Yards
UZD	Parking areas, including associated structures
UZE	Railway structures
UZF	Lifting gear structures
UZJ	Fencing and gates
UZK	Gardens, including structures
UZL	Noise abatement structures
UZM	Protective structures against external impact
UZN	Structure for flood protection
UZP	Structure for bank/shoreline stabilization
UZQ	Structure for river regulation
UZR	Jetty, quay
UZS	Shaft structure (shaft)
UZT	Piping and cable routes in outdoor area
UZU	Site security structure
UZW	Residential buildings, residential area
UZX	Special structure (plant-specific)
UZY	Bridge structure
UZZ	Ducting structure

Function Key X

X	Heavy machinery (not main machine sets)
XJ	Diesel engine plant
XJA	Engine
XJB	Turbocharger, blower
XJG	Liquid cooling system
XJH	Air intercooling system
XJK	Transmission gear between prime mover and driven machine
XJN	Fuel systems
XJP	Start-up unit (including flywheel)
XJQ	Air intake system
XJR	Exhaust gas system
XJV	Lubricant supply system
XJX	Fluid supply system for control and protection equipment
XJY	Control and protection equipment
XK	Generator plant
XKA	Generator frame, including stator, rotor, and all integral cooling equipment
XKB	Generator exciter set (use only if *MKC* is not sufficient for identification)
XKC	Generator exciter set
XKD	Bearings
XKF	Stator/rotor liquid cooling system, including coolant supply system
XKG	Stator/rotor gas cooling system, including coolant supply system
XKH	Stator/rotor nitrogen cooling system, including coolant supply system
XKQ	Exhaust gas system (if separate from *XKG* and *XKH*)
XKU	Stator/rotor cooling oil cooling system, including coolant supply system

XKV	Lubricant supply system (if separate system for generator)
XKW	Sealing oil system, including supply and treatment
XKY	Control and protection equipment
XP	Common installations for heavy machinery
XPA	Foundations
XPB	Sheating
XPG	Frame, support structure
XPR	Forced cooling system
XPS	Drying and layup system

Equipment Unit Key A/B

A	Mechanical equipment
AA	Valves, dampers, etc., including actuators, also manual, rupture disk equipment
AB	Isolating elements, air locks
AC	Heat exchangers, heat transfer surfaces
AE	Turning, driving, lifting, and slewing gear (also manipulators)
AF	Continuous conveyors, feeders (escalators)
AG	Generator units
AH	Heating, cooling, and air conditioning units
AJ	Size reduction equipment, only as part of process
AK	Compacting and packaging equipment, only as part of process
AM	Mixers, agitators
AN	Compressor units, fans
AP	Pump units
AS	Adjusting and tensioning equipment for non-electrical variables
AT	Cleaning, drying, filtering, and separating equipment, excluding *BT*
AU	Braking, gearbox, coupling equipment, non-electrical converters
AV	Combustion equipment
AW	Stationary tooling, treatment equipment
AX	Test and monitoring equipment for plant maintenance

B	Mechanical equipment
BB	Storage equipment (vessels, tanks)
BE	Shafts (for erection and maintenance only)
BF	Foundations
BN	Flow restrictors, limiters, orifices (not metering orifices)
BQ	Hangers, supports, racks, piping penetrations
BR	Piping, ductwork, chutes
BS	Silencers
BT	Flue gas catalytic converter modules
BU	Insulation, sheathing

Equipment Unit Key C/D

C	Direct measuring circuits
CB	Radiation variables, e.g., thermal radiation, flame monitoring (other than *CR* and *CQ*)
CD	Density

CE	Electrical variables (e.g., current, voltage, power, electrical frequency)
CF	Flow, rate
CG	Distance, length, position, direction of rotation
CH	Manual input (as manually operated sensor) (e.g., fire detector)
CJ	Power (mechanical, thermal)
CK	Time
CL	Level (also for dividing line)
CM	Moisture, humidity
CP	Pressure
CQ	Quality variables (analysis, material properties other than *CD*, *CM*, *CV*)
CR	Radiation variables
CS	Velocity, speed, frequency (mechanical), acceleration
CT	Temperature
CU	Combined and other variables
CV	Vicosity
CW	Weight, mass
CY	Vibration, expansion
D	Closed loop control circuits
DB	Radiation variables, e.g., thermal radiation, flame monitoring (other than *DR* and *DQ*)
DD	Density
DE	Electrical variables (e.g., current, voltage, power, electrical frequency)
DF	Flow, rate
DG	Distance, length, position, direction of rotation
DJ	Power (mechanical, thermal)
DK	Time
DL	Level (also for dividing line)
DM	Moisture, humidity
DP	Pressure
DQ	Quality variables (analysis, material properties other than *DD*, *DM*, *DV*)
DR	Radiation variables
DS	Velocity, speed, frequency (mechanical), acceleration
DT	Temperature
DU	Combined and other variables
DV	Viscosity
DW	Weight, mass
DY	Vibration, expansion

Equipment Unit Key E/F

E	Analog and binary signal conditioning
EA	Unit control
EB	Group control
EC	Subgroup control
EE	Subloop control, equipment unit changeover
EG	Alarm logics
EH	Hard-wired alarm annunciation system
EJ	Visualization of operation and monitoring displays
EK	Alarm logics

EM	Access control
EN	Status-display computer, criterion display
EP	Supervisory computer
EQ	Internal automation (signal processing)
ES	Internal automation (signal conditioning)
EU	Combined analog and binary signal conditioning
EV	Signal transmission, bus coupling
EY	Protective logics, priority, non-equipment-unit-related
EZ	Superordinate equipment unit protection
F	Indirect measuring circuits
FB	Radiation variables, e.g., thermal radiation, flame monitoring (other than *FR* and *FQ*)
FD	Density
FE	Electrical variables (e.g., electrical efficiency, power)
FF	Flow, rate
FG	Distance, length, position, direction of rotation
FJ	Power (mechanical, thermal)
FK	Time
FL	Level (also for dividing line)
FM	Moisture, humididty
FP	Pressure
FQ	Quality variables (analysis material properties other than *FD*, *FM*, *FV*)
FR	Radiation variables
FS	Velocity, speed, frequency (mechanical)
FT	Temperature
FV	Viscosity
FW	Weight, mass
FY	Vibration, expansion

Equipment Unit Key G/H

G	Electrical equipment
GA	Junction boxes for analog measured data
GB	Junction boxes for binary signals
GC	Junction boxes for analog safety I&C (SILT) measured data
GD	Junction boxes for power cables ≥ 1 kV
GE	Junction boxes for power cables < 1 kV
GF	Junction boxes for operational I&C (BELT) measured data
GH	Electrical and instrumentation and control installation units (e.g., cubicles, boxes)
GJ	Processing and storage equipment for process computers and automation systems (automation equipment and servers, etc.)
GK	Information display and operator control equipment for process computers and automation systems (keyboards, monitors, printers)
GM	Junction boxes for light-current systems of national telecommunications services
GN	Networking equipment (bus couplers and optic transceivers, etc.)
GP	Subdistribution/junction boxes for lighting
GQ	Subdistribution/junction boxes for power sockets

GR	DC generating equipment, batteries
GS	Switchgear equipment if not identified under process equipment
GT	Transformer equipment
GU	Converter equipment
GV	Structure-related earthing and lightning protection equipment, surge arrestors
GW	Cabinet power supply equipment
GX	Actuating equipment for electrical variables
GY	Junction boxes for light-current systems (not of national telecommunications services)
GZ	Hangers, supports, and racks for electrical and instrumentation and control equipment
H	Subassemblies of main and heavy machinery
HA	Machine stationary assembly
HB	Machine rotating assembly
HD	Bearing assembly

Component Key K/M/Q

K	Mechanical components
KA	Gate valves, globe valves, dampers, cocks, rupture disks, orifices
KB	Gates, doors, dam boards
KC	Heat exchangers, coolers
KD	Vessels/tanks, pools, surge tanks (fluid systems)
KE	Turning, driving, lifting, and slewing gear
KF	Continuous conveyors, feeders
KJ	Size reduction machines
KK	Compacting, packaging machines
KM	Mixers, agitators
KN	Compressors, blowers, fans
KP	Pumps
KT	Cleaning machines, dryers, separators, filters
KV	Burners, grates
KW	Stationary tooling and treatment machines for maintenance
M	Mechanical components
MB	Brakes
MF	Foundations
MG	Gearboxes
MK	Clutches, couplings
MM	Engines, not electrical
MR	Piping components, ductwork components
MS	Positioners, not electrical
MT	Turbines
MU	Transmission gear, non-electrical, converters, and boosters other than couplings and gearboxes
Q	Instrumentation and control components (non-electrical)
QA	Enclosures (for I&C component protection only)
QB	Sensors if not structurally integral with *QP*, metering orifices
QH	Signalling devices
QN	Controllers, flybolt governor

QP	Measuring instruments, testing equipment
QR	Instrument piping
QS	Condensation chambers (datum reservoir) in measuring circuits
QT	Thermowells and pockets for protection of sensors

Component Key

	Electrical components
-A	Assemblies and subassemblies
-B	Transducers for non-electrical to electrical variables and vice-versa
-C	Capacitors
-D	Binary elements, delay devices, memory devices
-E	Special components
-F	Protective devices
-G	Generators, power supplies
-H	Signalling devices
-K	Relays, contactors
-L	Inductors
-M	Motors
-N	Amplifiers, controllers
-P	Measuring instruments, testing equipment
-Q	Power switchgear
-R	Resistors
-S	Switches, selectors
-T	Transformers
-U	Modulators, convertors from electrical to other electrical variables
-V	Tubes, semiconductors
-W	Transmission paths, waveguides, aerials
-X	Terminals, plugs, sockets
-Y	Electrical positioners, e.g., solenoids (not motors)
-Z	Terminations, balancing equipment, filters, limiters, cable terminations, equalizers, hybird transformers

Component Key X/Y

X	Signal origins
XA	Functional group control/subloop control
XB	Control interface
XC	Closed-loop control
XG	Binary process signals conditioned by binary signal conditioning modules
XH	Binary limit signals derived from analog process signals
XJ	Signals from non-standard areas (e.g., black box, dedicated I&C)
XK	Equipment unit/component protection
XL	Control room and control stations, signals not assigned to specific control systems (e.g., control interface tiles)
XM	Higher-level static individual and group alarm signals
XN	Status-display computer/criterion display
XP	Supervisory computer (process computer)
XQ	Analog signals
XS	Functional group control step signals

XT	Turbine generator I&C, binary signals
XU	Higher-level dynamic individual and group alarm signals
XV	Signal gating (protective logic, alarm logic, etc.)
XW	Hardwired alarm annunciation system
Y	Signal applications
YA	Functional group control/subloop control
YB	Control interface
YC	Closed loop control
YJ	Signals from non-standard areas (e.g., black box, dedicated I&C)
YL	Control room and control stations, signals not assigned to specific control systems (e.g., control interface tiles)
YN	Status-display computer/criterion display
YP	Supervisory computer (process computer)
YQ	Analog signals
YT	Turbine generator I&C, binary signals
YV	Signal gating (protective logic, alarm logic, etc.)
YW	Hardwired alarm annunciation system

APPENDIX J
WELDING QUALIFICATION

The chain of events for welder qualifications are as follows:

1. The contractor's engineer is responsible for writing a Welding Procedure Specification (WPS) to ensure that the procedure qualification complies with the ASME Code, Section IX, I or VIII, Division I.1. The quality control manager or engineer is responsible for certifying all procedure records.
2. After the specification is written and certified, a Procedure Qualification Record (PQR) is written and the welding procedure is tested and qualified.
3. Once the WPS is qualified, the welders can be tested using the Manufacturer's Record of Welder or Welding Operator Qualification Test (WPQ).

Welding procedures use group numbers for materials (P), filler metals (F), and welding positions (numeric F and G numbers). These are provided by the code to reduce the number of procedures required by grouping similar procedures into one. A sample of a WPS, PQR, and WPQ are included in the following pages. Most companies write Welding Procedures Manuals as well as Welding Quality Control Manuals for personnel guidelines to ensure code compliance.

Welding symbols here are from the American Welding Society (AWS).

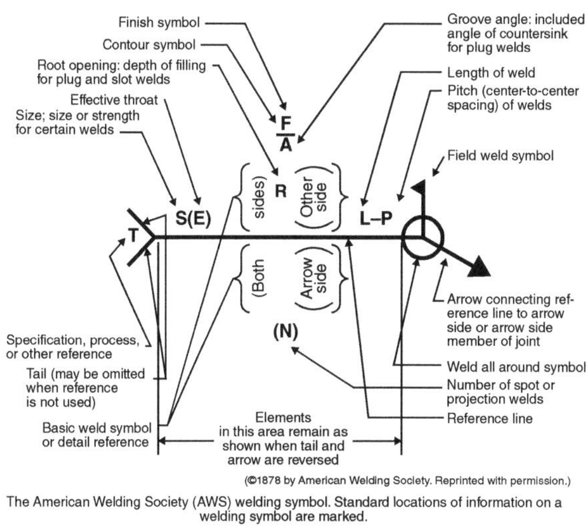

(©1878 by American Welding Society. Reprinted with permission.)

The American Welding Society (AWS) welding symbol. Standard locations of information on a welding symbol are marked.

Typical welding symbols

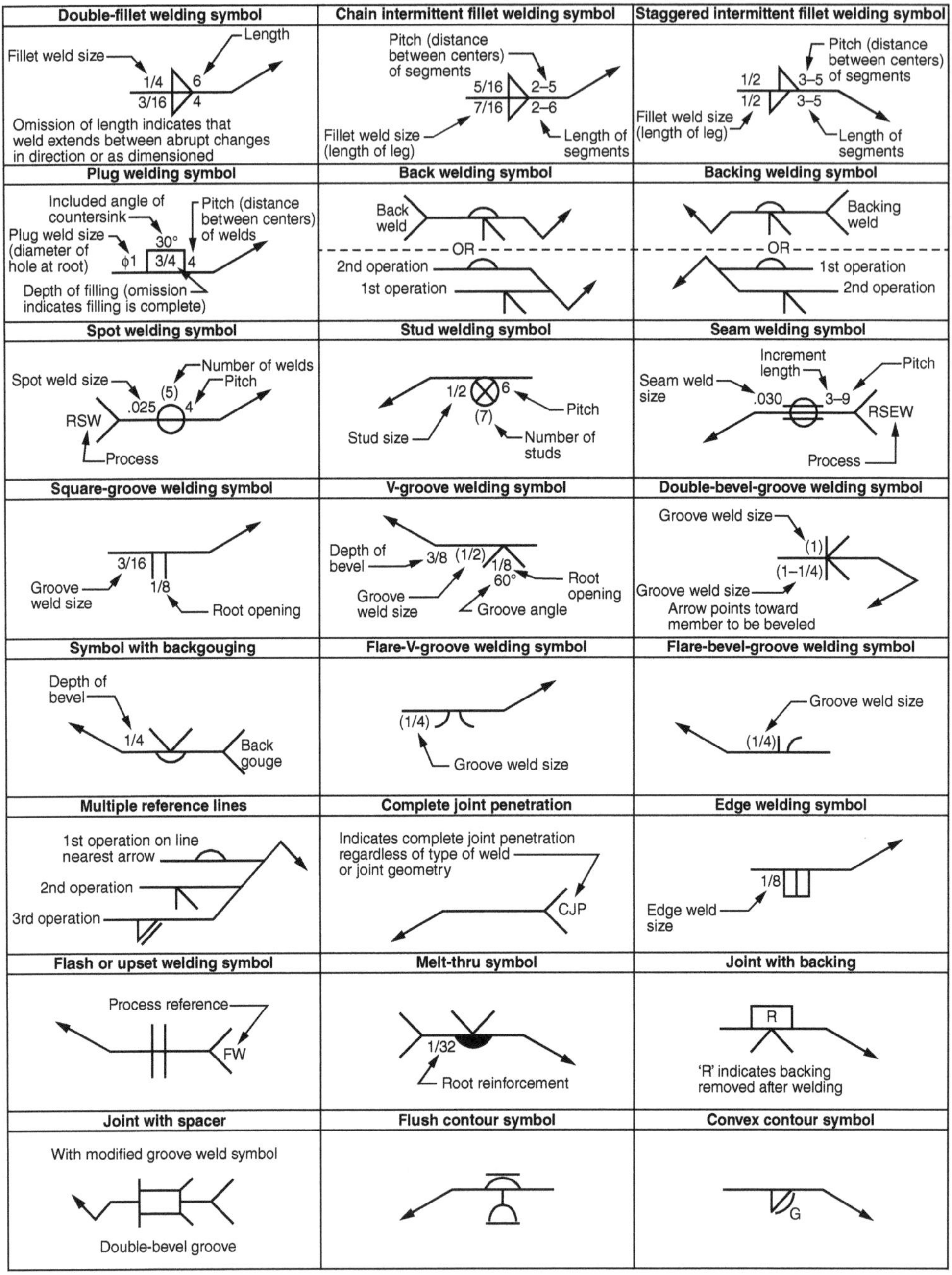

It should be understood that this chart is intended only as shop aids. The only complete and official presentation of the standard welding symbols is in AWS A2.4-98, standard symbols for welding, brazing, and nondestructive examination.

QW-482 SUGGESTED FORMAT FOR WELDING PROCEDURE SPECIFICATIONS (WPS)
(See QW-200.1, Section IX, ASME Boiler and Pressure Vessel Code)

Company Name DCR Engineers, Inc By D. C. Richardson

Welding Procedure Specification No. PP001 Date 10-10-14 Supporting PQR No. (s) PP001-1, 2, 3

Revision No. 1 Date 2-27-15

Welding Process(es) GTAW/SMAW Type(s) Manual
(Automatic, Manual, Machine, or Semi-Automatic)

JOINTS (QW-402)

Joint Design Single Vee Groove
Root Spacing NA
Backing: Yes ____ No X
Backing Material (Type) NA
(Refer to both backing and retainers)

☐ Metal ☐ Nonfusing Metal

☐ Nonmetallic ☐ Other

Sketches, Production Drawings, Weld Symbols, or Written Description should show the general arrangement of the parts to be welded. Where applicable, the details of weld groove may be specified.

[At the option of the manufacturer, sketches may be attached to illustrate joint design, weld layers, and bead sequence (e.g., for notch toughness procedures, for multiple process procedures, etc.)]

Details

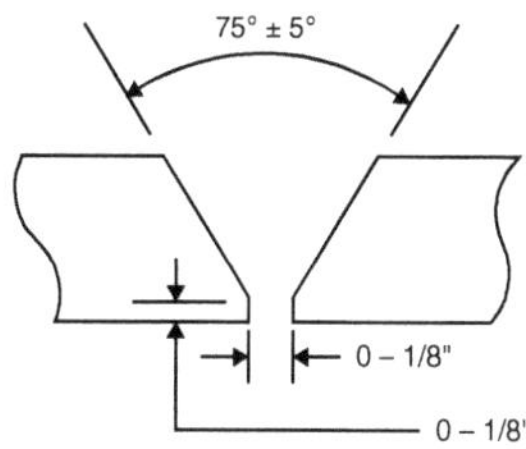

*BASE METALS (QW-403)

P-No. 1 Group No. 1 to P-No. 1 Group No. 1

OR

Specification and type/grade or UNS Number SA 106 Grade B
to Specification and type/grade or UNS Number SA 106 Grade B

OR

Chem. Analysis and Mech. Prop. NA
to Chem. Analysis and Mech. Prop. NA

Thickness Range:
Base Metal: Groove .280" - .864" Fillet All
Maximum Pass Thickness ≥1/2inch (13 mm) ☐ (Yes) ☐ (No)
Other

*FILLER METALS (QW-404)	GTAW	SMAW
Spec. No. (SFA)	5.18	5.1
AWS No. (Class)	E 70 S-2	E 7018
F-No.	6	4
A-No.	1	1
Size of Filler Metals	3/32, 1/8	3/32, 1/8, 5/32
Filler Metal Product Form		
Supplemental Filler Metal		
Weld Metal	1/16-3/16	1/16-3/16
Thickness Range:		
Groove		
Fillet		
Electrode-Flux (Class)		
Flux Type		
Flux Trade Name		
Consumable Insert		
Other		

*Each base metal-filler metal combination should be recorded individually.

QW-482 (Back)

WPS No. ________ Rev. ________

POSITIONS (QW-405)
Position(s) of Groove All
Welding Progression: Up All Down ________
Position(s) of Fillet ________
Other ________

POSTWELD HEAT TREATMENT (QW-407)
Temperature Range NA
Time Range NA
Other ________

PREHEAT (QW-406)
Preheat Temperature, Minimum 50°F
Interpass Temperature, Maximum 500°F
Preheat Maintenance As needed
Other ________
(Continuous or special heating, where applicable, should be recorded)

GAS (QW-408)

	Gas(es)	Percent Composition (Mixture)	Flow Rate
Shielding	Argon	WELD Grade	4-30 CFH
Trailing	NA		
Backing	None		
Other			

ELECTRICAL CHARACTERISTICS (QW-409)

Weld Pass(es)	Process	Filler Metal Classification	Filler Metal Diameter	Current Type and Polarity	Amps (Range)	Wire Feed Speed (Range)	Energy or Power (Range)	Volts (Range)	Travel Speed (Range)	Other (e.g., Remarks, Comments, Hot Wire Addition, Technique, Torch Angle, etc.)
Root	GTAW	E 70-S-2	3/32	Straight	70-100			7-11	1-11	
		E 70-S-2	1/8	Straight	70-115			7-11	1-11	
Fill	SMAW	E 7018	3/32	Reverse	70-100			20-24	2-11	
		E 7018	1/8	Reverse	100-175			16-28	2-11	
		E 7018	5/32	Reverse	125-225			20-29	2-11	

Amps and volts, or power or energy range, should be recorded for each electrode size, position, and thickness, etc.

Pulsing Current ________ Heat Input (max.) ________
Tungsten Electrode Size and Type 3/32 - 2% Toristed
(Pure Tungsten, 2% Thoriated, etc.)
Mode of Metal Transfer for GMAW (FCAW) ________
(Spray Arc, Short Circuiting Arc, etc.)
Other ________

TECHNIQUE (QW-410)
String or Weave Bead Both
Orifice, Nozzle, or Gas Cup Size 4-12
Initial and Interpass Cleaning (Brushing, Grinding, etc.) Chipping, Brushing, Grinding or any other means suitable for removing slag, scale or other deletrious materials
Method of Back Gouging NA
Oscillation -
Contact Tube to Work Distance NA
Multiple or Single Pass (Per Side) Multiple
Multiple or Single Electrodes Single
Electrode Spacing ________
Peening None
Other

QW-483 SUGGESTED FORMAT FOR PROCEDURE QUALIFICATION RECORDS (PQR)
(See QW-200.2, Section IX, ASME Boiler and Pressure Vessel Code)
Record Actual Variables Used to Weld Test Coupon

Company Name DCR Engineers, Inc
Procedure Qualification Record No. PP001-1 Date 10-10-14
WPS No. PP001-1
Welding Process(es) GTAW/SMAW
Types (Manual, Automatic, Semi-Automatic) Manual

JOINTS (QW-402)

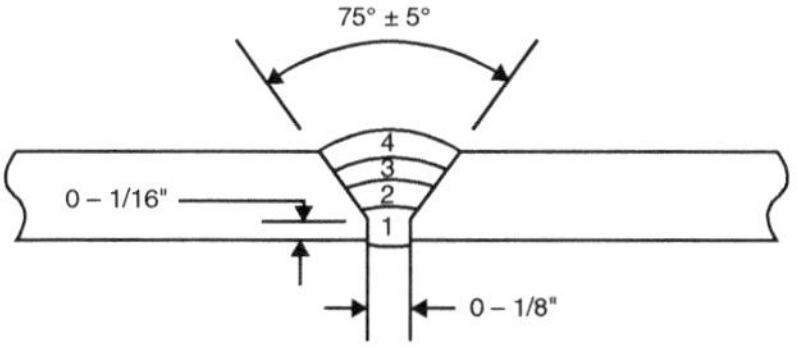

Passes 1 & 2: GTAW
Rest: SMAW

Groove Design of Test Coupon

(For combination qualifications, the deposited weld metal thickness shall be recorded for each filler metal and process used.)

BASE METALS (QW-403)
Material Spec. SA 106
Type/Grade, or UNS Number Grade B
P-No. 1 Group No. ____ to P-No. 1 Group No. ____
Thickness of Test Coupon .218
Diameter of Test Coupon 2"
Maximum Pass Thickness
Other

POSTWELD HEAT TREATMENT (QW-407)
Temperature NA
Time NA
Other

GAS (QW-408)

	Gas(es)	Percent Composition (Mixture)	Flow Rate
Shielding	Argon	WELD Grade	
Trailing			
Backing			
Other			

FILLER METALS (QW-404)

	GTAW	SMAW
SFA Specification	5.18	5.1
AWS Classification	E 70S-2	E 7018
Filler Metal F-No.	6	4
Weld Metal Analysis A-No.	1	1
Size of Filler Metal	3/32, 1/8	3/32, 1/8, 5/32
Filler Metal Product Form		
Supplemental Filler Metal		
Electrode Flux Classification		
Flux Type		
Flux Trade Name		
Weld Metal Thickness		
Other		

ELECTRICAL CHARACTERISTICS (QW-409)
Current Direct
Polarity GTAW-Straight SMAW-Reverse
Amps. 70-225 Volts 7-29
Tungsten Electrode Size
Mode of Metal Transfer for GMAW (FCAW)
Heat Input
Other

POSITION (QW-405)
Position of Groove 6G
Weld Progression (Uphill, Downhill) Uphill
Other

TECHNIQUE (QW-410)
Travel Speed 1-11 ipm
String or Weave Bead Both
Oscillation
Multipass or Single Pass (Per Side) Multipass
Single or Multiple Electrodes Single
Other

PREHEAT (QW-406)
Preheat Temperature 50°F
Interpass Temperature 500°F
Other

QW-483 (Back)

PQR No. PP001-1

Tensile Test (QW-150)

Specimen No.	Width	Thickness	Area	Ultimate Total Load	Ultimate Unit Stress, (psi or MPa)	Type of Failure and Location

Guided-Bend Tests (QW-160)

Type and Figure No.	Result
FACE No. 1 QW-462.3(a)	Satisfactory
FACE No. 2 QW-462.3(a)	Satisfactory
Roof No. 1 QW-462.3(a)	Satisfactory
Roof No. 2 QW-462.3(a)	Satisfactory

Toughness Tests (QW-170)

Specimen No.	Notch Location	Specimen Size	Test Temperature	Impact Values			Drop Weight Break (Y/N)
				ft-lb or J	% Shear	Mils (in.) or mm	

Comments ______

Fillet-Weld Test (QW-180)

Result — Satisfactory: Yes No Penetration into Parent Metal: Yes No

Macro — Results ______

Other Tests

Type of Test ______

Deposit Analysis ______

Other

....

Welder's Name John Doe Clock No. 507-56-7477 Stamp No. FDC

Tests Conducted by Industrial Testing Inter. Inc. Laboratory Test No. JN613.1

We certify that the statements in this record are correct and that the test welds were prepared, welded, and tested in accordance with the requirements of Section IX of the ASME Boiler and Pressure Vessel Code.

Manufacturer or Contractor DCR Engineers, Inc.

Date 3/5/14 Certified by Duncan C. Richardson

(Detail of record of tests are illustrative only and may be modified to conform to the type and number of tests required by the Code.)

03/08

FORM QW-484A SUGGESTED FORMAT A FOR WELDER PERFORMANCE QUALIFICATIONS (WPQ)
(See QW-301, Section IX, ASME Boiler and Pressure Vessel Code)

Welder's name John Doe Identification no. 525-94-2834

Test Description

Identification of WPS followed PP001 ☐ Test coupon ☐ Production weld

Specification and type/grade or UNS Number of base metal(s) ____ Thickness ____

Testing Variables and Qualification Limits

Welding Variables (QW-350)	Actual Values	Range Qualified
Welding process(es)	GTAW	GTAW
Type (i.e.; manual, semi-automatic) used	Manual	Manual
Backing (with/without)	Without	With or Without
☐ Plate ☐ Pipe (enter diameter if pipe or tube)		
Base metal P-Number to P-Number	P1 to P1	P1, P8 to P8
Filler metal or electrode specification(s) (SFA) (info. only)		
Filler metal or electrode classification(s) (info. only)		
Filler metal F-Number(s)	6	6
Consumable insert (GTAW or PAW)		
Filler Metal Product Form (solid/metal or flux cored/powder) (GTAW or PAW)		
Deposit thickness for each process		
Process 1 ____ 3 layers minimum ☐ Yes ☐ No		
Process 2 ____ 3 layers minimum ☐ Yes ☐ No		
Position qualified (2G, 6G, 3F, etc.)		
Vertical progression (uphill or downhill)	uphill	uphill
Type of fuel gas (OFW)		
Inert gas backing (GTAW, PAW, GMAW)		
Transfer mode (spray/globular or pulse to short circuit-GMAW)		
GTAW current type/polarity (AC, DCEP, DCEN)	DCEP	DCEP

RESULTS

Visual examination of completed weld (QW-302.4) ____

☐ Transverse face and root bends [QW-462.3(a)] ☐ Longitudinal bends [QW-462.3(b)] ☐ Side bends (QW-462.2)

☐ Pipe bend specimen, corrosion-resistant weld metal overlay [QW-462.5(c)]

☐ Plate bend specimen, corrosion-resistant weld metal overlay [QW-462.5(d)]

☐ Pipe specimen, macro test for fusion [QW-462.5(b)] ☐ Plate specimen, macro test for fusion [QW-462.5(e)]

Type	Result	Type	Result	Type	Result
FACE A QW-462.3(a)	Accept	FACE B QW-462.3(a)	Accept	ROOT A QW-462.3(a)	Accept
				ROOT B QW-462.3(a)	Accept

Alternative Volumetric Examination Results (QW-191): ____ RT ☐ or UT ☐ (check one)

Fillet weld — fracture test (QW-181.2) ____ Length and percent of defects ____

☐ Fillet welds in plate [QW-462.4(b)] ☐ Fillet welds in pipe [QW-462.4(c)]

Macro examination (QW-184) ____ Fillet size (in.) ____ × ____ Concavity/convexity (in.) ____

Other tests ____

Film or specimens evaluated by ____ Company ____

Mechanical tests conducted by Industrial Testing Inter. Inc. Laboratory test no. 613.2

Welding supervised by ____

We certify that the statements in this record are correct and that the test coupons were prepared, welded, and tested in accordance with the requirements of Section IX of the ASME BOILER AND PRESSURE VESSEL CODE.

Organization DCR Engineers, Inc.

Date 3-12-14 Certified by ____

APPENDIX K

TENSILE STRENGTH TABLE FOR CARBON STEEL

Spec. No.	Grade	Type or Class	Nominal Composition	P-No.	Notes	Specified Minimum Tensile, ksi	Specified Minimum Yield, ksi	*E* or *F*
Seamless Pipe and Tube								
A 53	A	S	C	1	(2)	48	30	1.00
	B	S	C-Mn	1	(2)	60	35	1.00
A 106	A	. . .	C-Si	1	(2)	48	30	1.00
	B	. . .	C-Si	1	(2)	60	35	1.00
	C	. . .	C-Si	1	(2)	70	40	1.00
A 179	. . .	. . .	C	1	(1)(2)(5)	(47)	26	1.00
A 192	. . .	. . .	C-Si	1	(2)(5)	(47)	26	1.00
A 210	A1	. . .	C-Si	1	(2)	60	37	1.00
	C	. . .	C-Mn-Si	1	(2)	70	40	1.00
A 333	1	. . .	C-Mn	1	(1)	55	30	1.00
	6	. . .	C-Mn-Si	1	. . .	60	35	1.00
A 369	FPA	. . .	C-Si	1	(2)	48	30	1.00
	FPB	. . .	C-Mn	1	(2)	60	35	1.00
API-5L	A	. . .	C	1	(1)(2)(14)	48	30	1.00
	B	. . .	C-Mn	1	(1)(2)(14)	60	35	1.00
Furnace Butt Welded Pipe								
A 53	. . .	F	C	1	(4)	48	30	0.60
API-5L	A25	I & II	C	1	(1)(4)(14)	45	25	0.60
Electric Resistance Welded Pipe and Tube								
A 53	A	E	C	1	(2)	48	30	0.85
	B	E	C-Mn	1	(2)	60	35	0.85
A 135	A	. . .	C	1	(1)(2)	48	30	0.85
	B	. . .	C-Mn	1	(1)(2)	60	35	0.85

TABLE A-1 Carbon Steel[a]

Maximum Allowable Stress Values in Tension, ksi, for Metal Temperature, °F, Not Exceeding												
−20 to 100	200	300	400	500	600	650	−20 to 650	700	750	800	Grade	Spec. No
...	...	...	...	...	...	...	12.0	11.7	10.7	9.0	A	A 53
...	...	...	...	...	...	...	15.0	14.4	13.0	10.8	B	
...	...	...	...	...	...	...	12.0	11.7	10.7	9.0	A	A 106
...	...	...	...	...	...	...	15.0	14.4	13.0	10.8	B	
...	...	...	...	...	...	...	17.5	16.6	14.8	12.0	C	
...	...	...	...	...	...	...	11.8	11.5	10.6	9.2	...	A 179
...	...	...	...	...	...	...	11.8	11.5	10.7	9.0	...	A 192
...	...	...	...	...	...	...	15.0	14.4	13.0	10.8	A1	A 210
...	...	...	...	...	...	...	17.5	16.6	14.8	12.0	C	
...	...	...	...	...	...	...	13.8	...	...	...	1	A 333
...	...	...	...	...	...	...	15.0	14.4	...	...	6	
...	...	...	...	...	...	...	12.0	11.7	10.7	9.0	FPA	A 369
...	...	...	...	...	...	...	15.0	14.4	13.0	10.8	FPA	
...	...	...	...	...	...	...	12.0	11.7	10.7	9.0	A	API-5L
...	...	...	...	...	...	...	15.0	14.4	13.0	10.8	B	
Furnance Butt Welded Pipe												
...	...	...	...	...	...	...	6.8	6.5	...	...	...	A 53
6.7	6.6	6.7	6.7	...	...	...	...	...	...	...	A 25	API-5L
Electric Resistance Welded Pipe and Tube												
...	...	...	...	...	...	...	10.2	9.9	9.1	7.7	A	A 53
...	...	...	...	...	...	...	12.8	12.2	11.0	9.2	B	
...	...	...	...	...	...	...	10.2	9.9	9.1	7.9	A	A 135
...	...	...	...	...	...	...	12.8	12.2	11.0	9.2	B	

APPENDIX L

PIPE PRESSURE–TEMPERATURE RATINGS

The pressure-temperature ratings for pipe shown in these tables are calculated using ANSI B31.1, 2004. Never use tables provided in handbooks that do not reflect the latest codes. There are some handbooks, including pipe fitting handbooks, that have not been updated in decades. Nevertheless, these tables are for reference only and should not be substituted for engineering calculations. Only ASTM Specification A53 and A106, Grades A and B, are provided here solely for the purpose of illustrating the effect temperature has on maximum operating pressure and should not be used in lieu of engineering calculations.

The following is a sample calculation used in the formulation of the tables in this appendix:

$$P = 2\,SE\,(t_m - A)/D_o - 2\,y\,(t_m - A)$$

where: P = internal design pressure, psig

SE = maximum allowable stress in material due to internal pressure and joint efficiency (or casting quality factor) at the design temperature, psi (values and definitions are provided in B31.1)

t_m = minimum required wall thickness, in

t_n = nominal wall thickness

$t_m = t_n(0.875)$, per ANSI B36.10, 12.5% mill tolerance

D_o = outside pipe diameter

y = coefficient having values as given in Table 104.1.2(A), B31.1

A = additional thickness, in, to compensate for material removed in threading, grooving, etc., required to make a mechanical joint. This value is zero for the tables in this appendix.

Given: A53/A106 Grade A, 6-in Schedule 40 pipe

$SE = 12{,}000$

$D_o = 6.625$

$t_m = 0.245$

$y = 0.400$

$A = 0$ (For these tables, A is assumed to be zero; that is fitting tolerances are not used.)

$T = -20$ to 650

$P = 2 \times 12{,}000 \times 0.245/(6.625 - 2 \times 0.4 \times 0.245)$

$= 915$ psig

It should be noted that all the ANSI B31 series codes do not have the same values for *SE;* therefore, the values for P in these tables only apply to power piping.

Pressure–Temperature Ratings

Plain End Seamless Carbon Steel Pipe
Grade A, ASTM Specifications A-53 and A-106, Minimum Tensile Strength 48,000 PSI

					Temperature Deg F	–20 to 650	700	750	800
					Allowable Stress (SE) PSI	12,000	11,700	10,700	9000
Pipe Size	Schedule Number	Weight Class	D_o	t_n, in	t_m, in	Maximum Working Pressure, psig			
½	40	STD	0.840	0.109	0.095	2984	2910	2661	2036
	80	XS		0.147	0.129	4202	4097	3747	2764
	160	—		0.188	0.165	5536	5397	4936	3536
	—	XXS		0.294	0.257	8858	8637	7899	5507
¾	40	STD	1.050	0.113	0.099	2447	2386	2182	1697
	80	XS		0.154	0.135	3439	3354	3067	2314
	160	—		0.219	0.192	5115	4987	4560	3291
	—	XXS		0.308	0.270	7419	7233	6615	4628
1	40	STD	1.315	0.133	0.116	2278	2221	2031	1588
	80	XS		0.179	0.157	3168	3089	2825	2149
	160	—		0.250	0.219	4611	4496	4112	2998
	—	XXS		0.358	0.313	6830	6660	6091	4284
1-¼	40	STD	1.660	0.140	0.123	1890	1843	1686	1384
	80	XS		0.191	0.167	2626	2560	2341	2405
	160	—		0.250	0.219	3540	3451	3156	3153
	—	XXS		0.382	0.334	5685	5542	5069	4809
1-½	40	STD	1.900	0.145	0.127	1695	1652	1511	1203
	80	XS		0.200	0.175	2386	2327	2128	1658
	160	—		0.281	0.246	3466	3380	3091	2330
	—	XXS		0.400	0.350	5156	5028	4598	3316
2	40	STD	2.375	0.154	0.135	1429	1393	1274	1023
	80	XS		0.218	0.191	2063	2011	1839	1447
	160	—		0.344	0.301	3385	3300	3018	2281
	—	XXS		0.436	0.382	4430	4320	3950	2895
2-½	40	STD	2.875	0.203	0.178	1563	1524	1394	1114
	80	XS		0.276	0.242	2166	2112	1931	1742
	160	—		0.375	0.328	3013	2938	2687	2053
	—	XXS		0.552	0.483	4656	4540	4152	3024
3	40	STD	3.500	0.216	0.189	1355	1321	1208	972
	80	XS		0.300	0.263	1919	1871	1711	1352
	160	—		0.438	0.383	2878	2806	2566	1969
	—	XXS		0.600	0.525	4091	3989	3648	2700
3-½	40	STD	4.000	0.226	0.198	1237	1206	1103	792
	80	XS		0.318	0.278	1766	1722	1575	1112
4	40	STD	4.500	0.237	0.207	1146	1118	1022	670
	80	XS		0.337	0.295	1660	1619	1481	954
	120	—		0.438	0.383	2192	2137	1954	1239
	160	—		0.531	0.465	2703	2636	2411	1504
	—	XXS		0.674	0.590	3515	3428	3135	1909
5	40	STD	5.563	0.258	0.226	1008	983	899	731
	80	XS		0.375	0.328	1485	1448	1324	1061
	120	—		0.500	0.438	2017	1966	1798	1417
	160	—		0.625	0.547	2561	2497	2284	1769
	—	XXS		0.750	0.656	3125	3047	2786	2122

					Temperature Deg F	–20 to 650	700	750	800
					Allowable Stress (SE) PSI	12,000	11,700	10,700	9000
Pipe Size	Schedule Number	Weight Class	D_o	t_n, in	t_m, in	Maximum Working Pressure, psig			
6	40	STD	6.625	0.280	0.245	915	892	816	665
	80	XS		0.432	0.378	1435	1399	1279	1027
	120	—		0.562	0.492	1895	1848	1690	1336
	160	—		0.719	0.629	2466	2404	2199	1708
	—	XXS		0.864	0.756	3014	2939	2687	2053
8	20	—	8.625	0.250	0.219	622	606	555	457
	30	—		0.277	0.242	689	672	614	505
	40	STD		0.322	0.282	806	786	718	588
	60	—		0.406	0.355	1021	996	911	741
	80	XS		0.500	0.438	1270	1239	1133	914
	100	—		0.594	0.520	1520	1482	1356	1085
	120	—		0.719	0.629	1859	1812	1657	1312
	140	—		0.812	0.711	2118	2065	1889	1483
	—	XXS		0.875	0.766	2294	2237	2046	1598
	160	—		0.906	0.793	2382	2322	2124	1654
10	20	—	10.750	0.250	0.219	497	485	443	367
	30	—		0.307	0.269	613	598	546	450
	40	STD		0.365	0.319	730	711	650	534
	60	XS		0.500	0.438	1011	986	901	733
	80	—		0.594	0.520	1208	1177	1077	870
	100	—		0.719	0.629	1473	1436	1314	1053
	120	—		0.844	0.739	1746	1702	1557	1237
	140	—		1.000	0.875	2090	2037	1863	1464
	160	—		1.125	0.984	2370	2311	2114	1647
12	20	—	12.75	0.250	0.219	418	408	373	309
	30	—		0.330	0.289	554	540	494	408
	—	STD		0.375	0.328	630	615	562	463
	40	—		0.406	0.355	683	666	609	501
	—	XS		0.500	0.438	848	827	756	618
	60	—		0.562	0.492	956	932	852	694
	80	—		0.688	0.602	1178	1148	1050	849
	100	—		0.844	0.739	1459	1422	1301	1043
	120	—		1.000	0.875	1743	1699	1554	1235
	140	—		1.125	0.984	1974	1925	1760	1388
	160	—		1.312	1.148	2329	2270	2076	1620
14	10	—	14	0.250	0.219	380	371	339	281
	20	—		0.312	0.273	475	464	424	351
	30	STD		0.375	0.328	573	559	511	421
	40	—		0.438	0.383	671	654	599	492
	—	XS		0.500	0.438	770	751	687	563
	60	—		0.594	0.520	919	896	819	668
	80	—		0.750	0.656	1168	1139	1042	843
	100	—		0.938	0.821	1477	1440	1317	1055
	120	—		1.094	0.957	1735	1692	1547	1230
	140	—		1.250	1.094	2000	1950	1784	1406
	160	—		1.406	1.230	2268	2211	2022	1580

Pipe Size	Schedule Number	Weight Class	D_o	t_n, in	t_m, in	−20 to 650	700	750	800
					Temperature Deg F	−20 to 650	700	750	800
					Allowable Stress (SE) PSI	12,000	11,700	10,700	9000
						Maximum Working Pressure, psig			
16	10	—	16	0.250	0.219	332	324	296	246
	20	—		0.312	0.273	415	405	370	307
	30	STD		0.375	0.328	500	488	446	369
	40	XS		0.500	0.438	672	655	599	492
	60	—		0.656	0.574	886	864	790	645
	80	—		0.844	0.739	1151	1122	1026	831
	100	—		1.031	0.902	1417	1381	1263	1014
	120	—		1.219	1.067	1691	1648	1508	1200
	140	—		1.438	1.258	2014	1963	1796	1414
	160	—		1.594	1.395	2249	2193	2006	1568
18	10	—	18	0.250	0.219	295	287	263	219
	20	—		0.312	0.273	368	359	329	273
	—	STD		0.375	0.328	444	433	396	328
	30	—		0.438	0.383	520	507	463	383
	—	XS		0.500	0.438	596	581	531	438
	40	—		0.562	0.492	671	654	598	492
	60	—		0.750	0.656	901	878	803	655
	80	—		0.938	0.821	1136	1108	1013	820
	100	—		1.156	1.012	1413	1378	1260	1011
	120	—		1.375	1.203	1695	1652	1511	1202
	140	—		1.562	1.367	1941	1892	1730	1366
	160	—		1.781	1.558	2232	2176	1990	1557
20	10	—	20	0.250	0.219	265	258	236	197
	20	STD		0.375	0.328	399	389	356	295
	30	XS		0.500	0.438	535	522	477	394
	40	—		0.594	0.520	637	621	568	468
	60	—		0.812	0.711	878	856	783	639
	80	—		1.031	0.902	1123	1095	1001	811
	100	—		1.281	1.121	1408	1373	1256	1008
	120	—		1.500	1.313	1663	1621	1483	1181
	140	—		1.750	1.531	1957	1908	1745	1377
	160	—		1.969	1.723	2221	2165	1980	1549
22	10	—	22	0.250	0.219	241	235	215	179
	20	STD		0.375	0.328	362	353	323	268
	30	XS		0.500	0.438	486	473	433	358
	60	—		0.875	0.766	860	838	766	626
	80	—		1.125	0.984	1113	1085	993	804
	100	—		1.375	1.203	1372	1338	1224	983
	120	—		1.625	1.422	1636	1595	1459	1162
	140	—		1.875	1.641	1904	1856	1698	1341
	160	—		2.125	1.859	2175	2121	1939	1520
24	10	—	24	0.250	0.219	221	215	197	164
	20	STD		0.375	0.328	332	323	296	246
	—	XS		0.500	0.438	444	433	396	328
	30	—		0.562	0.492	500	488	446	369
	40	—		0.688	0.602	614	599	548	451
	60	—		0.969	0.848	873	851	778	635
	80	—		1.219	1.067	1106	1079	986	799
	100	—		1.531	1.340	1403	1368	1251	1004
	120	—		1.812	1.586	1675	1633	1493	1188
	140	—		2.062	1.804	1919	1871	1711	1352
	160	—		2.344	2.051	2202	2146	1963	1537

Plain End Seamless Carbon Steel Pipe
Grade B, ASTM Specifications A-53 and A-106, Minimum Tensile Strength 60,000 PSI

					Temperature Deg F	−20 to 650	700	750	800
					Allowable Stress (SE) PSI	15,000	14,400	13,000	10,800
Pipe Size	Schedule Number	Weight Class	D_o	t_n, in	t_m, in	Maximum Working Pressure, psig			
½	40	STD	0.840	0.109	0.095	3730	3581	3233	2443
	80	XS		0.147	0.129	5252	5042	4552	3317
	160	—		0.188	0.165	6920	6643	5997	4243
	—	XXS		0.294	0.257	11073	10630	9597	6608
¾	40	STD	1.050	0.113	0.099	3059	2937	2651	2036
	80	XS		0.154	0.135	4299	4127	3726	2777
	160	—		0.219	0.192	6393	6137	5541	3950
	—	XXS		0.308	0.270	9273	8903	8037	5554
1	40	STD	1.315	0.133	0.116	2847	2733	2468	1905
	80	XS		0.179	0.157	3960	3802	3432	2579
	160	—		0.250	0.219	5764	5534	5541	3597
	—	XXS		0.358	0.313	8538	8197	8037	5141
1-¼	40	STD	1.660	0.140	0.123	2363	2268	2048	1660
	80	XS		0.191	0.167	3282	3151	2845	2886
	160	—		0.250	0.219	4425	4248	3835	3784
	—	XXS		0.382	0.334	7106	6821	6158	5771
1-½	40	STD	1.900	0.145	0.127	2119	2034	1836	1444
	80	XS		0.200	0.175	2983	2864	2585	1989
	160	—		0.281	0.246	4333	4160	3755	2796
	—	XXS		0.400	0.350	6446	6188	5586	3979
2	40	STD	2.375	0.154	0.135	1787	1715	1548	1228
	80	XS		0.218	0.191	2579	2475	2235	1737
	160	—		0.344	0.301	4231	4062	3667	2737
	—	XXS		0.436	0.382	5538	5316	4799	3474
2-½	40	STD	2.875	0.203	0.178	1954	1876	1694	1337
	80	XS		0.276	0.242	2708	2599	2347	2091
	160	—		0.375	0.328	3766	3616	3264	2464
	—	XXS		0.552	0.483	5820	5588	5044	3628
3	40	STD	3.500	0.216	0.189	1693	1625	1694	1166
	80	XS		0.300	0.263	2398	2303	2347	1623
	160	—		0.438	0.383	3598	3454	3264	2363
	—	XXS		0.600	0.525	5114	4909	5044	3240
3-½	40	STD	4.000	0.226	0.198	1546	1484	1340	1222
	80	XS		0.318	0.278	2208	2119	1913	1715
4	40	STD	4.500	0.237	0.207	1433	1375	1242	993
	80	XS		0.337	0.295	2076	1992	1799	1416
	120	—		0.438	0.383	2740	2630	2375	1838
	160	—		0.531	0.465	3379	3244	2929	2232
	—	XXS		0.674	0.590	4394	4218	3808	2832
5	40	STD	5.563	0.258	0.226	1260	1209	1092	877
	80	XS		0.375	0.328	1856	1782	1609	1273
	120	—		0.500	0.438	2521	2420	2185	1700
	160	—		0.625	0.547	3202	3074	2775	2123
	—	XXS		0.750	0.656	3906	3750	3385	2547

				Temperature Deg F		−20 to 650	700	750	800
				Allowable Stress (SE) PSI		15,000	14,400	13,000	10,800
Pipe Size	Schedule Number	Weight Class	D_o	t_n, in	t_m, in	Maximum Working Pressure, psig			
6	40	STD	6.625	0.280	0.245	1143	1098	991	799
	80	XS		0.432	0.378	1794	1722	1554	1232
	120	—		0.562	0.492	2369	2274	2053	1604
	160	—		0.719	0.629	3082	2959	2671	2050
	—	XXS		0.864	0.756	3767	3617	3265	2464
8	20	—	8.625	0.250	0.219	778	746	674	548
	30	—		0.277	0.242	861	827	746	606
	40	STD		0.322	0.282	1007	967	873	706
	60	—		0.406	0.355	1277	1226	1107	889
	80	XS		0.500	0.438	1588	1524	1376	1097
	100	—		0.594	0.520	1900	1824	1647	1302
	120	—		0.719	0.629	2323	2230	2014	1575
	140	—		0.812	0.711	2648	2542	2295	1780
	—	XXS		0.875	0.766	2868	2753	2486	1918
	160	—		0.906	0.793	2977	2858	2580	1985
10	20	—	10.750	0.250	0.219	621	596	538	440
	30	—		0.307	0.269	766	735	664	540
	40	STD		0.365	0.319	912	875	790	641
	60	XS		0.500	0.438	1264	1213	1095	880
	80	—		0.594	0.520	1510	1449	1308	1044
	100	—		0.719	0.629	1842	1768	1596	1263
	120	—		0.844	0.739	2182	2095	1891	1484
	140	—		1.000	0.875	2612	2507	2264	1757
	160	—		1.125	0.984	2963	2845	2568	1976
12	20	—	12.75	0.250	0.219	522	502	453	371
	30	—		0.330	0.289	693	665	600	489
	—	STD		0.375	0.328	788	756	683	555
	40	—		0.406	0.355	854	820	740	601
	—	XS		0.500	0.438	1060	1017	918	742
	60	—		0.562	0.492	1195	1147	1035	833
	80	—		0.688	0.602	1472	1413	1276	1019
	100	—		0.844	0.739	1823	1750	1580	1251
	120	—		1.000	0.875	2178	2091	1888	1482
	140	—		1.125	0.984	2468	2369	2139	1666
	160	—		1.312	1.148	2911	2794	2523	1944
14	10	—	14	0.250	0.219	475	456	412	338
	20	—		0.312	0.273	594	570	515	421
	30	STD		0.375	0.328	716	688	621	506
	40	—		0.438	0.383	839	806	727	591
	—	XS		0.500	0.438	963	924	834	675
	60	—		0.594	0.520	1148	1102	995	802
	80	—		0.750	0.656	1460	1402	1266	1012
	100	—		0.938	0.821	1846	1772	1600	1266
	120	—		1.094	0.957	2169	2083	1880	1476
	140	—		1.250	1.094	2501	2401	2167	1687
	160	—		1.406	1.230	2835	2722	2457	1897

					Temperature Deg F	−20 to 650	700	750	800
					Allowable Stress (SE) PSI	15,000	14,400	13,000	10,800
Pipe Size	Schedule Number	Weight Class	D_o	t_n, in	t_m, in	Maximum Working Pressure, psig			
16	10	—	16	0.250	0.219	415	399	360	295
	20	—		0.312	0.273	519	498	450	368
	30	STD		0.375	0.328	625	600	542	443
	40	XS		0.500	0.438	840	806	728	591
	60	—		0.656	0.574	1108	1064	960	774
	80	—		0.844	0.739	1439	1381	1247	997
	100	—		1.031	0.902	1771	1700	1535	1217
	120	—		1.219	1.067	2113	2029	1832	1440
	140	—		1.438	1.258	2517	2416	2181	1697
	160	—		1.594	1.395	2812	2699	2437	1882
18	10	—	18	0.250	0.219	369	354	319	263
	20	—		0.312	0.273	461	442	399	327
	—	STD		0.375	0.328	555	533	481	393
	30	—		0.438	0.383	649	623	563	459
	—	XS		0.500	0.438	744	715	645	525
	40	—		0.562	0.492	838	805	727	590
	60	—		0.750	0.656	1126	1081	976	787
	80	—		0.938	0.821	1420	1363	1231	985
	100	—		1.156	1.012	1766	1695	1531	1214
	120	—		1.375	1.203	2118	2034	1836	1443
	140	—		1.562	1.367	2426	2329	2102	1639
	160	—		1.781	1.558	2790	2678	2418	1868
20	10	—	20	0.250	0.219	331	318	287	236
	20	STD		0.375	0.328	499	479	432	354
	30	XS		0.500	0.438	669	642	580	473
	40	—		0.594	0.520	797	765	690	561
	60	—		0.812	0.711	1098	1054	951	767
	80	—		1.031	0.902	1404	1347	1216	973
	100	—		1.281	1.121	1760	1690	1526	1210
	120	—		1.500	1.313	2079	1996	1802	1417
	140	—		1.750	1.531	2446	2348	2120	1652
	160	—		1.969	1.723	2776	2665	2406	1859
22	10	—	22	0.250	0.219	301	289	261	215
	20	STD		0.375	0.328	453	435	392	322
	30	XS		0.500	0.438	607	583	526	430
	60	—		0.875	0.766	1074	1031	931	751
	80	—		1.125	0.984	1392	1336	1206	965
	100	—		1.375	1.203	1715	1647	1487	1180
	120	—		1.625	1.422	2045	1963	1772	1395
	140	—		1.875	1.641	2380	2285	2062	1610
	160	—		2.125	1.859	2719	2610	2356	1824
24	10	—	24	0.250	0.219	276	265	239	197
	20	STD		0.375	0.328	415	398	359	295
	—	XS		0.500	0.438	556	533	482	394
	30	—		0.562	0.492	625	600	542	442
	40	—		0.688	0.602	768	737	666	541
	60	—		0.969	0.848	1091	1047	945	763
	80	—		1.219	1.067	1383	1328	1199	959
	100	—		1.531	1.340	1753	1683	1520	1205
	120	—		1.812	1.586	2093	2009	1814	1426
	140	—		2.062	1.804	2399	2303	2079	1622
	160	—		2.344	2.051	2752	2642	2385	1844

APPENDIX M
VALVE PRESSURE–TEMPERATURE RATINGS

MANDATORY APPENDIX VII ASME B16.34-2004

Table VII-2-1.2 Ratings For Group 1.2 Materials

A 106 Gr. C (1)	A 203 Gr. E (2)	A 350 Gr. LF6 CI. 2 (3)	A 352 Gr. LC3 (4)
A 203 Gr. B (2)	A 216 Gr. WCC (2)	A 352 Gr. LC2 (4)	A 352 Gr. LCC (4)

NOTES:
(1) Not to be used over 800°F.
(2) Upon prolonged exposure to temperatures above 800°F, the carbide phase of steel may be converted to graphite. Permissible, but not recommended for prolonged use above 800°F.
(3) Not to be used over 500°F.
(4) Not to be used over 650°F.

A—Standard Class

Temperature, °F	Working Pressures by Class, psig						
	150	300	600	900	1500	2500	4500
–20 to 100	290	750	1500	2250	3750	6250	11,250
200	260	750	1500	2250	3750	6250	11,250
300	230	730	1455	2185	3750	6070	10,925
400	200	705	1405	2110	3640	5865	10,555
500	170	665	1330	1995	3520	5540	9965
600	140	605	1210	1815	3025	5040	9070
650	125	590	1175	1765	2940	4905	8825
700	110	555	1110	1665	2775	4630	8330
750	95	505	1015	1520	2535	4230	7610
800	80	410	825	1235	2055	3430	6170
850	65	320	640	955	1595	2655	4785
900	50	225	445	670	1115	1855	3345
950	35	135	275	410	685	1145	2055
1000	20	85	170	255	430	715	1285

B—Standard Class

Temperature, °F	Working Pressures by Class, psig						
	150	300	600	900	1500	2500	4500
−20 to 100	290	750	1500	2250	3750	6250	11,250
200	290	750	1500	2250	3750	6250	11,250
300	290	750	1500	2250	3750	6250	11,250
400	290	750	1500	2250	3750	6250	11,250
500	290	750	1500	2250	3750	6250	11,250
600	290	750	1500	2250	3750	6250	11,250
650	290	750	1500	2250	3750	6250	11,250
700	280	715	1425	2140	3565	5940	10,690
750	280	635	1270	1905	3170	5285	9515
800	255	515	1030	1545	2570	4285	7715
850	200	400	795	1195	1995	3320	5980
900	140	280	555	835	1395	2320	4180
950	85	170	345	515	855	1430	2570
1000	55	105	215	320	535	895	1605

ASME B16.34-2004 MANDATORY APPENDIX VII

Table VII-2-1.3 Ratings For Group 1.3 Materials

A 203 Gr. A (1)	A 352 Gr. LCE (2)	A 516 Gr. 65 (1)(3)	A 672 Gr. B65 (1)
A 203 Gr. D (1)	A 352 Gr. LC1 (2)	A 675 Gr. 65 (1)(4)(5)	A 672 Gr. C65 (1)
A 217 Gr. WC1 (6)(7)	A 515 Gr. 65 (1)		

NOTES:

(1) Upon prolonged exposure to temperatures above 800°F, the carbide phase of steel may be converted to graphite. Permissible, but not recommended for prolonged use above 800°F.

(2) Not to be used over 650°F.

(3) Not to be used over 850°F.

(4) Leaded grades shall not be used where welded or in any application above 500°F.

(5) For service temperatures above 850°F, it is recommended that killed steel containing not less than 0.10% residual silicon be used.

(6) Upon prolonged exposure to temperatures above 875°F, the carbide phase of steel of carbon-molybdennum steel may be converted to graphite. Permissible, but not recommended for prolonged use above 875°F.

(7) Use normalized and tempered material only.

A—Standard Class

Temperature, °F	Working Pressures by Class, psig						
	150	300	600	900	1500	2500	4500
−20 to 100	265	695	1395	2090	3480	5805	10,445
200	255	660	1320	1980	3300	5505	9905
300	230	640	1275	1915	3190	5315	9565
400	200	615	1230	1845	3075	5125	9225
500	170	585	1175	1760	2930	4885	8795
600	140	550	1150	1655	2755	4595	8270
650	125	535	1065	1600	2665	4440	7990
700	110	510	1025	1535	2560	4270	7685
750	95	475	955	1430	2385	3970	7150
800	80	390	780	1175	1955	3255	5865
850	65	300	595	895	1490	2485	4475
900	50	200	405	605	1010	1685	3035
950	35	135	275	410	685	1145	2055
1000	20	85	170	255	430	715	1285

B—Standard Class

Temperature, °F	Working Pressures by Class, psig						
	150	300	600	900	1500	2500	4500
−20 to 100	290	695	1395	2090	3480	5805	10,445
200	290	695	1395	2090	3480	5805	10,445
300	290	695	1395	2090	3480	5805	10,445
400	290	695	1395	2090	3480	5805	10,445
500	290	695	1395	2090	3480	5805	10,445
600	290	695	1395	2090	3480	5805	10,445
650	290	695	1390	2080	3470	5780	10,405
700	280	660	1320	1985	3305	5510	9915
750	280	595	1190	1785	2980	4965	8935
800	245	490	975	1465	2445	4070	7330
850	185	375	745	1120	1865	3105	5595
900	125	255	505	760	1265	2105	3795
950	85	170	345	515	855	1430	2570
1000	55	105	215	320	535	895	1605

MANDATORY APPENDIX VII ASME B16.34-2004

Table VII-2-1.4 Ratings For Group 1.4 Materials

A 106 Gr. B (1)	A 516 Gr. 60 (1)(2)	A 675 Gr. 65 (1)(3)(4)	A 672 Gr. B60 (1)
A 350 Gr. LF1 (1)	A 675 Gr. 60 (1)(2)(3)	A 696 Gr. B	A 672 Gr. C60 (1)
A 515 Gr. 60 (1)(2)			

NOTES:
(1) Upon prolonged exposure to temperatures above 800°F, the carbide phase of steel may be converted to graphite. Permissible, but not recommended for prolonged use above 800°F.
(2) Not to be used over 850°F.
(3) Leaded grades shall not be used where welded or in any application above 500°F.
(4) For service temperatures above 850°F, it is recommended that killed steels containing not less than 0.10% residual silicon be used.

A—Standard Class

Temperature, °F	Working Pressures by Class, psig						
	150	300	600	900	1500	2500	4500
−20 to 100	235	615	1235	1850	3085	5145	9255
200	215	565	1130	1695	2830	4715	8485
300	210	545	1090	1635	2725	4545	8175
400	200	525	1055	1580	2635	4390	7900
500	170	500	1005	1505	2510	4185	7530
600	140	475	945	1420	2365	3945	7095
650	125	455	915	1370	2285	3805	6850
700	110	440	885	1325	2210	3685	6635
750	95	430	855	1285	2140	3565	6420
800	80	370	740	1110	1850	3085	5555
850	65	300	595	895	1490	2485	4475
900	50	170	345	515	855	1430	2570
950	35	135	275	410	685	1145	2055
1000	20	85	170	255	430	715	1285

B—Standard Class

Temperature, °F	Working Pressures by Class, psig						
	150	300	600	900	1500	2500	4500
–20 to 100	245	645	1285	1930	3215	5355	9645
200	245	645	1285	1930	3215	5355	9645
300	245	645	1285	1930	3215	5355	9645
400	245	645	1285	1930	3215	5355	9645
500	245	645	1285	1930	3215	5355	9645
600	235	615	1230	1850	3080	5135	9240
650	230	595	1190	1785	2975	4955	8920
700	220	575	1150	1730	2880	4800	8640
750	215	555	1115	1670	2785	4645	8355
800	175	465	925	1390	2315	3855	6945
850	145	375	745	1120	1865	3105	5595
900	80	215	430	645	1070	1785	3215
950	65	170	345	515	855	1430	2570
1000	40	105	215	320	535	895	1605

APPENDIX N

THERMAL EXPANSION OF METALS

ASME B31.3-2004 Appendix C

TABLE C-1 Total Thermal Expansion, U.S. Units, for Metals

Total Linear Thermal Expansion Between 70° and Indicated Temperature, in/100 ft

	Material							
Temp., °F	Carbon Steel Carbon-Moly-Low-Chrome (Through 3Cr-Mo)	5Cr-Mo Through 9Cr-Mo	Austenitic Stainless Steels 18Cr-8Ni	12Cr, 17Cr, 27Cr	25Cr-20Ni	UNS NO4400 Monel 67Ni-30Cu	3½Ni	Copper and Copper Alloys
–450	—	—	—	—	—	—	—	-3.93
–425	—	—	—	—	—	—	—	-3.93
–400	—	—	—	—	—	—	—	-3.91
-375	—	—	—	—	—	—	—	-3.87
-350	—	—	—	—	—	—	—	-3.79
-325	-2.37	-2.22	-3.85	-2.04	—	-2.62	-2.25	-3.67
-300	-2.24	-2.10	-3.63	-1.92	—	-2.50	-2.17	-3.53
-275	-2.11	-1.98	-3.41	-1.80	—	-2.38	-2.07	-3.36
-250	-1.98	-1.86	-3.19	-1.68	—	-2.26	-1.96	-3.17
-225	-1.85	-1.74	-2.96	-1.57	—	-2.14	-1.86	-2.97
-200	-1.71	-1.62	-2.73	-1.46	—	-2.02	-1.76	-2.76
-175	-1.58	-1.50	-2.50	-1.35	—	-1.90	-1.62	-2.53
-150	-1.45	-1.37	-2.27	-1.24	—	-1.79	-1.48	-2.30
-125	-1.30	-1.23	-2.01	-1.11	—	-1.59	-1.33	-2.06
-100	-1.15	-1.08	-1.75	-0.98	—	-1.38	-1.17	-1.81
-75	-1.00	-0.94	-1.50	-0.85	—	-1.18	-1.01	-1.56
-50	-0.84	-0.79	-1.24	-0.72	—	-0.98	-0.84	-1.30
-25	-0.68	-0.63	-0.98	-0.57	—	-0.77	-0.67	-1.04
0	-0.49	-0.46	-0.72	-0.42	—	-0.57	-0.50	-0.77
25	-0.32	-0.30	-0.46	-0.27	—	-0.37	-0.32	-0.50
50	-0.14	-0.13	-0.21	-0.12	—	-0.20	-0.15	-0.22

TABLE C-1 Total Thermal Expansion, U.S. Units, for Metals (*Continued*)

Total Linear Thermal Expansion Between 70° and Indicated Temperature, in/100 ft

	Material							
Temp., °F	Carbon Steel Carbon-Moly-Low-Chrome (Through 3Cr-Mo)	5Cr-Mo Through 9Cr-Mo	Austenitic Stainless Steels 18Cr-8Ni	12Cr, 17Cr, 27Cr	25Cr-20Ni	UNS NO4400 Monel 67Ni-30Cu	3½ Ni	Copper and Copper Alloys
70	0	0	0	0	0	0	0	0
100	0.23	0.22	0.34	0.20	0.32	0.28	0.23	0.34
125	0.42	0.40	0.62	0.36	0.58	0.52	0.42	0.63
150	0.61	0.58	0.90	0.53	0.84	0.75	0.61	0.91
175	0.80	0.76	1.18	0.69	1.10	0.99	0.81	1.20
200	0.99	0.94	1.46	0.86	1.37	1.22	1.01	1.49
225	1.21	1.13	1.75	1.03	1.64	1.46	1.21	1.79
250	1.40	1.33	2.03	1.21	1.91	1.71	1.42	2.09
275	1.61	1.52	2.32	1.38	2.18	1.96	1.63	2.38
300	1.82	1.71	2.61	1.56	2.45	2.21	1.84	2.68
325	2.04	1.90	2.90	1.74	2.72	2.44	2.05	2.99
350	2.26	2.10	3.20	1.93	2.99	2.68	2.26	3.29
375	2.48	2.30	3.50	2.11	3.26	2.91	2.47	3.59
400	2.70	2.50	3.80	2.30	3.53	3.25	2.69	3.90
425	2.93	2.72	4.10	2.50	3.80	3.52	2.91	4.21
450	3.16	2.93	4.41	2.69	4.07	3.79	3.13	4.51
475	3.39	3.14	4.71	2.89	4.34	4.06	3.35	4.82
500	3.62	3.35	5.01	3.08	4.61	4.33	3.58	5.14
525	3.86	3.58	5.31	3.28	4.88	4.61	3.81	5.45
550	4.11	3.80	5.62	3.49	5.15	4.90	4.04	5.76

TABLE C-1 Total Thermal Expansion, U.S. Units, for Metals (*Continued*)

Total Linear Thermal Expansion Between 70° and Indicated Temperature, in/100 ft

Material								
Aluminum	Gray Cast Iron	Bronze	Brass	70Cu-30Ni	UNS N08XXX Series Ni-Fe-Cr	UNS N06XXX Series Ni-Fe-Cr	Ductile Iron	Temp., °F
—	—	—	—	—	—	—	—	−450
—	—	—	—	—	—	—	—	−425
—	—	—	—	—	—	—	—	−400
—	—	—	—	—	—	—	—	−375
—	—	—	—	—	—	—	—	−350
−4.68	—	−3.98	−3.88	−3.15	—	—	—	−325
−4.46	—	−3.74	−3.64	−2.87	—	—	—	−300
−4.21	—	−3.50	−3.40	−2.70	—	—	—	−275
−3.97	—	−3.26	−3.16	−2.53	—	—	—	−250
−3.71	—	−3.02	−2.93	−2.36	—	—		−225
−3.44	—	−2.78	−2.70	−2.19	—	—		−200
−3.16	—	−2.54	−2.47	−2.12	—	—		−175
−2.88	—	−2.31	−2.24	−1.95	—	—		−150
−2.57	—	−2.06	−2.00	−1.74	—	—		−125
−2.27	—	−1.81	−1.76	−1.53	—	—		−100
−1.97	—	−1.56	−1.52	−1.33	—	—		−75
−1.67	—	−1.32	−1.29	−1.13	—	—		−50
−1.32	—	−1.25	−1.02	−0.89	—	—		−25
−0.97	—	−0.77	−0.75	−0.66	—	—		0
−0.63	—	−0.49	−0.48	−0.42	—	—		25
−0.28	—	−0.22	−0.21	−0.19	—	—		50
0	0	0	0	0	0	0	0	70
0.46	0.21	0.36	0.35	0.31	0.28	0.26	0.21	100
0.85	0.38	0.66	0.64	0.56	0.52	0.48	0.39	125
1.23	0.55	0.96	0.94	0.82	0.76	0.70	0.57	150
1.62	0.73	1.26	1.23	1.07	0.99	0.92	0.76	175
2.00	0.90	1.56	1.52	1.33	1.23	1.15	0.94	200
2.41	1.08	1.86	1.83	1.59	1.49	1.38	1.13	225
2.83	1.27	2.17	2.14	1.86	1.76	1.61	1.33	250
3.24	1.45	2.48	2.45	2.13	2.03	1.85	1.53	275
3.67	1.64	2.79	2.76	2.40	2.30	2.09	1.72	300
4.09	1.83	3.11	3.08	2.68	2.59	2.32	1.93	325
4.52	2.03	3.42	3.41	2.96	2.88	2.56	2.13	350
4.95	2.22	3.74	3.73	3.24	3.18	2.80	2.36	375
5.39	2.42	4.05	4.05	3.52	3.48	3.05	2.56	400
5.83	2.62	4.37	4.38	—	3.76	3.29	2.79	425
6.28	2.83	4.69	4.72	—	4.04	3.53	3.04	450
6.72	3.03	5.01	5.06	—	4.31	3.78	3.28	475
7.17	3.24	5.33	5.40	—	4.59	4.02	3.54	500
7.63	3.46	5.65	5.75	—	4.87	4.27	3.76	525
8.10	3.67	5.98	6.10	—	5.16	4.52	3.99	550

TABLE C-1 Total Thermal Expansion, U.S. Units, for Metals (*Continued*)

Total Linear Thermal Expansion Between 70° and Indicated Temperature, in/100 ft

	Material							
Temp., °F	Carbon Steel Carbon-Moly-Low-Chrome (Through 3Cr-Mo)	5Cr-Mo Through 9Cr-Mo	Austenitic Stainless Steels 18Cr-8Ni	12Cr, 17Cr, 27Cr	25Cr-20Ni	UNS NO4400 Monel 67Ni-30Cu	$3^1/_2$Ni	Copper and Copper Alloys
575	4.35	4.02	5.93	3.69	5.42	5.18	4.27	6.07
600	4.60	4.24	6.24	3.90	5.69	5.46	4.50	6.09
625	4.86	4.47	6.55	4.10	5.96	5.75	4.74	—
650	5.11	4.69	6.87	4.31	6.23	6.05	4.98	—
675	5.37	4.92	7.18	4.52	6.50	6.34	5.22	—
700	5.63	5.14	7.50	4.73	6.77	6.64	5.46	—
725	5.90	5.38	7.82	4.94	7.04	6.94	5.70	—
750	6.16	5.62	8.15	5.16	7.31	7.25	5.94	—
775	6.43	5.86	8.47	5.38	7.58	7.55	6.18	—
800	6.70	6.10	8.80	5.60	7.85	7.85	6.43	—
825	6.97	6.34	9.13	5.82	8.15	8.16	6.68	—
850	7.25	6.59	9.46	6.05	8.45	8.48	6.93	—
875	7.53	6.83	9.79	6.27	8.75	8.80	7.18	—
900	7.81	7.07	10.12	6.49	9.05	9.12	7.43	—
925	8.08	7.31	10.46	6.71	9.35	9.44	7.68	—
950	8.35	7.56	10.80	6.94	9.65	9.77	7.93	—
975	8.62	7.81	11.14	7.17	9.95	10.09	8.17	—
1000	8.89	8.06	11.48	7.40	10.25	10.42	8.41	—
1025	9.17	8.30	11.82	7.62	10.55	10.75	—	—
1050	9.46	8.55	12.16	7.95	10.85	11.09	—	—
1075	9.75	8.80	12.50	8.18	11.15	11.43	—	—
1100	10.04	9.05	12.84	8.31	11.45	11.77	—	—
1125	10.31	9.28	13.18	8.53	11.78	12.11	—	—
1150	10.57	9.52	13.52	8.76	12.11	12.47	—	—
1175	10.83	9.76	13.86	8.98	12.44	12.81	—	—
1200	11.10	10.00	14.20	9.20	12.77	13.15	—	—
1225	11.38	10.26	14.54	9.42	13.10	13.50	—	—
1250	11.66	10.53	14.88	9.65	13.43	13.86	—	—
1275	11.94	10.79	15.22	9.88	13.76	14.22	—	—
1300	12.22	11.06	15.56	10.11	14.09	14.58	—	—
1325	12.50	11.30	15.90	10.33	14.39	14.94	—	—
1350	12.78	11.55	16.24	10.56	14.69	15.30	—	—
1375	13.06	11.80	16.58	10.78	14.99	15.66	—	—
1400	13.34	12.05	16.92	11.01	15.29	16.02	—	—
1425	—	—	17.30	—	—	—	—	—
1450	—	—	17.69	—	—	—	—	—
1475	—	—	18.08	—	—	—	—	—
1500	—	—	18.47	—	—	—	—	—

General Note: For code references to this Appendix, see para. 319.3.1. These data are for use in the absence of more applicable data. It is the designer's responsibility to verify that materials are suitable for the intended service at the temperatures shown.

APPENDIX O
PRESSURE TESTING REFERENCES

HYDROSTATIC TEST PROCEDURE

This sample procedure outlines hydrostatic testing of piping and/or pressure vessels. This procedure is to be used only as a guide to proven industry practice. All testing must be conducted using guidelines set forth in the appropriate codes. A list is provided to aid test personnel in locating the applicable codes. The information provided must be checked to ensure it is current and correct.

Preparation and Precautions

1. A piping and instrument drawing (P&ID) must be provided with the boundary "highlighted" for easy reference.
2. Access area around vents should be roped off during testing.
3. Safety and relief valves must be blanked off for testing.
4. All boundary valves must be danger tagged.
5. The first drain outside a boundary valve must be left open to ensure hydro pressure is limited to the test boundary.
6. Pressure vessel outer metal and water temperature must be greater than the nil-ductility temperature specified by the manufacturer and in no case below 70°F.
7. Water chemistry must be of the same quality as the process requirements if the system and/or vessel is not to be cleaned following hydrostatic testing.

General Procedure

1. Fill the system using the system in-line pumps, venting all high points.
2. Provide a recently calibrated (within 2 weeks) gauge with a range twice the test pressure.
3. Provide low-volume, high-head hydrostatic test pump and reservoir.
4. Pump the system or vessel pressure to test level while thoroughly examining for leaks.
5. Check drains outside the test boundary for leakage. If leakage occurs, record leakage rate of the boundary valve by timing the flow, with a stop watch, into a 5-gal bucket.
6. Check valves for packing leakage and tighten glands as necessary, using caution not to overtighten and render the valve inoperable.

Note

Leakage in valves is not cause for failing a hydrostatic test, but cause for initiation action to repair the valve in question. Valves may be repaired after the test. Also, test personnel may find it necessary to obtain a high-volume hydrostatic test pump if a number of valves are leaking beyond the capacity of the low-volume pump.

Caution

Do not use "cheaters" or otherwise overtighten valves to stop leakage. Isolation valves with leakage and control valves with excessive leakage must be disassembled and lapped or otherwise repaired.

7. Check weld joints and bolted flanges for leakage.
8. After system and/or vessel is determined "tight," ready the test for inspection and witness.
9. Bring the system to test pressure. If portions of the system cannot be inspected, test must be a decay test. All buried or insulated pipe or vessels must be decay rate tested.
10. Decay tests must be conducted during periods of constant ambient temperature to assure test pressure is not effected by temperature changes.
11. Record test data on installation record, and have inspector(s) witness and sign the record. Attach a P&ID indicating the test boundary.

    Test duration should be as follows:

    Decay test—30 min—Zero decay

    Visual test—10 min—Zero leakage

    Test pressure should be as follows:

Hydro	1.5 times design pressure
Pneumatic	1.5 times design piping pressure and no more than 100 psig for inspection
Pneumatic	1.25 times design vessel pressure and no more than 100 psig for inspection

Note

A system exposed to the sun may show a decay in pressure if the system or portion of the system becomes shaded during the test period.

Caution

A system may become overpressurized if shaded and, during the test period, becomes exposed to the sun.

Pressure Testing Reference Log

Code	Reference	Hydrostatic Test Pressure and Test Sequence	Pneumatic Test Pressure (If Permitted)	Visual Examination Pressure (After Test Pressure)	Gauges	AI Witness Required	Record Forms to Be Signed by AI	Water Temp. Controls	Notes
ASME Section I "Power Boilers"	PG-99	The tests shall be made in two stages in this sequence: 1. Hydrostatic pressure tests shall be applied by raising the pressure to 1.5 times the maximum allowable working pressure shown on the Data Report to be stamped on the boiler, but no less then 1.25 times the maximum allowable working pressure of any part of the boiler. The pressure shall be under proper control at all times so that the required test pressure is never exceeded by more than 6%. Close visual inspection for leakage is not required at this stage. 2. The hydrostatic test pressure may then be reduced to the maximum allowable working pressure, as shown on the Data Report, to be stamped on the boiler and maintained at this pressure while the boiler is carefully examined.	Hydrostatic test is only test method stated.	Maximum allowable working pressure is minimum pressure during examination.	An indicating gauge, visible to the operator controlling the pressure applied, shall be connected to the pressure parts. Hydrostatic head on the gauge shall be considered depending on the location of the gauge. Indicating pressure gauges used in testing shall preferably have dials graduated over their entire range of about double the intended maximum test pressure, but in no case shall the range be less than 1.5 times the pressure. The spacing between shall be such that the inspector and the operator controlling the test shall be able to determine when the required pressure has been applied.	Yes	P-2- All boilers except water and electric. P-2A- Electric Boilers. P-3- Water tube Boilers, Superheaters, Waterwalls and Economizers. P-3A- Engineering Contractor Data for a completed Boiler Unit. P-4- Partial Data Report. P-4A- Process Steam Generators. P-6- Supplemental Sheets.	Ambient, but in no case less than 70ºF. The metal temperature shall not exceed 120ºF during close examination.	At no time during hydrostatic test shall any part of the boiler be subjected to a stress greater than 90% of its yield strength (0.2% offset) at test temperature.
	PW-54	Except as modified above, all welded drums and other welded pressure parts shall be subjected to a hydrostatic test pressure of not less than 1.5 times the maximum allowable working pressure. The hydrostatic test may be made either in the Manufacturer's Shop or in the field.							

Pressure Testing Reference Log

Code	Reference	Hydrostatic Test Pressure and Test Sequence	Pneumatic Test Pressure (If Permitted)	Visual Examination Pressure (After Test Pressure)	Gauges	AI Witness Required	Record Forms to Be Signed by AI	Water Temp. Controls	Notes
ASME Section VIII "Pressure Vessels"	UG-99 and UG-100	1.5 times max. allowable working pressure times lowest ration of material stress value at design temp. 1.5 (S at design temp.)/S at test temp.	When using pneumatic pressure, 1.25 times maximum allowable working pressure, times lowest ratio of material stress value at test temp. to stress value at design temp. 1.25P (S at design temp.)/S at test temp. Raise pressure gradually to 1/2 test pressure, then increase in increments of 1/10 the test pressure not exceeding 1.25 times the basis for the calculated test pressure, defined in UA-60(e).	2/3 test pressure for hydromatic test of 4/5 test pressure for pneumatic test.	Connect indicating gauge directly to vessel. If visibility of gauge is limited, supply additional gauge. Recording gauge recommended for large vessels. Range: Double the test pressure, but never less than 1.5 times.	Yes	U-1- Pressure Vessels U-2-1 Partial Data U-3- Miniature Vessels U-4- Supplementary Sheets		Proof tests may be used to establish maximum working pressure. Pneumatic testing at 1.25 working pressure may be used if water is not feasible. Pneumatic testing is advised against because of hazards of compressed air in large volume.

Pressure Testing Reference Log

Code	Reference	Hydrostatic Test Pressure and Test Sequence	Pneumatic Test Pressure (If Permitted)	Visual Examination Pressure (After Test Pressure)	Gauges	AI Witness Required	Required Forms to Be Signed by AI	Water Temp. Controls
ANS1 B31.1 "Power Piping"	137.1 to 137.7	1.5 times the design pressure for a minimum of 10 minutes.	Not less than 1.20 nor more than 1.50 times design pressure for a minimum of 10 minutes. Increase pressure to not over 1/2 test pressure; then increase in 1/10 test pressure until test pressure is reached. Keep personnel at safe distance from pipe while at test pressure. It shall then be reduced to the lesser of the design pressure or 100 psig and held for such time as may be necessary to conduct the examination for leverage.	Hydro and Pneumatic: Examine at design pressure following holding for required time at test pressure.	Connect gauge where it is visible. Dial shall be graduated to approximately double the test pressure but not less than 1.5 times this pressure.	Yes.	P4A "Data Report Form for Fabricated Piping," ASME Section 1, Appendix X.	Temperature not stated.

Pressure Testing Reference Log									
Code	Reference	Shell Test	Valve Closure Test	Valve Gas Closure Test (Optional)	Visual Examination Pressure (After Test Pressure)	Gauges	AI Witness Required	Required Forms to Be Signed by AI	Water Temp. Controls
ANS1 B16.5 "Steel Pipe Flanges, Flanged Valves, and Fittings"	8.1 to 8.5 and Table 3.	1.5 times the pressure rating, minus nothing, but rounded off to next-higher 25 psi increment. Hold minimum of 15 seconds for valves 2 in and smaller, 1 minute for valves 2.5 in through 8 in, and 3 minutes for valves 10 in and larger.	110% of 100°F pressure rating for all valves over 10 in and valves 4 in to 8 in with pressure rating class 600 and higher.	Minimum of 80 psig on valves 4 in to 8 in under pressure rating class 600 and smaller valves of all pressure ratings.	Examine at test pressure.	Not stated.	As required for the system in which the component or part is used.	Rating marked on valve per MSS SP-25.	Not stated.

Pressure Testing Reference Log

Code	Reference	Valve Closure Test	Valve Gas Closure Test (Optional)	Visual Examination Pressure (After Test Pressure)	Gauges	AI Witness Required	Required Forms to Be Signed by AI	Water Temp. Controls
API 650 "Welded Steel Tanks for Oil Storage"	5.3.2 to 5.3.9	1st - Test tank bottom welds with soapsuds, linseed oil, or other suitable material for leak detection, using vacuum box at minimum of 2 psi vacuum (see 5.3.9). 2nd - Fill tank with water. Tanks with tight roofs are to be filled 2 in above the top leg of the top angle. 3rd - Vacuum test tank roof welds as in 1st step.	No pneumatic test other than vacuum box leak testing, except where water is not available. In that event, 2 psi air pressure is put in the tank and all welds are soapsuds inspected externally. This test or the vacuum box test are acceptable for roof weld inspection.	The tank shall be observed from a safe distance while filling. Thorough inspection of all welds shall be done starting 1 hour after completion of filling.	Air pressure or vacuum gauges shall have connections directly to the space to which the pressure or vacuum is applied.	Inspection required by Owner's Inspector. AI is not specified. The AI may act as Owner's Inspector, however.	API 650 "Storage Tank Specification Data Sheet," Appendix L.	Temperature not stated.

Pressure Testing Reference Log

Code	Reference	Mandatory Preliminaries	Visual Examination Pressure (After Test Pressure)	Gauges	AI Witness Required	Required Forms to Be Signed by AI	Water Temp. Controls
API 620 "Large Welded Low-Pressure Storage Tanks" (0–15 psi)	5.23.1 to 5.23.10.	1st - Magnetic Particle - examine all attachment welds around openings and reinforcements. 2nd - Pressurize space between wall and reinforcements with 80 to 100 psi air through telltale holes. Apply soapsuds to welds. 3rd - Apply soapsuds to tank bottom welds and pull partial vacuum of at least 3 psig.	Following all foregoing tests, and after release of pressure, thorough visual inspection shall be made of the entire tank with special attention to attachments to the shell. Additional tests may be made where it is advisable. Do not permit personnel near the tank while pressure is being applied for the first time.	The manometer for measuring vacuum must be attached to the vapor space. Air pressure gauges are also connected to the vapor space. Water pressure gauges at any point must take into account the vertical hydrostatic head of .433 psi per vertical foot. A record of all pressures shall be kept. All gauges shall be calibrated on deadweight tester before starting tests.	Inspector to be qualified per API 620, 5.02. Owner also shall provide his own inspection.	Manufacturer's Certificate (see Appendix M). Summary Report per 5.27.2 Weld Procedures and Welder Qualifications per 5.27.3 Tanks shall be stamped with API Standard 620 official symbol.	Temperature not stated. Fill rate shall not exceed 3 feet per hour.

Pressure Testing Reference Log		
Code	Reference	Hydrostatic Test Pressure and Test Sequence
API 620 "Large Welded Low-Pressure Storage Tanks" (0–15 psi)	5.23.1 to 5.23.10	4th - Alter any repairs, pressurize tank with 2 psig air of 1/2 the design pressure of the vapor space. Apply soapsuds to all welds above high-liquid design level. Combination Hydro-Pneumatic Test of Tanks Not Designed to Hold Liquid Above the Specified Capacity Level After Mandatory Preliminaries: 1st - Blind off pressure relief valves and vent top of tank. Fill with water to capacity design level. 2nd - Close vents. Pressurize with air to 1/2 the vapor space design pressure. Increase pressure gradually to 1.25 times the vapor space design pressure and hold for one hour, observing from a safe distance for any sign of distress. 3rd - Release pressure slowly and remove pressure valve blinds. Inject air until pressure equals vapor space design pressure at which relief valves shall release air. Hold this pressure while all welds are inspected visually for leaks. Examine all welds above liquid level with soapsuds at this pressure. Completely Hydrostatic Tests of Tanks Designed for Liquid to Roof Top After Mandatory Preliminaries: 1st - Blind off relief valves. Fill with water to roof top with top vented to atmosphere. Close vent. Increase water pressure slowly until top-most point of roof equals 1.25 times vapor space design pressure. Hold for one hour minimum. 2nd - Reduce pressure to vapor space design pressure and hold until all welds have been examined for leakage. 3rd - Vent tank to atmosphere. Lower water to below relief valve inlets. Remove blinds. Inject air until pressure reaches vapor space design pressure at which relief valves shall release air.

Pressure Testing Reference Log

Code	Reference	Hydrostatic Test Pressure and Test Sequence	Pneumatic Test Pressure (If Permitted)	Visual Examination Pressure (After Test Pressure)	Gauges	AI Witness Required	Record Forms to Be Signed by AI	Water Temp. Controls	Notes
API 620 "Large Welded Low-Pressure Tanks" (0–15 psi) (Concluded)		Partial Vacuum Test (Required after Either of the Preceding Two Tests) On tanks that are not vertical or do not have cylindrical sidewalls, release vapor space air pressure. Connect manometer to this space. Withdraw water, with vents closed, until the design partial vacuum relief valve(s) shall operate. With vents closed, continue to withdraw maximum partial vacuum until the liquid level reaches 1/2 the high liquid (capacity) level for which the pressure equals vapor space design pressure. Observe shape of tank for any appreciable change. On tanks which are vertical and do have cylindrical sidewalls, water is withdrawn from the tank until it is substantially empty. When the manometer indicates the design partial vacuum pressure, the vacuum relief valves shall operate. Then inject air until the pressure reaches the vapor space design pressure. Tank shall be carefully observed under both conditions for appreciable changes in shape.							

Pressure Testing Reference Log

Code	Reference	Hydrostatic Test Pressure and Test Sequence	Pneumatic Test Pressure (If Permitted)	Visual Examination Pressure (After Test Pressure)	Gauges	AI Witness Required	Record Forms to Be Signed by AI	Water Temp. Controls	Notes
NFPA “Foam-Water Sprinkler Systems”	Volume 2, 1977, Chapter 5	All piping, including yard piping and system piping, 200 psi for 2 hours, or 50 psi in excess of maximum static pressure if this pressure exceeds 150 psi.	Hydrostatic test is only test method stated.	Examine at test pressure.	Not stated.	Not stated.	Not stated.	Not stated.	
AWWA D100-73 “Welded Steel Elevated Tanks, Land Pipes, and Reservoirs for Water Storage”	Section 12.0	Before painting, fill tank with water to maximum working water level. Inspect for leakage.	Flat bottoms shall be vacuum box inspected at 2 psi minimum vacuum with soapsuds solution or linseed oil.	During and after filling.	Vacuum gauge directly on vacuum box.	Owner’s inspector AI is not specified.	Certificate of Compliance, Page 56.	Temperature not stated.	

APPENDIX P

FLANGE AND FITTING DIMENSIONS

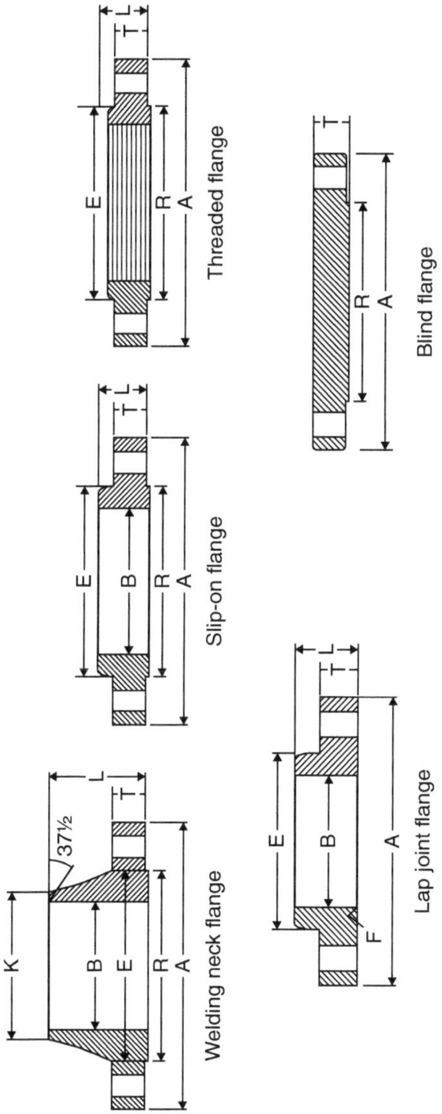

Source: Catalog 87, 1977, courtesy of Republic Supply Company.

Flange and Fitting Dimensions: 150-Pound Forged Steel Flanges

Nominal Pipe Size (in)	A Outside Diam. Flange (in)	†T Min. Flange Thick. (in)	R Outside Diam. Raised Face (in)	E Diam. Hub at Base (in)	K Diam. Hub at Pt. of Welding (in)	F Radius (in)	L. Length through Hub (in)			B. Diameter Bore (in)			Drilling Template			Bolt Size (in)	
							†Welding Neck	†Slip-on and Threaded	Lab Joint	*Welding Neck	Slip-on	Lap Joint	No. Holes	Diam. Holes (in)	Bolt Circle (in)	Diam.	‡Length Stud Bolt
½	3½	7/16	1 3/8	1 3/16	.84	1/8	1 7/8	5/8	5/8	.62	.88	.90	4	5/8	2 3/8	½	2¼
¾	3 7/8	½	1 11/16	1½	1.05	1/8	2 1/16	5/8	5/8	.82	1.09	1.11	4	5/8	2¾	½	2¼
1	4¼	9/16	2	1 15/16	1.32	1/8	2 3/16	11/16	11/16	1.05	1.36	1.38	4	5/8	3 1/8	½	2½
1¼	4 5/8	5/8	2½	2 5/16	1.66	3/16	2¼	13/16	13/16	1.38	1.70	1.72	4	5/8	3½	½	2½
1½	5	11/16	2 7/8	2 9/16	1.90	¼	2 7/16	7/8	7/8	1.61	1.95	1.97	4	5/8	3 7/8	½	2¾
2	6	¾	3 5/8	3 1/16	2.38	5/16	2½	1	1	2.07	2.44	2.46	4	¾	4¾	5/8	3
2½	7	7/8	4 1/8	3 9/16	2.88	5/16	2¾	1 1/8	1 1/8	2.47	2.94	2.97	4	¾	5½	5/8	3¼
3	7½	15/16	5	4¼	3.50	3/8	2¾	1 3/16	1 3/16	3.07	3.57	3.60	4	¾	6	5/8	3½
3½	8½	15/16	5½	4 13/16	4.00	3/8	2 13/16	1¼	1¼	3.55	4.07	4.10	8	¾	7	5/8	3½
4	9	15/16	6 3/16	5 5/16	4.50	7/16	3	1 5/16	1 5/16	4.03	4.57	4.60	8	¾	7½	5/8	3½
5	10	15/16	7 5/16	6 7/16	5.56	7/16	3½	1 7/16	1 7/16	5.05	5.66	5.69	8	7/8	8½	¾	3¾
6	11	1	8½	7 9/16	6.63	½	3½	1 9/16	1 9/16	6.07	6.72	6.75	8	7/8	9½	¾	3¾
8	13½	1 1/8	10 5/8	9 11/16	8.63	½	4	1¾	1¾	7.98	8.72	8.75	8	7/8	11¾	¾	4
10	16	1 3/16	12¾	12	10.75	½	4	1 15/16	1 15/16	10.02	10.88	10.92	12	1	14¼	7/8	4½
12	19	1¼	15	14 3/8	12.75	½	4½	2 3/16	2 3/16	12.00	12.88	12.92	12	1	17	7/8	4½
14	21	1 3/8	16¼	15¾	14.00	½	5	2¼	3 1/8	13.25	14.14	14.18	12	1 1/8	18¾	1	5
16	23½	1 7/16	18½	18	16.00	½	5	2½	3 7/16	15.25	16.16	16.19	16	1 1/8	21¼	1	5¼
18	25	1 9/16	21	19 7/8	18.00	½	5½	2 11/16	3 13/16	17.25	18.18	18.20	16	1¼	22¾	1 1/8	5¾
20	27½	1 11/16	23	22	20.00	½	5 11/16	2 7/8	4 1/16	19.25	20.20	20.25	20	1¼	25	1 1/8	6
22	29½	1 13/16	25¼	24¼	22.00	½	5 7/8	3 1/8	4¼	21.25	22.22	22.25	20	1 3/8	27¼	1¼	6½
24	32	1 7/8	27¼	26 1/8	24.00	½	6	3¼	4 3/8	23.25	24.25	24.25	20	1 3/8	29½	1¼	6¾

*Flanges will be furnished bored to these dimensions (Standard Wall Pipe) unless otherwise ordered.
†Raised face, 1/16-in, included in minimum thickness and length.
‡Lengths include thickness of two nuts but not height of crown.
Socket weld flanges also available.
Source: From Republic Supply Catalog 87, 1977.

Flange and Fitting Dimensions: 300-Pound Forged Steel Flanges

Nominal Pipe Size (in)	A Outside Diam. Flange (in)	†T Min. Flange Thick. (in)	R Outside Diam. Raised Face (in)	C Counter-bore (in)	E Diam. Hub at Base (in)	K Diam. Hub at Pt. of Welding (in)	F Radius (in)	L. Length through Hub (in) †Welding Neck	†Slip-on and Threaded	Lab Joint	B. Diameter Bore Inches *Welding Neck	Slip-on	Lap Joint	Drilling Template No. Holes	Diam. Holes (in)	Bolt Circle (in)	Bolt Size I(in) Diam.	‡Length Stud Bolt
½	3¾	9/16	1 3/8	.93	1½	.84	1/8	2 1/16	7/8	7/8	.62	.88	.90	4	5/8	2 5/8	½	2½
¾	4 5/8	5/8	1 11/16	1.14	1 7/8	1.05	1/8	2¼	1	1	.82	1.09	1.11	4	¾	3¼	5/8	2¾
1	4 7/8	11/16	2	1.41	2 1/8	1.32	1/8	2 7/16	1 1/16	1 1/16	1.05	1.36	1.38	4	¾	3½	5/8	3
1¼	5¼	¾	2½	1.75	2½	1.66	3/16	2 9/16	1 1/16	1 1/16	1.38	1.70	1.72	4	¾	3 7/8	5/8	3
1½	6 1/8	13/16	2 7/8	1.99	2¾	1.90	¼	2 11/16	1 3/16	1 3/16	1.61	1.95	1.97	4	7/8	4½	¾	3½
2	6½	7/8	3 5/8	2.50	3 5/16	2.38	5/16	2¾	1 5/16	1 5/16	2.07	2.44	2.46	8	¾	5	5/8	3¼
2½	7½	1	4 1/8	3.00	3 15/16	2.88	5/16	3	1½	1½	2.47	2.94	2.97	8	7/8	5 7/8	¾	3¾
3	8¼	1 1/8	5	3.63	4 5/8	3.50	3/8	3 1/8	1 11/16	1 11/16	3.07	3.57	3.60	8	7/8	6 5/8	¾	4
3½	9	1 3/16	5½	4.13	5¼	4.00	3/8	3 3/16	1¾	1¾	3.55	4.07	4.10	8	7/8	7¼	¾	4¼
4	10	1¼	6 3/16	4.63	5¾	4.50	7/16	3 3/8	1 7/8	1 7/8	4.03	4.57	4.60	8	7/8	7 7/8	¾	4¼
5	11	1 3/8	7 5/16	5.69	7	5.56	7/16	3 7/8	2	2	5.05	5.66	5.69	8	7/8	9¼	¾	4½
6	12½	1 7/16	8½	6.75	8 1/8	6.63	½	3 7/8	2 1/16	2 1/16	6.07	6.72	6.75	12	7/8	10 5/8	¾	4¾
8	15	1 5/8	10 5/8	8.75	10¼	8.63	½	4 3/8	2 7/16	2 7/16	7.98	8.72	8.75	12	1	13	7/8	5¼
10	17½	1 7/8	12¾	10.88	12 5/8	10.75	½	4 5/8	2 5/8	3¾	10.02	10.88	10.92	16	1 1/8	15¼	1	6
12	20½	2	15	12.94	14¾	12.75	½	5 1/8	2 7/8	4	12.00	12.88	12.92	16	1¼	17¾	1 1/8	6½
14	23	2 1/8	16¼	14.19	16¾	14.00	½	5 5/8	3	4 3/8	13.25	14.14	14.18	20	1¼	20¼	1 1/8	6¾
16	25½	2¼	18½	16.19	19	16.00	½	5¾	3¼	4¾	15.25	16.16	16.19	20	1 3/8	22½	1¼	7¼
18	28	2 3/8	21	18.19	21	18.00	½	6¼	3½	5 1/8	17.25	18.18	18.20	24	1 3/8	24¾	1¼	7½
20	30½	2½	23	20.19	23 1/8	20.00	½	6 3/8	3¾	5½	19.25	20.20	20.25	24	1 3/8	27	1¼	8
22	33	2 5/8	25¼	22.19	25¼	22.00	½	6½	4	5¾	21.25	22.22	22.25	24	1 5/8	29¼	1½	8¾
24	36	2¾	27¼	24.19	27 5/8	24.00	½	6 5/8	4 3/16	6	23.25	24.25	24.25	24	1 5/8	32	1½	9

†Raised face, 1/16-in, included in minimum thickness and length.
‡Lengths include thickness of two nuts but not height of crown.
Socket weld flanges also available.
Source: From Republic Supply Catalog 87, 1977.

Flange and Fitting Dimensions: 400-Pound Forged Steel Flanges

Nominal Pipe Size (in)	A Outside Diam. Flange (in)	T Min. Flange Thick. (in)	R Outside Diam. Raised Face (in)	C Counter-bore (in)	E Diam. Hub at Base (in)	K Diam. Hub at Pt. of Welding (in)	F Radius (in)	L. Length through Hub (in)			[†]B. Diameter Bore (in)		Drilling Template			Stud Bolt	
								Welding Neck	Slip-on and Threaded	Lab Joint	Slip-on	Lap Joint	No. Holes	Diam. Holes (in)	Bolt Circle (in)	Diam.	Length [§](in)
[‡]½	3¾	9/16	1 3/8	.93	1½	.84	1/8	2 1/16	7/8	7/8	.88	.90	4	5/8	2 5/8	½	3
[‡]¾	4 5/8	5/8	1 11/16	1.14	1 7/8	1.05	1/8	2¼	1	1	1.09	1.11	4	¾	3¼	5/8	3¼
[‡]1	4 7/8	11/16	2	1.41	2 1/8	1.32	1/8	2 7/16	1 1/16	1 1/16	1.36	1.38	4	¾	3½	5/8	3½
[‡]1¼	5¼	13/16	2½	1.75	2½	1.66	3/16	2 5/8	1 1/8	1 1/8	1.70	1.72	4	¾	3 7/8	5/8	3¾
[‡]1½	6 1/8	7/8	2 7/8	1.99	2¾	1.90	¼	2¾	1¼	1¼	1.95	1.97	4	7/8	4½	¾	4
[‡]2	6½	1	3 5/8	2.50	3 5/16	2.38	5/16	2 7/8	1 7/16	1 7/16	2.44	2.46	8	¾	5	5/8	4
[‡]2½	7½	1 1/8	4 1/8	3.00	3 15/16	2.88	5/16	3 1/8	1 5/8	1 5/8	2.94	2.97	8	7/8	5 7/8	¾	4½
‡3	8¼	1¼	5	3.63	4 5/8	3.50	3/8	3¼	1 13/16	1 13/16	3.57	3.60	8	7/8	6 5/8	¾	4¾
[‡]3½	9	1 3/8	5½	4.13	5¼	4.00	3/8	3 3/8	1 15/16	1 15/16	4.07	4.10	8	1	7¼	7/8	5¼
4	10	1 3/8	6 3/16	4.63	5¾	4.50	7/16	3½	2	2	4.57	4.60	8	1	7 7/8	7/8	5¼
5	11	1½	7 5/16	5.69	7	5.56	7/16	4	2 1/8	2 1/8	5.66	5.69	8	1	9¼	7/8	5½
6	12½	1 5/8	8½	6.75	8 1/8	6.63	½	4 1/16	2¼	2¼	6.72	6.75	12	1	10 5/8	7/8	5¾
8	15	1 7/8	10 5/8	8.75	10¼	8.63	½	4 5/8	2 11/16	2 11/16	8.72	8.75	12	1 1/8	13	1	6½
10	17½	2 1/8	12¾	10.88	12 5/8	10.75	½	4 7/8	2 7/8	4	10.88	10.92	16	1¼	15¼	1 1/8	7¼
12	20½	2¼	15	12.94	14¾	12.75	½	5 3/8	3 1/8	4¼	12.88	12.92	16	1 3/8	17¾	1¼	7¾
14	23	2 3/8	16¼	14.19	16¾	14.00	½	5 7/8	3 5/16	4 5/8	14.14	14.18	20	1 3/8	20¼	1¼	8
16	25½	2½	18½	16.19	19	16.00	½	6	3 11/16	5	16.16	16.19	20	1½	22½	1 3//8	8½
18	28	2 5/8	21	18.19	21	18.00	½	6½	3 7/8	5 3/8	18.18	18.20	24	1½	24¾	1 3/8	8¾
20	30½	2¾	23	20.19	23 1/8	20.00	½	6 5/8	4	5¾	20.20	20.25	24	1 5/8	27	1½	9½
22	33	2 7/8	25¼	22.19	25¼	22.00	½	6¾	4¼	6	22.22	22.25	24	1¾	29¼	1 5/8	10
24	36	3	27¼	24.19	27 5/8	24.00	½	6 7/8	4½	6¼	24.25	24.25	24	1 7/8	32	1¾	10½

[*]The ¼-in raised face is not included in thickness *T* or length *L*.
[†]Specify desired I.D. for welding neck flanges when ordering.
[‡]Dimensions for these sizes are the same as for 600-lb flanges.
[§]Lengths include thickness of two nuts, but not height of crown.
Socket weld flanges also available.
Source: From Republic Supply Catalog 87, 1977.

Flange and Fitting Dimensions: 600-Pound Forged Steel Flanges

Nominal Pipe Size (in)	A Outside Diam. Flange (in)	T Min. Flange Thick. (in)	R Outside Diam. Raised Face (in)	C Counter-bore (in)	E Diam. Hub at Base (in)	K Diam. Hub at Pt. of Welding (in)	F Radius (in)	L. Length through Hub (in)			†B. Diameter Bore (in)		Drilling Template			Stud Bolt	
								Welding Neck	Slip-on and Threaded	Lab Joint	Slip-on	Lap Joint	No. Holes	Diam. Holes (in)	Bolt Circle (in)	Diam.	Length (in)
½	3¾	9/16	1 3/8	.93	1½	.84	1/8	2 1/16	7/8	7/8	.88	.90	4	5/8	2 5/8	½	3
¾	4 5/8	5/8	1 11/16	1.14	1 7/8	1.05	1/8	2¼	1	1	1.09	1.11	4	¾	3¼	5/8	3¼
1	4 7/8	11/16	2	1.41	2 1/8	1.32	1/8	2 7/16	1 1/16	1 1/16	1.36	1.38	4	¾	3½	5/8	3½
1¼	5¼	13/16	2½	1.75	2½	1.66	3/16	2 5/8	1 1/8	1 1/8	1.70	1.72	4	¾	3 7/8	5/8	3¾
1½	6 1/8	7/8	2 7/8	1.99	2¾	1.90	¼	2¾	1¼	1¼	1.95	1.97	4	7/8	4½	¾	4
2	6½	1	3 5/8	2.50	3 5/16	2.38	5/16	2 7/8	1 7/16	1 7/16	2.44	2.46	8	¾	5	5/8	4
2½	7½	1 1/8	4 1/8	3.00	3 5/16	2.88	5/16	3 1/8	1 5/8	1 5/8	2.94	2.97	8	7/8	5 7/8	¾	4½
3	8¼	1¼	5	3.63	4 5/8	3.50	3/8	3¼	1 13/16	1 13/16	3.57	3.60	8	7/8	6 5/8	¾	4¾
3½	9	1 3/8	5½	4.13	5¼	4.00	3/8	3 3/8	1 15/16	1 15/16	4.07	4.10	8	1	7¼	7/8	5¼
4	10¾	1½	6 3/16	4.63	6	4.50	7/16	4	2 1/8	2 1/8	4.57	4.60	8	1	8½	7/8	5½
5	13	1¾	7 5/16	5.69	7 7/16	5.56	7/16	4½	2 3/8	2 3/8	5.66	5.69	8	1 1/8	10½	1	6¼
6	14	1 7/8	8½	6.75	8¾	6.63	½	4 5/8	2 5/8	2 5/8	6.72	6.75	12	1 1/8	11½	1	6½
8	16½	2 3/16	10 5/8	8.75	10¾	8.63	½	5¼	3	3	8.72	8.75	12	1¼	13¾	1 1/8	7½
10	20	2½	12¾	10.88	13½	10.75	½	6	3 3/8	4 3/8	10.88	10.92	16	1 3/8	17	1¼	8¼
12	22	2 5/8	15	12.94	15¾	12.75	½	6 1/8	3 5/8	4 5/8	12.88	12.92	20	1 3/8	19¼	1¼	8½
14	23¾	2¾	16¼	14.19	17	14.00	½	6½	3 11/16	5	14.14	14.18	20	1½	20¾	1 3/8	9
16	27	3	18½	16.19	19½	16.00	½	7	4 3/16	5½	16.16	16.19	20	1 5/8	23¾	1½	9¾
18	29¼	3¼	21	18.19	21½	18.00	½	7¼	4 5/8	6	18.18	18.20	20	1¾	25¾	1 5/8	10½
20	32	3½	23	20.19	24	20.00	½	7½	5	6½	20.20	20.25	24	1¾	28½	1 5/8	11¼
22	34¼	3¾	25¼	22.19	26¼	22.00	½	7¾	5¼	6 7/8	22.22	22.25	24	1 7/8	30 5/8	1¾	12
24	37	4	27¼	24.19	28¼	24.00	½	8	5½	7¼	24.25	24.25	24	2	33	1 7/8	12¾

*Raised face, ¼-in, not included in minimum thickness and length.
Lengths include thickness of two nuts but not height of crown.

†Specify desired I.D. for welding neck flanges when ordering.
Socket weld flanges also available.

Source: From Republic Supply Catalog 87, 1977.

Flange and Fitting Dimensions: 900-Pound Forged Steel Flanges

Nominal Pipe Size (in)	A Outside Diam. Flange (in)	T Min. Flange Thick. (in)	R Outside Diam. Raised Face (in)	C Counter-bore (in)	E Diam. Hub at Base (in)	K Diam. Hub at Pt. of Welding (in)	F Radius (in)	L. Length through Hub (in)			†B. Diameter Bore (in)		Drilling Template			Stud Bolt	
								Welding Neck	Slip-on and Threaded	Lab Joint	Slip-on	Lap Joint	No. Holes	Diam. Holes (in)	Bolt Circle (in)	Diam.	Length §(in)
‡½	4¾	7/8	1 3/8	.93	1½	.84	1/8	2 3/8	1¼	1¼	.88	.90	4	7/8	3¼	¾	4
‡¾	5 1/8	1	1 11/16	1.14	1¾	1.05	1/8	2¾	1 3/8	1 3/8	1.09	1.11	4	7/8	3½	¾	4¼
‡1	5 7/8	1 1/8	2	1.41	2 1/16	1.32	1/8	2 7/8	1 5/8	1 5/8	1.36	1.38	4	1	4	7/8	4¾
‡1¼	6¼	1 1/8	2½	1.75	2½	1.66	3/16	2 7/8	1 5/8	1 5/8	1.70	1.72	4	1	4 3/8	7/8	4¾
‡1½	7	1¼	2 7/8	1.99	2¾	1.90	¼	3¼	1¾	1¾	1.95	1.97	4	1 1/8	4 7/8	1	5¼
‡2	8½	1½	3 5/8	2.50	4 1/8	2.38	5/16	4	2¼	2¼	2.44	2.46	8	1	7/8	7/8	5½
‡2½	9 5/8	1 5/8	4 1/8	3.00	4 7/8	2.88	5/16	4 1/8	2½	2½	2.94	2.97	8	1 1/8	1	1	6
3	9½	1½	5	3.63	5	3.50	3/8	4	2 1/8	2 1/8	3.57	3.60	8	1	7/8	7/8	5½
4	11½	1¾	6 3/16	4.63	6¼	4.50	7/16	4½	2¾	2¾	4.57	4.60	8	1¼	1 1/8	1 1/8	6½
5	13¾	2	7 5/16	5.69	7½	5.56	7/16	5	3 1/8	3 1/8	5.66	5.69	8	1 3/8	1¼	1¼	7¼
6	15	2 3/16	8½	6.75	9¼	6.63	½	5½	3 3/8	3 3/8	6.72	6.75	12	1¼	12½	1 1/8	7½
8	18½	2½	10 5/8	8.75	11¾	8.63	½	6 3/8	4	4½	8.72	8.75	12	1½	15½	1 3/8	8½
10	21½	2¾	12¾	10.88	14½	10.75	½	7¼	4¼	5	10.88	10.92	16	1½	18½	1 3/8	9
12	24	3 1/8	15	12.94	16½	12.75	½	7 7/8	4 5/8	5 5/8	12.88	12.92	20	1½	21	1 3/8	9¾
14	25¼	3 3/8	16¼	14.19	17¾	14.00	½	8 3/8	5 1/8	6 1/8	14.14	14.18	20	1 5/8	22	1½	10½
16	27¾	3½	18½	16.19	20	16.00	½	8½	5¼	6½	16.16	16.19	20	1¾	24¼	1 5/8	11
18	31	4	21	18.19	22¼	18.00	½	9	6	7½	18.18	18.20	20	2	27	1 7/8	12¾
20	33¾	4¼	23	20.19	24½	20.00	½	9¾	6¼	8¼	20.20	20.25	20	2 1/8	29½	2	13½
24	41	5½	27¼	24.19	29½	24.00	½	11½	8	10½	24.25	24.25	20	2 5/8	35½	2½	17

*The ¼-in raised face is not included in thickness *T* or length *L*.
†Specify desired I.D. for welding neck flanges when ordering.
‡Dimensions for these sizes are the same as for 1500-lb flanges.
§Lengths include thickness of two nuts, but not height of crown.
Socket weld flanges also available.
Source: From Republic Supply Catalog 87, 1977.

Flange and Fitting Dimensions: 1500-Pound Forged Steel Flanges

Nominal Pipe Size (in)	A Outside Diam. Flange (in)	§T Min. Flange Thick. (in)	R Outside Diam. Raised Face (in)	C Counter-bore (in)	E Diam. Hub at Base (in)	K Diam. Hub at Pt. of Welding (in)	F Radius (in)	L. Length through Hub (in) §Welding Neck	L. Length through Hub (in) §Slip-on and Threaded	L. Length through Hub (in) Lab Joint	*B. Diameter Bore (in) Slip-on	*B. Diameter Bore (in) Lap Joint	Drilling Template No. Holes	Drilling Template Diam. Holes (in)	Drilling Template Bolt Circle (in)	Stud Bolt Diam.	Stud Bolt ‡Length (in)
½	4¾	7/8	1 3/8	.93	1½	.84	1/8	2 3/8	1¼	1¼	.88	.90	4	7/8	3¼	¾	4
¾	5 1/8	1	1 11/16	1.14	1¾	1.05	1/8	2¾	1 3/8	1 3/8	1.09	1.11	4	7/8	3½	¾	4¼
1	5 7/8	1 1/8	2	1.41	2 1/16	1.32	1/8	2 7/8	1 5/8	1 5/8	1.36	1.38	4	1	4	7/8	4¾
1¼	6¼	1 1/8	2½	1.75	2½	1.66	3/16	2 7/8	1 5/8	1 5/8	1.70	1.72	4	1	4 3/8	7/8	4¾
1½	7	1¼	2 7/8	1.99	2¾	1.90	¼	3¼	1¾	1¾	1.95	1.97	4	1 1/8	4 7/8	1	5¼
2	8½	1½	3 5/8	2.50	4 1/8	2.38	5/16	4	2¼	2¼	2.44	2.46	8	1	6½	7/8	5½
2½	9 5/8	1 5/8	4 1/8	3.00	4 7/8	2.88	5/16	4 1/8	2½	2½	2.94	2.97	8	1 1/8	7½	1	6
3	10½	1 7/8	5	3.63	5¼	3.50	3/8	4 5/8	2 7/8	2 7/8	3.57	3.60	8	1¼	8	1 1/8	6¾
4	12¼	2 1/8	6 3/16	4.63	6 3/8	4.50	7/16	4 7/8	3 9/16	3 9/16	4.57	4.60	8	1 3/8	9½	1¼	7½
5	14¾	2 7/8	7 5/16	5.69	7¾	5.56	7/16	6 1/8	4 1/8	4 1/8	5.66	5.69	8	1 5/8	11½	1½	9½

*Slip-on flanges are covered by the Standard in sizes through 2½-in and threaded flanges only through 12 in.
†Lengths include thickness of two nuts but not height of crown.
‡¼-in raised face not included in minimum thickness and length.
§Specify desired I.D. for welding neck flanges when ordering.
Socket weld flanges also available.
Source: From Republic Supply Catalog 87, 1977.

Flange and Fitting Dimensions: 2500-Pound Forged Steel Flanges

Nominal Pipe Size (in)	A Outside Diam. Flange (in)	‡T Min. Flange Thick. (in)	R Outside Diam. Raised Face (in)	C Counter-bore (in)	E Diam. Hub at Base (in)	K Diam. Hub at Pt. of Welding Inches	F Radius (in)	L. Length through Hub (in)		§B. Diameter Bore (in)		Drilling Template			Stud Bolt	
								‡Welding Neck	‡Slip-on ‡Threaded Lap Joint	Slip-on	Lap Joint	No. Holes	Diam. Holes (in)	Bolt Circle (in)	Diam.	†Length (in)
½	5¼	1 3/16	1 3/8	.93	1 11/16	.84	1/8	2 7/8	1 9/16	.88	.90	4	7/8	3½	¾	4¾
¾	5½	1¼	1 11/16	1.14	2	1.05	1/8	3 1/8	1 11/16	1.09	1.11	4	7/8	3¾	¾	4¾
1	6¼	1 3/8	2	1.41	2¼	1.32	1/8	3½	1 7/8	1.36	1.38	4	1	4¼	7/8	5¼
1¼	7¼	1½	2½	1.75	2 7/8	1.66	3/16	3¾	2 1/16	1.70	1.72	4	1 1/8	5 1/8	1	5¾
1½	8	1¾	2 7/8	1.99	3 1/8	1.90	¼	4 3/8	2 3/8	1.95	1.97	4	1¼	5¾	1 1/8	6½
2	9¼	2	3 5/8	2.50	3¾	2.38	5/16	5	2¾	2.44	2.46	8	1 1/8	6¾	1	6¾
2½	10½	2¼	4 1/8	3.00	4½	2.88	5/16	5 5/8	3 1/8	2.94	2.97	8	1¼	7¾	1 1/8	7½
3	12	2 5/8	5	3.63	5¼	3.50	3/8	6 5/8	3 5/8	3.57	3.60	8	1 3/8	9	1¼	8½
4	14	3	6 3/16	4.63	6½	4.50	7/16	7½	4¼	4.57	4.60	8	1 5/8	10¾	1½	9¾
5	16½	3 5/8	7 5/16	5.69	8	5.56	7/16	9	5 1/8	5.66	5.69	8	1 7/8	12¾	1¾	11½
6	19	4¼	8½	6.75	9¼	6.63	½	10¾	6	6.72	6.75	8	2 1/8	14½	2	13½
8	21¾	5	10 5/8	8.75	12	8.63	½	12½	7	8.72	8.75	12	2 1/8	17¼	2	15
10	26½	6½	12¾	10.88	14¾	10.75	½	16½	9	10.88	10.92	12	2 5/8	21¼	2½	19
12	30	7¼	15	12.94	17 3/8	12.75	½	18¼	10	12.88	12.92	12	2 7/8	24 3/8	2¾	21

†Lengths include thickness of two nuts but not height of crown.
‡¼-in raised face not included in minimum thickness and length.
§Specify desired I.D. for welding neck flanges when ordering.
Socket weld flanges also available.
Source: From Republic Supply Catalog 87, 1977.

APPENDIX Q

PIPE DIMENSION AND WEIGHT TABLE

$$wt.ft = 10.69 \times (OD - \text{Wall Thickness}) \times \text{Wall thickness}$$

Standard Chart

Steel Pipe Sizes, Weights, Dimensions

Size	Outer Diameter	Schedule 5	Schedule 10	Schedule 20	Schedule 30	Schedule 40	Standard	Schedule 60	Schedule 80	X-Heavy	Schedule 100	Schedule 120	Schedule 140	Schedule 160	XX-Heav
1/8	0.405	.035 .138	.049 .1863			.068 2.447	.068 2.447		.095 .3145	.095 .3145					
1/4	0.540	.049 .2570	.065 .3297			.088 .4248	.088 .4248		.119 .5351	.119 .5351					
3/8	0.675	.049 .3276	.065 .4235			.091 .5676	.091 .5676		.126 .7388	.126 .7388					
1/2	0.840	.065 .5383	.083 .6710			.109 .8510	.109 .8510		.147 1.088	.147 1.088				.187 1.304	.29 1.71
3/4	1.050	.065 .6838	.083 .8572			.113 1.131	.113 1.131		.154 1.474	.154 1.474				.218 1.937	.30 2.44
1	1.315	.065 .8678	.190 1.404			.133 1.679	.133 1.679		.179 2.172	.179 2.172				.250 2.844	.35 3.65
1 1/4	1.660	.065 1.107	.109 1.806			.140 2.273	.140 2.273		.191 2.997	.191 2.997				.250 3.765	.38 5.21
1 1/2	1.900	.065 1.274	.109 2.638			.145 2.718	.145 2.718		.200 3.631	.200 3.631				.281 4.859	.40 6.40
2	2.375	.065 1.604	.109 2.638			.154 3.653	.154 3.653		.218 5.022	.218 5.022				.343 7.444	.43 9.02
2 1/2	2.875	.083 2.475	.120 3.531			.203 5.793	.203 5.793		.276 7.661	.276 7.661				.375 10.01	.55 13.7
3	3.500	.083 3.029	.120 4.332			.216 7.576	.216 7.576		.300 10.25	.300 10.25				.437 14.32	.60 18.5
3 1/2	4.000	.083 3.472	.120 4.973			.226 9.109	.226 9.109		.318 12.51	.318 12.51					.63 22.8
4	4.500	.083 3.915	.120 5.613			.237 10.79	.237 10.79	.281 12.66	.337 14.98	.337 14.98		.437 19.01		.531 22.51	.67 27.5
4 1/2	5.000						.247 12.53			.355 17.61					.71 32.5
5	5.563	.109 6.349	.134 7.770			.258 14.62	.258 14.62		.375 20.78	.375 20.78		.500 27.04		.625 32.96	.75 38.5
6	6.625	.109 7.585	.134 9.289			.280 18.97	.280 18.97		.432 28.57	.432 28.57		.562 36.39		.718 45.30	.86 53.1
7	7.625						.301 [illegible]			.500 [illegible]					.87 [illegible]

Source: All Steel Pipe, www.allsteelpipe.com

8	**8.625**	**.109** 9.914	**.148** 13.40	**.250** 22.36	**.277** 24.70	**.322** 28.55	**.322** 28.55	**.406** 35.64	**.500** 43.39	**.500** 43.39	**.593** 50.87	**.718** 60.63	**.812** 67.76	**.906** 74.69	**.875** 72.42
9	**9.625**						**.342** 33.90			**.500** 48.72					
10	**10.750**	**.134** 15.19	**.165** 18.65	**.250** 28.04	**.307** 34.24	**.365** 40.48	**.365** 40.48	**.500** 54.74	**.593** 64.33	**.500** 54.74	**.718** 76.93	**.843** 89.20	**1.000** 104.1	**1.125** 115.7	
11	**11.750**						**.375** 45.55			**.500** 60.07					
12	**12.750**	**.156** 21.07	**.180** 24.20	**.250** 33.38	**.330** 43.77	**.406** 53.53	**.375** 49.56	**.562** 73.16	**.687** 88.51	**.500** 65.42	**.843** 107.2	**1.000** 125.5	**1.125** 139.7	**1.312** 160.3	
14	**14.000**	**.156** 23.06	**.250** 36.71	**.312** 45.68	**.375** 54.57	**.437** 63.37	**.375** 54.57	**.593** 84.91	**.750** 106.1	**.500** 72.09	**.937** 130.7	**10.93** 150.7	**1.250** 170.2	**1.406** 189.1	
16	**16.000**	**.165** 27.90	**.250** 42.05	**.312** 52.36	**.375** 62.58	**.500** 82.77	**.375** 62.58	**.656** 107.5	**.843** 136.5	**.500** 82.77	**1.031** 164.8	**1.218** 192.3	**1.437** 223.5	**1.593** 245.1	
18	**18.000**	**.165** 31.43	**.250** 47.39	**.312** 59.03	**.437** 82.06	**.562** 104.8	**.375** 70.59	**.750** 138.2	**.937** 170.8	**.500** 93.45	**1.156** 208.0	**1.375** 244.1	**1.562** 274.2	**1.781** 308.5	
20	**20.000**	**.188** 39.78	**.250** 52.73	**.375** 78.60	**.500** 104.1	**.593** 122.9	**.375** 78.60	**.812** 166.4	**1.031** 208.9	**.500** 104.1	**1.280** 256.1	**1.500** 296.4	**1.750** 341.1	**1.968** 379.0	
24	**24.000**	**.218** 55.37	**.250** 63.41	**.375** 94.62	**.562** 140.8	**.687** 171.2	**.375** 94.62	**.968** 238.1	**1.218** 296.4	**.500** 125.5	**1.531** 367.1	**1.812** 429.4	**2.062** 483.1	**2.343** 541.9	
26	**26.000**		**.312** 85.60	**.500** 136.2			**.375** 102.6			**.500** 136.2					
28	**28.000**		**.312** 92.26	**.500** 146.8	**.625** 182.7		**.375** 110.6			**.500** 146.8					
30	**30.000**	**.250** 79.43	**.312** 98.93	**.500** 157.5	**.625** 196.1		**.375** 118.6			**.500** 157.5					
32	**32.000**		**.312** 105.6	**.500** 168.2	**.625** 209.4	**.688** 230.1	**.375** 126.7			**.500** 168.2					
34	**34.000**		**.344** 123.7	**.500** 178.9	**.625** 222.8	**.688** 244.8	**.375** 134.7			**.500** 178.9					
36	**36.000**		**.312** 118.9	**.500** 189.6	**.625** 236.1	**.750** 282.3	**.375** 142.7			**.500** 189.6					
42	**42.000**						**.375** 166.7			**.500** 221.6					
48	**48.000**						**.375** 190.7			**.500** 253.6					

Bold = nominal wall thickness

APPENDIX R
TRIGONOMETRIC TABLE

↓

Rad	Deg	Sin	Cos	Tan	Csc	Sec	Cot		
.0000	00	.0000	1.0000	.0000	—	1.0000	—	90	1.5707
.0175	01	.0175	.9998	.0175	57.2987	1.0002	57.2900	89	1.5533
.0349	02	.0349	.9994	.0349	28.6537	1.0006	28.6363	88	1.5359
.0524	03	.0523	.9986	.0524	19.1073	1.0014	19.0811	87	1.5184
.0698	04	.0698	.9976	.0699	14.3356	1.0024	14.3007	86	1.5010
.0873	05	.0872	.9962	.0875	11.4737	1.0038	11.4301	85	1.4835
.1047	06	.1045	.9945	.1051	9.5668	1.0055	9.5144	84	1.4661
.1222	07	.1219	.9925	.1228	8.2055	1.0075	8.1443	83	1.4486
.1396	08	.1392	.9903	.1405	7.1853	1.0098	7.1154	82	1.4312
.1571	09	.1564	.9877	.1584	6.3925	1.0125	6.3138	81	1.4137
.1745	10	.1736	.9848	.1763	5.7588	1.0154	5.6713	80	1.3953
.1920	11	.1908	.9816	.1944	5.2408	1.0187	5.1446	79	1.3788
.2094	12	.2079	.9781	.2126	4.8097	1.0223	4.7046	78	1.3614
.2269	13	.2250	.9744	.2309	4.4454	1.0263	4.3315	77	1.3439
.2443	14	.2419	.9703	.2493	4.1336	1.0306	4.0108	76	1.3265
.2618	15	.2588	.9659	.2679	3.8637	1.0353	3.7321	75	1.3090
.2793	16	.2756	.9613	.2867	3.6280	1.0403	3.4874	74	1.2915
.2967	17	.2924	.9563	.3057	3.4203	1.0457	3.2709	73	1.2741
.3142	18	.3090	.9511	.3249	3.2361	1.0515	3.0777	72	1.2566
.3316	19	.3256	.9455	.3443	3.0716	1.0576	2.9042	71	1.2392
.3491	20	.3420	.9397	.3640	2.9238	1.0642	2.7475	70	1.2217
.3665	21	.3584	.9336	.3839	2.7904	1.0711	2.6051	69	1.2043
.3840	22	.3746	.9272	.4040	2.6695	1.0785	2.4751	68	1.1868
.4014	23	.3907	.9205	.4245	2.5593	1.0864	2.3559	67	1.1694
.4189	24	.4067	.9135	.4452	2.4586	1.0946	2.2460	66	1.1519
.4363	25	.4226	.9063	.4663	2.3662	1.1034	2.1445	65	1.1345
.4538	26	.4384	.8988	.4877	2.2812	1.1126	2.0503	64	1.1170

.4712	27	.4540	.8910	.5095	2.2027	1.1223	1.9626	63	1.0996
.4887	28	.4695	.8829	.5317	2.1301	1.1326	1.8807	62	1.0821
.5061	29	.4848	.8746	.5543	2.0627	1.1434	1.8040	61	1.0647
.5236	30	.5000	.8660	.5774	2.0000	1.1547	1.7321	60	1.0472
.5411	31	.5150	.8572	.6009	1.9416	1.1666	1.6643	59	1.0297
.5585	32	.5299	.8480	.6249	1.8871	1.1792	1.6003	58	1.0123
.5760	33	.5446	.8387	.6494	1.8361	1.1924	1.5399	57	.9948
.5934	34	.5592	.8290	.6745	1.7883	1.2062	1.4826	56	.9774
.6109	35	.5736	.8192	.7002	1.7434	1.2208	1.4281	55	.9599
.6283	36	.5878	.8090	.7265	1.7013	1.2361	1.3764	54	.9425
.6458	37	.6018	.7986	.7536	1.6616	1.2521	1.3270	53	.9250
.6632	38	.6157	.7880	.7813	1.6243	1.2690	1.2799	52	.9076
.6807	39	.6293	.7771	.8098	1.5890	1.2868	1.2349	51	.8901
.6981	40	.6428	.7660	.8391	1.5557	1.3054	1.1918	50	.8727
.7156	41	.6561	.7547	.8693	1.5243	1.3250	1.1504	49	.8552
.7330	42	.6691	.7431	.9004	1.4945	1.3456	1.1106	48	.8378
.7505	43	.6820	.7314	.9325	1.4663	1.3673	1.0724	47	.8203
.7679	44	.6947	.7193	.9657	1.4396	1.3902	1.0355	46	.8029
.7854	45	.7071	.7071	1.0000	1.4142	1.4142	1.0000	45	.7854
		Cos	Sin	Cot	Sec	Csc	Tan	Deg	Rad

↑

APPENDIX S

USEFUL EQUATIONS

$A + B = C = 90^\circ$
$a^2 + b^2 = c^2$ (Pythagorean theorem)
$c = (a^2 + b^2)^{1/2}$
Area = ½ base × height

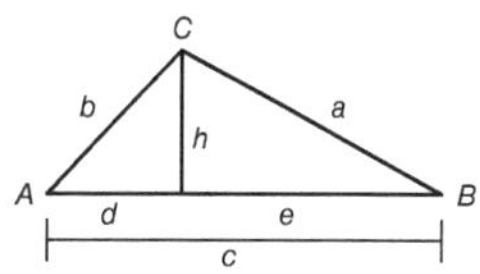

$C = 2\,\pi\,r = \pi D$
$A = \pi\,r^2 = \frac{1}{4}\,\pi\,D^2$
$C = 2\,(\pi A)^{1/2}$

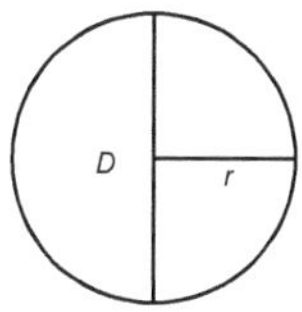

$s = (R^2 + h^2)^{1/2}$
$A_{\text{Lateral}} = (R^2 + h^2)^{1/2}$
$A_{\text{Total}} = \pi\,Rs$
$V = 1/3\,\pi\,R^2\,h$

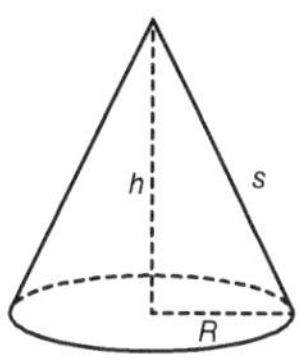

$s = \{(R_1 + R_2)^2 + h^2\}^{1/2}$
$A_{\text{Lateral}} = \pi\,s\,(R_1 + R_2)$
$A_{\text{Total}} = \pi\,\{R^2{}_1 + R^2{}_2 + (R_1 + R_2)s\}$
$V = 1/3\,\pi\,h\,(R^2{}_1 + R^2{}_2 + R_1R_2)$

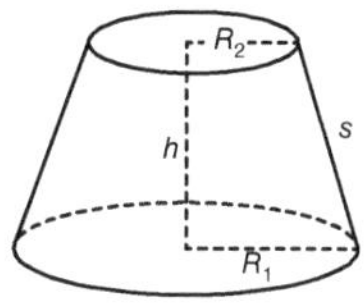

$A_{\text{lateral}} = 2\pi Rh$
$A_{\text{total}} = 2\pi R(R + h)$
$V = \pi R^2 h$

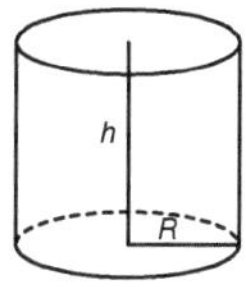

Sphere
$A = 4\pi R^2 = \pi D^2$
$V = 4.189\,R^3$
Zone and Segment of Base
$A = 2\pi Rh$
$V = 1/3\pi h^2(3R - h)$

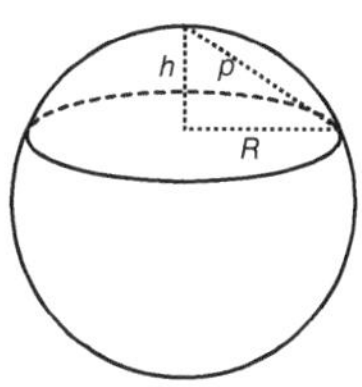

ELECTRICAL

E = volts
V = electrical pressure, volts
I = current flow, amperes
R = resistance, ohms
f = frequency = PN/120
PF = power factor = cos θ
N = speed = rpm
T = period = sec/cycle
P = number of poles
P = power, watts
emf = electromotive force
cemf = counter emf

Ohm's Law

$$E = IR$$

Kirchhoff's Law

$$i_1 + i_2 + i_3 + i_4 + i_5 = 0$$

$$v_{ab} + v_{bc} + v_{cd} + v_{da} = 0$$

$$E_{\text{RMS}} = E_{\text{MAX}} / 0.707$$

$$= E_{\text{MAX}} \times 1.414$$

$$\sqrt{2} = 1.414$$

$$2 / \pi = 0.636$$

$$E_{\text{AVE}} = 0.636 \times E_{\text{MAX}}$$

All handheld multimeters read effective power, E_{RMS}; therefore, for a meter reading 110 V, $E_{\text{MAX}} = 1.414 \times 110 = 155.54$ V.

A continuous cemf is always present in AC/DC circuits.

$R = E/I$, where R is one component of opposition and is pure resistance.

Reactance opposition to AC current flow is called inductive reactance, X_L.

$$X_L = 2\,\pi f L,\ L = \text{inductance in Henrys}$$

$$E = I\,X_L$$

Reactance opposition to source AC voltage and current is capacitive reactance, X_C.

$$X_C = ½\,\pi f C,\ C = \text{capacitance in ohms}$$

In an inductive circuit, voltage leads current. In a capacitance circuit, current leads voltage. Remember, ELI the ICE man.

Impedance, Z, is the total opposition to current flow.

Z is a combination of X_C, X_L, and R.

$Z = (R^2 + X^2)½$, series circuit
$X = X_L - X_C$, L dominant
$X = X_C - X_L$, C dominant

X is never negative.

$P = I^2 R, E I, E^2/R$
$P_{\text{APPARENT}} = P_A$
$P_{\text{TOTAL}} = P_T$
$P_{\text{TRUE}} = P_t$
$P_A = E I$, volt-amps, VA

DC

HP = $E I \eta/746$, η = efficiency

AC

HP = $E I \eta$ PF/746, single phase
HP = $2 E I \eta$ PF/746, two phase
HP = $1.73 E I \eta$ PF/746, three phase

Determine Indicated Circulating Water Pump HP

Cable data: 3-1/C 350 MCM
$L = 1920$ ft
$R = 0.0308$

Motor data: $E = 4.16$ kV
$\eta = 0.942$
$P_M = 1129$ kW at MCC

Indicated HP

$I = P_M / 3½ E \cos \theta$
$= 1129$ kW/ 1.732(4.16 kV)(0.1771)
$= 203$ amps
$P_L = 3 I^2 R$
$= 203^2$ (0.0308)(1920)71000
$= 2.443$ KW
$P = (1129 - 2.443)0.942$
$= 1084$ KW (1.341)
$=$ 1453 HP ←

Power Equations

$P = I R$, DC circuit

$P = I R \cos \theta$, single phase AC circuit

$P = 2\, I R \cos \theta$, two phase AC circuit

$P = 3^{1/2} I R \cos \theta$, 3 phase AC circuit

PF = true power/apparent power

$P_t = I^2 R$

$P_a = I^2 Z$

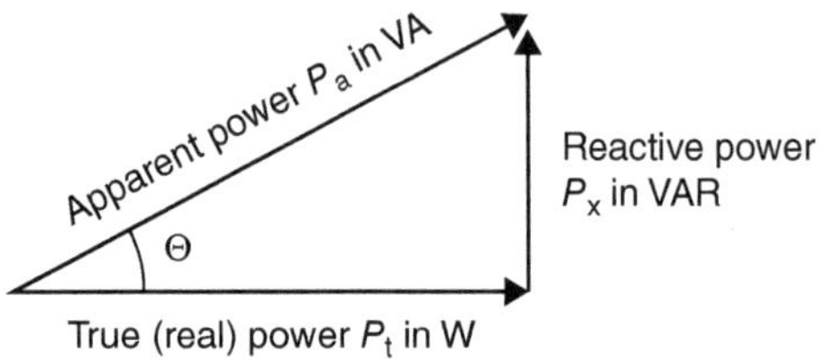

Low-Voltage Tips for Mechanical Engineers

- Lighting circuits are rated as 110–120 V. That is, during limited-load periods, such as at night, the voltage will be 120 V. At peak load periods, during the day, the voltage will approach 110. Two phase is rated at 220–240 V, and three phase 440–480 V for the same reason. The 480-V system is often called a 460-V system, which is the average voltage.
- Lighting fixtures use the same wiring for 120 as well as 240, but the light bulbs must be compatible. *Caution:* The bulb threaded receptacle is "hot" for 240-V circuits. It is the neutral for 120 V.
- In 120-V systems, the hot wire is always Black. The neutral is always White and the ground bare or Green. Traveler wires for three way switches, which use a Romex white wire, must be recoded with black tape in the lighting fixture outlet boxes.
- All 120-V neutrals are grounded, if not at the building ground, at the utility ground. All fixtures and boxes, etc., must also be wired in a continuous network to a solid ground.
- Two-phase 240-V conductors are not grounded; therefore, care must be taken to run an extra ground wire to ensure all fixtures, conduits, boxes, etc., are wired in a continuous network to a solid ground.
- Metal conduit can serve as the grounding conductors, but wire jumpers must be installed at non-conductor boxes, etc.
- In 120-V circuits, only "load" outlets have neutrals such as lamp fixtures. Switches are always wired with single hot wires for both single- and two-phase circuits.
- Current in the neutral and hot leads of single-phase and both hot leads of two-phase circuits have the same current flow. Ground fault circuit interrupters trip if there is a differential between the wires.
- Clamp-on-ampmeter current measurements across both two-phase wires will read zero because the alternating current cancels out.

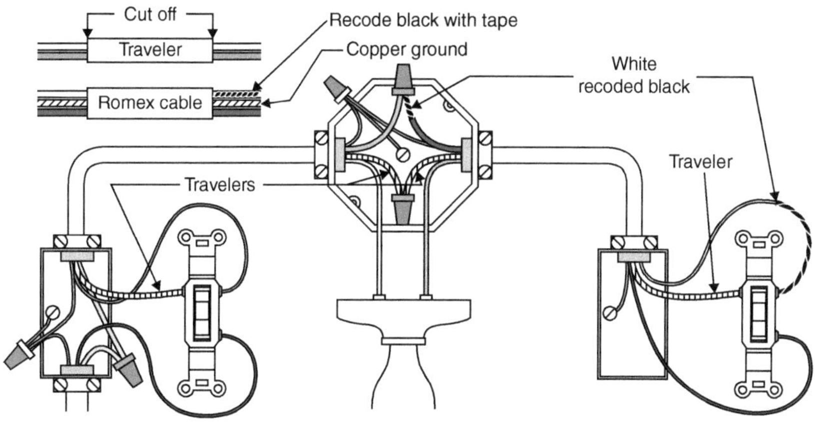

The switches shown here are wired with a four-wire Romex cable, which has a bare copper ground and white, black, and red wires.

A 240-V circuit has red and black hot wires. In the United States, if the circuit is for a clothes dryer, it will also have a white neutral wire for the 120-V timer control. A four-wire plug is best because it also has a ground wire to connect the dryer directly to the house ground. If you have ever touched a washer or dryer and felt a current, that meant the appliance was not grounded with a wire directly from the appliance to the house ground. Only the three-plug (without a neutral) and the four-plug, 240-V receptacles have grounds.

Home Improvement Tips for Electrical Engineers

Square, Level, and Plumb

Square, level, and plumb is the journeyman's rule for good workmanship. All corners should be square; runs must be level and plumb. No duct bank, conduit, or piping run should cut across a space. They must run vertically or horizontally with the structure, which is also square, level, and plumb.

For example, if you build a deck or a fence, your friends from work will automatically line up your work with an existing structure to see if you followed the "rule."

Decks

Attach, as an anchor point, the deck floor joist header to the house baseplate. For CMU[a] (block) structures, attach a header with wedge anchors to the structure. Place pier blocks for the columns level but not to the final elevation. That is, do not excavate so that all piers are at the same elevation. Final elevation is fixed; therefore, for each pier, which are at different grade elevations, a column will be cut to make up the difference. (See the "Fences." section) Run your lightning conduit underneath square with your headers or joists; don't cut across a space.

Fences

Fence posts, like deck columns, do not have to be the same length. That is, the post holes don't have to be exactly the same depth. Line up square, level and plumb the posts, and back-fill or concrete-fill the post holes. Using a surveyor's level, mark all posts at the same height.

[a]Concrete masonry unit (the proper name for "block").

Cut the tops off, which will be different lengths, leaving a perfectly level top fence rail. Your friends will "eyeball" your work, so keep to the "rule."

MECHANICAL

Classical Mechanics

Newton's Law, $F = ma$

Force = Mass × Acceleration is the foundation of Newtonian or "Classical" mechanics.

The *First Law of Thermodynamics* is the conservation of energy. For example, all energy flowing into and out of a control volume must be accounted for when calculating heat balance problems.

The *Second Law of Thermodynamics* is the concept of entropy. The concept can be used to quantify irreversibilities. That is, the entropy (S) of a system always increases and that increase can determine efficiency. Entropy cannot be less than unity. Impossible "perpetual motion" can be identified using this concept.

Energy equation = NRG

NRG: $Q_{in} = Q_{out}$

Continuity equation = mass going in is equal to mass going out. $W_{in} = W_{out}$

Feedwater Heater Heat Balance

We want to find the steam to, and drain flow from, a feedwater heater.

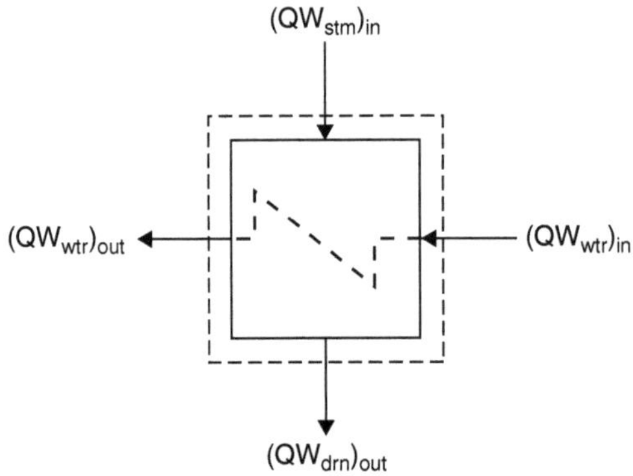

Continuity:

$$W_{in} = W_{out}$$

NRG:

$$Q = W\Delta h$$

$$\therefore\ W_{stm} = (W_{fw}\ \Delta h_{fw})/\Delta h_{stm}$$

In this heat exchanger–heat balance, we have assumed an adiabatic process (zero heat losses in the process control volume). This is a reasonable and universally accepted assumption.

Equations of state are equations that depict state changes for a process with either a constant volume or pressure. For example:

$$P = \rho RT$$
$$H = U + \rho V$$

Temperature Drop Calculation

For a main steamline, the calculation is as follows:

HEAT LOSS

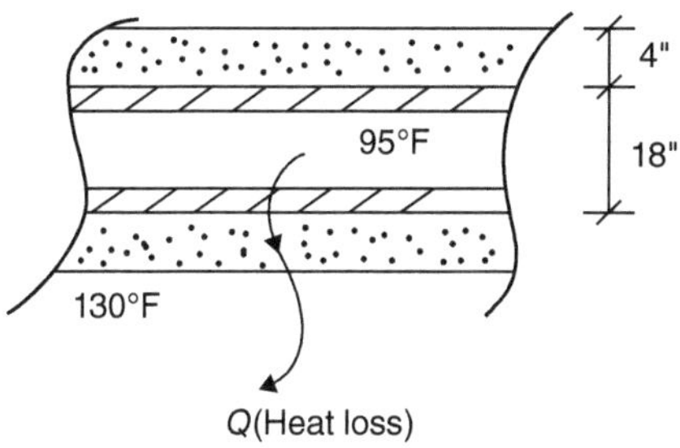

Given:

$$L = 300 \text{ ft}$$
$$K_i = 0.75, \text{ calcium silicate}$$
$$K_p = 1.04, \text{ carbon steel}$$
$$D_i = 8 \text{ in}$$
$$D_{pod} = 18.0 \text{ in}$$
$$D_{pid} = 14.44 \text{ in}$$
$$D_{io} = 26.0 \text{ in}$$
$$Q = (2\pi L\Delta T)/(P + I)$$

where:

$$P = 1/K_p \log_e r_0/r_i$$
$$I = 1/K_i \log_e r_{i0}/r_i$$
$$\therefore P = 1/1.04 \log_e 9.00/7.22 = 0.212$$
$$I = 1/0.75 \log_e 13.0/9.0 = 0.490$$

where:

$$Q_L = 2\pi\ 300\ (950 - 130)/(0.212 + 0.490)$$
$$= 2.20 \times 10^6 \text{ Btu/hr}$$

Heat Balance

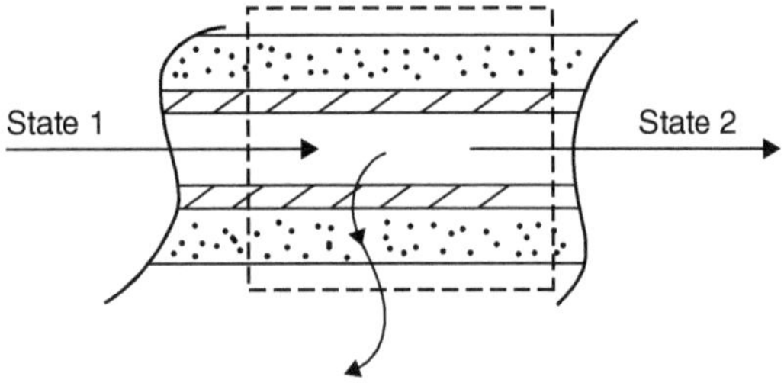

Assumptions:

1. Steam obeys the perfect gas laws.
2. Process is irreversible isentropic.

$$T = 950°\text{F}$$
$$P \approx 2780 \text{ psia}$$
$$W_1 = 4.904 \times 10^6 \text{ lb/hr}$$
$$h_1 = 1411 \text{ Btu/lb}$$

Continuity:

$$W_1 = W_2$$

NRG:

$$Q_1 = W_1 C_p T_1$$
$$Q_2 = W_2 C_p T_2$$

where:

$$h_1 W_1 = h_2 W_2 + Q_L$$
$$h_2 = 1411 - 2.20/4.904 = 1410.6 \text{ Btu/lb}$$
$$Q_1 = Q_2 + Q_L$$
$$Q_L = (h_1 - h_2)W$$
$$\mathbf{\Delta T = (h_1 - h_2)/C_p = 0.6/0.24 = 2.5°F}$$

Condenser Performance Evaluation

Steam flow and heat load to the condenser were found by a total system heat balance of the feedwater heaters and turbine:

Heat in = 566,652,060
Heat out = 43,315,777
Heat loss to CW = 523,336,283
CW flow (lb/hr)

$$W_{cw} = \text{Heat load (loss)}/C_P T_{RISE}$$
$$= 523{,}336{,}283/(1 \times 20.172)$$

CW flow (gpm)

$$W_{cw} = W_{LB/HR}\,(0.1247)/(\text{lb/ft}^3)$$
$$= 108{,}597$$
$$\text{LMTD} = (T_2 - T_1)/[(\ln(T_S - T_1)/(T_S - T_2)]$$
$$\text{LMTD} = 11.83$$

Heat Transfer Coefficient (Test)

$$Q = U\,A\,\Delta T$$
$$U = 566.7 \times 10^6/(70{,}000)(11.83)$$
$$= 684.34$$

Heat Transfer Coefficient (HEI)[2]

Coefficient (uncorrected) = 670

Temperature correction = 1.05

Tube material correction = 0.90

Coefficient (corrected) = 633.15

Apparent Cleanliness Factor

C.F. = U (test)/U (HEI)

= 684.34/633.15

= 1.081

Expected Turbine Backpressure

$$Q = U\,A\,\Delta T$$
$$\Delta T = 566.7 \times 10^6/633.15 \times 70{,}000$$
$$= 12.786$$
$$\text{LMTD} = (T_2 - T_1)/[(\ln(T_S - T_1)/(T_S - T_2)]$$
$$\text{LMTD} = 108.48°\text{F}$$
$$P_S = 2.48 \text{ inHg (HEI)}$$
$$P = 2.43 \text{ inHg (Test)}$$

Gas Turbine Heat Balance

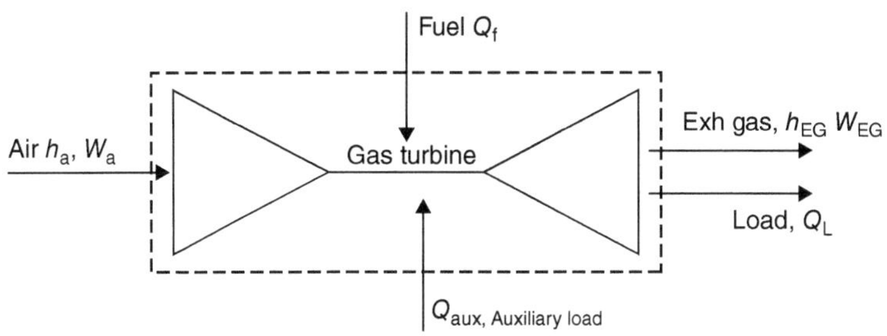

Find: Heat input to HRSG
Known: T_{AMB}, P_{BARO}, T_E, P_E, Q_L, Q_{AUX}, Q_F

1. Continuity: $W_{EG} = W_A + W_F$
2. NRG: $h_A W_A + Q_F - \Sigma Q_L = h_{EG} W_{EG}$
3. Simultaneous Equations: $h_A W_A + Q_F - \Sigma Q_L = h_{EG}(W_A + W_F)$

$$\therefore W_A = (h_{EG} W_F + \Sigma Q_L - Q_F)/(h_F - h_{EG}) \text{ and}$$

4. $O_{EG}\left[\left(\dfrac{h_{EG} W_F + \Sigma Q_L - Q_F}{h_A - h_{EG}}\right) + W_F\right] h_{EG}$

$T_A = 83.19 + 460 = 543.2$
$T_{EG} = 959 + 460 = 1419$

From Ref. 3:

$h_A = 129.8$, $h'_{EG} = 10{,}242.9$, $M' \approx 28.58$
$h_{EG} = h'/M' = 359$ Btu/lb
$W_F = 29{,}100$ lb/hr
$Q_F = 650 \times 10^6$ Btu/hr

$$Q_{EG} = \left[\left(\frac{359 \times 29.1 + 1765 - 650 \times 10^2}{1298 - 359}\right)\right] + 29{,}100 \times 359$$
$$= 743 \times 10^6 \text{Btu/lb}$$

Constituent Analysis

	%	C	H
CO_2	0.20		
N_2	0.26		
CH_4	97.70	0.977	3.908
C_2H_6	1.66	0.033	0.099
C_3H_8	0.15	0.005	0.012
C_4H_8	0.02	0.001	0.002
C_4H_{10}	0.01	0.000	0.001
	100	1.016	4.022

Combustion Gas Analysis (Products)

	%
CO_2	2.96
O_2	15.42
CO	0
N_2	81.62

Stoichiometric Reaction

$$C_{1.016}\,H_{4.022} + bO_2 + 3.76\,bN_2 \rightarrow aCO_2 + bO_2 + 3.76bN_2 + eH_2O$$

$$a = 1.016$$

$$e = 4.022/2 = 2.011$$

$$b = 1.016 + 2.011/2$$

$$= 2.022$$

Actual Reaction

$$aC_{1.016}\,H_{4.022} + bO_2 + 3.76\,bN_2 \rightarrow 2.96CO_2 + 15.42O_2 + 81.62N_2 + eH_2O$$

$$a = 2.96/1.016 = 2.91$$

$$b = 81.62/3.76 = 21.71$$

$$e = (2.91)(4.022)/2 = 5.86$$

Excess Air

$$100 \times [(21.71/2.91) - 2.022]/\,2.022 = 269\%$$

From Ref. 3:

where: 200% excess air $T = 1419$, $h' = 10402.1$
400% excess air $T = 1419$, $h' = 10242.9$
Interpolating for 269% excess air $h' = 10347.2$

$$\therefore\ h_{COR} = 10347.2/28.58 \text{ lb/ lb-mole}$$

$$h_{COR} = 363 \text{ Btu/lb}$$

Subsituting $h_{EG} = 359$ for h_{COR} in equation 4)

$$\mathbf{W_{GAS\,(CORRECTED)} = 2.035 \times 10^6\ lb/hr} \leftarrow$$

Pump Total Dynamic Head

$$TDH = Vel_{HEAD} + \Delta H_{GAGE} + \Delta H_{FRICTION}\ \Delta H_{ELEVATION}$$

Given: $\Delta H_{GAGE} = (41.746 - 3.9)$psi

$$\Delta H_{ELEVATION} = 0$$

$$\Delta H_{FRICTION} = 0$$

where: $Vel_{HEAD} = (Q/\rho\, A\, 3600)^2/64.4$

where: $A_{IN} = 12.6$ ft^2
$A_{OUT} = 9.63$ ft^2

$$Vel_{DICH} = [(41.746 - 3.9)/9.63 \times 3600]/64.4 = 3.13 \text{ ft}$$

$$Vel_{SUCT} = [(41.746 - 3.9)/12.6 \times 3600]/64.4 = 1.83 \text{ ft}$$

$$\mathbf{TDH = 1.3 + (37.046 \times 2.31) = 88.7\ ft} \leftarrow$$

Pump Water HP

$$\text{Given: } Q = 25.946 \times 10^6 \text{ lb/hr}$$
$$\text{TDH} = 88.9 \text{ ft}$$
$$\text{HP}_{\text{MOTOR}} = 1453 \text{ hp}$$
$$P = Q \times \text{TDH} \times 0.0505 \times 10^{-5}$$
$$\mathbf{P = 1165 \text{ hp}}$$
$$\eta = \mathbf{1165/1453 = 0.80} \leftarrow$$

Losses Throttling Main Steam

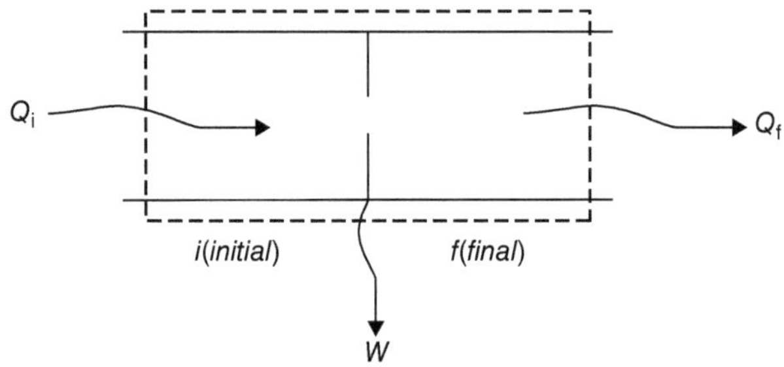

$$P_i = 800,\ T_i = 860$$
$$P_f = 400,\ T_f = 830$$
$$h_i = h_f = 1433 \text{ Btu/lm}$$
$$\text{Mollier diagram: } s_i = 1.625 \text{ Btu/°F-lbm}$$
$$s_f = 1.7 \text{ Btu/°F-lbm}$$

Equation of State:

$$v_i = v_i\,(800, 860) = 0.9212 \text{ ft}^3\text{/lbm}$$
$$v_f = v_f\,(400, 830) = 0.9010 \text{ ft}^3\text{/lbm}$$

NRG Equation

$$(W)_u = -d\,(E + P_o\,v - T_o\,s),\ ds = \delta\,Q/T$$

$$W = -\,[(u_i + P_o\,v_i - T_o\,s_i) - (u_f + P_o\,v_f - T_o\,s_f)]$$

where: $u_i = h_i - P_i\,v_i = 1433 - 800\,(144/778)\,0.9010 = 1366$

$u_f = 1433 - 400\,(144/778)\,0.9212 = 1297$

$$\therefore\ W = 1297 + 14.7\,(144/778)\,0.9212 - 520\,(1.625) - 1366 - 14.7\,(144/778)\,0.9010 + 520\,(1.7) = 30 \text{ Btu/lb}$$

Rated Flow: 495,000 lb/hr

$$\therefore\ \mathbf{W = (30 \times 495{,}000)/3413 = 4.33\ MW} \leftarrow$$

Gas Flow Test Orifice Sizing[4]

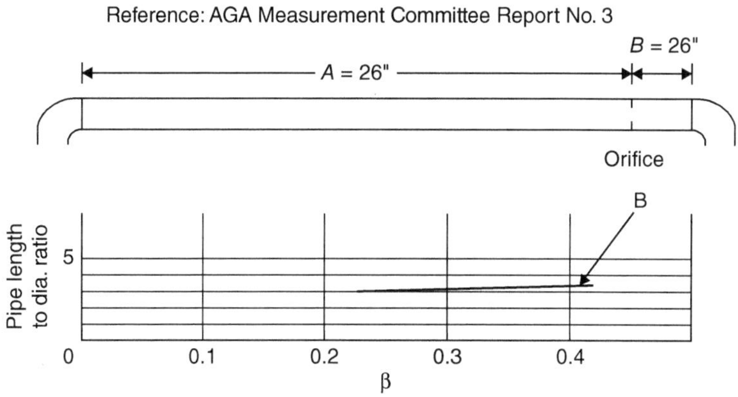

Iteration

$$B = 26/8 = 3.25D$$
$$\therefore \beta = d_o/D_P = 0.38 \text{ from Fig. 3}$$
$$d = 3'' \text{ orifice dia}$$

Performance Test Data

$$P_f = 228 \text{ psig} + 29.84 = 242.6 \text{ psia}$$
$$F_{pv} = 1.0016 \text{ @ estimated } h_w = 40 \text{ in-Hg}$$
$$F_t = 1.0053 \text{ @ } 54.5°\text{F}$$
$$F_g = 1.3327 \text{ @ } 0.563 \text{ sg gr Table 15}$$
$$y = 0.9734 \text{ where } h_w/P_f = 2.25 \text{ @ estimated } h_w = 40 \text{ in-Hg Table 7}$$
$$F_R = 1.00, \text{ Table 5}$$
$$F_m = 0.99$$
$$F_d = F_R F_m y = 0.976$$
$$F_b = 1843, \text{ Table 4}$$
$$C' = 1843 \times 1.0016 \times 1.0053 \times 1.3327 \times 0.976$$
$$= 2370$$
$$q = C'(h_w \times P_f)^{1/2}$$
$$h_w = (8/C')^2 (1/P_f)$$
$$= (686 \times 10^3/2370)^2 (1/242.6)$$
$$= 345 \text{ in } H_2O$$
$$= 26 \text{ in Hg}$$

Referring back to the figure at the top of this page choose $d = 3.1$ max by dimension "B" limitation.

$\therefore$ ***d* = 3.1**

> **Note**
>
> Data for flow measurement orifices cannot be used to size pressure reduction (break down) orifices. Tap pressure for flow orifices is created by fluid velocity and is mostly recovered downstream.

Boiler Air Blow Calculations

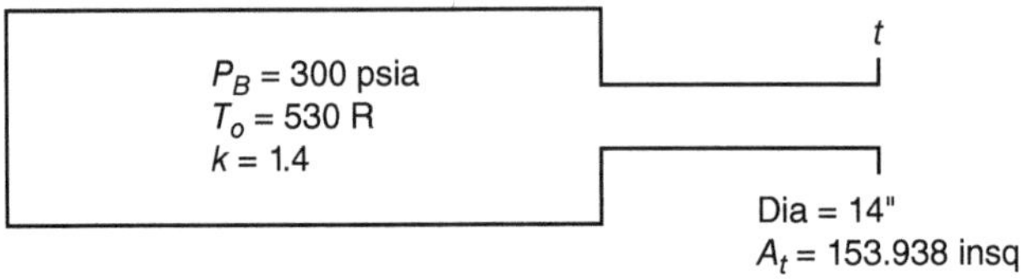

Assume: Final P = atm = 15 psia

Continuity:

$$dm/dt = \dot{m} = \rho VA$$

Sonic Velocity:

$$v = (kg_c RT)^{1/2}$$

Equation of State:

$$\rho = P/RT$$

$$\therefore \dot{m} = (A_t P_t / RT_t)(kg_c RT_t)^{1/2}$$

$$P_o/P_t = (T_o/T_t) \text{ where } T_t = T_o\,[(k+1)/2]$$

$$\therefore \dot{m} = A_t(k/g_c\,R)^{1/2}\,[(k+1)/2]\,[(2/(k+1)]^{k/k-1}\,(P_o/\,T_o^{\,1/2})$$

$$= 153.938\,(1.4/32.2 \times 53.3)^{1/2}(1.2)^{1/2}\,(0.833)^{3.5}\,(300/530^{1/2})$$

$$= 33 \text{ lbm/sec}$$

$$\dot{m} = 119{,}000 \text{ lb/hr}$$

Rated HRSG Steam Flow

$$\dot{m} = 238{,}000 \text{ lb/hr}$$

Force Ratio:

$$R = 119{,}000/238{,}000 = 0.50$$

$$= 164{,}340/238{,}000 = 0.69$$

Time Duration of Blow

Volume estimates:

Main steam = 906 ft^3

Boiler feed = 166 ft^3

$$\text{HRSG} = 3123 \text{ ft}^3$$
$$\text{Total} = 4200 \text{ ft}^3$$
$$dm/dt = \dot{m} = A_t\,(k/g_c\,R)^{1/2}\,[(k+1)/2]\,[(2/(k+1)]^{k/k-1}\,(P_o/T_o^{1/2})$$
$$C_1 = A_t\,(k/g_c\,R)^{1/2}\,[(k+1)/2]\,[(2/(k+1)]^{k/k-1}\,(1/T_o^{1/2})$$
$$C_2 = v/C_1RT$$
$$dm/dt = C_1\,P_o$$
$$dt = dm/C_1\,P_o \text{ and } P_o = \text{mRT}/v$$
$$t = \int C_2\,dm/m$$
$$t = C_2 \ln m_1/m_2, \text{ where } C_2 = 194.63 \text{ @ } T_o = 70°\text{F}$$
$$P_i = 600$$
$$P_f = 400$$
$$m_i = P_i\,v/RT = [(600)(4200)(144)]/[(53.3)(530)]$$
$$= 12{,}846 \text{ lbm}$$
$$m_f = P_f\,v/RT = [(400)(4200)(144)]/[(53.3)(530)]$$
$$= 8{,}564 \text{ lbm}$$
$$\therefore t = C_2 \ln m_1/m_2 = 194.63 \ln 12{,}846/8564$$
$$= 34 \text{ sec}$$

Recharge Time:

Compressor capacity = 4800 cfm

$$\dot{m} = [(15)(4800)(144)]/[(53.3)(530)] = 367 \text{ lbm/min}$$

Air Blow Table

	Blowdown	
t = 34 sec	P_i = 600	P_f = 400
t = 2.5 min	P_i = 600	P_f = 100
t = 2.1 min	P_i = 500	P_f = 100
t = 1.9 min	P_i = 400	P_f = 100
t = 4.6 min	P_i = 400	P_f = 15.0
	Recharge	
t = 11.6 min	P_i = 600	P_f = 400
t = 29 min	P_i = 600	P_f = 100
t = 23 min	P_i = 500	P_f = 100
t = 17.5 min	P_i = 400	P_f = 100
t = 22.5 min	P_i = 400	P_f = 15.0

Generator Air Leakage Test

$$L_A = 238\,V/H\,[(P_i + \text{Pb}_i)/(273 + T_i) - (P_f + \text{Pb}_f)/(273 + T_f)]$$

where: V = generator volume
H = hours of testing

P_i = initial generator pressure, in-Hg
P_f = final generator pressure, in-Hg
Pb_i = initial barometric pressure, in-Hg
Pb_f = final barometric pressure, in-Hg
T_i = initial generator air temperature
T_f = final generator air temperature

Maximum Suction Screen ΔP

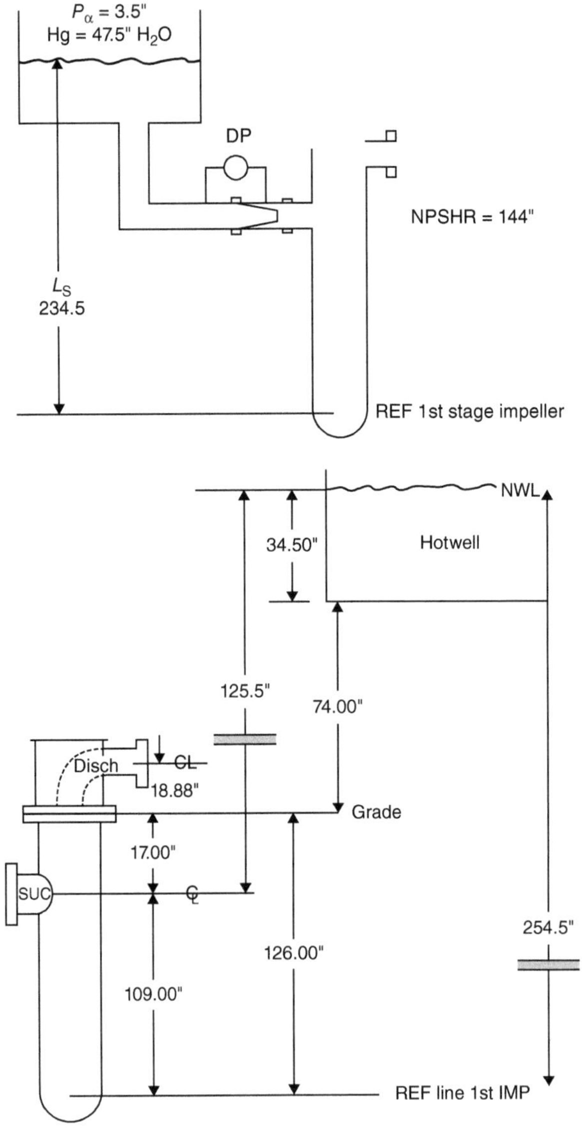

Flow Equations

Bernoulli Equation

$$P_1 + 1/2\ \rho\ v^2 = P_2 + 1/2\ \rho\ v^2$$

where:

p = pressure
ρ = density
v = flow velocity

Continuity Equation

$$q = v_1\, A_1 = v_2\, A_2$$

where:

q = flow rate
A = flow area

Simultaneous Equation

$$q = A_2\, [2(P_1 - P_2)/\rho(1 - (A_2/A_1)^2)]^{1/2}$$

Flow Coefficients

Find the flow constant k for a feedwater flow element.
Given:

$$D_o = 10.74 \text{ in}$$
$$\beta = 0.70159$$
$$h = 100 \text{ in } H_2O$$
$$T = 290°\text{F}$$
$$W = 3.6 \times 10^6 \text{ lb/hr}$$
$$\therefore\ W = k\, (\text{h}/\acute{\upsilon})^{1/2},\ \acute{\upsilon} = 0.01736$$
$$k = 3.6 \times 10^6/(100/0.01736)^{1/2}$$
$$\mathbf{k = 47432.6}$$

If the flow coefficient for any section of pipe or duct is known, the flow can be determined by the differential pressure across that section of pipe or duct. For performance test purposes, standard flow elements are used because the flow coefficients and expansion factors have been determined by rigorous laboratory controlled tests.

However, for commissioning tests used to anticipate, detect, or troubleshoot problems, coefficients and expansion factors determined by laboratory testing are not necessary because the uncertainties associated with the use of plant instrumentation negate any uncertainties associated with the use of approximate coefficients and expansion factors.

On the other hand, if plant instrumentation is used and the flow element data sheet is available, it will provide crucial information that can be used to determine the flow constants.

Now the flow can be determined for any flow differential pressure across an element with this β ratio and D_o. The same principles can be applied to any constant geometry element. For example, during flue gas testing of the air quality control equipment, the gas flow will be measured by duct traverse. It is a good idea to take differential pressure measurements across all boiler passes, air preheaters, and air quality equipment. Once the flow and differentials are determined, a flow constant can be calculated and used to determine the flow at any other differential pressure. This is a helpful tool to determine fouling. The flow can be determine by measuring the differential across a clean section, and the differential across other sections will determine their cleanliness.

Coefficients vary according to the β ratio (D_o/D_P) and R_d (Reynolds number). Close approximation for field tests can be used as illustrated in the following table.

Flow Coefficients for Square Edged Orifice

Pipe Diameter (in)	2	4	8	16
$\beta = 0.5, R_d = 10^5$	0.61	0.61	0.60	0.60
$\beta = 0.7, R_d = 10^5$	0.61	0.61	0.60	0.60
$\beta = 0.5, R_d = 10^6$	0.61	0.61	0.60	0.60
$\beta = 0.7, R_d = 10^6$	0.61	0.61	0.60	0.60

Values for temporary square edged orifices fabricated in the field with flanged taps.

Note

Flow coefficients for ASME flow nozzles are close to unity (0.98–0.99). Functions of expansion, and process gas to ideal gas ratios all approximate unity (0.99–1.00). For precise flow coefficients and expansion factors, ASME MFC-3M-1989, Measurement of Fluid Flow in Pipes Using Orifice, Nozzle, and Venturi;[5] ASME Application Part II of Fluid Meters;[6] or AGA Gas Measurement Committee Report No. 3[4] can be used.

Moody Diagram[7]

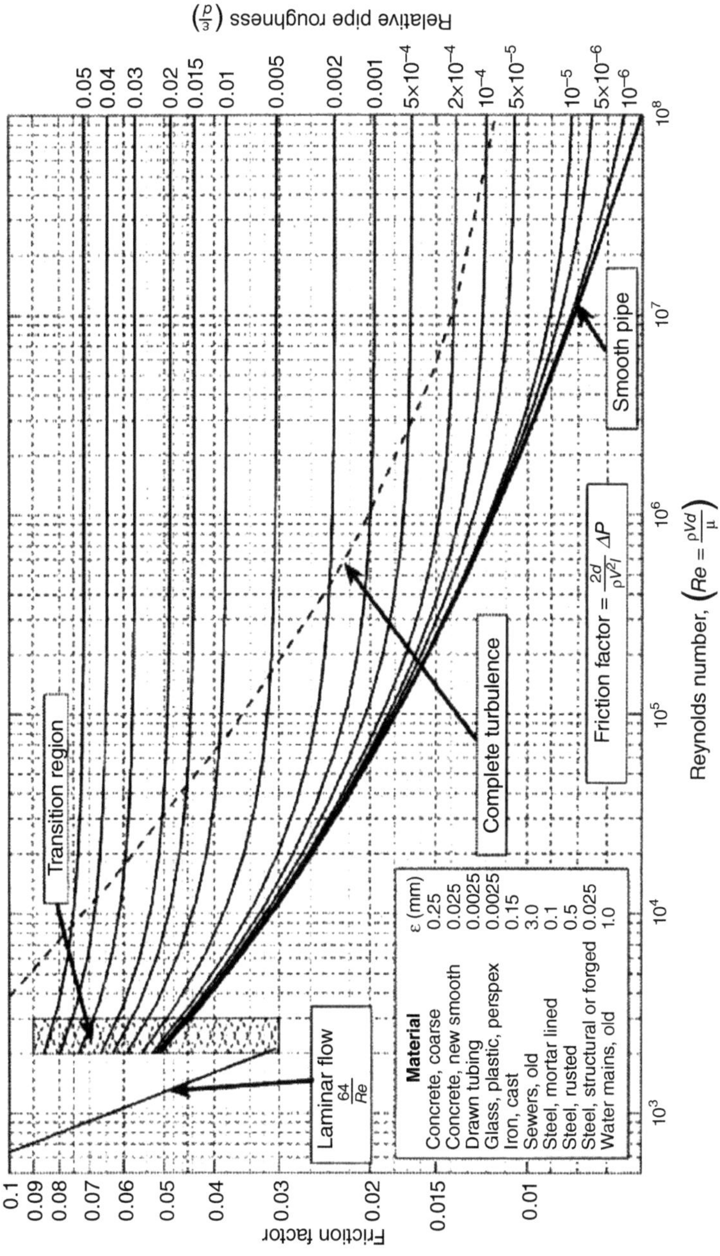

BOLT TORQUE TABLE[8]

$\frac{1}{4}$–$\frac{3}{4}$ INCH

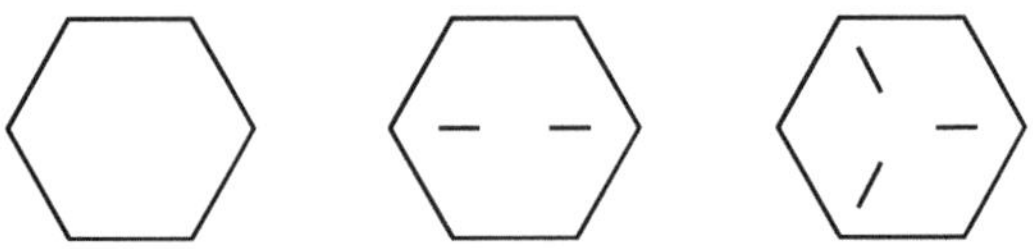

SEA 0,1,2	SEA 3	SEA 5
74,000 psi	100,000 psi	120,000 psi med
low carbon steel	med carbon steel	carbon heat treated

Bolt Size	Grade 0, 1, 2	Grade 3	Grade 5
Dry Torque (ft-lb)			
¼	6	9	10
5/16	12	17	19
3/8	20	30	33
7/16	32	47	54
½	47	69	78
9/16	69	103	114
5/8	96	145	154
¾	155	234	257

REFERENCES

1. Standards for Steam Surface Condensers, Heat Exchange Institute, Cleveland Ohio.
2. Keenan, J. F. and Kaye, J., Gas Tables , John Wiley and Sons, 1948.
3. American Gas Association, "Orifice metering of natural gas and other related hydrocarbon fluids," AGA Report No. 3.
4. ASME MFC-3M-1989.
5. ASME Research Committee on Fluid Meters, Part II, Application.
6. Beck, S. and Collins, R., University of Sheffield, Wikimedia Commons, 2008.
7. Glover, T. J., *Pocket Reference*, Sequoia Publishing, 1993.
8. Thomas J. Glover, *Pocket Reference*, Sequoia Publishing, 1993.

APPENDIX T
FLUSHING TOOLS

Sieve Mesh Number	Sieve Opening Millimeters	Sieve Opening Inches	Wire Diameter Inches
2½	8.00	0.315	0.073
3	6.72	0.265	0.065
3½	5.66	0.223	0.057
4	4.76	0.187	0.050
5	4.00	0.157	0.044
6	3.36	0.132	0.040
7	2.83	0.111	0.036
8	2.38	0.094	0.033
10	2.00	0.079	0.030
12	1.68	0.066	0.027
14	1.41	0.056	0.024
16	1.19	0.047	0.021
18	1.00	0.039	0.019
20	0.84	0.033	0.017
25	0.71	0.029	0.015
30	0.59	0.023	0.013
35	0.50	0.020	0.011
40	0.42	0.017	0.010
45	0.35	0.014	0.009
50	0.30	0.012	0.007
60	0.25	0.010	0.006
70	0.21	0.008	0.006
80	0.18	0.007	0.005
100	0.15	0.006	0.004
120	0.13	0.005	0.003
140	0.11	0.004	0.003
170	0.09	0.004	0.003
200	0.07	0.003	0.002
230	0.06	0.003	0.002
270	0.05	0.002	0.002
325	0.04	0.002	0.001

Screen Mesh Sizes
0.001 mm = 1 micron = 1 μm

Screens and Baskets
"Y" and Basket Type Strainers

1/32" Openings	1/50" Openings	1/64" Openings	.009 Openings	.007 Openings	.005 Openings

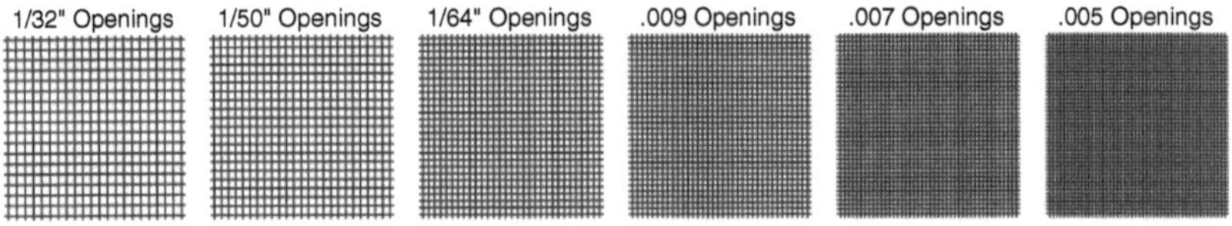

20 Mesh (52% Open Area)	30 Mesh (41% Open Area)	40 Mesh (36% Open Area)	60 Mesh (31% Open Area)	80 Mesh (31% Open Area)	100 Mesh (30% Open Area)

.027 Dia.	.033 Dia. (1/32")	.045 Dia. (3/64")	.057 Dia. (1/16")

400 Holes/SQ. IN. (23% Open Area)	.324 Holes/SQ. IN. (28% Open Area)	.225 Holes/SQ. IN. (36% Open Area)	.144 Holes/SQ. IN. (37% Open Area)

1/10" Dia.	1/8" Dia.	5/32" Dia.	1/16" Dia.	1/4" Dia.

.49 Holes/SQ. IN. (39% Open Area)	.32 Holes/SQ. IN. (40% Open Area)	.33 Holes/SQ. IN. (62% Open Area)	.18 Holes/SQ. IN. (50% Open Area)	.8 Holes/SQ. IN. (40% Open Area)

Perforated metal and mesh

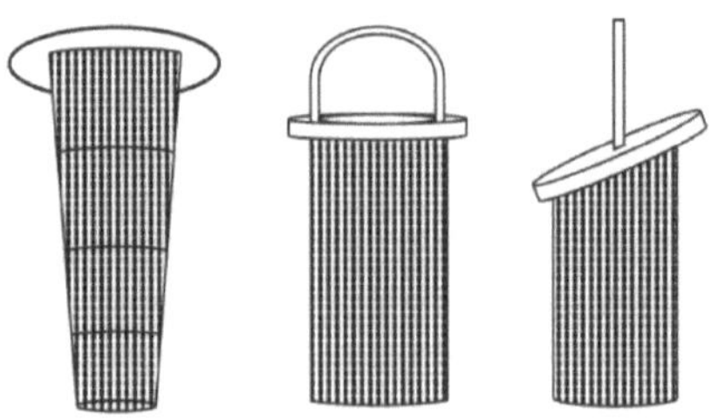

Basket Types

Temporary L.O. heaters

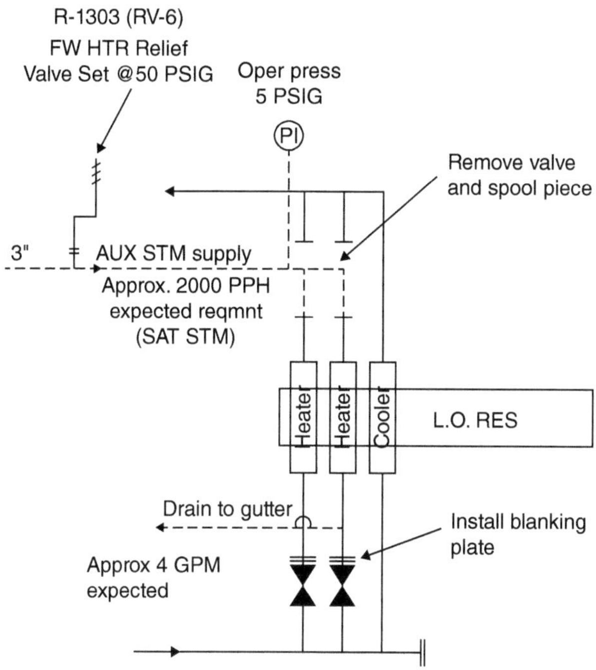

AUX STM for L.O. heating

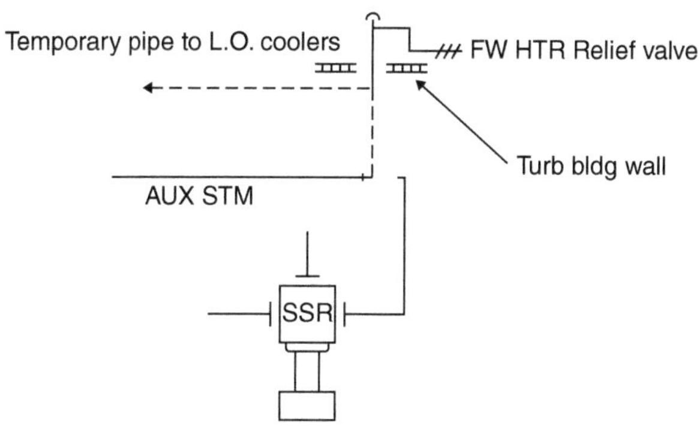

STEAM REQUIREMENT FOR LO HEATING

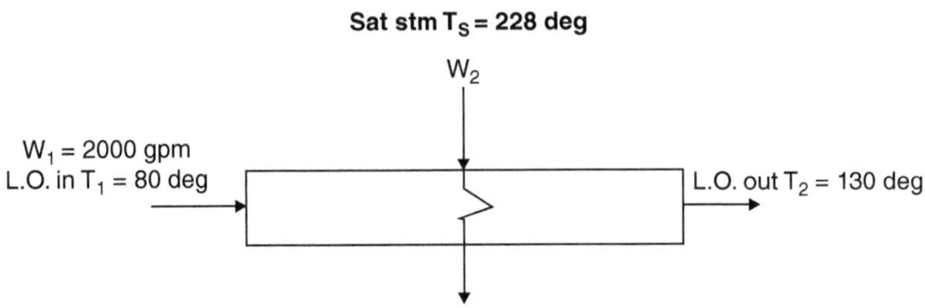

$W_1 = (h_1 - h_2)W_2^{1/Sg}/C_p\ (T_2 - T_1)$; Where for oil $C_p = 0.51$, $Sg = 0.79$

$10^6 = (1156.3 - 196.3)\ W_2^{1/Sg}/0.511\ (130 - 80)$

$W_2 = 2100$ PPH

$= 42$ gpm

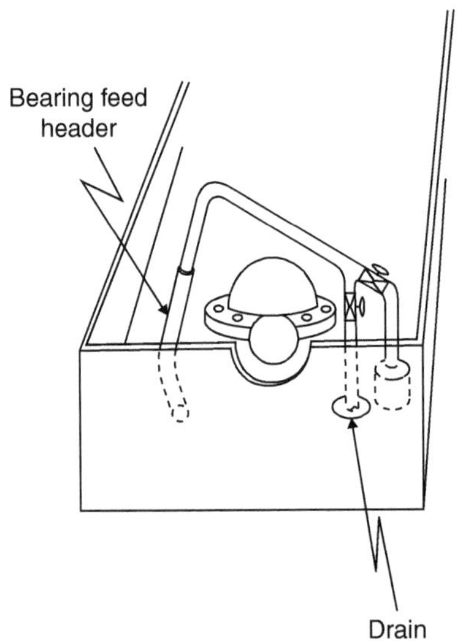

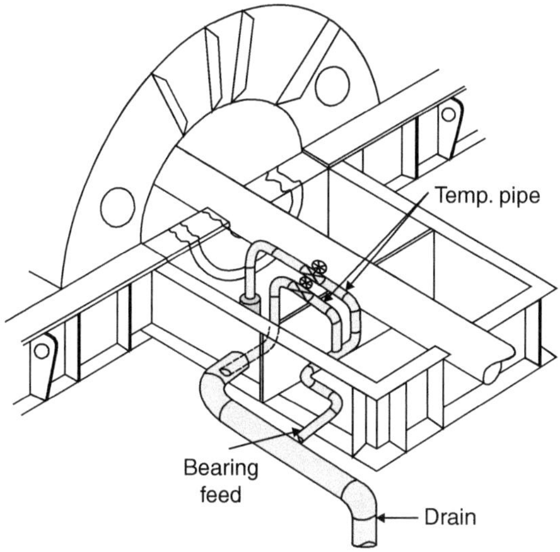

Typical temporary flush pipe arrangement
No. 2,3, & 4 BRNG

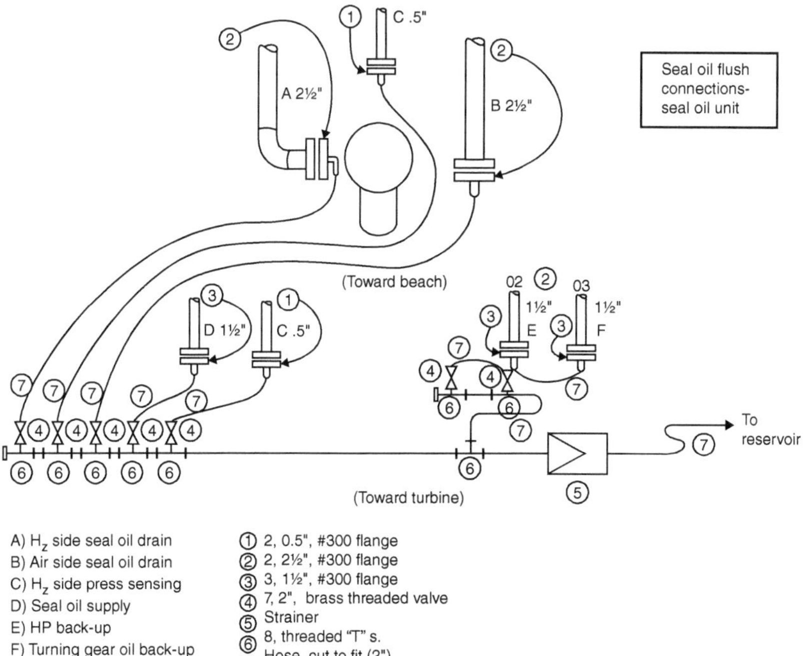

Seal oil unit connections

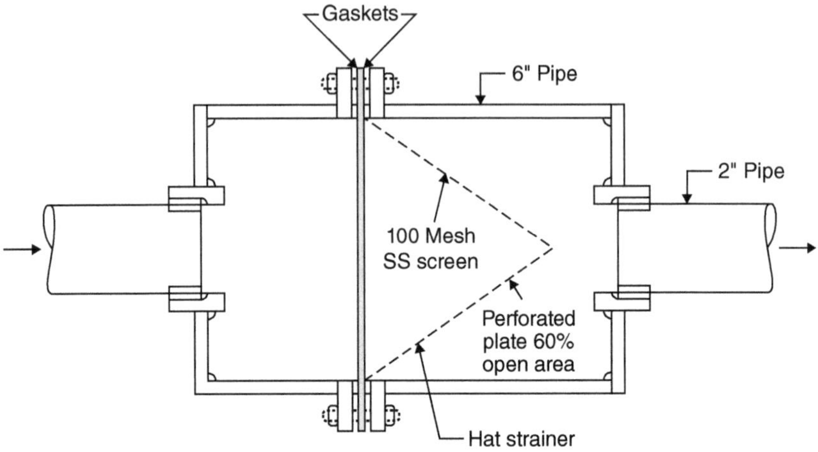

Strainer basket

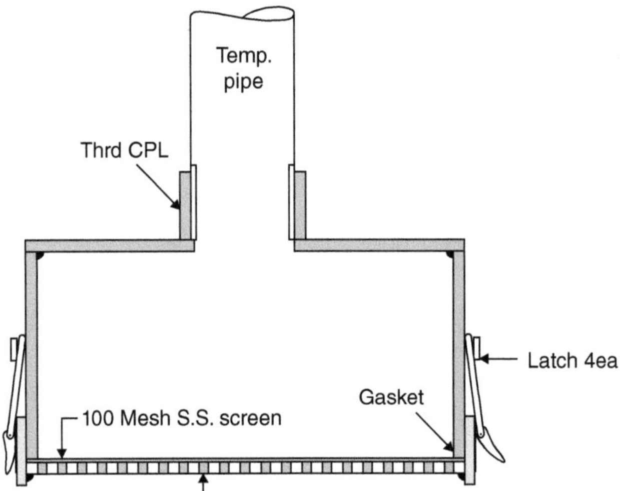

Typical basket screen
6 REQ

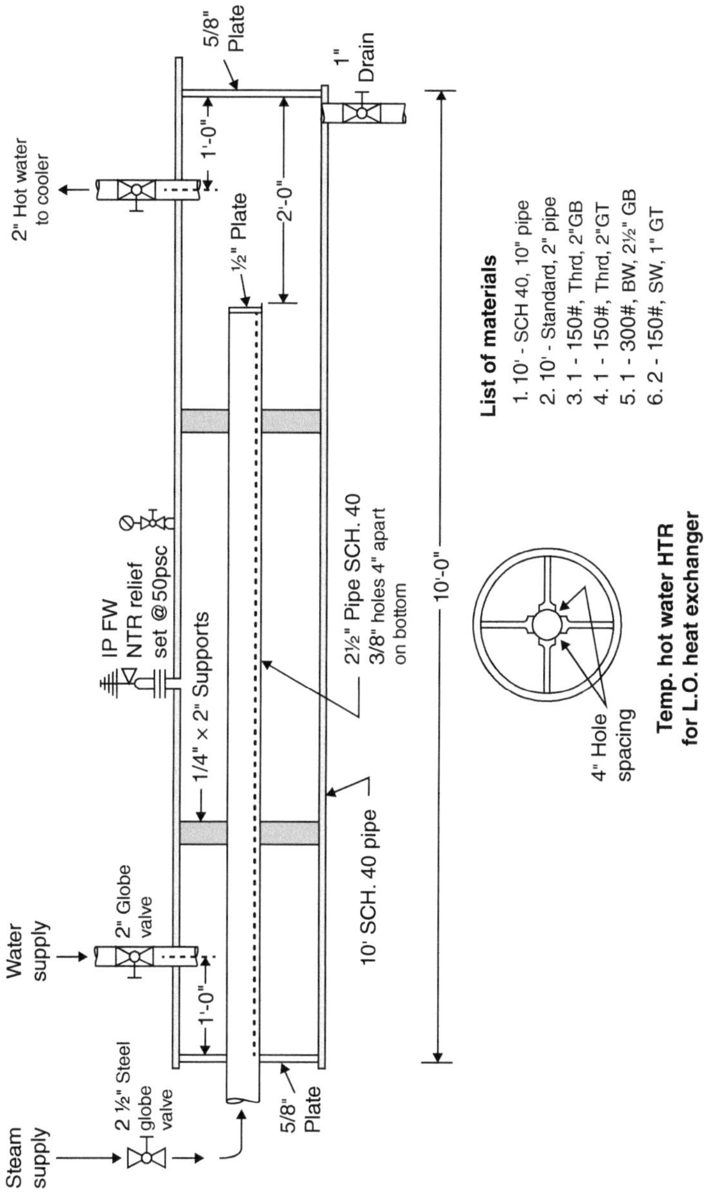
Steam supply
2 ½" Steel globe valve
Water supply
2" Globe valve
1'-0"
5/8" Plate
10' SCH. 40 pipe
1/4" × 2" Supports
IP FW NTR relief set @50psc
2½" Pipe SCH. 40 3/8" holes 4" apart on bottom
10'-0"
½" Plate
2'-0"
2" Hot water to cooler
1'-0"
5/8" Plate
1" Drain
List of materials
1. 10' - SCH 40, 10" pipe
2. 10' - Standard, 2" pipe
3. 1 - 150#, Thrd, 2"GB
4. 1 - 150#, Thrd, 2"GT
5. 1 - 300#, BW, 2½" GB
6. 2 - 150#, SW, 1" GT
4" Hole spacing
Temp. hot water HTR for L.O. heat exchanger

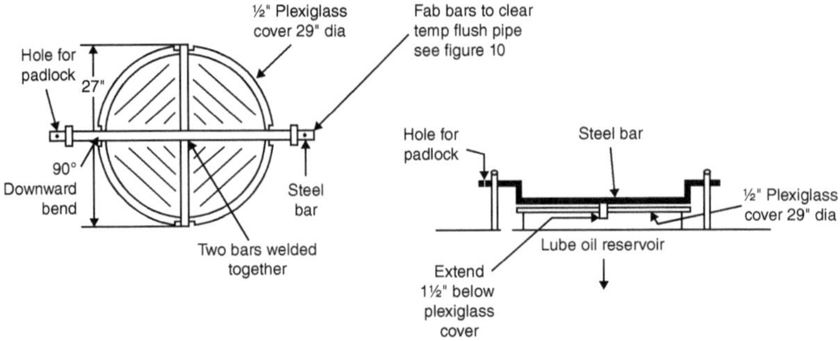

Reserovir covers

PEDESTAL COVERS

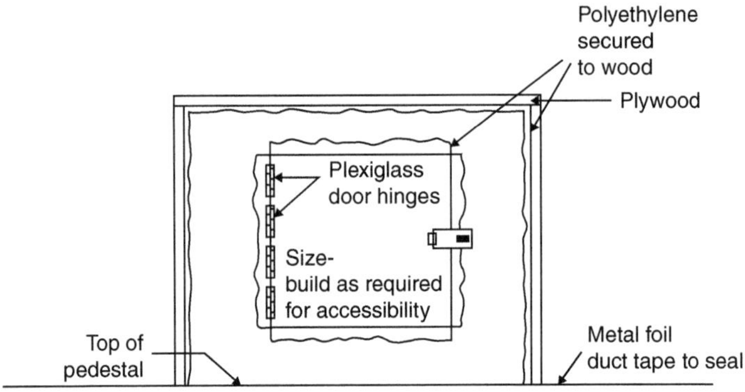

ISO/NAS Correlation Table

ISO 4460	Particles/Milliliter			NAS 1636 1964
23/21/18	80,000	20,000	2500	12
22/20/18	40,000	10,000	2500	
22/20/17	40,000	10,000	1300	11
22/20/16	40,000	10,000	640	
21/19/16	20,000	5000	640	10
20/18/15	10,000	2500	320	9
19/17/14	5000	1300	160	8
18/16/13	2500	640	80	7
17/15/12	1300	320	40	6
16/14/12	640	160	40	
16/14/11	640	160	20	5
15/13/10	320	80	10	4
14/12/9	160	40	5	3
13/11/8	80	20	2.5	2
12/10/8	40	10	2.5	
12/10/7	40	10	1.3	1
12/10/6	40	10	0.64	

APPENDIX U

TYPICAL STARTUP SCHEDULE SEQUENCE AND QUANTITY SURVEY FORM

1. Diesel generator in service
2. Energize 480-V bus
3. Batteries, DC system energized
4. Check vital AC and battery charger
5. Startup transformer in service
6. 6.9-kV Buses energized
7. 480-V control center energized
8. DCS energized
9. Service water system in service
10. Fire protection system hydro and city inspection
11. Fire protection system in service
12. Clean intake pit
13. Operate screenwash pumps, traveling screens
14. Start large motor rotation checks
15. Demineralized water available
16. Operate condensate makeup pumps
17. Cooling water pump operation
18. Closed cooling water system in service
19. Operate instrument air compressors
20. Instrument air dryers in service
21. Operate service air compressors
22. Service and instrument air systems in service
23. Elevator in service
24. Inspect intake tunnel
25. Condenser waterbox priming pump operation
26. Operate circulating water pumps
27. Reverse circulating water system in service
28. Clean and flush condenser steam side
29. Flush condenser pump suction
30. Condenser hydro
31. Operate condensate pumps
32. Operate chemical feed system
33. Flush preboiler cycle
34. Nitrogen system complete
35. Clean lube oil reservoir
36. Turbine oil pump operation
37. Turbine oil flush
38. Turbine oil purifier in service
39. EHC system flush
40. Decontamination tanks ready
41. Chemical clean preboiler cycle
42. Seal oil system in service
43. Turbine generator on turning gear
44. Boiler feed pump turbine oil flush
45. HVAC systems operation
46. Preboiler operation
47. Fuel heaters and oil recirculation
48. Auxiliary boiler operation
    a. Fans
    b. Feedwater pumps
    c. Fuel oil system
    d. Control system
    e. Interlocks
    f. Boilout
49. Steam—seal—piping blowout
50. Steam seals in service
51. Vacuum pump operation

52. Blowout steam lines to boiler feed pump turbines
53. Pull condenser vacuum check leaks
54. Operate boiler feed pump turbines
55. Boiler feed pump seal water system in service
56. Operate boiler feed pumps—suction screens installed
57. Hypochlorinator operation
58. Ignition and main gas system blowout
59. Operate air preheaters
60. Clean and inspect all boiler passes
61. Air preheater coil operation
62. Operate induced draft (ID) fans
63. Operate forced draft (FD) fans
64. Air dilution fan operation
65. Manual control for water oil air flow
66. Boiler interlocks checked
67. Laboratory complete
68. Pilot burner operation
69. Main oil burner operation
69. Boiler boilout
70. Boiler acid cleaning
71. Main steam blows
72. Cathodic protection installed
73. Operation sootblowers
74. Stator cooling water system flush
75. Generator air test
76. Oxygen analyzer in service
77. Blowout main heater and boiler feed pump turbine (BFPT) steam-lines
78. Set reheat safety valves
79. Revert steam piping to normal
80. All controls, annunciators complete in control room
81. Loop checks complete
82. Remove condensate pump suction screens
83. Set safety valves
84. Main turbine generator control checks
85. Generator phase check
86. Fill generator with hydrogen
87. Final CT and PT checks
88. Hi-pot tests
89. Functional relay check
90. Hydrogen cooler operation
91. Water induction system in service
92. Reverse current valve check
93. Heater level control check
94. Fire boiler for turbine roll
95. Overspeed tests
96. Synchronize generator—trial operation
97. Operation of unit at loads up to full load
98. Check hangers and expansion
99. Check all equipment operation
100. Adjust boiler excess air
101. Obtain data for combustion control auto operation
102. Combustion control on auto
103. Full-load run
104. Shutdown unit
105. Remove fine screens
106. Bearing inspection
107. Remove boiler feed pump screens
108. Make repairs as necessary
109. Operate unit up to full load
110. Check variable-pressure operation
111. Commercial operation
112. Equipment performance testing

Quantity Survey Form

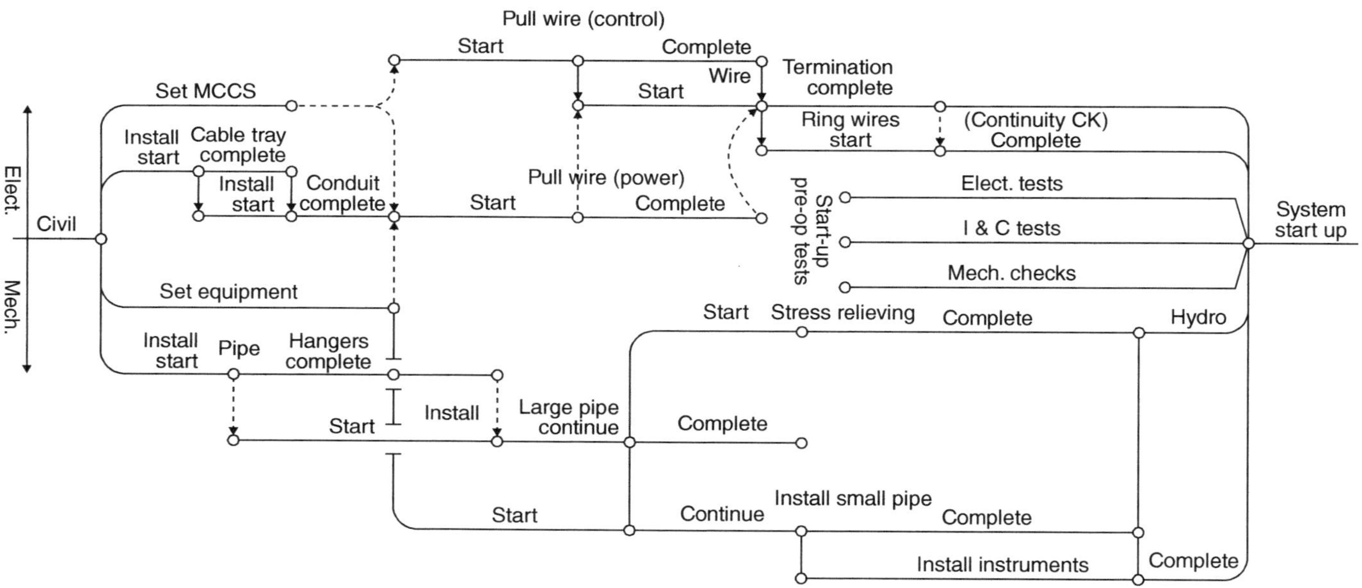

	Item			Quantity	Unit
	Estimated			MH unit	
Description	Qty.	MHS		Used	
Large pipe					
SM pipe & valves					
Instruments					
Tubing					
Hydro					
Tray					
Conduit					
Cable pull					
1. Power					
2. Control					
Cable term					
1. Power					
2. Control					

APPENDIX V
ELECTRICAL SUPPLIES

The purpose of this appendix is to provide proper nomenclature for fittings to narrow searches for components using a vendor's online catalog. The selected electrical fittings in this appendix are among the most commonly used in construction for residential, commercial, and industrial facilities.

A limited selection of electrical supplies are redrawn from Catalog No. 83; courtesy of Consolidated Electrical Distributors Inc. For identification reference only.

Octagon Boxes

K.O.

Clamps

Extension

Box Covers

Blank

K.O.

Shallow Ceiling Boxes

Concrete Rings and Knockouts

Ring

Plain Plate

Stud Plate

OCTAGON BOXES
With Brackets

"J" Bracket   Flat Bracket

SQUARE BOXES

½" K.O.   Romex

OCTAGON BOX COVERS

Flat Blank   Plaster Ring   Dup. Receptical

SQUARE BOX COVERS

Flat Blank   Plaster Ring   Switch Ring

EXPANDABLE BAR SETS

Octagon Boxes

SQUARE BOXES
With Brackets

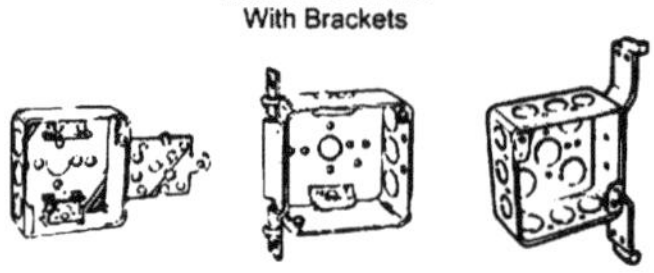

Long   Flat   Offset

SQUARE BOXES
With 1/2" K.O.

SQUARE SURFACE COVERS

Toggle Switch   Single Receptacle   Duplex Receptacle

Toggle & Duplex   Two Duplex Receptacle

SQUARE BOX COVERS

Flat Blank   Switch Ring

SQUARE RAISED SURFACE COVERS

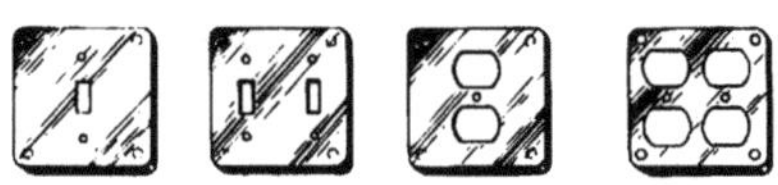

Toggle   2 Toggle   Duplex   2 Duplex

Toggle & Duplex   Single   Blank

HANDY BOXES

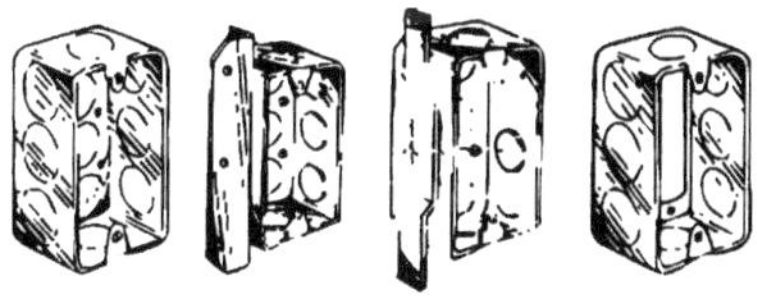

½" K.O. | Angle Bracket | Flet Bracket | Extension

HANDY BOX COVERS

Blank | Duplex Receptacle | Toggle Switch

PLENUM BOXES & COVERS

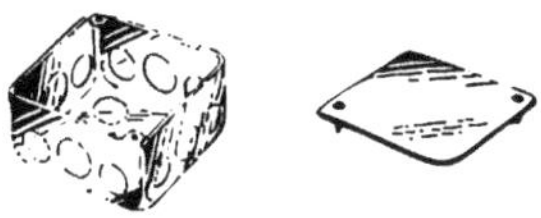

SMALL SIZE BOX
Without Brackets

SWITCH BOX
With Conduit Knockouts

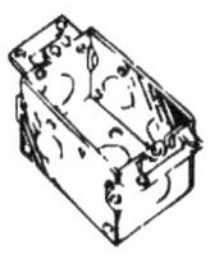

SWITCH BOXES
With Brackets

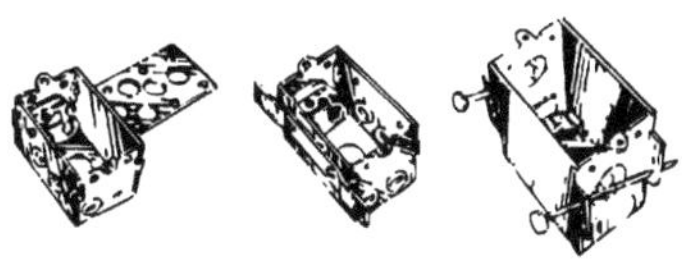

Long | Flat | Nail

SWITCH BOX
With BX Clamps
and with Ears

SWITCH BOX
With Romex Clamps
and with Square Corners

SWITCH BOX
With Romex Clamps
and with Beveled Corners

SWITCH BOX EXTENSIONS
With Screws

SWITCH BOX SUPPORTS

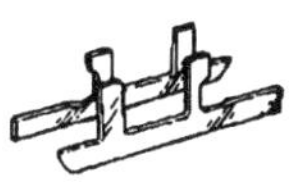

MASONRY BOXES

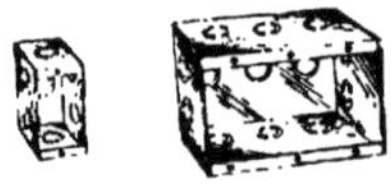

1 Gang 3 Gang

SOLID GANG SWITCH BOXES

LOW VOLTAGE PARTITIONS
For Masonry Boxes

SOLID GANG BOX COVERS

MASONRY
THRU-THE-WALL BOXES

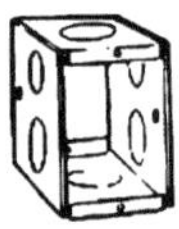

SOLID GANG
FLAT BLANK COVERS

GROUNDING ACCESSORIES

Clip Screw Screw With Pigtail

BAR HANGERS

Adjustable

NON-METALLIC
ROUND OUTLET BOXES

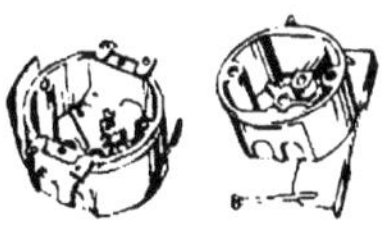

NON-METALLIC
SQUARE "N" BRACKET BOX
With Nails

NON-METALLIC
SWITCH BOX
With Nails / No Clamps

NON-METALLIC
ROUND OUTLET BOXES

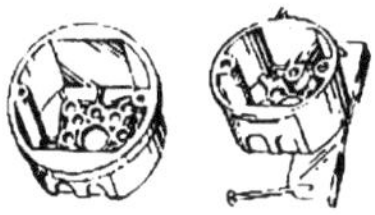

No Bracket "N" Bracket

NON-METALLIC
ROUND OUTLET BOXES

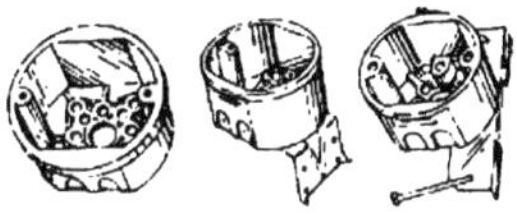

NON-METALLIC
2-GANG SWITCH BOX
With "N" Bracket

NON-METALLIC SWITCH BOX
No Clamps / #2 Bracket

NON-METALLIC
2 GANG SWITCH BOX
With Nail Bracket / No Clamps

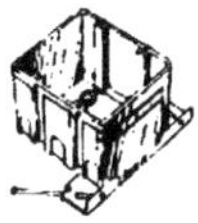

NON-METALLIC
SWITCH BOX
For Old and New Work
No Clamps

NON-METALLIC
ROUND OUTLET BOX
With Bar Hanger / No Clamps

NON-METALLIC
ROUND OUTLET BOX COVERS

Pendant or Blank

Plaster Ring

NON-METALLIC
SQUARE OUTLET BOX COVERS

Flat Blank

Switch Ring

Plaster Ring

PEDESTAL COVERS

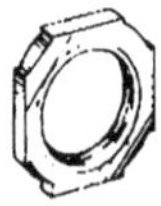

CONDUIT LOCKNUTS

CONDUIT BUSHINGS

METAL CONDUIT SEALS (PENNIES)

PLASTIC PULL CAPS

PLASTIC INSULATING BUSHINGS

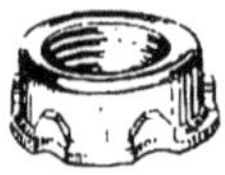

INSULATED METALLIC BUSHINGS

GROUNDING BUSHINGS

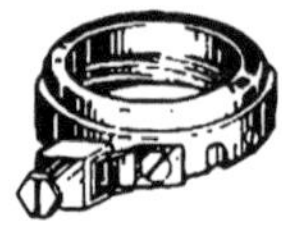

INSULATED GROUNDING BUSHINGS

CAPPED BUSHINGS

GROUNDING LOCKNUTS

KNOCKOUT SNAP-IN BLANKS

REDUCING BUSHINGS

REDUCING WASHERS

BUSHED CONDUIT NIPPLES

INSULATED BUSHED CONDUIT NIPPLES

3/4" OFFSET NIPPLES

1-1/2" OFFSET NIPPLES

CONDUIT UNIONS

SET SCREW CONNECTOERS

INSULATED THROAT SET SCREW CONNECTORS

SET SCREW COUPLINGS

THREADLESS
CONDUIT CONNECTORS

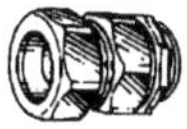

INSULATED THROAT
THREADLESS CONNECTORS

THREADLESS
CONDUIT COUPLINGS

90°THREADLESS
SHORT ELBOWS

INSULATED THROAT 90°
THREADLESS SHORT ELBOWS

NO-BOLT FIXTURE STUDS

3/8" FIXTURE EXTENSIONS

COMPRESSION TYPE
EMT CONNECTORS

INSULATED THROAT
COMPRESSION
CONNECTORS

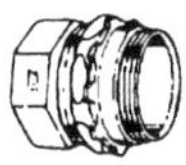

COMPRESSION TYPE
EMT COUPLINGS

COMPRESSION TYPE
EMT SHORT ELBOWS

INSULATED THROAT
EMT SHORT ELBOWS

THREADED
ENTRANCE CAPS
FOR RIGID CONDUIT

CLAMP TYPE
ENTRANCE CAPS

SLIP FITTER TYPE
ENTRANCE CAPS

SLIP FITTER
TYPE FOR MAST
INSTALLATIONS

STRAIGHT ROUND
ENTRANCE CAPS

SET SCREW EMT
CONNECTORS

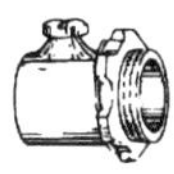

INSULATED THROAT
SET SCREW CONNECTORS

SET SCREW EMT
COUPLINGS

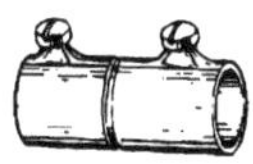

EMT SPLIT
ADAPTERS

INDENTER EMT
CONNECTORS

INSULATED THROAT
INDENTER CONNECTORS

INDENTER EMT
COUPLINGS

SET SCREW EMT
OFFSET CONNECTORS

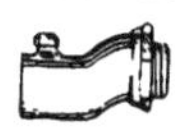

COMBINATION EMT
TO RIGID COUPLINGS

Compression Type

Set Screw Type

ONE HOLE MALLEABLE
EMT STRAPS

ONE HOLE STEEL
EMT STRAPS

TWO HOLE
EMT STRAPS

ENTRANCE ELLS

CAPPED ELLS

GALVANIZED HEAVY WALL
RIGID STEEL CONDUIT

GALVANIZED
CONDUIT NIPPLES

CORNER PULLING ELBOWS
Female to Female

CORNER PULLING ELBOWS
Male to Female

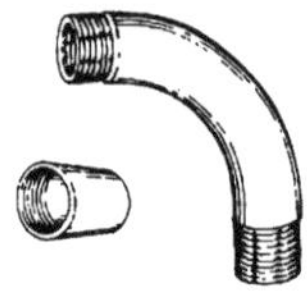

GALVANIZED
COUPLINGS
& ELBOWS

SHORT ELBOWS

INSULATED
SHORT ELBOWS

BUSHED ELBOWS

TWO HOLE PIPE STRAPS

ONE HOLE MALLEABLE
IRON PIPE STRAPS

ONE HOLE STAMPED
STEEL PIPE STRAPS

CLAMP BACKS

CONDUIT BODIES

TYPE C

TYPE E

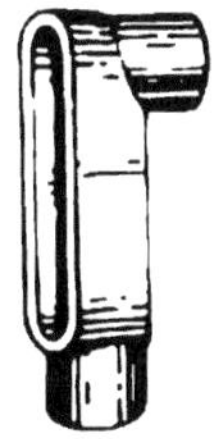

TYPE LB

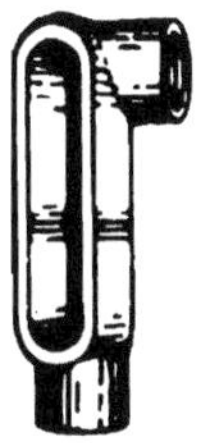

TYPE LL

TYPE LR

TYPE 1

TYPE X

CONDUIT BODIES

One Gang
TYPE FS

TYPE FSA

TYPE FSC

TYPE FSL

TYPE FSR

TYPE FSS

TYPE FSCA

TYPE FSCC

TYPE FSCT

TYPE FST

TYPE FSX

Two Gang
TYPE FS

TYPE FSC

Type F (S) 1- 5/8-in deep, type F (D) 2-½-in deep

CONDUIT SEALS

EYS1

EYS16

EYS11

EYS116

EYS4

EYS46

EZS1

EZS16

EYS29

WEATHER-RESISTANT
RECEPTACLES

With
Spring Door
Vaportight

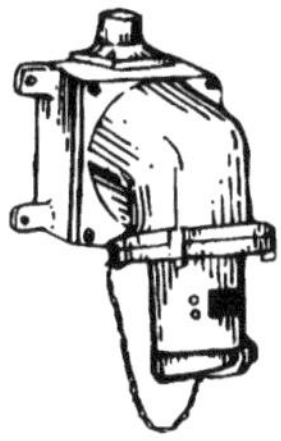
With
Screw Cap
Raintight

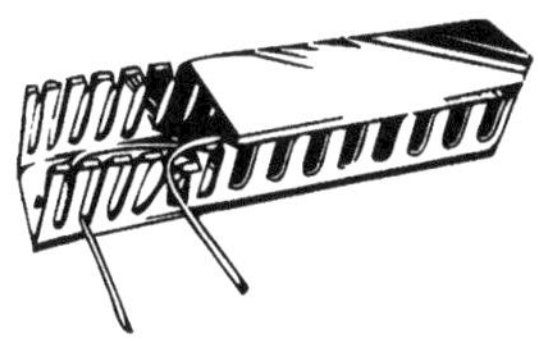
PLASTIC WIRING DUCT & COVERS

SPIRAL WRAPPING

ADHESIVE BACKED
LATCHING DUCT

ADHESIVE BACKED
LATCHING CLIP

SCRU-TITE HUBS

ROMEX CABLE
CONNECTORS

SILVEY
HICKEY

SCREW-IN TYPE
FLEX CONNECTORS

THINWALL BENDER

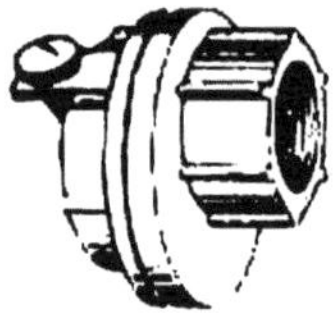

SCRU-TITE GROUND
OFFSET HUBS

EMT CONNECTORS
& COUPLINGS TAP-ON

HICKEYS
FOR RIGID CONDUIT

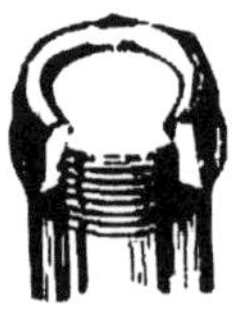

MISENER
SAFE-T-SEALS
(Paper Pennies)

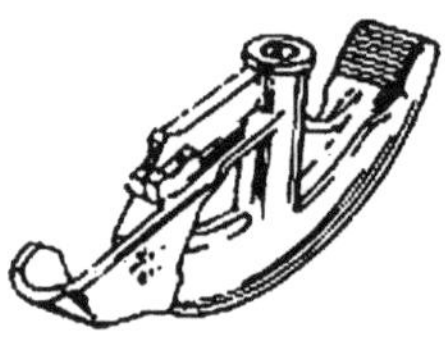

RIGID ALUMINUM
CONDUIT BENDERS

INDEX

Note: Page numbers followed by *f* denote figures; page numbers followed by *t* denote tables.

Printed by Libri Plureos GmbH in Hamburg,
Germany

9 780071 809894